ELECTRICITY
AND
MAGNETISM

ELECTRICITY AND MAGNETISM

P. F. Kelly

CRC Press
Taylor & Francis Group
Boca Raton London New York

CRC Press is an imprint of the
Taylor & Francis Group, an **informa** business

CRC Press
Taylor & Francis Group
6000 Broken Sound Parkway NW, Suite 300
Boca Raton, FL 33487-2742

First issued in paperback 2020

© 2015 by Taylor & Francis Group, LLC
CRC Press is an imprint of Taylor & Francis Group, an Informa business

No claim to original U.S. Government works

ISBN-13: 978-1-4822-0635-7 (hbk)
ISBN-13: 978-0-367-78369-3 (pbk)

Visit the Taylor & Francis Web site at
http://www.taylorandfrancis.com

and the CRC Press Web site at
http://www.crcpress.com

Contents

List of Examples

Preface

My original motivation for writing this series of books, *Elements of Mechanics*, *Properties of Materials*, and *Electricity and Magnetism*, was to provide students attending my classes with advance copies of notes for each lecture. Successive accretions of explanatory material [much of it generated in response to student questions] transformed this project into something more substantial, which could be read as

- A sole source for a sequence of introductory physics classes

- A student supplement to a standard textbook

- A review in preparation for graduate/professional/comprehensive examinations

The provenance of these volumes as notes for lecture is manifest throughout in style and content. Physics is, in this telling, an engaging endeavour rather than a spectator sport. Also, attempts are made to ensure that the reasons for studying physics:

    LEARN   to acquire knowledge about nature,

    CONTROL   to exercise dominion over nature,

    CREATE   to experience the joys of invention and artistry,

are not lost entirely under the weight of detailed investigation of specific models for particular phenomena. Finally, while occasional mention is made of *au courant* topics, our primary concern is with the foundations of classical physics. Readers who wish to more fully appreciate the "modern" developments in physics are encouraged to work through these notes first.

Very early on, while each book in the trilogy was scarcely more than an inkling, they were dubbed $\mathcal{MAPS}$, $\mathcal{SPAM}$, and $\mathcal{AMPS}$. These names derived fuller significance by interpretation as acronyms obtained from permutations of the four Latin words

$$\textbf{\textit{Ars}} \qquad \textit{Mechanica} \qquad \textit{Physica} \qquad \textit{Scientia}.$$

This artifice is a nod toward the deeper mathematical structures hinted at in these notes. The words themselves have import too.

$\mathcal{A\,M}$ The *Art of Mechanics* conveys the notion that the practice of physics is essentially creative. In another riff on this pair of words, the *Artes Mechanicae*, as traditionally understood, are practical skills[1] such as those employed by artisans in the production of useful goods and decorative artwork. Throughout these notes, we shall craft many mathematical models providing serviceable approximations to [pertinent aspects of] physical systems.

---

[1] In the past, *Artes Mechanicae* suffered invidious comparison with *Artes Liberales* or "Liberal Arts."

$\mathcal{P S}$ The *Science of Physics* is the ordering of knowledge[2] of the physical world. This ordering culminates in the discernment of fundamental symmetries and the formulation of conservation laws and theorems which characterise [some would dare to say "govern"] the behaviour of physical systems.

The three volumes in the series are partially sequenced: $\mathcal{MAPS}$ is propaedeutic, while $\mathcal{SPAM}$ and $\mathcal{AMPS}$ may be subsequently read in either order. There is also a gradual shift in writing style from a more discursive tone [with text and mathematical syntactic redundancy] in $\mathcal{MAPS}$ to a somewhat terser one in $\mathcal{SPAM}$ and $\mathcal{AMPS}$. An aim is to be welcoming to those new to the study of physics. Our intent is that upon completing this series students be prepared for specialised-subject upper-level undergraduate physics courses and textbooks.

A burden of thanks is owed to all of the students whom I have taught, and especially to those who greatly enjoyed it.

[This came as a surprise to some!]

Special credit is owed to those who have assisted me in the preparation of figures, especially Andy Geyer, and Carl and Jaspar von Buelow.

---

[2]This aspect of the courses hews more closely to traditional notions of liberal education.

Preface for Volume III

The acronym for this volume, AMPS, is the colloquial form of the SI unit for electric current. Here, we shall present one of science's greatest intellectual achievements: the formulation of the classical laws governing electromagnetism.

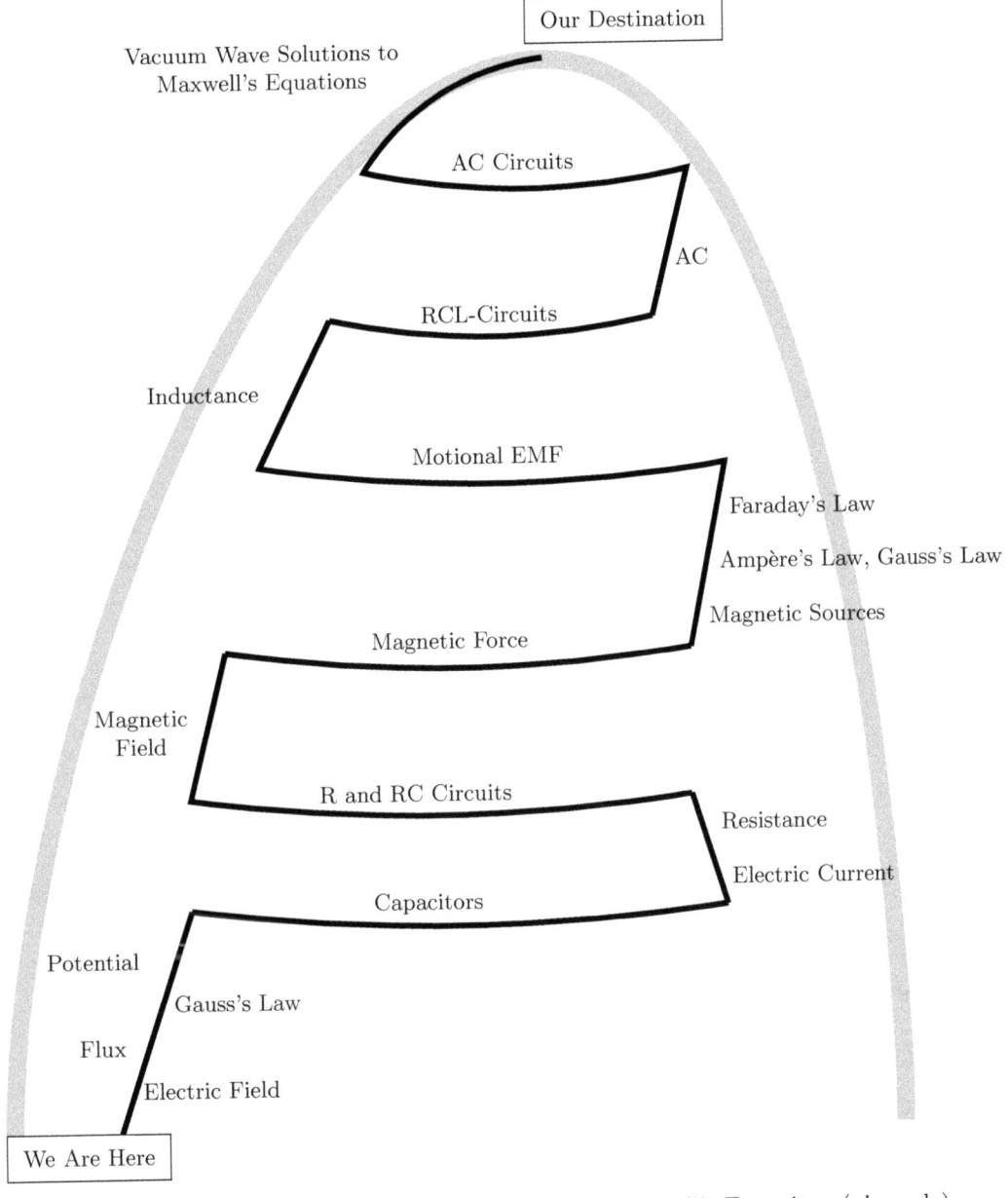

FIGURE 1 Our Route from Mechanics (base camp) to Maxwell's Equations (pinnacle)

We shall follow the path illustrated in Figure 1. The features named on the map will be explained in detail as we encounter them. Our ascent up this particular "Seven Storey

Mountain"[3] shall have periods of stiff climbing [studying aspects of electric and magnetic fields] punctuated by brief respites [analysing assorted circuits].

A potential pitfall along this route is that one might falsely conclude that the climbs and the switchbacks are disjoint. That they are indeed proper to the same mountain is illustrated in the Epilogue—after we have reached the summit. All of the various phenomena we study find their places in the single, common, unified structure identified by Maxwell in the 1860s.

---

[3]With apologies to both Thomas Merton and Dante Alighieri.

Part I

Electricity and Magnetism

Chapter 1

Electric Charge, Coulomb's Law, Electric Field

Electric charge is an observed [*i.e.*, phenomenological] property of matter. Two types[1] of charge exist: POSITIVE (+) and NEGATIVE (−). There is no compelling theoretical explanation for the existence of electric charge.

Electric charge is underlyingly responsible for nearly all of the structure and dynamics that we observe in the world around us. The chemical bonding of elements into molecules, and elements and molecules into compounds, forming the building blocks of observed matter, is essentially[2] of electric origin. The cohesion of matter into liquid and solid states and the approximate rigidity of solids [discussed in VOLUME II] result from residual electromagnetic effects. Normal forces and forces of contact [whose phenomenology was studied in VOLUME I] are also [distantly] of electromagnetic origin. [These residual forces also prevent the unlimited collapse of gravitating bodies comprised of normal matter,[3] like the Sun and planets.]

> ASIDE: With only slight exaggeration [residuals of the strong nuclear force bind the protons and neutrons into atomic nuclei characteristic of chemical isotopic elements], we can tease our philosopher friends by averring that electric forces generate order and form.

Investigations into electricity and magnetism began with the Greek philosopher-scientists who realised that, by rubbing silk or fur on glass or amber, they could do *party tricks*.

> ASIDE: The history of science makes for fascinating study and, as is so often the case with human endeavour, it is suffused with drama, serendipity, nobility, concupiscence, inspiration and perspiration. While the development of the classical theory of electromagnetism is a rich tale and much deserves to be told, we shall mostly eschew the historical telling in favour of a presentation of the finished models. Still, one cannot avoid the "accidents of convention." A sterling example of such is the definition of electric current, promoted by Ben Franklin and accepted by the community of eighteenth century scientists, as the flux of positive charge. Much later, in the latter part of the nineteenth century, it was realised that, in most cases, the microscopic origin of electric current is the coherent drift of negatively charged electrons in the diametrically opposite direction.

Ordinary matter is electrically neutral. Any net electric charge attributed to a material object [on scales ranging from the sub-nuclear to the galactic] is derived from all of the fundamental particles which together comprise[4] the charged object. Generation of **static electric charges** [net non-zero amounts which persist for long time scales] in bulk matter is usually accomplished by the transfer of electrons **to** or **from** neutral matter.

$$\text{Net transfer of electrons} \begin{bmatrix} \text{TO} \\ \text{FROM} \end{bmatrix} \text{an object}$$

$$\text{results in the accumulation of net} \begin{bmatrix} \text{NEGATIVE} \\ \text{POSITIVE} \end{bmatrix} \text{charge by the object.}$$

---

[1] Alternative words, if one prefers, include species or varieties.

[2] Quantum mechanical and magnetic effects are responsible for some particulars.

[3] Neutron stars must rely on a quantum mechanical property to forestall further compression.

[4] This is yet another instance of the whole's being precisely equal to the sum of its parts.

Three terms which we must define (qualitatively) are conductor, insulator, and ground.

CONDUCTOR   A conductor is a type of material within which electric charge is easily transported from one point in space to another. That is, charge may be made to flow without the application of appreciable effort.

INSULATOR   An insulator is a type of material in which charge is not easily transported. That is, electric charge flows with great difficulty, if at all.

GROUND   Electric ground is a source of charge which is large enough to be considered infinite.

There are two paramount aspects of electric charge.

  • **Electric charge is CONSERVED.**

   Electric charge is neither created nor destroyed. This does not prevent transfer of charge from one location to another, nor the production of equal and opposite charges from neutral matter by polarisation.

  • **Electric charge is QUANTISED.**

   Observable electric charge is quantised in units of the fundamental[5] charge.

Elementary Particle	Charge
PROTON	$+e$
ELECTRON	$-e$
NEUTRON	zero

ASIDE: All forms of nuclear matter are composed of **quarks** and **antiquarks**. In the STANDARD MODEL OF PARTICLE PHYSICS, there are six known flavours of quark, grouped into three families.

UP	STRANGE	TOP/TRUTH	electric charge $= +2\,e/3$
DOWN	CHARM	BOTTOM/BEAUTY	electric charge $= -e/3$

Quark confinement ensures that in all but the most extremely energetic conditions single ["bare"] quarks can never be isolated, and only quark–antiquark pairs or quark–quark–quark (or antiquark–antiquark–antiquark) triplets can be directly observed. The net charges borne by these aggregates of quarks are found in the set: $\{-2e, -e, 0, +e, +2e\}$. All of the observations made in terrestrial and astrophysical experiments have been consistent with this confined quark model.

It follows that the exact amount of electric charge borne by any physical object must be an integer multiple of the fundamental [*a.k.a.* elementary] charge,

$$q = \mathcal{N}\,e, \quad \text{for some } \mathcal{N} \in \mathbb{Z}, \text{ the set of integers}.$$

For large values of $\mathcal{N}$, the quantum nature of the charge may be difficult to discern.

---

[5]The fundamental charge is equal to that borne by an electron [NEGATIVE] or proton [POSITIVE]. It is a mystery why the magnitudes of the proton and electron charges are the same [to the precision with which we are able to compare them]. However, if this were not the case, then agglomerations of matter would have little chance of being electrically neutral, and this would be disruptive to the formation of structure in the universe.

The SI unit of charge is the coulomb [C]. The coulomb is not fundamental; rather, it is a derived unit. The elementary charge, expressed in coulombs, is

$$e = 1.609 \times 10^{-19} \text{ C} \simeq 1.6 \times 10^{-19} \text{ C}.$$

There are approximately $1/(1.6 \times 10^{-19}) \simeq 6.25 \times 10^{18}$ elementary charges per coulomb. The immense size of this number illustrates why the granularity of electric charge is not readily apparent, and justifies the classical model[6] of charge as a continuous fluid.

All of this is well and good. However, one cannot resist asking the following question.

Q: How do (two) charges interact?

A: Perhaps unsurprisingly, there are multiple model-dependent answers to this question.

COLLOQUIAL   The colloquial answer is: **like charges repel**, while **unlike charges attract**.

CLASSICAL   The classical answer is that the interaction between two point charges is described mathematically by the **Coulomb Force Law**, expounded upon below.

MODERN   In the modern theory of QUANTUM ELECTRODYNAMICS [QED], electromagnetic interactions are mediated by the exchange of virtual photons.

COULOMB'S LAW (for Two Point Charges)   The expression for the force exerted **by** a point[7] particle with electric charge q_1, located at spatial position $\vec{r}_1$, **on** another point particle bearing electric charge q_2, and located at $\vec{r}_2$, is quoted in Figure 1.1.

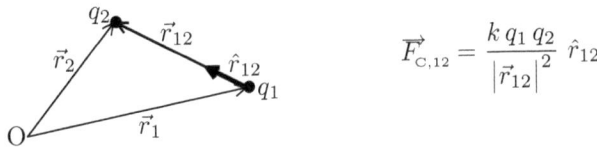

$$\vec{F}_{\text{C},12} = \frac{k\, q_1\, q_2}{\left|\vec{r}_{12}\right|^2}\, \hat{r}_{12}$$

FIGURE 1.1   Coulomb's Law for Point Charges

In the force expression, the **Coulomb constant**, k, equals 8.99×10^9 N $\cdot$ m^2/C^2, while $\vec{r}_{12} = \vec{r}_2 - \vec{r}_1$ is the relative position vector of the point charge being acted upon, with respect to the charge exerting the Coulombic force. The unit vector, $\hat{r}_{12}$, indicates the direction of $\vec{r}_{12}$, while $\left|\vec{r}_{12}\right|$ represents its magnitude. For brevity and concision, we continue the longstanding custom of labelling particles by their salient features, and simply refer to them as q_1 and q_2. **The Coulomb Force Law is an inverse-square Law.**

ASIDE: Newton's Law of Universal Gravitation (*circa* 1666) possesses the pulchritudinous properties [conservative force, integrability, associated potential energy] elucidated in VOLUME I, on account of its inverse-square form. It followed immediately upon Coulomb's discovery of his eponymous Law (*circa* 1785) that it too would have these properties.

---

[6]The same ideas pertain to the atomic/molecular granularity of matter.
[7]Not to worry, we'll get to point-like charges anon.

The colloquial description of the interaction of electric charges is readily seen to be consistent with Coulomb's Force Law.

- **like** charges   IF $q_1 q_2 > 0$, *i.e.*, the charges are of like sign, THEN the Coulombic force produced **by** q_1 **on** q_2 acts in the direction of $\hat{r}_{12}$, and thus they **repel**.

- **unlike** charges   IF $q_1 q_2 < 0$, *i.e.*, the charges have opposite signs, THEN the Coulombic force produced **by** q_1 **on** q_2 acts in the direction of $-\hat{r}_{12}$, and thus they **attract**.

Having determined the force produced by the point particle bearing charge q_1, acting on another with q_2, it is eminently reasonable to ask the following question.

Q: What is the force, $\vec{F}_{C,21}$, exerted **by** q_2, acting **on** q_1?

A: The result, obtained by merely permuting the labels, $1 \leftrightarrow 2$, in the expression of Coulomb's Law, is:

$$\vec{F}_{C,21} = \frac{k\, q_2\, q_1}{\left|\vec{r}_{21}\right|^2}\, \hat{r}_{21}\,.$$

The reversed vector, $\vec{r}_{21}$, is of precisely the same magnitude as $\vec{r}_{12}$, while pointing in the diametrically opposite direction. Hence, $\left|\vec{r}_{21}\right| = \left|\vec{r}_{12}\right|$ and $\hat{r}_{21} = -\hat{r}_{12}$. Thus, it follows that

$$\vec{F}_{C,21} = -\vec{F}_{C,12}\,,$$

in perfect accord with Newton's Third Law.

Extending the analysis, we pose another question.

Q: What happens when more than two charged particles interact?

A: Invoke the PRINCIPLE OF LINEAR SUPERPOSITION.

The net force exerted upon a charge, q_0, by a set of point charges, denoted by the collective symbol q', $q' = \{q_i, i = 1, 2, 3, \ldots, \mathcal{N}\}$, is the vector sum of the forces exerted by each of the q_i charges taken to act individually. That is,

$$\vec{F}_{q'q_0} = \vec{F}_{q_1 q_0} + \vec{F}_{q_2 q_0} + \vec{F}_{q_3 q_0} + \ldots \vec{F}_{q_{\mathcal{N}} q_0} = \sum_{i=1}^{\mathcal{N}} \vec{F}_{q_i q_0}\,.$$

Q: This appears daunting. Are there any helpful tricks that might spare us tedious computations?

A: Yes, perhaps. Exploit symmetries wherever possible.

EXAMPLE [*Net Coulombic Force in a Symmetrical Situation*]

Two equal, $1\,\mu\text{C}$ charges reside $4\,\text{m}$ apart. Without loss of generality, the charges may be labelled q_1 and q_2, and they may be assigned symmetric positions along the x-axis, *viz.*, $\vec{r}_1 = (-2,\, 0,\, 0)\,\text{m}$, while $\vec{r}_2 = (2,\, 0,\, 0)\,\text{m}$. Our intent is to determine the net Coulombic force exerted upon a charge, q_0, located at the origin, $\vec{r}_0 = (0,\, 0,\, 0)\,\text{m}$.

The symmetry of the system ensures that the net force on q_0 is identically zero. In fact, the x-component of the net force vanishes whenever q_0 lies anywhere in the yz-plane.

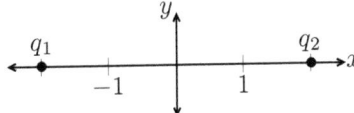

FIGURE 1.2 Coulombic Force in a Symmetric Situation

Coulomb's Law is plagued by the same ACTION AT A DISTANCE difficulties that infest Newton's Law of Universal Gravitation. Amelioration of the philosophical problems associated with this notion is accomplished by the introduction of an **electric field** permeating all of space.

ELECTRIC FIELD   The electric field, $\vec{E}_{q'}$ is formally defined via a limit:

$$\vec{E}_{q'}(\mathcal{P}) = \lim_{q_0 \to 0} \frac{\vec{F}_{q'q_0}}{q_0}\,.$$

This seemingly simple vector-valued function of position requires close explication.

q' represents the set of charges which together act as the **source** of the field.

q_0 denotes a **test charge** residing at the point in space, $\mathcal{P}$, where the field is to be determined.

$q_0 \to 0$ is taken so that the charges which comprise the source of the field are not themselves affected by Coulombic forces produced by q_0.

The electric field has SI units of newtons per coulomb [N/C]. Such units are consistent with the interpretation of the electric field as the **electric force per unit charge** at a particular point in space.

FIELD POINT   A field point is a location in space at which the value of a field [scalar, vector, or tensor[8]] is of interest.

....

EXAMPLE [*Electric Field of a Single Point Charge*]

The electric field produced by a single point charge, q_1, at a field point, $\mathcal{P}$, may be straightforwardly ascertained. Let $\vec{r}_{10}$ represent the relative position of $\mathcal{P}$ with respect to the location of the point charge, as is shown in Figure 1.3. The expression for the Coulombic force exerted on a test-charge at $\mathcal{P}$ appears alongside the sketch.

Thus, the electric field produced by q_1 at the point $\mathcal{P}$ is

$$\vec{E}_1(\vec{r}_{10}) = \lim_{q_0 \to 0} \frac{\vec{F}_{10}}{q_0} = \lim_{q_0 \to 0} \frac{\frac{k\,q_1\,q_0}{|\vec{r}_{10}|^2}\,\hat{r}_{10}}{q_0} = \frac{k\,q_1}{|\vec{r}_{10}|^2}\,\hat{r}_{10}\,.$$

The electric field, at an arbitrary point in space, produced by a single point charge is completely determined by the strength of the charge and the relative position of the field point.

---

[8]Appealing to [post-]Newtonian gravity for examples: the gravitational potential [the gravitational potential energy per unit mass at a point in space] is a scalar field; the gravitational field, $\vec{g}$, is vector-valued; in General Relativity, the spacetime metric tensor is the dynamical object.

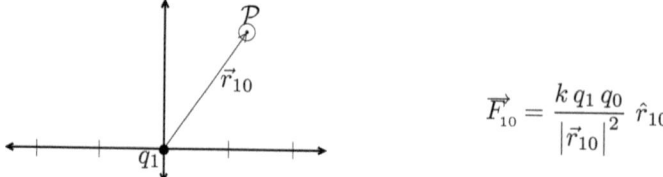

$$\vec{F}_{10} = \frac{k\,q_1\,q_0}{\left|\vec{r}_{10}\right|^2}\,\hat{r}_{10}$$

FIGURE 1.3   Electric Field of a Single Point Charge at the Field Point $\mathcal{P}$

EXAMPLE [*Electric Field of a Collection of Point Charges*]

The electric field at $\mathcal{P}$ may be produced by a collection of $\mathcal{N}$ point charges, q'. In Figure 1.4, there are four constituent charges, so $\mathcal{N} = 4$.

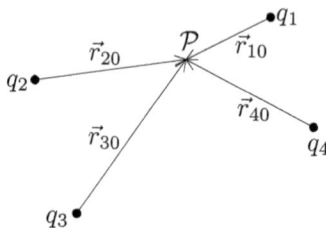

FIGURE 1.4   Four Charges Acting to Produce an Electric Field at $\vec{r}_0$

The Coulombic force obeys the LAW OF LINEAR SUPERPOSITION. Consequently, the electric field must do so, too. Thus, the net electric field is given by

$$\vec{E}_{q'}(r_p) = \lim_{q_0 \to 0}\frac{\vec{F}_{q'\,q_0}}{q_0} = \lim_{q_0 \to 0}\frac{\sum_{i=1}^{\mathcal{N}}\vec{F}_{q_i q_0}}{q_0} = \sum_{i=1}^{\mathcal{N}}\lim_{q_0 \to 0}\frac{\vec{F}_{q_i q_0}}{q_0} = \sum_{i=1}^{\mathcal{N}}\vec{E}_i(r_p)\,.$$

In these trivial examples, Coulomb's Law is evidently just as useful a calculational tool as the electric field. This observation inspires an existential enquiry.

Q: Why bother to introduce the electric field?

A: There are situations in which it is too hard to use Coulomb's Law directly, but we may still determine the electric field. In addition, it is always the case that the electric force exerted **on** a point particle, with charge q, resident at $\mathcal{P}$, in an electric field $\vec{E}$, is $\vec{F}_q = q\,\vec{E}(\mathcal{P})$, irrespective of how the electric field is produced!

ASIDE: The notation can be a barrier to understanding. Often, we write the electric field at a field point as $\vec{E}(\mathcal{P})$, sometimes as $\vec{E}(\vec{r}_p)$, and even, abstractly, as $\vec{E}$. In all cases, however, the precise meaning should be clear from the context.

Chapter 2

Electric Dipole, Motion of Charged Particles

Prior to beginning the discussion of dipoles, let's introduce a bit of broader terminology.

1   A **monopole** field is that produced by a single isolated point[-like] charge.

2   A **dipole** field is equivalent to the net field produced by a pair of isolated point[-like] **equal magnitude** AND **opposite sign** charges.

4   **Quadrupole** field configurations are produced by particular arrangements of two identical dipoles.

8   **Octopoles** ensue from certain superpositions of two quadrupoles.

⋮   ⋮

Any assemblage of point-like charges can be described in terms of its collective monopole, dipole, quadrupole, ... moments. Continuous [and spatially bounded] distributions of charge are also amenable to the multi-pole expansion.

ASIDE: For **electrostatics**, the simplest field structure is the electric monopole, while for [classical] **magnetostatics**, the elementary field configuration [readily] found in nature is the magnetic dipole.

ELECTRIC DIPOLE   The archetypical electric dipole consists of two equal and opposite point electric charges, $\{+q, -q\}$, separated by spatial distance $2\,a$, as illustrated in the leftmost panel of Figure 2.1.

FIGURE 2.1   Defining the Electric Dipole and Dipole Moment Vector

ELECTRIC DIPOLE MOMENT VECTOR   The electric dipole is characterised by its dipole moment vector, $\vec{p}$. In the case of Figure 2.1,

$$\vec{p} = 2\,a\,q\,\left[- \rightarrow +\right].$$

The magnitude of the dipole moment is equal to the distance between the centres of the point[-like] charges, multiplied by the magnitude of the charges. The SI units associated with the dipole moment are coulomb·metres $[\mathtt{C} \cdot \mathtt{m}]$. The dipole moment vector lies along the axis formed by the centres of charge, oriented **from** the [centre of] negative charge **toward** the [centre of] positive charge.

EXAMPLE [*The Field Produced by an Electric Dipole*]

Q: What is the electric field in the vicinity of a dipole?

A: We will provide a partial and approximate answer by explicitly constructing the electric dipole field along the dipole axis [in (i) and (ii) below], and everywhere in the plane which is perpendicular to the dipole axis and passing through the midpoint of the charges [in (iii)].

> ASIDE: In truth, one can formally determine the electric field anywhere in space by exactly superposing the Coulombic fields of the point(-like) charges. Such a precise and accurate computation is ultimately less informative and useful than the imprecise (and yet still accurate) computations of the far field behaviour of dipole electric fields which follow below.

Recalling Figure 2.1, let's extend the line joining the charges to form the x-axis. The perpendicular bisector of the line joining the charges, restricted to the plane of the page, shall constitute the y-axis. With coordinates chosen, we can unambiguously[1] state the classes of field points, $\mathcal{P} = (x_p\,,\,y_p)$, for which the form of the electric field produced by this dipole arrangement of charges shall be ascertained. These are indicated in Figure 2.2.

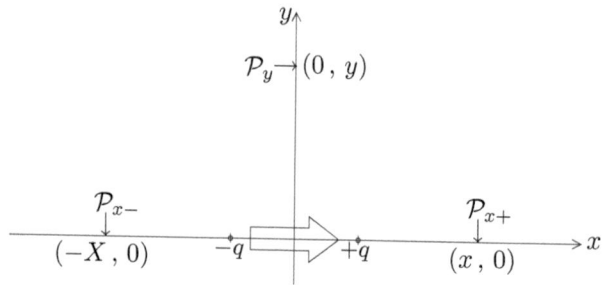

FIGURE 2.2   The Dipole Field at Particular Points

(i)   The first field point is found on the positive x-axis, far away from the charges, *i.e.*, $\mathcal{P}_{x+} = (x,\,0)$ with $x \gg a$. The scale on which the distance to the field point is deemed "far" is set by the size of the dipole.

The negative and positive charges each contribute to the net field at the location of the field point. Employing the definition of the constituent fields,

$$\vec{E}_{\mathrm{net}}(\mathcal{P}_{x+}) = \vec{E}_{-}(\mathcal{P}_{x+}) + \vec{E}_{+}(\mathcal{P}_{x+}) = \frac{k\,(-q)}{(x+a)^2}\,\hat{\imath} + \frac{k\,(+q)}{(x-a)^2}\,\hat{\imath}\,.$$

In each case, the numerator includes the charge of the source as a multiplicative factor, and thus is acutely sensitive to its sign. The denominators consist of the respective squared distances from source to field point, and thence the net field acquires crucial geometrical dependence. The unit vectors, always directed from the source to the field point, happen to coincide exactly with $\hat{\imath}$. The vector superposition, in this instance, yields

$$\vec{E}_{\mathrm{net}}(\mathcal{P}_{x+}) = k\,q\left[\frac{1}{(x-a)^2} - \frac{1}{(x+a)^2}\right]\hat{\imath}\,.$$

---

[1] The z-coordinates are elided in recognition of the axial rotation symmetry of the dipole configuration.

While this is exact, we shall invoke the far field limit, $x \gg a$, and employ a [truncated] binomial expansion of the terms in the bracket, to achieve a marked simplification:

$$\frac{1}{(x-a)^2} = \frac{1}{x^2\left(1-\frac{a}{x}\right)^2} = \frac{1}{x^2}\left(1-\frac{a}{x}\right)^{-2} = \frac{1}{x^2}\left(1+2\frac{a}{x}+\ldots\right),$$

$$\frac{1}{(x+a)^2} = \frac{1}{x^2\left(1+\frac{a}{x}\right)^2} = \frac{1}{x^2}\left(1+\frac{a}{x}\right)^{-2} = \frac{1}{x^2}\left(1-2\frac{a}{x}+\ldots\right).$$

Armed with these expansions, we may simplify the expression for the net electric field,

$$\vec{E}_{\text{net}}(\mathcal{P}_{x+}) = \frac{kq}{x^2}\left[1+2\frac{a}{x}+\ldots-1+2\frac{a}{x}+\ldots\right]\hat{\imath} \simeq \frac{4kqa}{x^3}\hat{\imath} = \frac{2k\vec{p}}{x^3}.$$

Q: Can we make sense of this result?

A: You betcha. Far away, the superposed fields are of nearly the same magnitude, possess opposite directions, and hence mostly cancel. One would expect the net field to be weaker, and more rapidly diminishing, than that which would be produced by either charge alone. In the expression derived above, the net electric field falls off as the cube of the distance from the dipole to the field point, in contrast to the inverse-square behaviour of the field proper to each monopole.

(ii) This field point is located along the negative x-axis, far away from the charges, *i.e.*, $\mathcal{P}_{x-} = (-X, 0)$ with $X \gg a$. In this analysis, X is a positive quantity representing how far the field point lies from the origin in the negative x-direction.

Once again, the negative and positive charges contribute monopolar fields which together yield the net field at the point of interest,

$$\vec{E}_{\text{net}}(\mathcal{P}_{x-}) = \vec{E}_{-}(\mathcal{P}_{x-}) + \vec{E}_{+}(\mathcal{P}_{x-}) = \frac{k(-q)}{(X-a)^2}(-\hat{\imath}) + \frac{k(+q)}{(X+a)^2}(-\hat{\imath}).$$

Employing the same conventions as in part (i),

$$\vec{E}_{\text{net}}(\mathcal{P}_{x-}) = kq\left[\frac{1}{(X-a)^2} - \frac{1}{(X+a)^2}\right](-\hat{\imath}).$$

In the far field limit, $X \gg a$, and binomial expansions analogous to those above,

$$\frac{1}{(X-a)^2} = \frac{1}{X^2}\left(1+2\frac{a}{X}+\ldots\right),$$

$$\frac{1}{(X+a)^2} = \frac{1}{X^2}\left(1-2\frac{a}{X}+\ldots\right),$$

simplify the approximation to the net electric field at $\mathcal{P}_{x-}$,

$$\vec{E}_{\text{net}}(\mathcal{P}_{x-}) = \frac{kq}{X^2}\left[1-2\frac{a}{X}+\ldots-1-2\frac{a}{X}+\ldots\right](-\hat{\imath}) \simeq \frac{-4kqa}{X^3}(-\hat{\imath}) = \frac{2k\vec{p}}{X^3}.$$

The dipole field far along the dipole axis has the same form in both directions.

(iii) The field point along the perpendicular bisector of the line joining the centres of the two point charges has coordinates $\mathcal{P}_y = (0, y)$. No *a priori* assumptions about the relative size of y and a shall be made, although later we will focus attention on the field

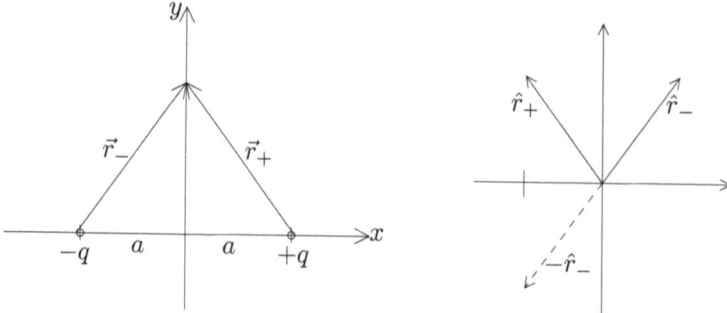

FIGURE 2.3 The Electric Field of a Dipole along Its Perpendicular Bisector

far from the dipole source. The geometry of the superposition is illustrated in Figure 2.3, which shows the vectors from the sources to the field point along with their directions.

Owing to the axial and reflection symmetries of the charge configuration,

$$|\vec{r}_+|^2 = |\vec{r}_-|^2 = a^2 + y^2 \,,$$

in the present case. Thus, the electric field at $\mathcal{P}_y$ is

$$\vec{E}_{\text{net}}(\mathcal{P}_y) = \vec{E}_-(\mathcal{P}_y) + \vec{E}_+(\mathcal{P}_y) = \frac{k\,(-q)}{a^2 + y^2}\,\hat{r}_- + \frac{k\,(+q)}{a^2 + y^2}\,\hat{r}_+ = \frac{k\,q}{a^2 + y^2}\left[\hat{r}_+ - \hat{r}_-\right].$$

The magnitude and direction of the difference of unit vectors, $\hat{r}_+ - \hat{r}_-$, must be determined. The rightmost panel of Figure 2.3 makes evident that, under this vector subtraction, the vertical components cancel, while the horizontal components add. Explicitly, the $(x,\,y)$ components of the unit vectors are

$$\hat{r}_+ = \frac{(-a\,,\,r)}{|\vec{r}_+|} = \left(\frac{-a}{\sqrt{a^2 + y^2}}\,,\,\frac{r}{\sqrt{a^2 + y^2}}\right)$$

and

$$\hat{r}_- = \frac{(a\,,\,r)}{|\vec{r}_-|} = \left(\frac{a}{\sqrt{a^2 + y^2}}\,,\,\frac{r}{\sqrt{a^2 + y^2}}\right).$$

Thus,

$$\hat{r}_+ - \hat{r}_- = \left(\frac{-2\,a}{\sqrt{a^2 + y^2}}\,,\,0\right) = \frac{2\,a}{\sqrt{a^2 + y^2}}\,(-1\,,\,0),$$

confirming our expectations. Hence,

$$\vec{E}_{\text{net}}(\mathcal{P}_y) = \frac{k\,q}{a^2 + y^2}\left[\hat{r}_+ - \hat{r}_-\right] = \frac{2\,k\,q\,a}{\left(a^2 + y^2\right)^{3/2}}\,(-1,\,0) = \frac{-k\,\vec{p}}{\left(a^2 + y^2\right)^{3/2}}\,.$$

In the far-field limit, $y \gg a$, the net dipole field along the perpendicular bisector falls off as the cube of the distance to the field point.

Cubic falling off with distance is characteristic of dipole fields.

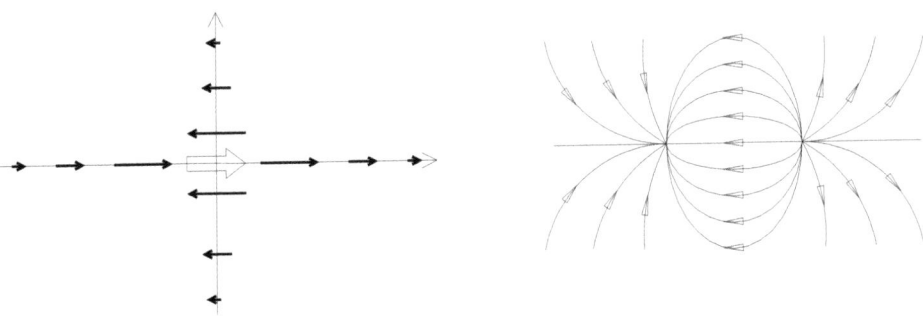

FIGURE 2.4   The Electric Field of a Dipole: Results (i → iii) and Their Extension

The electric [vector] field can be visualised by drawing arrows indicating its direction and relative magnitude at a finite set of discrete points in space. We do this for the portions of the dipole field along the dipole axis, (i - ii), and perpendicular bisector, (iii), in the left panel of Figure 2.4.

The iconic image of the dipole in the right panel displays the vector field along curves to which it is everywhere tangent. These are the **electric field lines** associated with the dipole charge distribution.

ELECTRIC FIELD LINES   Electric field lines are obtained by joining together the electric field vectors at adjacent points in space.

> ASIDE: This "integrability" is non-trivial.

The tangent to the field line at a point approximates the trajectory that a test charge would follow if it were released from rest at that point into the local electric field.

[Field lines are not the actual trajectories of charged particles, as no account is taken of inertia.]

Drawing field lines helps one to acquire knowledge about the magnitude and direction of electric fields in particular cases. The process of sketching such diagrams is codified by a set of six basic rules.

(i)   **SYMMETRY**   Exploit symmetries.

(ii) **SOURCE/SINK**   Field lines **originate** from (+) charges and **terminate** on (−) charges.

$$\begin{bmatrix} \text{Positive} \\ \text{Negative} \end{bmatrix} \text{charge acts as the} \begin{bmatrix} \text{source} \\ \text{sink} \end{bmatrix} \text{of electric field lines.}$$

(iii) **#**   The number of field lines originating from, or terminating on, a particular charge is proportional to its magnitude.

(iv) **DENSITY**   The density of the field lines in the neighbourhood of a field point corresponds to the local field strength.

(v) **FAR FIELD**   At large distances away from isolated charges [or bounded distributions of charge] the pattern of field lines simplifies. Field lines may even end at infinity.

(vi) ×   **Field lines are not permitted to intersect!**

PROOF that the Field Lines May Not Cross

Suppose that two field lines intersect at some particular point in space.

Q: In what direction is the electric field pointing at the intersection of the field lines?

A: The direction is ambiguous. However, this is completely unphysical, since it would imply that Newton's Laws could not be employed to predict the acceleration of a charged test particle released from rest at the point in question. The only way to avoid this *reductio ad absurdum* is to deny the premise upon which it is founded. Thus, we infer that crossed lines may never occur.

Motion of Charged Point[-like] Particles in a Uniform Electric Field

Before determining the influence of uniform electric fields upon the motion of charged particles, one might ask the logically prior and pertinent question of existence.

Q: How does one produce a uniform electric field?

A$_{snarky}$**:** WHO CARES?
All that is needed to determine the acceleration of a particle, of mass m and bearing charge q, is the net external force acting upon it. The magnitude and direction of the electric force is obtained from the electric field via $\vec{F}_E = q\,\vec{E}$. IF the electric force is the only force acting [or conditions allow for application of the Impulse Approximation], THEN

$$ m\,\vec{a} = \vec{F}_{\text{NET}} = \vec{F}_E = q\,\vec{E} \quad \Longrightarrow \quad \vec{a} = \frac{q}{m}\,\vec{E}. $$

IF the electric field is uniform, THEN the acceleration of the charged particle is constant.[2]

A$_{serious}$**:** In later chapters we will realise how one might produce a [relatively] uniform electric field within bounded regions of space. The uniform electric field is akin to the uniform gravitational field in its simplicity.

Whether we choose to be serious or snarky, our first inescapable inference is that the acceleration is parallel/antiparallel to the field for particles bearing positive/negative charge, and zero for those without net charge.

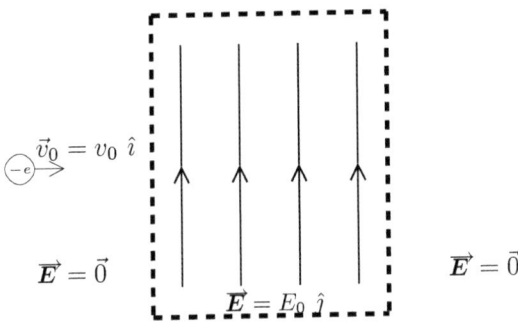

FIGURE 2.5 An Electron Is Injected into a Region of Non-Zero Electric Field

EXAMPLE [*Electron Deflection by an Electric Field*]

An electron is fired horizontally, with speed v_0, into a region, L metres wide, of constant vertical electric field. Elsewhere, the electric field vanishes. Three questions of import are listed (and answered) below.

Q: (i) What is the trajectory of the electron?

(ii) At what point does the electron exit the region of non-zero field?

(iii) With what velocity does the electron exit the field?

A: As is customary, we make a sketch (see Figure 2.5) to establish context and coordinates. The physical model developed to answer the questions includes the following assumptions.

⊖→ The electron is a point particle with mass m_e and charge $-e$.

⊖→ The electric field vanishes everywhere in space, except for a rectangular region in the xy-plane [of width L and sufficient height to accommodate the electron's trajectory] in which the field is $\vec{E} = E_0\,\hat{\jmath}$, where E_0 is constant. WLOG, the field boundaries occur at $x = 0$ and $x = L$.

⊖→ Prior to encountering the field region, the electron moves inertially with speed v_0 parallel to the x-axis.

⊖→ The initial conditions for the electron entering the field may be chosen to be $\vec{r}_0 = (0\,,0)$ and $\vec{v}_0 = (v_0\,,0)$.

⊖→ Notwithstanding that the force of gravity acts on the electron throughout its motion, the electric force is almost certainly of sufficient strength that the Impulse Approximation justifies the neglect of the electron's weight in the determination of its motion. Hence, the acceleration endured by the electron, once it enters the field, is $\vec{a} = (0\,, -e\,E_0/m_e)$.

⊖→ Thus, the constant acceleration kinematical formulae [adapted to the particular boundary conditions enumerated above] apply:

$$\vec{v}(t) = \left(v_0\,, -\frac{e\,E_0\,t}{m_e}\right) \quad \text{and} \quad \vec{r}(t) = \left(v_0\,t\,, -\frac{e\,E_0\,t^2}{2\,m_e}\right).$$

---

[2]Constancy of mass and electric charge is implicitly assumed throughout these notes.

⊖→   Subsequent to its passage through the [non-zero region of the] field, the electron's motion is, once again, inertial.

The portions of the trajectory of the electron prior to entering and within the non-zero field region have been described. We have not yet determined the post-field trajectory, nor have we answered the other questions posed at the start of this example. *En route* to completing the answer to the first, we must knock off the other two.

A$_{ii}$: Parametric inversion of the horizontal component of position, $t = x/v_0$, allows us to write

$$y(x) = -\frac{e\,E_0\,x^2}{2\,m_e\,v_0^2}\,.$$

The electron emerges from the field when its x-component is equal to L, *i.e.*,

$$\vec{r}_{\text{emergence}} = \left(L\,,\ -\frac{e\,E_0\,L^2}{2\,m_e\,v_0^2}\right).$$

A$_{iii}$: The exit velocity of the electron may be computed by a number of means.

METHOD ONE: [*A timely approach*]

The constant horizontal motion of the electron fixes the time at which it reaches the rightmost edge of the field region to be $t_L = L/v_0$. At this instant, its velocity is

$$\vec{v}(t_L) = \left(v_0\,,\ -\frac{e\,E_0\,L}{m_e\,v_0}\right).$$

This answers the third question and, along with the answer to the second, specifies the initial conditions for the inertial trajectory in the post-field region.

METHOD TWO: [*Tricky invocation of the kinematical formulae*]

Let's ignore the horizontal motion and consider only the vertical motion.

[One may adopt the inertial frame in which the electron is initially at rest.]

The third constant acceleration kinematical formula [applicable in 1-d] reads: $v_f^2 = v_i^2 + 2\,a\,(\Delta y)$. In the present context, the various quantities on the RHS are known,

$$v_i = 0\,, \qquad a = -\frac{e\,E_0}{m_e}\,, \qquad \Delta y = -\frac{e\,E_0\,L^2}{2\,m_e\,v_0^2}\,,$$

leading to an expression of the LHS,

$$v_f^2 = \frac{e^2\,E_0^2\,L^2}{m_e^2\,v_0^2} \qquad \Longrightarrow \qquad v_f = -\frac{e\,E_0\,L}{m_e\,v_0}\,,$$

where the vertical velocity component was assigned to be negative to comport with our understanding that the electron is deflected downward by the upward-directed electric field.

Upon exiting the field, the particle's motion is again inertial, and thus its trajectory is given parametrically by

$$x\left(\tilde{t}\right) = L + v_0\,\tilde{t} \qquad \text{and} \qquad y\left(\tilde{t}\right) = -\left[\frac{e\,E\,L}{m_e\,v_0}\right]\left(\frac{L}{2\,v_0} + \tilde{t}\right),$$

where we have introduced a split time parameter, $\tilde{t}$, for this third dynamical regime.

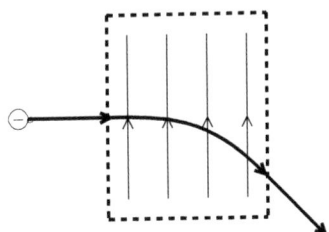

FIGURE 2.6 The Trajectory of an Electron Passing through a Uniform Field

---

The complete trajectory of the electron is sketched in Figure 2.6.

> ASIDE: This is, [very] roughly speaking, how the picture is formed on the screen of an old-fashioned computer monitor or television set. The electron gun and fluorescent screen assembly is called a Cathode Ray Tube [CRT]. Energetic electrons hitting phosphor dots on the screen excite atoms, whose subsequent de-excitation is accompanied by the emission of light of a particular colour/wavelength. The controlled spray of energetic electrons across the scintillator-coated screen is called a "**raster scan.**"

EXAMPLE [*Braking Fields*]

Q: What is the minimum initial speed needed for a proton to break through a specified region of opposing electric field?

A: Figure 2.7 provides context for the model whose attributes are listed below.

⊕→   The proton is a point[-like] particle bearing charge $q = +e$.

⊕→   A uniform non-zero electric field occupies a region of space with thickness, L, as shown in Figure 2.7. Everywhere else, the field vanishes.

⊕→   A mechanism fires the proton, directly against the field, with a calibrated [controlled and determined] initial speed, v_0.

⊕→   This problem is effectively one-dimensional. Choose the x-direction to be increasing towards the right, with its origin at the left edge of the field region. The initial velocity of the proton is $\vec{v}_i = v_0\,\hat{\imath}$.

⊕→   The Impulse Approximation applies. Prior to entering the non-zero field, the proton moves inertially. Inside the field, the force exerted on the particle and the acceleration it endures are

$$\vec{F}_{\text{NET}} = (+e)\,\vec{E} \qquad \Longrightarrow \qquad \vec{a} = \frac{e\,\vec{E}}{m_p} = \frac{e\,E_0}{m_p}\,(-\hat{\imath})\,.$$

Three possibilities arise.

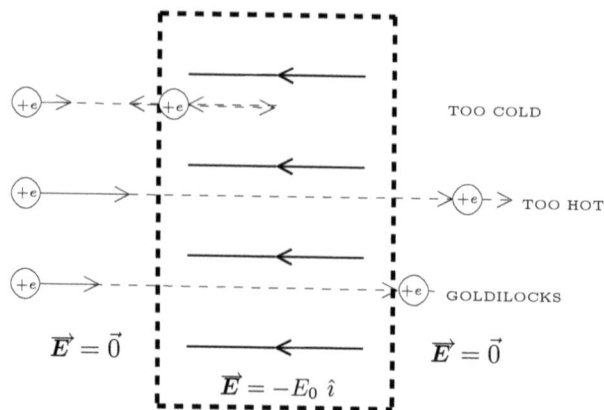

FIGURE 2.7   A Proton Propelled into an Opposing Electric Field

TOO COLD	The proton is stopped, and reversed, by the field.
TOO HOT	The proton sails through the field and continues in motion.
GOLDI-LOCKS	The proton comes to rest just outside of the non-zero field region at $x = L$. This is the critical case.

The third constant acceleration kinematical formulae, $v_f^2 = v_i^2 + 2\,a\,(\Delta x)$, adapted to the critical case, along with

$$a = -\frac{e\,E_0}{m_p}\,, \qquad v_f = 0\,, \qquad \text{and} \qquad \Delta x = L\,,$$

yields

$$v_{0,\text{min}}^2 = 2\,a\,(\Delta x) \qquad \Longrightarrow \qquad v_{0,\text{min}} = \sqrt{\frac{2\,e\,E_0\,L}{m_p}}\,.$$

The expression for the minimal breakthrough speed has dependences that seem eminently reasonable. For instance, it increases with the strength and thickness of the opposing field. Furthermore, the minimum speed is smaller for a more massive particle, *e.g.*, a deuteron or a singly ionised atom.

Later we'll see that this [very cute] result admits an energetic interpretation.

Chapter 3

Continuous Charge Distributions

Thus far, we have considered point charges only.

Q: What if the charges are distributed throughout a region of space?

[Charge is always associated with matter. Hence, the formalism must accommodate distributions.]

A: There are several ways of answering this sort of question.

    METHOD ONE: [*Getting perspective*]

Viewed from a sufficiently great distance, the extended distribution may appear point-like.

Problem solved,[1] eh?

    METHOD TWO: [*Getting a better perspective*]

LINEAR SUPERPOSITION enabled determination of the net field produced by a collection of isolated point charges. This militates for applying the PARTITION, COMPUTE, and SUM approach to continuous distributions.

    P   PARTITION the solid body into $\mathcal{N}$ [little] chunks. The chunk [*a.k.a.* subvolume, or charge element] labelled by the index, i, bears electric charge Δq_i. [The explicit use of "Δ" presages the eventual infinite refinement of the partition.]

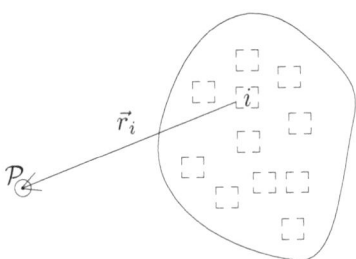

FIGURE 3.1   Distributed Charge Partitioned into Chunks

    The chunks are assumed to be boundable and small enough to make meaningful the association of a unique position vector with each chunk. The vector extending from the ith chunk to the field point, $\mathcal{P}$, is labelled $\vec{r}_i$.

    C   COMPUTE the electric field produced at the field point by the ith charge element, Δq_i, which we treat as though it were a point charge. *I.e.,*

$$\Delta \vec{E}_i(\mathcal{P}) = \frac{k \, \Delta q_i}{\left| \vec{r}_i \right|^2} \, \hat{r}_i \, ,$$

where each of the symbols in the above expression has its usual meaning.

---

[1] Unfortunately, this "solution" is inadequate.

S   The total or net electric field at $\mathcal{P}$ is the SUM of the field contributions from each of the charge elements comprising the entire distribution.

$$\vec{E}_{\text{net}}(\mathcal{P}) = \sum_{i=1}^{\mathcal{N}} \Delta \vec{E}_i(\mathcal{P}) = \sum_{i=1}^{\mathcal{N}} \frac{k\,\Delta q_i}{\left|\vec{r}_i\right|^2}\,\hat{r}_i\,.$$

Although this expression for the net field is daunting to contemplate, it is [yet] another triumph for the iterative PARTITION, COMPUTE, and SUM technique. One can, in principle and in practice,[2] determine $\vec{E}_{\text{net}}(\mathcal{P})$ to any desired degree of accuracy, subject to the assumptions made in the model for the distribution of charge.

For purposes of explication, we desire exact analytical results. To obtain these, one must infinitely REFINE the partition by taking, simultaneously and self-consistently, the limits:

$$\lim_{\mathcal{N}\to\infty} \qquad \text{AND} \qquad \lim_{\forall\, i,\ \Delta q_i \to 0}\,.$$

The sum over the charge elements passes over into an integral over the charge distribution:

$$\vec{E}(\mathcal{P}) = \int \frac{k\,dq}{\left|\vec{r}\right|^2}\,\hat{r}\,.$$

Despite its syntactic simplicity, this expression is a complicated vector integration over the charge elements distributed in space.

CAVEAT: This passage from sum to integral is sometimes presented as though it were merely a matter of symbolic replacements,

$$\Delta q_i \to dq \qquad \text{and} \qquad \sum_{i=1}^{N} \to \int\,.$$

The infinitesimal charge elements, dq, when taken together, comprise the entire distribution. In many cases, the charge elements must be re-expressed in terms of the geometry [size and shape] of the distribution in order to effect the integration(s). Similarly confounding vector integrations arose in the CofM computations performed in Chapter 30 of VOLUME I. In parallel with that discussion, we shall closely examine the electric fields generated by **lineal**, **areal**, and **volume** distributions of electric charge.

1-d   In a one-dimensional distribution of charge,

$$dq = \lambda\,ds\,,$$

where λ denotes the **local lineal charge density** at each point, and ds represents the line element along the segment of the charge distribution at the same point. The charge density typically depends on position. However, IF the charge is uniformly distributed, THEN λ is constant. In such cases, its value, $\lambda_0 = Q_{\text{Total}}/L$, is fixed by the total charge, Q_{Total}, and the [total] length, L.

2-d   In two dimensions,

$$dq = \sigma\,dA\,,$$

where σ and dA denote the **local areal charge density** and the local area element lying tangent to the surface at each point on the charge-bearing surface. IF an amount of charge, Q_{Total}, is uniformly distributed over a surface of area A, THEN the areal charge density is constant, $\sigma_0 = Q_{\text{Total}}/A$.

---

[2]It may take a large partition and a great deal of computer time, but it can always be done.

3-d   In 3-d,
$$dq = \rho \, dV \, ,$$

where ρ and dV represent the **local volume charge density** and the local volume element within the charge-bearing region, respectively. IF the charge is uniformly distributed through volume V, THEN $\rho = \rho_0 = Q_{\text{Total}}/V$.

Consistent SI units for the various charge densities are as follows:

$$[\lambda] = \frac{\text{C}}{\text{m}} \, , \qquad\qquad [\sigma] = \frac{\text{C}}{\text{m}^2} \, , \quad \text{and} \qquad\qquad [\rho] = \frac{\text{C}}{\text{m}^3} \, .$$

EXAMPLE [*Electric Field along the Axis and External to a Uniformly Charged Rod*]

A thin rod of length L has uniform positive lineal charge density, $\lambda_0 > 0$. Let us choose coordinates such that the x-axis lies along the rod, with the origin of coordinates at its centre, as illustrated in Figure 3.2.

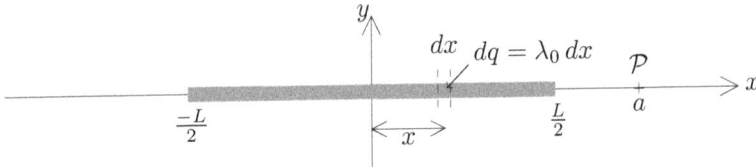

FIGURE 3.2   On-Axis Electric Field due to a Uniformly Charged Rod

We intend to determine the magnitude and direction of the electric field produced by the charge on the rod at any field point, $\vec{r}_p = (a, 0)$, with $a > L/2$.

ASIDE: Symmetry ensures that only $a > L/2$ need be considered. Pathologies might be expected for $|a| \leq L/2$.

As a warm-up,[3] let's compute the total charge on the rod,

$$Q_{\text{Total}} = \int dq = \int_{-\frac{L}{2}}^{\frac{L}{2}} \lambda_0 \, dx = \lambda_0 \, [x] \Big|_{-\frac{L}{2}}^{\frac{L}{2}} = \lambda_0 \, L \, .$$

It is no surprise to find that the total amount of electric charge on the rod is equal to its [constant] charge-per-unit-length multiplied by its total length.

The differential contribution to the vector electric field at a field point, $\mathcal{P}$, is

$$d\vec{E}(\mathcal{P}) = \frac{k \, dq}{|\vec{r}|^2} \, \hat{r} \, .$$

As shown in Figure 3.2, the field point is located at $\vec{r}_p = (a, 0)$, while the infinitesimal charge element shown, $dq = \lambda_0 \, dx$, is at $(x, 0)$. The vector from each differential source to the field point has magnitude $|\vec{r}| = a - x$, and in all cases its direction is $\hat{r} = \hat{\imath}$. Combining

---

[3]This is true, but is by no means the sole motivation.

these particulars into the general expression for $d\vec{E}$, and integrating over the extent of the rod, yields:

$$\vec{E}(\mathcal{P}) = (E_x, 0) = \int_{-\frac{L}{2}}^{\frac{L}{2}} \frac{k\lambda_0\,dx}{(a-x)^2}\,\hat{\imath} = k\lambda_0 \left[\frac{1}{a-x}\Big|_{-\frac{L}{2}}^{\frac{L}{2}}\right]\hat{\imath} = k\lambda_0 \left[\frac{1}{a-\frac{L}{2}} - \frac{1}{a+\frac{L}{2}}\right]\hat{\imath}$$

$$= \frac{k\lambda_0 L}{a^2 - \left(\frac{L}{2}\right)^2}\,\hat{\imath} = \frac{kQ_{\text{Total}}}{a^2 - \left(\frac{L}{2}\right)^2}\,\hat{\imath}.$$

In the last equality, the expression for the net, or total, field was rewritten in terms of the total amount of charge on the rod.

Q: Does this result make sense?

A: Yes, insofar as when $a \gg \frac{L}{2}$, the electric field produced by the charge distributed throughout the length of the rod looks like that of a point charge, Q_{Total}, situated at the point occupied by the centre of the rod.

EXAMPLE [*Field along the Perpendicular Bisector of a Uniformly Charged Rod*]

Considering the same thin rod as in the previous example, we now seek to determine the electric field that it produces along its perpendicular bisector. WLOG, we place the field point along the positive y-axis, *i.e.*, $\mathcal{P} = (0, a)$ with $a > 0$. The axial symmetry of the situation suggests[4] that only the y-component of the electric field will be non-zero.

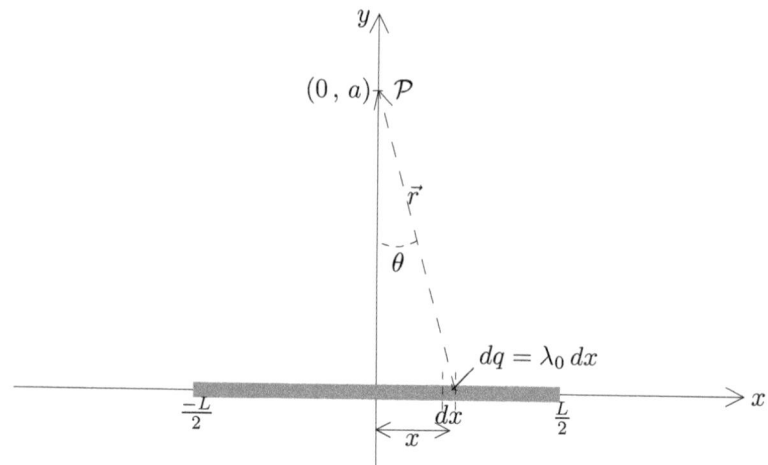

FIGURE 3.3 Field along the Perpendicular Bisector of a Uniformly Charged Rod

At the field point $\mathcal{P}$, illustrated in Figure 3.3, the [differential] contributions to the net electric field produced by different charge elements, dq, vary in both magnitude and direction. Consider the magnitude of $d\vec{E}$ for a particular dq:

$$|d\vec{E}| = \frac{k\,dq}{r^2}.$$

---

[4] In deference to the skeptical, we shall soon verify that the x-component of the net electric field vanishes.

The Cartesian components of the differential field are:

$$dE_x = \hat{\imath} \cdot d\vec{E}(\mathcal{P}) = -\left|d\vec{E}\right| \sin(\theta) \qquad \text{and} \qquad dE_y = \hat{\jmath} \cdot d\vec{E}(\mathcal{P}) = \left|d\vec{E}\right| \cos(\theta),$$

where θ is the angular position of the charge element as discerned from the field point. The unit vector aligned with the differential electric field lies parallel to the vector extending from dq to the field point, *i.e.*,

$$\hat{r} = \left(-\sin(\theta),\, \cos(\theta) \right).$$

The y-component of the net field is

$$E_y = \int \frac{k\, dq}{|\vec{r}|^2} \cos(\theta).$$

In the case at hand, $dq = \lambda_0\, dx$, where $\lambda_0 = Q_{\text{Total}}/L$ is constant; the denominator is $|\vec{r}|^2 = a^2 + x^2$; and the cosine factor may be written as

$$\cos(\theta) = \frac{a}{r} = \frac{a}{\sqrt{a^2 + x^2}}.$$

When we combine these results and recall that $\{x \in [-L/2, L/2]\,,\ y = 0\}$ describes the extent of the rod, we obtain an integral expression for the y-component of the electric field:

$$E_y = k\,\lambda_0\, a \int_{-\frac{L}{2}}^{\frac{L}{2}} \frac{dx}{\left(a^2 + x^2\right)^{\frac{3}{2}}}.$$

The indefinite integral appearing here has a closed-form solution,

$$\int \frac{dx}{\left(a^2 + x^2\right)^{\frac{3}{2}}} = \frac{1}{a^2}\, \frac{x}{\sqrt{a^2 + x^2}}.$$

Thus,

$$E_y = k\,\lambda_0\, \frac{1}{a}\, \frac{x}{\sqrt{a^2 + x^2}}\Bigg|_{-\frac{L}{2}}^{\frac{L}{2}} = \frac{k\,\lambda_0\, L}{a\,\sqrt{a^2 + \left(\frac{L}{2}\right)^2}} = \frac{k\,Q_{\text{Total}}}{a\,\sqrt{a^2 + \left(\frac{L}{2}\right)^2}}.$$

In the last equality, the net field is re-expressed in terms of the total charge on the rod.

Q: Does this result make sense?

A: Good question! Let's take some limits.

FAR   IF the field point is very far away from the rod, *i.e.*, $a \gg L$, THEN the rod appears small when viewed from the perspective of an observer at the field point, and hence the electric field should approach that of a point source [*i.e.*, a monopole]. In this limit, the denominator in the expression for E_y becomes

$$a\,\sqrt{a^2 + \left(\frac{L}{2}\right)^2}\,\Bigg|_{a \gg L} = a^2\,\sqrt{1 + \left(\frac{L}{2a}\right)^2}\,\Bigg|_{a \gg L} \simeq a^2,$$

so the force law exhibits [approximate] inverse-squared dependence at large distances, just like a point charge.

NEAR   IF the field point is very near to the rod, *i.e.*, $a \ll L$, THEN the rod subtends a large angle from the perspective of an observer standing at the field point. The electric field should resemble that produced by an infinitely long straight wire bearing uniform charge density. In the limit in which the rod is much longer than the off-axis distance to the field point,

$$a\sqrt{a^2 + \left(\frac{L}{2}\right)^2}\,\Bigg|_{a \ll L} = \frac{aL}{2}\sqrt{\left(\frac{2a}{L}\right)^2 + 1}\,\Bigg|_{a \ll L} \simeq a\,\frac{L}{2}\,,$$

and thus the field approaches

$$E_y \simeq \frac{2\,k\,Q_{\text{Total}}}{a\,L} = \frac{2\,k\,\lambda_0}{a}\,.$$

The robustness of the argument leading to this result inspires the realisation that **this is an accurate expression for the net electric field produced by an infinite line of uniform charge density**.

Two clarifying comments are helpful.

- It is, of course,[5] physically impossible to produce a truly infinite line of uniform charge density.

- The idea of SCALE is at work here, *viz.*, IF the rod is much longer than the [perpendicular] distance to the field point, THEN the infinite rod model provides a reasonable approximation to the actual field.

    As one gets closer and closer to any finite line segment, it begins to "look" infinite, as one must turn one's head by almost π radians to survey its extent.

———————

To placate the skeptics, let's work out the net x-component of the electric field in order to show explicitly that it vanishes. From the trigonometric analysis conducted above,

$$E_x = \int \frac{k\,dq}{|\vec{r}\,|^2}\,\sin(\theta)\,,$$

where, according to our previous determinations,

$$|\vec{r}\,|^2 = a^2 + x^2\,, \qquad \text{and} \qquad \sin(\theta) = \frac{x}{r} = \frac{x}{\sqrt{a^2 + x^2}}\,.$$

Hence,

$$E_x = k\,\lambda_0 \int_{-\frac{L}{2}}^{\frac{L}{2}} \frac{x\,dx}{\left(a^2 + x^2\right)^{\frac{3}{2}}}\,.$$

There are [at least] two clever ways to verify that this integral vanishes.

METHOD ONE: [*Symmetric Integration*]
The integral of an odd integrand over an even domain vanishes.

———————

[5] **Q:** Why "of course"?   **A:** There are intractable energy consequences.

METHOD TWO: [*Compute first, ask questions later!*]
Performing the indefinite integration yields:

$$\int \frac{x\,dx}{\left(a^2+x^2\right)^{\frac{3}{2}}} = \frac{1}{\left(a^2+x^2\right)^{\frac{1}{2}}}\,,$$

whereupon it is evident that the contributions at each endpoint are of exactly the same magnitude, and subtract to yield ZERO.

ASIDE: Recall that the electric field is defined through the electric force acting upon a test charge in the limit in which the test charge vanishes.

Q: What might happen were the test charge to remain finite?

A: A **back-reaction** can occur in which the test charge affects the putative source of the electric field. Suppose that a uniformly charged rod is made of a conducting material. If a nearby test charge, q_0, is vanishingly small, as shown in the left panel of the figure below, then the rod can remain uniformly charged, whereas if the test charge does not vanish, as is the case in the right-hand panel, then the electric field produced by the test charge disrupts the distribution of charge on the rod, and thus also influences its electric field.

FIGURE 3.4   (Non-)Effects of Vanishing and Non-Vanishing Test Charges

· ·

EXAMPLE [*On-Axis Electric Field Produced by a Uniformly Charged Ring*]

Let us now determine the electric field along the symmetry axis of a thin ring, with radius a, bearing uniform charge per unit length λ_0. In Figure 3.5, the field point [located along the axis perpendicular to, and through the centre of, the ring] has coordinates $\vec{r}_p = (x\,,0\,,0)$. The charge element, dq, gives rise to the differential contribution

$$d\vec{E}(\mathcal{P}) = \frac{k\,dq}{|\vec{r}|^2}\,\hat{r}\,,$$

to the field at $\mathcal{P}$.

The collection of $d\vec{E}$ at $\mathcal{P}$ forms a symmetric cone about the x-axis. The x-component of the field contribution from the particular charge element, dq, is

$$dE_x = \left|d\vec{E}\right|\cos(\theta) = \left|d\vec{E}\right|\frac{x}{r}\,.$$

Once again, $r^2 = a^2 + x^2$, and upon assembling all of the various factors,

$$dE_x = \frac{k\,dq}{a^2+x^2}\frac{x}{\sqrt{a^2+x^2}} = \frac{k\,x}{\left(a^2+x^2\right)^{\frac{3}{2}}}\,dq\,.$$

Integrating dq around the ring is trivial, as both x and a are constant for the entire set of charge elements. The total charge is $Q_{\text{Total}} = \int dq$. Thus, the x-component of the electric field at $\mathcal{P}$ is

$$E_x = k\,Q_{\text{Total}}\frac{x}{\left(a^2+x^2\right)^{\frac{3}{2}}}\,.$$

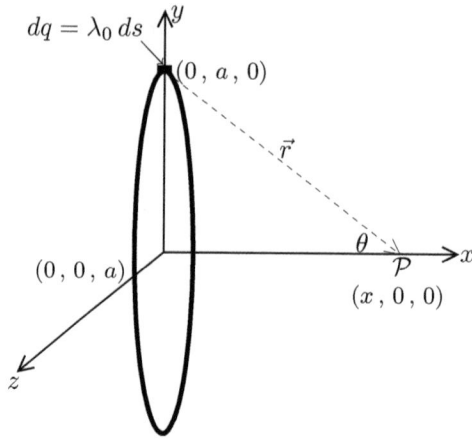

FIGURE 3.5   A Uniformly Charged Thin Ring

ASIDE: It is edifying to recognise that $Q_{\text{Total}} = \lambda_0\,2\pi\,a$.

Q: Does this result make sense?

A: Yes. We'll show this by taking some limits.

FAR   IF the field point is very far away from the ring, *i.e.*, $x \gg a$, THEN

$$E_x = k\,Q_{\text{Total}}\ \left.\frac{x}{x^3\left(1 + \frac{a^2}{x^2}\right)^{3/2}}\right|_{x \gg a} \simeq \frac{k\,Q_{\text{Total}}}{x^2}\,.$$

From far away, the field produced by the ring looks like that of a point charge.

NEAR   IF the field point is very close to the centre of the ring, *i.e.*, $x \to 0$, THEN $E_x \to 0$, as one might expect from the forward–backward symmetry of the ring.

EXAMPLE [*On-Axis Electric Field Produced by a Uniformly Charged Disk*]

Having computed the electric field produced along the axis of a uniform ring of charge, we are equipped to generalise the result to a uniform disk of charge. As illustrated in Figure 3.6, the disk is assembled by summing over thin rings.

The axial symmetry of each ring ensures that only E_x is non-zero for the disk. The differential amount of charge borne by a ring of radius a and width da is

$$dq = \sigma_0\,2\pi\,a\,da,$$

where σ_0 is the [constant] charge per unit area on the surface of the disk.

The differential contribution to the electric field at $\mathcal{P}$ from the thin charged ring of radius a is inferred from the previous example to be

$$dE_x = \frac{k\,\sigma_0\,2\pi\,a\,x\,da}{\left(a^2 + x^2\right)^{3/2}}\,.$$

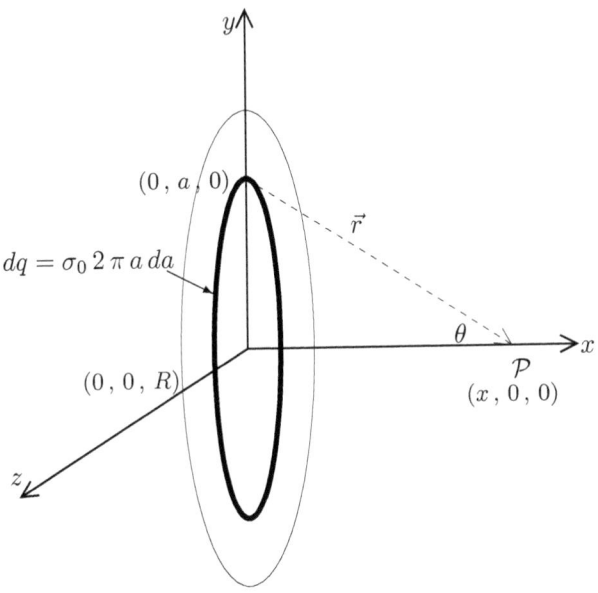

FIGURE 3.6   A Uniformly Charged Thin Disk

Integrating the field contributions over a from $a = 0$ at the centre of the disk, to $a = R$ at its edge, [assuming $x > 0$] yields

$$E_x = k\,2\,\pi\,\sigma_0 \int_0^R \frac{x\,a\,da}{\left(a^2 + x^2\right)^{3/2}} = -k\,2\,\pi\,\sigma_0 \left[\frac{x}{\sqrt{a^2 + x^2}}\right]\Bigg|_0^R = k\,2\,\pi\,\sigma_0 \left(1 - \frac{x}{\sqrt{x^2 + R^2}}\right).$$

To corroborate this formula, let's look at the far- and near-field asymptotics.

FAR   In preparation for taking this limit, we simplify the expression and implement a binomial expansion:

$$\frac{x}{\sqrt{x^2 + R^2}} = \frac{x}{x\left(1 + \frac{R^2}{x^2}\right)^{1/2}} = \left(1 + \frac{R^2}{x^2}\right)^{-\frac{1}{2}} = 1 - \frac{1}{2}\frac{R^2}{x^2} + \dots.$$

IF the field point is far from the disk, *i.e.*, $x \gg R$, THEN the truncated expansion yields a useful approximation. Therefore,

$$\left(1 - \frac{x}{\sqrt{x^2 + R^2}}\right)\Bigg|_{x \gg R} \simeq 1 - 1 + \frac{1}{2}\frac{R^2}{x^2} - \dots \simeq \frac{1}{2}\frac{R^2}{x^2},$$

and thus

$$E_x\Big|_{x \gg R} \simeq k\,2\,\pi\,\sigma_0 \left(\frac{1}{2}\frac{R^2}{x^2}\right) = \frac{k\left(\sigma_0\,\pi\,R^2\right)}{x^2} = \frac{k\,Q_{\text{Total}}}{x^2},$$

for Q_{Total} equal to the charge per unit area multiplied by the area of the disk. This expression looks exactly like the electric field produced by a point charge!

NEAR   IF the field point is near the disk, *i.e.*, $x \ll R$, THEN

$$\frac{x}{R} \to 0,$$

and thus,

$$\lim_{\frac{x}{R} \to 0} \left(1 - \frac{x}{\sqrt{x^2 + R^2}} \right) = \lim_{\frac{x}{R} \to 0} \left(1 - \frac{x/R}{\sqrt{1 + (x/R)^2}} \right) \to 1 \, .$$

In this limit, the electric field approaches

$$E_x \Big|_{x \ll R} \simeq k \, 2 \, \pi \, \sigma_0 \, ,$$

a constant value.

> ASIDE: The problematic aspects of attempting to realise a truly infinite plane of uniform charge density notwithstanding, this result is applicable whenever the [perpendicular] distance to the field point is much smaller than the radius of the disk.

This was great fun, and it shall prove worthwhile to bear these [canonical] cases in mind as we continue our investigations.

- Next, we'll revisit ideas of SCALING, which will [almost] ensure that the electric field is constant throughout the space adjacent to a uniform planar distribution of charge.

- We'll derive rigorously that (and how and why) the electric field above an infinite uniform planar distribution of charge is constant.

- We will, in Chapter 6, begin to learn a powerful (and less laborious) way to determine the behaviour of electric fields in highly symmetric [idealised] cases.

Chapter 4

Above a Uniformly Charged Rectangular Plate

EXAMPLE [*Field above the Centre of a Rectangular Uniform Distribution of Charge*]

This entire chapter will be concerned with the detailed examination of the net electric field produced by a rectangular plate carrying a uniform distribution of electric charge, at a field point, $\mathcal{P}$, located directly above the middle of the plate. The rectangular plate is deemed to have edge lengths $2\,a$ and $2\,b$, respectively, and to bear constant charge-per-unit-area, σ_0. The point, $\mathcal{P}$, at which we wish to compute the field is located a distance L above the centre of the plate, as illustrated in Figure 4.1. Let's choose Cartesian coordinates aligned naturally with the system, so that

$$z = 0\,, \qquad x \in [-b, b]\,, \qquad y \in [-a, a]\,,$$

together serve to define the plate, while $\vec{r}_{\mathcal{P}} = (0\,,0\,,L)$, as shown.

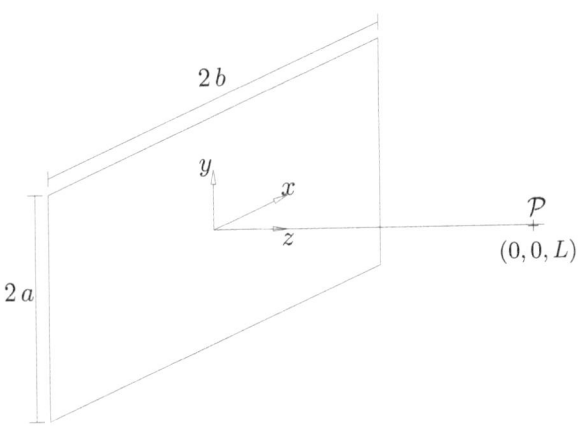

FIGURE 4.1 The Uniformly Charged Rectangular Plate

The first realisation is that we cannot pass quickly and easily to the solution. Instead we must PARTITION, COMPUTE, and SUM our way to determining the electric field at $\mathcal{P}$.

Partition Divide the planar rectangle into [little] rectangular chunks. Anticipating the eventual refinement of the partition, we'll attribute to the chunks length dx and width dy. Each chunk admits a[n increasingly well-defined] characteristic position vector [as the partition is refined]: $(x\,,y\,,0)$.

Compute   The differential electric field at $\mathcal{P}$, produced by the charge element corresponding to one such chunk, is

$$d\vec{E} = \frac{k\,dq}{|\vec{r}|^2}\,\hat{r}\,,$$

where dq is the infinitesimal charge on a particular chunk, $\vec{r}$ extends **from** the location characteristic of the chunk **to** the field point, $\hat{r}$ indicates the direction of $\vec{r}$, and k is Coulomb's constant. The charge element can be re-expressed as

$$dq = \sigma_0\,d[\text{AREA}] = \sigma_0\,dx\,dy\,,$$

since the chunks are rectangular with edge-lengths dx and dy. IF the chunks are small enough,[1] THEN each chunk acts as though it were a point charge, dq, located at the source point, $(x\,,\,y\,,\,0)$.

> ASIDE: If one is performing this computation numerically, one might wish to assign the point charge to the centre of the rectangular chunk. In the exact analysis, both $dx \to 0$ and $dy \to 0$, and it doesn't matter.

The vector from the chunk to the field point, $\vec{r}$, has components

$$\vec{r} = (0\,,\,0\,,\,L) - (x\,,\,y\,,\,0) = (-x\,,\,-y\,,\,L)\,,$$

magnitude, $|\vec{r}| = r = \sqrt{x^2 + y^2 + L^2}$, and direction $\hat{r} = \frac{\vec{r}}{|\vec{r}|} = \left(\frac{-x}{r}\,,\,\frac{-y}{r}\,,\,\frac{L}{r}\right)$. Thus, the Cartesian components of the differential electric field are

$$dE_x = -\frac{k\,x}{(x^2 + y^2 + L^2)^{3/2}}\,dq\,, \qquad dE_y = -\frac{k\,y}{(x^2 + y^2 + L^2)^{3/2}}\,dq\,, \qquad \text{and}$$

$$dE_z = \frac{k\,L}{(x^2 + y^2 + L^2)^{3/2}}\,dq\,.$$

Sum   Computing the sum of each component of the field requires an integration over the planar domain. Fortunately, the symmetry of the situation ensures that the net x- and y-components are both ZERO.

The vanishing of E_x is verified by the following computation.

$$E_x = \int_{\text{Area}} dE_x = \int_{-b}^{b} dx \int_{-a}^{a} dy\; -\frac{k\,\sigma_0\,x}{(x^2 + y^2 + L^2)^{3/2}}\,.$$

Performing the x-integration over the domain $[-b, b]$, along a strip of width dy, with y held constant, yields zero by SYMMETRIC INTEGRATION.[2] By an analogous argument, the y-component, E_y, vanishes, too.

---

[1] This condition will surely be met when the edge dimensions are of differential magnitude.
[2] Under inversion, $x \leftrightarrow -x$, the integrand is odd [antisymmetric], while the domain of integration is even [symmetric].

Thus, only the z-component of the overall electric field is non-zero:

$$E_z = \int_{\text{Area}} dE_z = \int_{-b}^{b} dx \int_{-a}^{a} dy \, \frac{k \, \sigma_0 \, L}{(x^2 + y^2 + L^2)^{3/2}}$$

$$= k \, \sigma_0 \, L \int_{-a}^{a} dy \int_{-b}^{b} \frac{dx}{(x^2 + y^2 + L^2)^{3/2}} \,.$$

Let's perform the x-integration first, holding y fixed at $y_0 \in [-a, a]$. Introduce a new variable, u, such that

$$u = \frac{x}{\sqrt{y_0^2 + L^2}} \,,$$

with the properties: $(x^2 + y_0^2 + L^2) = (y_0^2 + L^2)(1 + u^2)$ and $dx = \sqrt{y_0^2 + L^2} \, du$. The new variable has limits of integration

$$u_\pm = \pm \frac{b}{\sqrt{y_0^2 + L^2}} \,,$$

corresponding to the edges of the plate at $x_\pm = \pm b$. Written in terms of u, the expression for E_z reads

$$E_z = k \, \sigma_0 \, L \int_{-a}^{a} dy \, \frac{1}{y_0^2 + L^2} \int_{u_-}^{u_+} \frac{du}{(1 + u^2)^{3/2}} \,.$$

The hyperbolic trig substitution,

$$u = \sinh(\theta) \,, \qquad \text{with} \qquad du = \cosh(\theta) \, d\theta \qquad \text{and} \qquad 1 + u^2 = \cosh^2(\theta) \,,$$

leads to a tractable elementary integration. Performing this change of variables,

$$I_u = \int \frac{du}{(1 + u^2)^{3/2}} = \int \frac{\cosh(\theta) \, d\theta}{\cosh^3(\theta)} = \int \frac{d\theta}{\cosh^2(\theta)} = \int \operatorname{sech}^2(\theta) = \tanh(\theta) \,.$$

Re-expressing this result in terms of u, so as to more easily supply the upper and lower limits of integration, yields

$$I_u = \frac{\sinh(\theta)}{\cosh(\theta)} = \frac{u}{\sqrt{1 + u^2}}$$

$$I_u \Big|_{u_-}^{u_+} = \frac{u}{\sqrt{1 + u^2}} \Big|_{u_-}^{u_+} = \frac{b/\sqrt{y_0^2 + L^2}}{\sqrt{\frac{b^2 + y_0^2 + L^2}{y_0^2 + L^2}}} - \frac{-b/\sqrt{y_0^2 + L^2}}{\sqrt{\frac{(-b)^2 + y_0^2 + L^2}{y_0^2 + L^2}}} = \frac{2b}{\sqrt{b^2 + y_0^2 + L^2}} \,.$$

With the x-integration thus completed, the assumption of constancy for y_0 may be relaxed, $y_0 \to y$, and the integral over $y \in [-a, a]$ performed.

$$E_z = k \, \sigma_0 \, L \int_{-a}^{a} dy \, \frac{1}{y^2 + L^2} \, \frac{2b}{\sqrt{b^2 + y^2 + L^2}}$$

$$= 2 \, k \, \sigma_0 \, b \, L \int_{-a}^{a} dy \, \frac{1}{y^2 + L^2} \, \frac{1}{\sqrt{b^2 + y^2 + L^2}} \,.$$

The trick[3] which we next employ is introducing yet another new variable,

$$w = \frac{y}{\sqrt{L^2 + b^2 + y^2}} \,,$$

---

[3] Integration is a tricky business, eh?

with y derivative,

$$\frac{dw}{dy} = \frac{1}{\sqrt{L^2 + b^2 + y^2}} - \frac{y^2}{\left(L^2 + b^2 + y^2\right)^{3/2}} = \frac{L^2 + b^2}{\left(L^2 + b^2 + y^2\right)^{3/2}}.$$

Cleverly rewriting the integrand,

$$I_y = \frac{1}{y^2 + L^2} \frac{1}{\sqrt{b^2 + y^2 + L^2}} = \frac{dw}{dy} \frac{L^2 + b^2 + y^2}{\left(L^2 + y^2\right)\left(L^2 + b^2\right)},$$

and recognising that the complicated-looking fraction simplifies when expressed in terms of w,

$$L^2 + b^2 w^2 = L^2 + \frac{b^2 y^2}{L^2 + b^2 + y^2}$$

$$= \frac{L^4 + L^2 b^2 + y^2 L^2 y^2 + b^2 y^2}{L^2 + b^2 + y^2} = \frac{\left(L^2 + y^2\right)\left(L^2 + b^2\right)}{L^2 + b^2 + y^2},$$

gives rise to the following intermediate form:

$$\int_{-a}^{a} I_y \, dy = \int_{-a}^{a} dy \, \frac{dw}{dy} \frac{1}{L^2 + b^2 w^2} = \frac{1}{L^2} \int_{-a}^{a} dy \, \frac{dw}{dy} \frac{1}{1 + \left(\frac{b\,w}{L}\right)^2}.$$

Combining these intermediate results into the expression for E_z, retaining the y-parameterisation, and simplifying, yields

$$E_z = 2\,k\,\sigma_0 \int_{-a}^{a} dy \, \frac{b}{L} \frac{dw}{dy} \frac{1}{1 + \left(\frac{b\,w}{L}\right)^2}.$$

Finally, let's introduce yet another dummy variable,[4]

$$u = \frac{b\,w}{L}, \qquad \text{and} \qquad du = \frac{b}{L} \, dw.$$

Gathering these disparate threads together,

$$E_z = 2\,k\,\sigma_0 \int_{-a}^{a} dy \, \frac{du}{dy} \frac{1}{1 + u^2}.$$

The integral immediately above is in a standard form, *viz.*, $\int \frac{du}{1+u^2} = \tan^{-1}(u)$. Recalling that $u = \frac{b\,w}{L} = \frac{b\,y}{L\sqrt{L^2 + b^2 + y^2}}$, and that the y-limits of integration are $\pm a$, along with the realisation that $-\tan^{-1}(-\theta) = +\tan^{-1}(\theta)$, enables simple expression of the z-component of the electric field at height L above the centre of a rectangular plate of dimensions $(2\,a) \times (2\,b)$:

$$E_z = 4\,k\,\sigma_0 \tan^{-1}\left[\frac{a\,b}{L\sqrt{L^2 + b^2 + y^2}}\right].$$

This is the sought-after result. It provides an explicit expression for the non-zero component of the electric field at the point $\mathcal{P}$, directly above the centre of the plate, as a function of its height, L, and the plate geometry, parameterised by a and b.

---

[4]To emphasise that it is truly a dummy variable, we are reusing the symbol "u."

Q: Should we trust this result?

A: Let's test its formal properties and investigate its predictions in physical cases or limits where we have an intuition as to the nature of the field.

SYM   The expression for the field is symmetric in $a \leftrightarrow b$. This is a necessary condition for physical quantities associated with the rectangular plate.

$L \gg a, b$ IF the perpendicular distance to the field point is much larger than both dimensions of the plate, *i.e.*, $L \gg \{a, b\}$, THEN the electric field should approximate that of a point source.

Comparison with the expression just derived starts with the square-root factor in the denominator of the argument of the inverse tangent function:

$$\lim_{L \gg a,b} \sqrt{L^2 + a^2 + b^2} \ \to \ L + \text{ small corrections}.$$

Hence, the argument of the inverse tangent is well-approximated by

$$\left[\frac{a\,b}{L\,\sqrt{L^2 + a^2 + b^2}} \right] \sim \frac{a\,b}{L^2},$$

in this limit. Furthermore, for small θ, $\tan^{-1}(\theta) \sim \theta$, and thus,

$$\lim_{L \gg a,b} E_z \ \to \ 4\,k\,\sigma_0 \left[\frac{a\,b}{L^2} \right] = \frac{k\,(\sigma_0\,(2\,a)(2\,b))}{L^2} = \frac{k\,Q_{\text{Total}}}{L^2},$$

where the total charge is the [constant] charge per unit area, σ_0, integrated over the area, $(2\,a) \times (2\,b)$, of the plate. The field in this limit evinces the expected monopolar structure.

Far above a small plate, the field resembles that of a point charge.

$a \gg L \gg b$ IF the perpendicular distance to the field point is small with respect to one of the dimensions of the plate, while remaining large relative to the other, *i.e.*, $a \gg L \gg b$, THEN the electric field should resemble that of an infinite line source.

In this limiting case, the argument in square brackets reduces to

$$\lim_{a \gg L \gg b} \frac{a\,b}{L\,\sqrt{L^2 + a^2 + b^2}} \ \to \ \lim_{L \gg b} \frac{b}{L} + \text{ very small corrections}.$$

The value of $\tan^{-1}(\theta)$, when θ is small, is well-approximated by θ, and thus

$$E_z \sim 4\,k\,\sigma_0 \left[\frac{b}{L} \right] = \frac{2\,k\,(\sigma_0\,2\,b)}{L}.$$

Looking at the numerator appearing just above motivates introduction of an effective lineal charge density associated with this [long and narrow] plate,

$$\lambda_0 = 2\,b\,\sigma_0,$$

equal to the surface charge density multiplied by its width. When this is done, the expression for the field may be written as

$$E_z \sim \frac{2\,k\,\lambda_0}{L},$$

precisely matching the result obtained for a uniformly charged infinite straight wire.

This electric field resembles that of a long straight wire.

$a, b \gg L$ IF both dimensions of the plate are much larger than the distance to the field point, *viz.*, $\{a, b\} \gg L$, THEN the field should be not unlike that above an infinite uniformly charged plane.

Here, the argument of the inverse-tangent function exhibits limiting behaviour which is qualitatively and quantitatively distinct from the prior two cases:

$$\lim_{\{a,b\} \gg L} \frac{a\,b}{L\,\sqrt{L^2 + a^2 + b^2}} \to \infty\,!$$

There is no pathology associated with this, because

$$\lim_{x \to \infty} \tan^{-1}(x) \to \frac{\pi}{2}\,.$$

In this limit, the electric field approaches a constant value:

$$E_z \sim 4\,k\,\sigma_0 \left[\frac{\pi}{2}\right] = 2\,\pi\,k\,\sigma_0\,.$$

This is just like the infinite plane of uniform charge encountered in Chapter 3.

This electric field is like that of a uniformly charged infinite plane.

In consideration of the syntactic symmetry of the expression for the electric field and its exhibiting correct behaviour in the three distinct physical limits considered above, we can employ this derived result with some measure of confidence.

Three final comments finish off this example.

INFINITE
LIMITS   Consideration of the three special cases was only possible once the general result had been computed. Recall that the proper way to effect an infinite limit when integrating is via a limiting process,

$$\int_{-\infty}^{\infty} f(x)\,dx \equiv \lim_{\substack{a \to -\infty \\ b \to \infty}} \int_{a}^{b} f(x)\,dx\,,$$

exactly as was done here.

GEOMETRY   The infinite limits, $\{a, b\} \gg L$ for the plate, and $r \to \infty$ for the disk, coincide because the distinction between rectangular and circular geometry is lost. One must remember, though, that for any finite-sized situation it can be excruciatingly hard to analyse a rectangular system using polar coordinates and *vice versa*.

CAVEAT   This analytical result, while consistently derived, is only valid when its assumptions are reliable. For actual charge-bearing rectangular plates, any non-uniformity of the charge distribution will result in the field deviating from the form predicted here.

Chapter 5

Electric Flux and Gauss's Law

The concepts of FLUX and INTENSITY have many applications in physics.

- **Flux** is the rate of flow [of something] through a prescribed imaginary surface. The dimensions of flux are "stuff" per unit time.

- **Intensity** is the flux per unit cross-sectional area on the surface. The dimensions of intensity are "stuff" per unit area per unit time.

As an introduction to ELECTRIC FLUX, we'll first re-acquaint ourselves[1] with MASS FLUX.

UNIFORM FLOW OF IDEAL FLUID IN A PIPE

An **ideal** fluid[2] with constant mass density ρ_0 flows uniformly [with constant velocity $\vec{v}_0$] through a pipe with constant cross-sectional area A.

ASIDE: The assumed uniformity of the velocity field has several implications.

- The fluid undergoes LAMINAR rather than TURBULENT flow.
- The flow is IRROTATIONAL.

Uniformity is in contradistinction to the usual circumstance in which the fluid near the centre of the pipe moves with greater speed than that nearer to the edge.

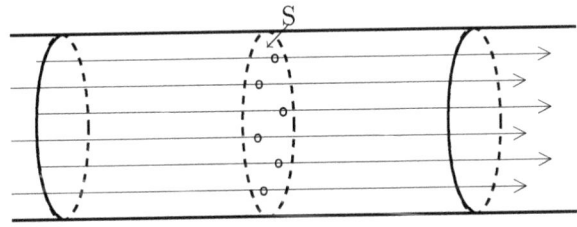

FIGURE 5.1   Streamlines in Ideal Uniformly Flowing Fluid

The situation is illustrated in Figure 5.1, in which **streamlines** showing the flow of [localised] mass/volume elements are represented by the arrows. The dashed circle is the boundary of a particular [imaginary] disk-like surface, S, through which all of the fluid in the pipe

---

[1] Material fluxes were discussed in VOLUME II, Chapter 6 *et seq.*

[2] Ideal fluid is **incompressible** [its volume is fixed and unchanging] and flows without **viscosity** [dissipation due to internal friction]. Also, ideal fluids do not experience friction with the walls of vessels in which they happen to be confined.

passes. The tiny "perforations" encircle points at which the identified streamlines intersect the surface.

The mass of fluid flowing through the indicated surface per unit of time is the MASS FLUX,

$$\Phi_{m,\mathrm{S}} = \rho_0\, v_0\, A\,.$$

A priori, the sign of the flux is indeterminate; here, it was chosen to be positive.

The following questions shall guide our recollection and review of flux.

Q: What if the [imaginary] surface through which the fluid passes lies at an oblique angle with respect to the edge of the pipe, as does S′ in Figure 5.2?

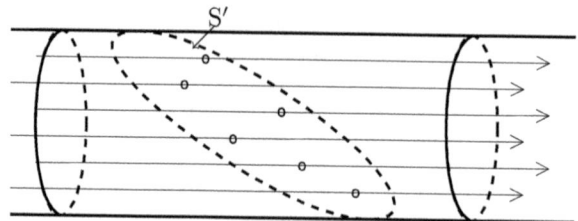

FIGURE 5.2   Flux through an Oblique Planar Imaginary Surface

A: The flux through S′ must be the same as that through S, shown in Figure 5.1, in order that the LAW OF CONSERVATION OF FLUID be observed.

Three generalisations flow from conservation considerations.

NON-PLANAR   It is not necessary, for conservation, that the imaginary surface be flat. To compute the flux accurately through a curved surface, $\widetilde{\mathrm{S}}$, one must employ the [local] integral formulation

$$\Phi_{m,\widetilde{\mathrm{S}}} = \int_{\widetilde{\mathrm{S}}} \rho\, \vec{v} \cdot d\vec{A}\,,$$

which can also accommodate variable densities and non-uniform velocity fields.

NON-SPANNING   IF two surfaces, S_1 and S_2, do not span the same section [or equivalent sections] of pipe, THEN nothing can be conclusively said about the relative sizes of their respective fluxes, Φ_{m,S_1} and Φ_{m,S_2}.

CLOSED SURFACE   IF a particular surface, S_o, is CLOSED [AND non-self-intersecting] and it bounds a region with volume V, AND IF there are no **sources** or **sinks** of fluid lying within the bounded region, THEN the net,[3] or total, flux through S_o vanishes identically:

$$\Phi_{m,\mathrm{S}_o} = \oint_{\mathrm{S}_o} \rho\, \vec{v} \cdot d\vec{A} \equiv 0\,.$$

A single closed surface, S_o, may be thought of as consisting of two complementary open surfaces, $\{\mathrm{S}_1, \mathrm{S}_{\bar{1}}\}$, along with their common boundary.[4]

---

[3]The ring appearing on the integral sign indicates that the surface over which the integral is to be performed is closed.

[4]The lawyers insist that we mention the boundary, even though it is one-dimensional.

ASIDE: Suppose that the closed surface is a spherical shell.

- The northern hemisphere [latitude $> 0°$ N, or polar angle $0 \leq \theta < \pi/2$] is an open surface within the shell. The northern hemisphere's open complement is the southern hemisphere [latitude $> 0°$ S, or $\pi/2 < \theta \leq \pi$]. The equator [latitude $= 0°$, or $\theta = \pi/2$] is the boundary of both hemispheres.
- Another open surface is the Arctic region, with latitudes greater than $66° \, 33'$ north, *i.e.*, above the Arctic Circle. [The Arctic Circle marks the southernmost extent of the north polar region in which the sun is visible for 24 hours (all day) on the summer solstice and fails to rise on the winter solstice.] The complementary open surface consists of the northern temperate and tropical regions and all of the southern hemisphere. The Arctic Circle is the common boundary.

In order that the net flux through the closed surface vanish, the fluxes through the two open surfaces must exactly cancel.

This ends our recapitulation of mass flux associated with an ideal fluid moving uniformly in a pipe. Now, we'll consider the electric flux associated with an electric field.

ELECTRIC FLUX (Simple Case) The electric flux associated with a uniform electric field, $\vec{E_0}$, passing through S, the planar surface with area vector $\vec{A_S}$, is

$$\Phi_{e,S} = \vec{E_0} \cdot \vec{A_S} \, .$$

Several comments are warranted.

UNITS   The SI units of electric flux are newtons·square-metres per coulomb,

$$[\, \Phi_{e,S} \,] = \frac{N \cdot m^2}{C} \, .$$

AREA   The vector describing the planar surface has magnitude, $|\vec{A_S}|$, equal to the area of the surface, and is directed along the surface normal.

DOT   The DOT PRODUCT of two vectors, $\{\vec{a}, \vec{b}\}$, may be expressed geometrically, in terms of their respective magnitudes and the angle θ, which lies between them; or algebraically, as the sum of products of their respective Cartesian components, *viz.*,

$$\vec{a} \cdot \vec{b} = |\vec{a}| \, |\vec{b}| \, \cos(\theta) \qquad \text{OR} \qquad \vec{a} \cdot \vec{b} = a_x \, b_x + a_y \, b_y + a_z \, b_z \, .$$

FIELD LINES   The electric flux may be interpreted as a measure of the number of electric field lines which penetrate a surface.

[Electric field lines are analogous to fluid streamlines.]

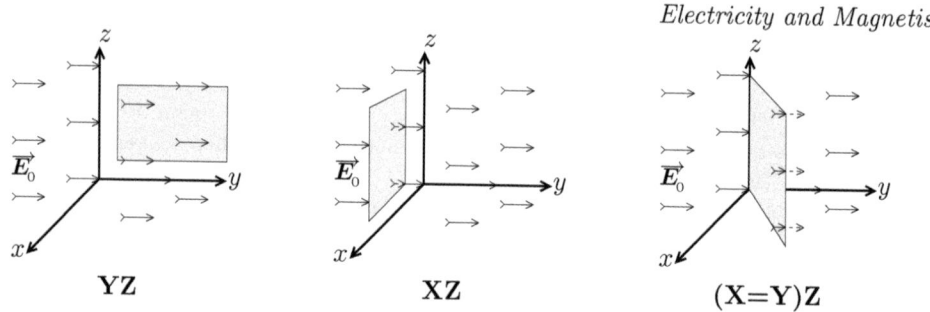

FIGURE 5.3   Electric Fluxes through Various Planar Surfaces

EXAMPLE [*Electrifying Fluxes*]

A uniform electric field, $\vec{E}_0 = E_0\,\hat{\jmath}$, is present in the three scenarios shown in Figure 5.3. We shall compute the electric fluxes through each of the [planar] shaded surfaces.

YZ   The surface S_{yz} lies in a portion of the yz-plane. Therefore,

$$\vec{E}_0 \cdot d\vec{A} = 0$$

everywhere on the surface over which the flux is computed. Thus, $\Phi_{e,S_{yz}} = 0$. Two comments are:

- The flux is zero, because no electric field lines pierce S_{yz}.
- Although the direction of the surface area vector is ambiguous, $\pm\hat{\imath}$, the electric flux vanishes in either case.

XZ   The surface S_{xz} lies in a portion of the xz-plane. Therefore,

$$\vec{E}_0 \cdot d\vec{A} = E_0\,dA\,\cos(0) = E_0\,dA$$

everywhere on this surface. The orientation ambiguity has been resolved [by *fiat*] by choosing the integrand to be positive. As E_0 is [assumed to be] constant across the surface, the integration trivially yields

$$\Phi_{e,S_{xz}} = E_0\,A_{xz}\,,$$

where A_{xz} is the area of the planar surface S_{xz}.

(X=Y)Z   The surface $S_{(x=y)z}$ lies in a plane which includes both the z-axis and the line $y = x$. Again exercising our freedom to render the flux positive,

$$E_0\,\hat{\jmath} \cdot d\vec{A} = E_0\,dA\,\cos(\pi/4) = \frac{1}{\sqrt{2}}\,E_0\,dA\,,$$

everywhere on the surface of interest. Thus,

$$\Phi_{e,S_{(x=y)z}} = \frac{1}{\sqrt{2}}\,E_0\,A_{(x=y)z}\,,$$

where $A_{(x=y)z}$ is the area of the planar surface.

The magnitude of the flux through $\left\{ \begin{array}{c} S_{xz} \\ S_{(x=y)z} \\ S_{yz} \end{array} \right\}$ is $\left\{ \begin{array}{c} \text{MAXIMAL} \\ \text{INTERMEDIATE} \\ \text{MINIMAL} \end{array} \right\}$ for this electric field.

The essential point here is that the flux through a planar surface depends strongly on the relative angle between the velocity field and the surface normal.

Q: How might one calculate the flux when the surface, $\tilde{S}$, is curved, and the electric field varies from point to point in space?

A: As is customary in such cases, we PARTITION, COMPUTE, and SUM.

Partition the curved surface into $\mathcal{N}$ non-overlapping approximately planar patches, $\{S_i, i \in [1, \ldots, \mathcal{N}]\}$, such that the local electric field, $\vec{E}_i$, is effectively uniform on each patch. The entire surface is reconstituted by assembling the patches:

$$\tilde{S} = \bigcup_{i=1}^{\mathcal{N}} S_i \,.$$

An area vector, $\Delta\vec{A}_i$, may be associated with each patch.[5]

Compute the electric flux through each of the patches, employing the expression for uniform fields and planar surfaces. The flux through the ith patch, produced by the local nearly constant field, is

$$\Delta\Phi_{m,S_i} = \vec{E}_i \cdot \Delta\vec{A}_i \,.$$

Sum the electric fluxes through all of the area elements in the partition. The net flux thus obtained is

$$\Phi_{e,\tilde{S},\mathcal{N}} = \sum_{i=1}^{\mathcal{N}} \Delta\Phi_{m,S_i} = \sum_{i=1}^{\mathcal{N}} \vec{E}_i \cdot \Delta\vec{A}_i \,.$$

The absence of overlap in the patches prevents double-counting in the sum.

The computed value of the electric flux depends on the electric field, $\vec{E}$, on the surface, $\tilde{S}$, and on the details of the partition.

ASIDE: In the particular instance of a globally constant electric field, the expression for the flux further simplifies to the form:

$$\Phi_{e,\tilde{S},\mathcal{N}} = \vec{E}_0 \cdot \left(\sum_{i=1}^{\mathcal{N}} \Delta\vec{A}_i \right) \,.$$

The answer to the preceding question is fine and dandy,[6] but we desire exact results and hence INFINITELY refine the partition by taking the simultaneous limits

$$\lim_{\mathcal{N} \to \infty} \quad \text{AND} \quad \left| \Delta\vec{A}_i \right| \to 0 \,, \ \forall \, i \,,$$

to transform the sum into an integral.

---

[5] The explicit incorporation of the Δ presages the eventual refinement of the partition.
[6] As an added bonus, these expressions are amenable to numerical evaluation.

ELECTRIC FLUX (General Case)   The electric flux, $\Phi_{e,\mathrm{S}}$, passing through S, an imaginary [non-self-interacting] surface, is

$$\Phi_{e,\mathrm{S}} = \int_{\mathrm{S}} \vec{E} \cdot d\vec{A}\,,$$

where $\vec{E}$ is the local electric field. A detailed parsing of the RHS of this expression is undertaken below.

$\vec{E}$ The electric field is a vector-valued function of position. [The electric field integrates to form field lines. It is often efficacious to think in terms of field lines.]

$d\vec{A}$ The area element at a particular location on the imaginary surface has as its magnitude the [infinitesimal] size of a patch of area on the surface. Its direction is perpendicular [*a.k.a.* normal] to the imaginary surface. Barring pathologies in the geometry of the surface,[7] the direction of $d\vec{A}$ is determined up to ORIENTATION.[8]

$\vec{E} \cdot d\vec{A}$ The DOT PRODUCT yields the projection of the electric field onto the infinitesimal local area element. **Only those electric field lines passing through the surface contribute to the flux.**

$\int_{\mathrm{S}}$ The net flux is the sum of the local fluxes through all points on the surface.

> Most often the surface is bounded and of finite extent, but this isn't necessary. The domain of the integrand may be restricted to the surface to simplify the analysis.

The electric flux, $\Phi_{e,\mathrm{S}}$, typically depends on the surface, S, and the electric field, $\vec{E}$, in ways that appear complicated. In its favour, the electric flux is a scalar quantity, possessing magnitude but not direction, affording the promise that it might be less onerous to work with than the vector-valued electric field.

We are finally prepared[9] to address the orientation ambiguity described previously. It is resolved by **restricting analysis to closed surfaces** [non-self-intersecting], and **establishing a convention.**

CLOSED   A **closed** surface envelops a 3-d volume while possessing no boundary itself. The surfaces of a sphere, a football, and a cube are all closed, whereas the surface of a disk, or one face of a cube might not be closed.

> [Topological considerations are important in Physics!]

NON-SELF-INTERSECTING   There exist closed surfaces which are not orientable.[10] It turns out that all such non-orientable closed surfaces self-intersect when embedded in 3-d space. Thus, insisting that the imaginary surfaces employed in the computation of the flux be non-self-intersecting ensures that this type of pitfall is avoided.

---

[7]Specific instances of pathology include self-intersection, creases, and cones.
[8]We will address this ambiguity momentarily.
[9]"Finely prepared," too, one might dare to hope.
[10]The most commonly known of these is the *Klein Bottle*, which is the 2-d generalisation of the 1-d *Möbius Strip*.

CONVENTION   Closed surfaces which do not self-intersect are orientable because the inside and outside volumes are distinguishable.[11] The orientation ($\pm$) ambiguity is resolved by the following rule:

On a closed surface, $d\vec{A}$ points OUTWARD.

ELECTRIC FLUX (Closed Surfaces)   The flux associated with an electric field, $\vec{E}$, passing through a [non-self-interacting] closed surface, S_o, is

$$\Phi_{e,S_o} = \oint_{S_o} \vec{E} \cdot d\vec{A}\,.$$

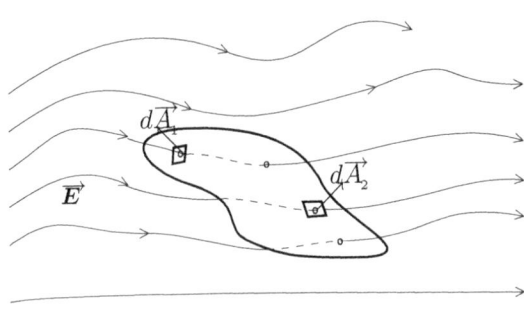

FIGURE 5.4   Schematic Illustration of Electric Flux through a Closed Surface

---

EXAMPLE [*Electric Flux: Point Charge and Concentric Spherical Surface*]

An isolated point charge, q, is fixed at a location in space. The electric field associated with the charge permeates all of space, is isotropic about the location of the charge, and has [Coulombic] inverse-square radial dependence. An imaginary spherical shell, S_{o_R}, of radius R, is centred on the point charge.

The electric field produced by the charge at points on S_{o_R} is

$$\vec{E}_q = \frac{k\,q}{R^2}\,\hat{R}\,.$$

The outward unit normal associated with each of the area elements is also $\hat{R}$, and thus

$$d\vec{A} = dA\,\hat{R}$$

holds locally everywhere on the surface of integration. The flux integrand simplifies,

$$\vec{E}_q \cdot d\vec{A} = \frac{k\,q}{R^2}\,dA\left(\hat{R} \cdot \hat{R}\right) = \frac{k\,q}{R^2}\,dA\,,$$

---

[11] A snarky person might object that this is NOT strictly true IF the universe itself is closed.

because the local electric field and the area vector at each point on the surface are everywhere [locally] parallel.

The collection of area elements, taken together, spans the entirety of the spherical surface forming the domain of integration. The spherical surface is characterised by the constancy of R, the radius. It happens that, in this case, the integrand has the same [constant] value everywhere throughout the domain of integration. The flux integral then is trivially performed,[12]

$$\Phi_{e,S_{o_R}} = \oint_{S_{o_R}} \frac{k\,q}{R^2}\, dA = \frac{k\,q}{R^2} \oint_{S_{o_R}} dA = \frac{k\,q}{R^2}\left(4\,\pi\,R^2\right) = 4\,\pi\,k\,q\,.$$

The electric flux computed here depends only on a mathematical constant, $4\,\pi$; Coulomb's constant, k; and the quantity of charge, q; and does not depend on the radius of the spherical surface. This is, it must be emphasised, a profound and general result, not to be obscured by its serendipitous discovery within this very symmetric special case.

GAUSS'S LAW for ELECTRIC FIELDS   The net electric flux through any closed [non-self-interacting] surface, S_o, which bounds a volume, V, and encloses a net amount of electric charge, $q_{\text{encl.}}$, is

$$\Phi_{e,S_o} = \oint_{S_o} \vec{E}\cdot d\vec{A} \equiv \frac{q_{\text{encl.}}}{\epsilon_0}\,,$$

where ϵ_0 is the **permittivity of free space**, with value

$$\epsilon_0 = \frac{1}{4\,\pi\,k} = 8.85 \times 10^{-12}\ \frac{\text{C}^2}{\text{N}\cdot\text{m}^2}\,.$$

GAUSS'S LAW is a fundamental LAW of NATURE.

---

[12]The area of a spherical surface of radius R is $A_R = 4\,\pi\,R^2$.

Chapter 6

More Gauss's Law

Recall the example in the previous chapter, in which an isolated point charge, q, is fixed at a particular point in space which is taken to be the origin of coordinates. At a field point, $\mathcal{P}$, described by the vector $\vec{r}$, the electric field produced by the point charge is

$$\vec{E}_q(\mathcal{P}) = \lim_{q_0 \to 0} \frac{\vec{F}_{qq_0}}{q_0} = \frac{k\,q_0}{\left|\vec{r}\right|^2}\,\hat{r}\,.$$

Now imagine a spherical shell [a.k.a. a Gaussian surface], S_o, of radius $R = \left|\vec{r}\right|$, concentric with the charge. We shall recapitulate and extend our computation [performed in Chapter 5] of the electric flux through this closed surface. The fluxes through infinitesimal patches of the surface, $d\Phi_e = \vec{E} \cdot d\vec{A}$, integrate to yield

$$\Phi_{e,S_o} = \oint_{S_o} \vec{E} \cdot d\vec{A}\,.$$

By convention, the vector differential area element is parallel to the unit outward normal[1] on the surface. The flux integral is easily performed, because the electric field and the differential area element are everywhere parallel, and the electric field has constant magnitude, throughout the entire surface [the domain of integration]. Hence,

$$\Phi_{e,S_o} = \oint_{S_o} \vec{E} \cdot d\vec{A} = \oint_{S_o} \frac{k\,q}{R^2}\,dA = \frac{k\,q}{R^2}\left(4\,\pi\,R^2\right) = 4\,\pi\,k\,q = \frac{q}{\epsilon_0}\,.$$

In the third equality, we've used the fact that the area of the spherical surface of radius R is equal to $4\,\pi\,R^2$. In the ultimate expression, the permittivity of free space,

$$\epsilon_0 = \frac{1}{4\,\pi\,k} = 8.85 \times 10^{-12}\,\frac{\mathrm{C}^2}{\mathrm{N} \cdot \mathrm{m}^2}\,,$$

has been reintroduced.

The net electric flux produced by a point charge, through a spherical surface centred on the charge, is equal to the magnitude of the charge divided by the permittivity of free space.

Let's generalise this result in three ways.

SIZE   The computed flux does not depend on the radius of the spherical surface. To understand better why this is so, it is instructive to think of the flux as proportional to the number of field lines penetrating the surface. Changing the radius of the spherical Gaussian surface affects the surface density of the field lines [the magnitude of the electric field on the surface], but not the total number of field lines [the flux].

---

[1]This is globally consistent, provided that we specialise to closed and non-self-intersecting surfaces.

Size doesn't matter!

SHAPE   In fact, the flux doesn't depend on the geometry of the surface either! To prove this requires high-powered mathematical techniques, so we will instead satisfy ourselves with the following heuristic argument.

Consider an isolated point charge surrounded by an ellipsoidal[2] Gaussian surface. Next, consider two spherical surfaces, an INSCRIBING sphere which lies completely inside the ellipsoid, and an EXSCRIBING sphere completely encompassing the ellipsoid. The electric fluxes through the two spherical surfaces must be exactly the same, because [as was shown immediately above] the size of the spherical surface does not enter into the expression for the flux. Alternatively, any field lines which pass through one sphere must also pass through the other, because there are no charges in the region between the spheres upon which the field lines might terminate.

> ASIDE: On the inner spherical surface, the electric field is stronger than, and the area is smaller than, its corresponding value on the outer sphere.

The fluxes through the inscribed and exscribed spheres are equal, and no electric charges are present in the interstitial space. Therefore, the same amount of flux passes through the ellipsoidal surface.

Shape doesn't matter!

SYMMETRY   The flux doesn't depend on the symmetry of the situation either. We shall again advance a heuristic argument, in favour of this new claim.

Consider the situation in which the point charge is located off-centre inside a spherical Gaussian surface. The electric field is stronger on the near side and weaker on the far side of the surface. Next, imagine two more spherical surfaces, INSCRIBED and EXSCRIBED, both centred on the charge, with the inscribed sphere lying entirely within the original Gaussian surface, and the exscribed one completely enclosing the other two. According to our previous arguments, the fluxes through the inscribed and exscribed surfaces are precisely the same, and their common value must be equal to the flux through the non-concentric sphere.

Symmetry doesn't matter!

Gauss's Law is a very powerful statement about the behaviour of electric flux.

**Whatever the distribution of charge,
whatever the shape of the Gaussian surface,
the net electric flux depends only on
the net amount of charge enclosed by the surface.**

**Gauss's Law is of practical importance because,
in cases which are highly symmetric,
knowledge of the flux can determine the electric field!**

---

[2]To visualise this, think of a somewhat squashed sphere or an American-style football.

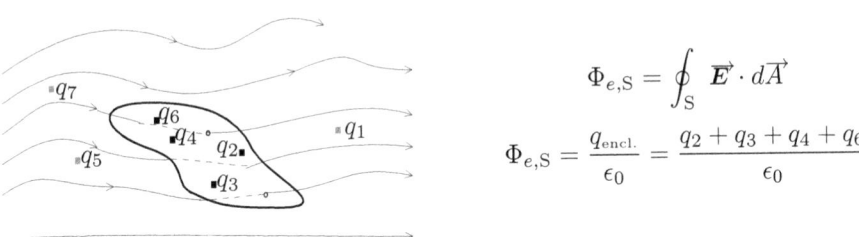

$$\Phi_{e,S} = \oint_S \vec{E} \cdot d\vec{A}$$

$$\Phi_{e,S} = \frac{q_{encl.}}{\epsilon_0} = \frac{q_2 + q_3 + q_4 + q_6}{\epsilon_0}$$

FIGURE 6.1   Only the Enclosed Charge Contributes to the Net Flux

EXAMPLE [*Gauss's Law: Point Charge to Flux to Field*]

Here we shall reconsider the isolated point charge case. This time, we will apply our knowledge of the flux, computed using Gauss's Law, to infer the form of the electric field. For this analysis, let the location of the charge be the origin of the system of coordinates.

By Gauss's Law, the electric flux through a spherical Gaussian surface, S_R, of radius R, centred on [an isolated point-like] charge, q, is

$$\Phi_{e,S_R} = \frac{q}{\epsilon_0} \,.$$

As a consequence of the **isotropy** of the space surrounding the single point charge,[3] its electric field must be radially directed,[4] and possess constant magnitude, everywhere on S_R. These two properties have very strong implications on the form of the flux integrand, *viz.*,

$$\vec{E} \cdot d\vec{A} = |\vec{E}| \, |d\vec{A}| \, \cos(0) = E_R \, dA \,,$$

where E_R is the particular [constant] value assumed by the magnitude of the field everywhere on the spherical Gaussian surface. Thus, the electric flux through the sphere is

$$\Phi_{e,S_R} = \oint_{S_R} \vec{E} \cdot d\vec{A} = \oint_{S_R} E_R \, dA = E_R \oint_{S_R} dA = E_R \times (4\pi R^2) \,.$$

Invoking Gauss's Law fixes the magnitude of the electric field, *i.e.*,

$$E_R \, 4\pi R^2 = \frac{q}{\epsilon_0} \qquad \Longrightarrow \qquad E_R = \frac{q}{4\pi\epsilon_0 R^2} = \frac{k\,q}{R^2} \,.$$

The final equality rests on the definition of Coulomb's constant in terms of the permittivity of free space, $k \equiv (4\pi\epsilon_0)^{-1}$. Recalling that the field must be radially directed, we determine it to be

$$\vec{E} = E_R \, \hat{R} = \frac{k\,q}{R^2} \, \hat{R} \,,$$

or, in other words, **we have just derived Coulomb's Law from Gauss's Law!**

ASIDE: One might wonder how it is possible to determine the field, a vector quantity with three components, from the flux, which is a scalar quantity. Gauss's Law can be "reversed," IFF the physical situation possesses sufficient symmetry to completely fix the field's direction. Subsequent determination of the flux then fixes the magnitude of the field.

---

[3] An ISOTROPIC system "looks the same" irrespective of the direction from which you choose to view it.
[4] The orientation of the field is OUTWARD for $q > 0$ and INWARD for $q < 0$.

Let's now use Gauss's Law to analyse a case in which direct application of Coulomb's Law would be onerous. Afterward, we shall rederive and extend the uniformly charged infinite plane results that were previously obtained.

EXAMPLE [*Electric Field Produced by a Uniformly Charged Spherical Source*]

A spherical volume of insulating material[5] has uniform charge density ρ_0 and radius a. It bears total charge

$$Q_{\text{Total}} = \rho_0 \, \frac{4\,\pi}{3} \, a^3 \,.$$

Our goal is to determine the electric field everywhere in space: both **outside** and **inside** of the sphere.

METHOD ONE: [*Partition, Compute, Sum, and Refine*]

P   Partition the charge-containing volume into chunks. The amount of charge on a chunk of differential volume is $dq = \rho_0 \, d[\text{Volume}]$.

C   Compute the differential electric fields produced by the charge elements in the partition, treating each as though it were a point source of electric field,

$$d\vec{E} = \frac{k \, dq}{\left| \vec{r} \right|^2} \,,$$

where $\vec{r}$ is the vector from the chunk to the field point.

S   Sum the field contributions from each chunk [linear superposition] to obtain the net field. This step requires integrating over the 3-d volume of the spherical region, a standard but thorny problem in Calculus classes. [In Chapter 48 of VOLUME I, we almost did this analysis.]

There has got to be a better (read *"easier"*) *way than this, eh?*

―――――――――――――

[5]In the next chapter, we shall come to realise that it is necessary for the material to be insulating.

METHOD TWO: [*Appeal to Gauss's Law*]

EXTERNAL   First, let's determine the electric field outside of the charged sphere.

Imagine a Gaussian sphere of radius R, with $R > a$, concentric with the spherical charge distribution. According to Gauss's Law,

$$\oint_{S_R} \vec{E} \cdot d\vec{A} = \frac{q_{\text{encl.}}}{\epsilon_0} \,.$$

Under the conditions of isotropy which apply to this situation, the LHS evaluates to $E(R) \times 4\pi R^2$, where $E(R)$ is constant.[6] Meanwhile, all of the charge on the ball of insulating material is enclosed in the enveloping Gaussian surface, and therefore the RHS is $Q_{\text{Total}}/\epsilon_0$.

Invoking Gauss's Law [by setting its LHS equal to its RHS] shows that

$$E(R) = \frac{Q_{\text{Total}}}{4\pi \epsilon_0 R^2}$$

outside the volume of charge-bearing insulating material, *viz.*, for $R > a$.

**The electric field outside the distribution
is identical to that which would occur
if the total charge were
concentrated at the central point of the distribution.**

> ASIDE: **Q:** Why should the effects of a distribution of charge appear to be exactly like those of a point charge?
>
> **A:** There is no compelling *a priori* reason why this must be so. Nevertheless, it (1) is observed to hold in nature, (2) flows from the inverse-squared dependence of the force/field, and (3) has other profound implications, as we'll see anon.

INTERNAL   Second, let's determine the electric field inside of the charged sphere.

Imagine a Gaussian sphere of radius R, with $R < a$, concentric with the spherical charge distribution.

Application of Gauss's Law and the analysis of its LHS proceeds exactly as above, because nothing in the flux derivation depends on the relative sizes of R and a. The RHS meanwhile is changed on account of the fact that only a portion of the charge-bearing volume is enclosed within the Gaussian surface.

The actual amount of charge enclosed is straightforwardly computed in this case,

$$q_{\text{encl.}} = \rho_0 \frac{4\pi}{3} R^3 \,,$$

which is best rewritten as

$$q_{\text{encl.}} = \rho_0 \frac{4\pi}{3} a^3 \left(\frac{R}{a} \right)^3 = Q_{\text{Total}} \frac{R^3}{a^3} \,.$$

---

[6] Although spherical symmetry and isotropy force the magnitude of the field to be constant everywhere on the spherical surface, the value is expected to vary for differing radii, and hence the parametric dependence on R is made explicit. In previous computations, the R-dependence was indicated by a subscript.

Application of Gauss's Law yields

$$E(R)\, 4\,\pi\, R^2 \equiv \frac{Q_{\text{Total}}\, \frac{R^3}{a^3}}{\epsilon_0} = \frac{Q_{\text{Total}}}{\epsilon_0\, a^3}\, R^3 \qquad \Longrightarrow \qquad E(R) = \frac{Q_{\text{Total}}}{4\,\pi\,\epsilon_0\, a^3}\, R = \frac{k\, Q_{\text{Total}}}{a^3}\, R\,.$$

Thus, within the insulating material, the electric field grows linearly from zero at the centre to its value at the surface.

To convince ourselves that the exterior and interior solutions are consistent, let's consider Figure 6.2 and the following comments.

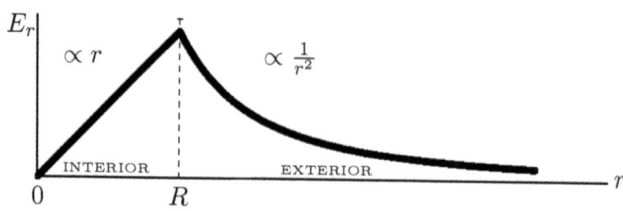

FIGURE 6.2   Radial Field of a Spherically Symmetric Uniformly Charged Ball

⊙ The electric field vanishes at the centre, as it must, in accord with the isotropy of the spherically symmetric distribution.

The only vector that points in all directions simultaneously is the ZERO vector.

○ The electric field is continuous [but not differentiable] at the surface of the insulator [*i.e.*, the boundary separating the interior and exterior solutions].

$$\lim_{R\to a^-} E(R) = \lim_{R\to a^-} \frac{k\, Q_{\text{Total}}}{a^3}\, R \qquad\qquad \lim_{R\to a^+} E(R) = \lim_{R\to a^+} \frac{k\, Q_{\text{Total}}}{R^2}$$

$$\text{AND}$$

$$= \frac{k\, Q_{\text{Total}}}{a^2} \qquad\qquad\qquad\qquad\qquad = \frac{k\, Q_{\text{Total}}}{a^2}$$

EXAMPLE [*Electric Field of an Infinite Uniformly Charged Planar Source*]

Consider an infinite plane surface with uniform charge density σ_0, as exhibited in Figure 6.3. The direction of the electric field at every point must be perpendicular to the charged plane owing to the symmetry of the source.

ASIDE: The field cannot tilt "sideways," because that would signal a preferred direction in the plane which is incompatible with the symmetry of the situation. Nor can the field depend on the position of the point on the plane directly beneath the field point. We might also suspect that, because there is no intrinsic scale to the infinite plate, the field cannot depend on the height of the field point above the plane. Although we shall prove momentarily that this is indeed the case, *a priori*, we suppose that the field does vary.

Imagine a Gaussian CYLINDER, S_c, with height $2\,a$ and face area A, centred on and bisected by the plane as shown in Figure 6.4. The closed cylinder is comprised of two open disk-shaped end caps [upper, S_u, and lower, S_l], and a single side, S_s, wrapping all the way around:

$$S_c = S_u \cup S_l \cup S_s\,.$$

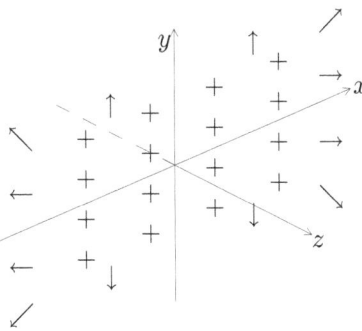

FIGURE 6.3 An Infinite Uniformly Charged Plane Surface

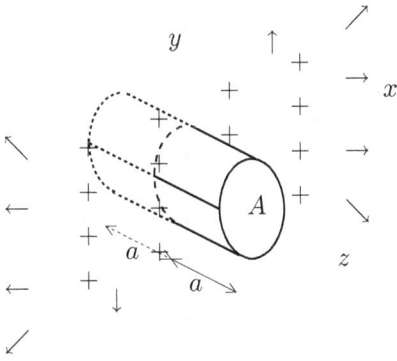

FIGURE 6.4 Gaussian Cylinder Intersecting an Infinite Uniformly Charged Plane

The electric flux through the Gaussian cylinder is the sum of the fluxes through the respective portions of the closed surface.

SIDE   Since the field is tangent to the wall of the cylinder, no flux goes through the side of the cylinder, and $\Phi_{S_s} = 0$.

UPPER   The upper face, with area A, is at constant height a above the plane. The unit normal on this surface points perpendicularly away from the plane, as must the electric field. Everywhere on S_u, the electric field has the same magnitude, which we designate $E(a)$. The flux through this surface is computed to be

$$\Phi_{S_u} = E(a)\,A\,.$$

LOWER   The lower and upper faces are symmetric. Though the area vectors are oppositely oriented, so too are the electric fields. The net effect is that the flux through each surface is exactly the same:

$$\Phi_{S_l} = E(a)\,A\,.$$

Therefore, the total electric flux passing through the Gaussian cylinder is $\Phi_{S_c} = 2\,E(a)\,A$, while the amount of charge enclosed within its confines is $q_{\text{encl.}} = \sigma_0\,A$. Invoking Gauss's Law enables determination of the manner in which the magnitude of the electric field varies with a, the height above the plane. *I.e.*,

$$2\,E(a)\,A \equiv \frac{\sigma_0\,A}{\epsilon_0} \quad \Longrightarrow \quad E(a) = \frac{\sigma_0}{2\,\epsilon_0} = 2\,\pi\,k\,\sigma_0\,.$$

Hence, the expression for the (constant) electric field above an infinite plane of uniform charge density was successfully rederived using Gauss's Law.   [*Cool, eh?*]

One final word: The electric field in this case is discontinuous at the [vanishingly thin] surface of the charged plane. Since the field points outward from the plane on both sides,

$$\vec{E}(z) = \begin{cases} + \dfrac{\sigma_0}{2\,\epsilon_0}\,\hat{k}\,, & \text{for } z > 0\,, \text{ and} \\[2mm] - \dfrac{\sigma_0}{2\,\epsilon_0}\,\hat{k}\,, & \text{for } z < 0\,. \end{cases}$$

EXAMPLE [*Two Infinite Uniformly Opposite-Charged Planar Surfaces*]

Consider the pair of parallel infinite sheets separated by distance L, bearing opposite uniform surface charge densities $\pm\sigma_0$, illustrated in the left panel of Figure 6.5.

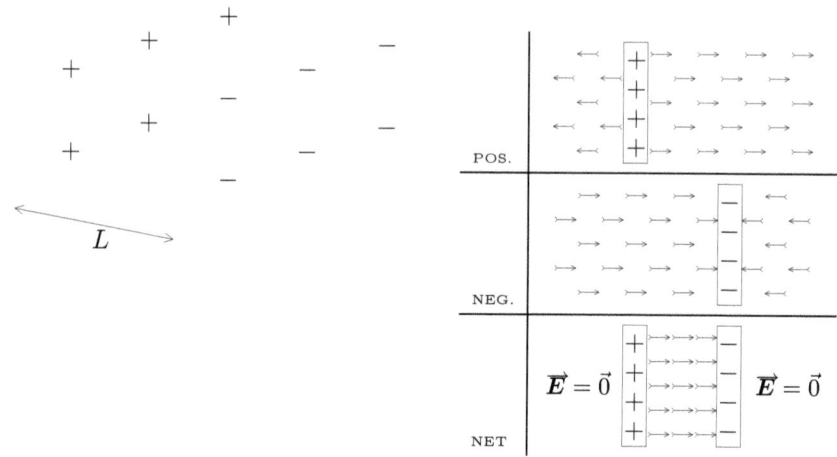

FIGURE 6.5   Two Nearby Oppositely Charged Infinite Parallel Plates and the Superposition of Their Electric Fields

The fields produced by the plates alone, along with the net field obtained by linear superposition, are shown in the rightmost panel of the figure. For this **parallel plate** configuration,

$$\vec{E}_{\parallel} = \begin{cases} \vec{0}\,, & \text{outside the volume enclosed by the plates,} \\[2mm] \dfrac{\sigma_0}{\epsilon_0}\,[+\rightarrow-]\,, & \text{within the volume enclosed.} \end{cases}$$

ASIDE: In Chapter 2, concern was expressed as to the provenance of uniform electric fields. Now we have a robust means of producing uniform electric fields of finite extent and sharply defined boundaries.

Chapter 7

Electrostatic Implications and Potential Energy

ELECTROSTATIC EQUILIBRIUM   Electrostatic equilibrium is attained within a region of space when the prevailing charge distribution remains constant in time.

A couple of general qualifications are in order before we consider the specifics pertinent to conductors in electrostatic equilibrium. First, the charge distribution need not be uniform.

Uniformity is antithetical to equilibrium in many circumstances.

Second, electrostatic equilibrium does not preclude the motion of particular charges. Rather, it requires that a compensatory counterflow of other charges neutralise the net effect on the overall distribution.

Individual charges may move, but the net charge flux vanishes.

For a charged conductor[1] to be in electrostatic equilibrium, three conditions must be met.

- **The electric field must vanish everywhere within the bulk of the conductor.**

   IF the electric field were not everywhere equal to zero, THEN any free charges present would accelerate and flow from one position to another in a manner incompatible with the idea of electrostatic equilibrium. Thus, the electric field must vanish within the body of a conducting material in electrostatic equilibrium.

- **Any excess charge must reside on the surface of the conductor.**

   IF a net amount of electric charge were present at some location entirely within the bulk of the material, THEN there would have to be a non-zero electric field there as well, as a consequence of Gauss's Law.

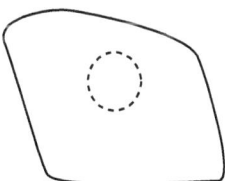

FIGURE 7.1   A Gaussian Surface Enclosed within a Lump of Conductor

Consider a closed Gaussian surface completely embedded in the conductor as indicated by the dotted circle appearing in Figure 7.1. IF the enclosed volume contains non-zero net charge, THEN the electric flux through the Gaussian surface is non-zero according to the RHS of Gauss's Law.

---

[1]Recall that charge flows easily within a conductor.

When the RHS of Gauss's Law is non-zero, so too must be the LHS!

For the LHS to be non-zero, it is necessary that the electric field be non-vanishing in some region or regions which intersect the Gaussian surface. And yet, precisely this circumstance has been ruled out above, as being incompatible with the conditions imposed by electrostatic equilibrium.

- **The field just outside a charged conductor must be orthogonal to the surface.**

IF the electric field on the surface has a non-zero projection onto the tangent plane, THEN charges on the surface will accelerate in response to the surface field in a manner completely at odds with the notion of electrostatic equilibrium

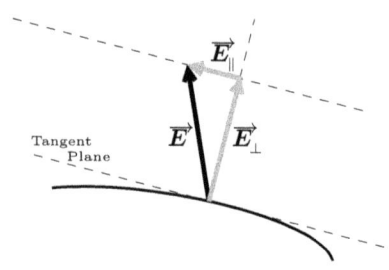

FIGURE 7.2   The Electric Field Just outside of a Charged Conductor

---

ASIDE: Let's revisit these three conditions in alternative formulations.

- Suppose that a non-zero electric field is applied to the bulk of a neutral conductor. In response, the conductor will polarise: the POSITIVE charges tending "downfield," while the NEGATIVE charges are drawn "upfield." This separation of charge within the neutral matter gives rise to an **induced field** oriented to precisely oppose the externally applied field. The polarisation increases until the applied field is completely cancelled within the bulk of the conductor.

- Suppose that some amount of net electric charge is dispersed within the bulk of a conductor. The interior charge must give rise to a non-zero local electric field in the interior, by Gauss's Law. Any charges bathed in such local fields will respond so as to distance themselves as much as possible from the other LIKE charges present. This continues until their *backs are to the wall* [*i.e.,* they reach the edge of the conductor], and they are maximally spread out upon the surface.

- A non-zero projection of the electric field onto the surface of the conductor will inevitably cause a redistribution of the charges resident on the surface. This proceeds until the redistributed charges give rise to an induced field which precisely cancels the in-surface components of the original electric field everywhere on the surface.

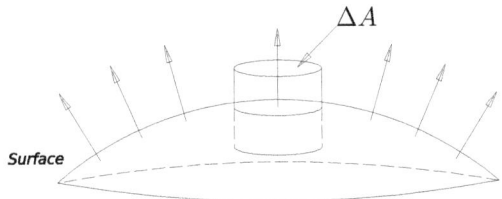

FIGURE 7.3 Conductor Surface with Embedded Gaussian Cylinder

THE ELECTRIC FIELD AT THE SURFACE OF A CHARGED CONDUCTOR

An isolated lump of conductor bearing a non-zero net amount of charge lies in electrostatic equilibrium. Imagine a tiny Gaussian cylinder, S_c, poking through a small section of its surface, as illustrated in Figure 7.3.

The upper and lower faces of the cylinder have area ΔA, and are effectively parallel to the small patch of surface through which the Gaussian cylinder pokes.

[The Δ informs us that the face area is small on all scales set by the dimensions of the lump.]

The electric flux through the lower face is exactly zero because $\vec{E} \equiv \vec{0}$ everywhere inside the bulk of the conductor. The flux through the side of the cylinder is also zero on account of the perpendicularity of the local field and the unit normals on the cylinder wall. For sufficiently short cylinders, the electric field on the top face is effectively constant and approximately equal to its surface value. The flux through the top face then is well-approximated by the product of the magnitude of the surface field with the face area: $\Phi_{e,S_u} = E(0) \Delta A$. Thus,

$$\Phi_{e,S_c} = \Phi_{e,S_u} + \Phi_{e,S_l} + \Phi_{e,S_s} = E(0) \Delta A .$$

The amount of charge enclosed within S_c is the local surface charge density multiplied by the cylinder's tiny face area: $q_{\text{encl.}} = \sigma \Delta A$. Gauss's Law relates the net flux to the enclosed charge. Cancelling the common factor of area yields an expression for the surface field strength in terms of the local charge density:

$$E(0) = \frac{\sigma}{\epsilon_0} .$$

No mention whatsoever has been made of the manner in which the areal electric charge density, σ, varies on the surface except to say that it is reasonably constant over [very tiny] local patches.[2] Remarkably, this formula [determining the magnitude of the electric field at the surface of a charged conductor in electrostatic equilibrium in terms of the local value of the surface net electric charge density] holds generally.

> ASIDE: The charge density is almost certain to be non-uniform over the surface. [In fact, excess charge tends to accumulate where the surface curvature is large.]
>
> Electrical probes, fabricated from conducting material and shaped to a sharp point, make use of this effect. The concentration of charge toward the tip gives rise to a stronger field and more tightly bunched electric field lines.

---

[2]Mathematically speaking, σ is continuous.

Lightning rods work on this principle. Concentrations of separated electric charges built up in the atmosphere are slowly dissipated [a.k.a. grounded] before they become sufficiently large to produce a lightning strike.

ELECTRIC FIELD QUENCHING

IF a neutral conductor is immersed in an externally applied electric field, THEN the conductor **polarises** so as to produce zero net electric field within its bulk.

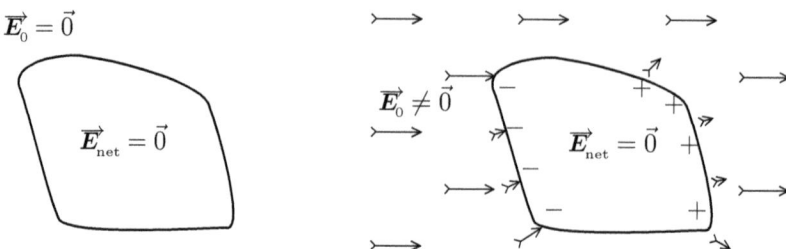

FIGURE 7.4 A Conductor Quenches an Applied Electric Field

Inside the conductor, the external field is opposed/cancelled out by the field induced by the separation of charge on the external surfaces of the conducting body:

$$\vec{E}_{\text{net}} = \vec{E}_0 + \vec{E}_i.$$

One says that the external field has been **quenched** within the volume of the conductor. We will expound further on this subject when we encounter dielectrics in Chapter 15.

ELECTROSTATIC POTENTIAL ENERGY

A point particle situated at $\mathcal{P}$, bearing charge q_0, experiences an electric force,

$$\vec{F}_{\text{E}} = q_0\,\vec{E}(\mathcal{P}),$$

determined by the local electric field. Suppose that the particle moves from an initial position, $\vec{r}_i$, to a final position, $\vec{r}_f$, along a specified path, γ, during the time interval $\Delta t = t_f - t_i$. The mechanical work done by the electric force on the particle as it traverses the path is

$$W_{if}\left[\vec{F}_{\text{E}}\right]_\gamma = \int_{\vec{r}_i}^{\vec{r}_f} \vec{F}_{\text{E}} \cdot d\vec{s} = q_0 \int_{\vec{r}_i}^{\vec{r}_f} \vec{E} \cdot d\vec{s}.$$

That the electric force is CONSERVATIVE has the following implications.

- The work done by the electric force depends only upon the initial and final positions and not on the path taken.

- An electrostatic potential energy function depending on position, U_E, may be associated with the electric force in the customary manner:

$$U_E(\vec{r}_f) - U_E(\vec{r}_i) \equiv -W_{if}\left[\vec{F}_{\text{E}}\right]_\gamma = -q_0 \int_{\vec{r}_i}^{\vec{r}_f} \vec{E} \cdot d\vec{s}.$$

- The electrostatic potential energy is not uniquely defined. Only changes in potential energy, rather than the [value of the] potential energy itself, have significance.

  Potential energy at a single point is meaningless; context is everything!

  > ASIDE: Recall from VOLUME I, Chapter 25, that Mathilde added a constant value, C, to PK's already serviceable potential energy function, U, thereby obtaining $\widetilde{U} = U + C$. When Mathilde and PK compared potential energy differences computed with her function to those computed with his, they found themselves always in agreement:
  > $$\Delta \widetilde{U} = \widetilde{U}_f - \widetilde{U}_i = [U_f + C] - [U_i + C] = U_f - U_i = \Delta U.$$

- The units of electrostatic potential energy are $[\text{N} \cdot \text{m}] = [\text{J}] = \text{joules}$.

ELECTRIC POTENTIAL

The electric potential function, $V(\mathcal{P})$, may be thought of as the electrostatic potential energy per unit charge, *i.e.*,

$$V(\vec{r}) = \frac{U_{E,q_0}(\vec{r})}{q_0}.$$

Electric potential is more precisely defined by

$$V(\vec{r}) = -\int_{\vec{r}_0}^{\vec{r}} \vec{E} \cdot d\vec{s}.$$

The reference point, $\vec{r}_0$, is that for which the potential is deemed to vanish. As the electrostatic potential energy is only defined up to an additive constant, so too is the potential function, and the reference point may be freely chosen.

Electric potential is measured in **volts**,

$$[\text{V}] = \left[\frac{\text{N} \cdot \text{m}}{\text{C}} \right] = \left[\frac{\text{J}}{\text{C}} \right],$$

honouring Alessandro Volta (1745–1827), the inventor of the electrochemical battery. In this reckoning, one joule of work is required to increase the potential of one coulomb of charge by one volt.

[*Potential* confusion between the potential and its unit exists, but is seldom actualised.]

Potential differences are of greater import, since these are more truly physical and consequential. Provided that a common reference point is chosen,

$$\Delta V_{12} = V(\vec{r}_2) - V(\vec{r}_1) = \left(-\int_{\vec{r}_0}^{\vec{r}_2} \vec{E} \cdot d\vec{s} \right) - \left(-\int_{\vec{r}_0}^{\vec{r}_1} \vec{E} \cdot d\vec{s} \right) = -\int_{\vec{r}_1}^{\vec{r}_2} \vec{E} \cdot d\vec{s}.$$

That is, the difference in potential depends only[3] on the two field points. The conservative nature of the field ensures that the potential function and potential differences depend only on the endpoints and not [explicitly] on details of the path taken from the reference point or from one point to the other.

All measurements of potential are, properly speaking, potential differences, and in common usage the "Δs" are elided.

[CAVEAT: The *potential* for confusion between the potential and its differences is ever-present.]

---

[3] Well, *almost* only, since it is also necessary that there exist a path from one point to the other.

Chapter 8

Potentially Fun!

Let's start with a brief recapitulation and one augmentation.

V   The **electric potential**, V, at a point in space, admits interpretation as the electrostatic potential energy per unit charge there [with respect to a chosen reference point]:

$$V(\vec{r}) = \frac{U_{E,q_0}(\vec{r})}{q_0} = -\int_{\vec{r}_0}^{\vec{r}} \vec{E} \cdot d\vec{s} \,.$$

V   The units of potential are volts, V, where $\mathtt{V} = \left[\dfrac{\mathtt{N} \cdot \mathtt{m}}{\mathtt{C}}\right] = \left[\dfrac{\mathtt{J}}{\mathtt{C}}\right]$.

ΔV The **potential difference**, ΔV_{12}, existing between the points $\vec{r}_2$ and $\vec{r}_1$, is

$$\Delta V_{12} = V(\vec{r}_2) - V(\vec{r}_1) = -\int_{\vec{r}_1}^{\vec{r}_2} \vec{E} \cdot d\vec{s} \,.$$

The potential difference is independent of the path employed in its computation.

eV   The **electron-volt** is the amount of energy gleaned or consumed in the passage of a proton, bearing elementary charge, $+e$, through a potential difference of $1\,\mathtt{V}$.

$$\left[\,\mathtt{eV}\,\right] = 1.602 \times 10^{-19} \left[\,\mathtt{C}\,\right] \times \left[\,\mathtt{V}\,\right] = 1.602 \times 10^{-19} \left[\,\mathtt{C}\,\frac{\mathtt{J}}{\mathtt{C}}\,\right] \simeq 1.6 \times 10^{-19}\,\mathtt{J} \,.$$

With these reminiscences over, let's have fun with potentials.

EXAMPLE [*Electric Potential Associated with a Uniform Field*]

A uniform electric field, $\vec{E} = E_0\,\hat{\imath}$, fills a region of space. A particle bearing charge q_0 is moved from the point $\vec{r}_1 = (x_1\,,\,y_0\,,\,z_0)$ to the point $\vec{r}_2 = (x_2\,,\,y_0\,,\,z_0)$, along a straight path parallel to the x-axis.

The electrostatic potential difference between the initial and the final points is

$$\Delta V_{12} = -\int_{\vec{r}_1}^{\vec{r}_2} \vec{E} \cdot d\vec{s} \,.$$

Every step along this particular path proceeds purely in the x-direction,

$$d\vec{s} = (dx\,,\,dy\,,\,dz)\Big|_{\substack{dy=0 \\ dz=0}} = (dx\,,\,0\,,\,0) \,,$$

and thus

$$\vec{E} \cdot d\vec{s} = E_0\,\hat{\imath} \cdot dx\,\hat{\imath} = E_0\,dx \,.$$

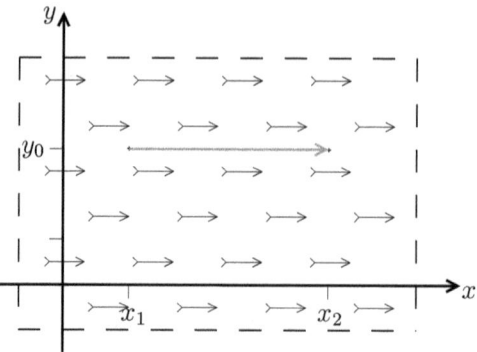

FIGURE 8.1   Electric Potential (1-d case)

Hence, the potential difference is

$$\Delta V_{12} = -\int_{x_1}^{x_2} E_0 \, dx = -E_0 \int_{x_1}^{x_2} dx = -E_0 \, x \Big|_{x_1}^{x_2} = -E_0 \left(x_2 - x_1 \right) = -E_0 \, L \, ,$$

where L is the directed magnitude of the displacement,

$$L \, \hat{\imath} = \vec{r}_2 - \vec{r}_1 \, .$$

The imputed change in the electrostatic potential energy of the particle which was moved from $\vec{r}_1$ to $\vec{r}_2$ is

$$\Delta U_E = q_0 \, \Delta V_{12} = -q_0 \, E_0 \, L \, .$$

This is reasonable, since moving a positively charged particle forward in the field, $L > 0$, lowers its potential energy, $\Delta U_E < 0$. Conversely, moving it backward, $L < 0$, leads to an increase in its electrostatic potential energy, $\Delta U_E > 0$.

Example   [*Potential Associated with Uniform Field, Encore*]

Let's again consider a point particle bearing charge q_0 immersed in a uniform electric field, $\vec{E} = E_0 \, \hat{\imath}$. This time the particle is moved from an initial point, $\vec{r}_1 = (x_1, y_1, z_0)$, to a final point, $\vec{r}_1 = (x_2, y_2, z_0)$, along a straight line path. Henceforth, the z-coordinate shall be self-consistently ignored, and we'll work exclusively in 2-d.

The potential difference, computed in the usual way, is

$$\Delta V_{12} = -\int_{\vec{r}_1}^{\vec{r}_2} \vec{E} \cdot d\vec{s} \, .$$

The novelty in this instance is that the path is at an angle to the field. We'll perform the requisite analysis two different ways.

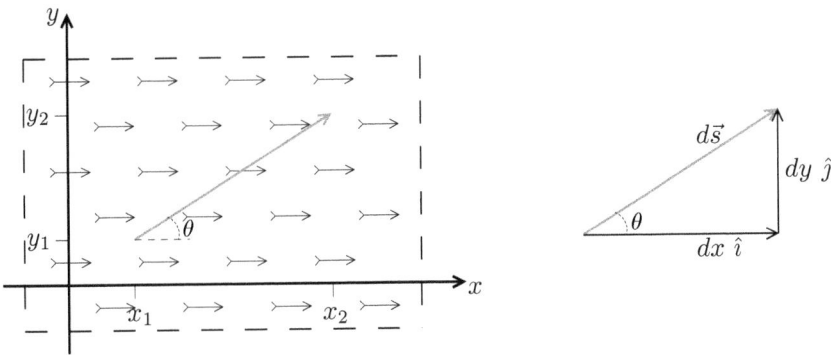

FIGURE 8.2 Electric Potential (2-d case)

METHOD ONE: [*Expand the integrand and then specialise to constant field*]
The differential path element, expressed in Cartesian coordinates, is

$$d\vec{s} = \big(\cos(\theta)\,\hat{\imath} + \sin(\theta)\,\hat{\jmath} \big)\,ds\,,$$

where the fixed angle, θ, is constrained to satisfy

$$\tan(\theta) = \frac{y_2 - y_1}{x_2 - x_1}\,,$$

and ds is the differential arc length of the segment. Thus, the integrand reads

$$\vec{E} \cdot d\vec{s} = \big(E_0\,\hat{\imath} \big) \cdot \big(\cos(\theta)\,\hat{\imath} + \sin(\theta)\,\hat{\jmath} \big)\,ds = E_0\,\cos(\theta)\,ds\,.$$

As both E_0 and θ are everywhere constant along the path, the integral leading to the potential difference simplifies:

$$-\int_{\vec{r}_1}^{\vec{r}_2} \vec{E} \cdot d\vec{s} = -E_0\,\cos(\theta) \int_{\vec{r}_1}^{\vec{r}_2} ds = -E_0\,L\,\cos(\theta)\,,$$

where L is the arc length of the path proceeding straight from the initial to the final point.

METHOD TWO: [*Specialise to constant field and then integrate*]
That the field is constant in this particular case can be exploited to simplify the computations as follows:

$$-\int_{\vec{r}_1}^{\vec{r}_2} \vec{E} \cdot d\vec{s} = -\vec{E} \cdot \int_{\vec{r}_1}^{\vec{r}_2} d\vec{s} = -\vec{E} \cdot \vec{L} = -E_0\,L\,\cos(\theta)\,.$$

The first equality arises because the electric field is constant throughout the domain of integration, *i.e.*, everywhere along the path. The second equality introduces the displacement vector, $\vec{L} = \vec{r}_2 - \vec{r}_1$, while the third stems from the geometric definition of the dot product.

Whichever method is employed to compute the potential difference, its expected path independence becomes manifest with realisation that $L\cos(\theta) = x_2 - x_1$. Thus,

$$\Delta V_{12} = -E_0\left(x_2 - x_1\right)$$

depends only on the net distance moved *uphill* or *downhill* with respect to the field and not at all on the lateral distance traversed. In this instance, **all** points with a particular value of x have the same potential.

EQUIPOTENTIAL REGION   An equipotential region is a contiguous subdomain of space within which the electric potential has everywhere the same value. These regions may be line-like [a rare occurrence], sheet-like [these are called "equipotential surfaces"], or fully three-dimensional. Their topology may be complicated. Disjoint equipotentials may happen to have equal values of potential.

[The collection of equipotentials reconstitutes 3-d space up to a set of measure zero.]

Important properties of equipotential surfaces are:

⊥   The local electric field is everywhere perpendicular to equipotential surfaces and lines. Within equipotential volumes, the electric field must vanish.

0   It takes ZERO work to move charges from one point to another within an equipotential region.

PROOF: The electric work done on a particle bearing charge q, moved from $\vec{r}_i$ to $\vec{r}_f$, in the presence of an electric field, $\vec{E}$, is

$$W_{if}\left[\vec{F_{\text{E}}}\right] = \int_{\vec{r}_i}^{\vec{r}_f} q\,\vec{E}\cdot d\vec{s}\,.$$

For an external agent to move the particle, starting and ending at rest, from $\vec{r}_i$ to $\vec{r}_f$, requires that the external agent provide an amount of work which exactly compensates for the electric work.

- Within an equipotential region the electric field vanishes, $\vec{E} = \vec{0}$, and thus the electric work is identically zero, and no compensatory external work is required.

- Upon an equipotential surface, the electric field is everywhere parallel to the local normal, and perpendicular to the local tangent plane. Thus, $\vec{E}\cdot d\vec{s} = 0$, everywhere along the path, and zero external work is required to accomplish the movement of the charge.

The bulk of a conductor in electrostatic equilibrium forms an equipotential region.

Chapter 9

Electrostatic Potential Energy

EXAMPLE [*Electric Potential Associated with an Isolated Point Charge*]

An isolated point charge, q, lies at the origin of coordinates. Two field points lie at $\vec{r}_1$ and $\vec{r}_2$, respectively. The electric potential difference between these points, ΔV_{12}, is

$$\Delta V_{12} = V(\vec{r}_2) - V(\vec{r}_1) = -\int_{\vec{r}_1}^{\vec{r}_2} \vec{E}_q \cdot d\vec{s}.$$

The electric field produced by the isolated point charge is directed radially outward, with its magnitude falling off as the square of the distance from the charge,

$$\vec{E}_q = \frac{k\,q}{r^2}\,\hat{r}.$$

Accordingly, only the radial components of the path survive the projection onto the field in the integrand.

> ASIDE: The equipotential surfaces in this case are spherical shells, as these are everywhere orthogonal to the radial field. A path from $\vec{r}_1$ to $\vec{r}_2$ first proceeds along an equipotential [spherical shell] to the point, $\vec{r}_c$, which lies along the semi-infinite ray from the origin passing through $\vec{r}_2$. The second portion of the path is purely radial from $\vec{r}_c$ to $\vec{r}_2$.

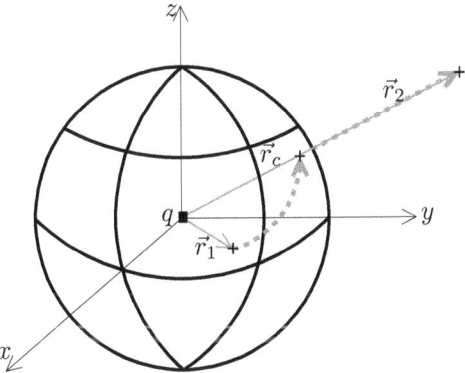

FIGURE 9.1   A Point Charge, Field Points, and an Equipotential Surface

Since the first portion of the path begins and ends on the equipotential shell,[1] $\Delta V_{1c} = 0$. Given that $\Delta V_{12} = \Delta V_{1c} + \Delta V_{c2}$, we may write

$$\int_{\vec{r}_1}^{\vec{r}_2} \vec{E}_q \cdot d\vec{s} = \int_{\vec{r}_1}^{\vec{r}_c} \vec{E}_q \cdot d\vec{s} + \int_{\vec{r}_c}^{\vec{r}_2} \vec{E}_q \cdot d\vec{s} = 0 + \int_{\vec{r}_c}^{\vec{r}_2} \vec{E}_q \cdot d\vec{s}. = \int_{r_1}^{r_2} E_{q,r}\,dr.$$

---

[1]Stated more strongly: the entirety of this portion remains within the equipotential shell.

However one prefers to think about the reduction of the integral, one obtains

$$\Delta V_{12} = -\int_{r_1}^{r_2} \frac{k\,q\,dr}{r^2} = k\,q\left[\frac{1}{r}\bigg|_{r_1}^{r_2}\right] = k\,q\left[\frac{1}{r_2} - \frac{1}{r_1}\right].$$

Remaining with this example for a wee bit longer, one is inclined to establish the zero of electrostatic potential at infinite radial distance from the isolated point charge, *i.e.*, $r \to \infty$.

[Although the notion of a single point at infinity might seem odd, it is rigorous.]

With the point of reference at infinite radial distance, the potential function associated with an isolated point charge, q, located at the origin admits a particularly simple form:

$$V(\vec{r}) = \frac{k\,q}{|\vec{r}|} = \frac{k\,q}{r}.$$

This potential function is scalar and spherically symmetric.

Q: What electrostatic potential is produced by a collection of point charges?

A: It is the superposition of the potentials produced by each of the charges individually. Recall that electric fields obey the PRINCIPLE OF LINEAR SUPERPOSITION in that the total field produced at a point arising from the presence of $\mathcal{N}$ contributors[2] is

$$\vec{E}_{\text{net}} = \sum_{i=1}^{\mathcal{N}} \vec{E}_i,$$

where $\vec{E}_i$ is the electric field produced by the ith contributor. Realising that the dot product is distributive over vector addition, we may write

$$\vec{E}_{\text{net}} \cdot d\vec{s} = \left(\sum_{i=1}^{\mathcal{N}} \vec{E}_i\right) \cdot d\vec{s} = \sum_{i=1}^{\mathcal{N}} \vec{E}_i \cdot d\vec{s}.$$

Returning to the definition of electrostatic potential,

$$V_{\text{net}}(\vec{r}) = -\int_{\vec{r}_0}^{\vec{r}} \vec{E}_{\text{net}} \cdot d\vec{s} = \sum_{i=1}^{\mathcal{N}} \left(-\int_{\vec{r}_0}^{\vec{r}} \vec{E}_i \cdot d\vec{s}\right) = \sum_{i=1}^{\mathcal{N}} V_i(\vec{r}),$$

where $V_i(\vec{r})$ is the electric potential [with respect to a common reference point] associated with the ith contributing electric source. In the course of the mini-derivation quoted above, the sum over sources and the integral along the path were interchanged.

[This switcheroo can only be effected when both the sum and integral converge.]

The above analysis works readily when it is assumed that the source distributions are bounded and the field point lies external to all sources. These conditions are not always essential and can be relaxed in particular instances, as we shall later see.

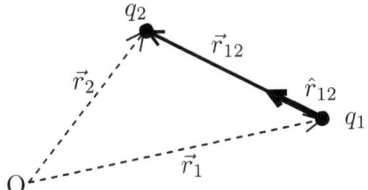

FIGURE 9.2 A Two Point Charge System

THE ELECTROSTATIC POTENTIAL ENERGY OF A TWO-CHARGE SYSTEM

ASIDE: **Q:** What about the electric potential energy of a one-charge system?

A: There is no proper answer to this question. It is like asking:

"What is the sound of one hand clapping?"

Consider two point charges, q_1 and q_2, held fixed in space with separation vector $\vec{r}_{12}$ extending from the first to the second, as illustrated in Figure 9.2.

12 The electric potential produced by q_1, at the location of q_2, is

$$V_1(\vec{r}_{12}) = \frac{k\,q_1}{|\vec{r}_{12}|}\,.$$

Thus, the electrostatic potential energy of q_2 lying in the electric field and potential produced by q_1 is

$$U_{E,12} = q_2\,V_1(\vec{r}_{12}) = \frac{k\,q_1\,q_2}{|\vec{r}_{12}|}\,.$$

21 The electric potential produced by q_2, at the location of q_1, is

$$V_2(\vec{r}_{21}) = \frac{k\,q_1}{|\vec{r}_{21}|}\,.$$

Thus, the electrostatic potential energy of q_1 lying in the potential of q_2 is

$$U_{E,21} = q_1\,V_2(\vec{r}_{21}) = \frac{k\,q_1\,q_2}{|\vec{r}_{21}|} = \frac{k\,q_1\,q_2}{|\vec{r}_{12}|} = U_{E,12}\,.$$

The computed potential energies are equal: $U_{E,21} = U_{E,12}$. This is necessary, as the electrostatic potential energy is a property of the **system of charges**[3] and not attributed to one or the other individually. The potential energy may be positive [supplied to] or negative [given up by] the system, depending on the relative signs of q_1 and q_2.

To assemble $\begin{bmatrix} \text{LIKE} \\ \text{UNLIKE} \end{bmatrix}$ charges, $\begin{bmatrix} \text{POSITIVE} \\ \text{NEGATIVE} \end{bmatrix}$ external work is required,

and the system has $\begin{bmatrix} \text{POSITIVE} \\ \text{NEGATIVE} \end{bmatrix}$ electrostatic potential energy.

---

[2] These contributors may have as their ultimate sources point-, line-, areal-, or volume-charges.

[3] The potential energy corresponds to the amount of energy which must be supplied to [or liberated by], the system to bring the two charges from infinite initial separation to their final separation of $|\vec{r}_{12}|$.

Q: Does this last bit comport with our developing intuitions?

A: Yes, as can readily be seen in the following *Gedanken* experiment.

Imagine that you, an external agent, assemble the system of two charges by first placing q_1 at its desired location [in a region of space devoid of other charges]. As the ambient electric field vanishes, there is no net electric force acting on q_1 to be overcome. Once q_1 is in place, affix it to that spot, and prepare to bring q_2 [from far away] to its ultimate location. The field produced by q_1 bathes q_2, and the ensuing electric force must be countered in order to place q_2 at its appointed spot.

+ IF q_1 and q_2 are of LIKE sign, *i.e.*, $q_1 q_2 > 0$, THEN the two charges repel. The applied force and the displacement [from infinity to its final position in the vicinity of q_1] are sufficiently aligned that the net work done by the external force is positive. This positive amount of work has been converted into electrostatic potential energy proper to the system.

> ASIDE: Here's another way to ensure that we are thinking correctly.
>
> In bringing q_2 from infinity to its ultimate destination, at distance r from q_1, it must first attain distances r_a and r_b, with
>
> $$r < r_a < r_b \,.$$
>
> If instead q_2 were brought to rest and released at r_a, then the repulsive electric force would propel it outward so that at some later time it would again be at distance r_b, and there it would be moving with some speed v_b. Acquisition of kinetic energy takes place by conversion of electrostatic potential energy, and thus the potential energy at smaller distances must be greater than that at larger distances.

− IF q_1 and q_2 are of UNLIKE sign, *i.e.*, $q_1 q_2 < 0$, THEN the two charges attract. To bring q_2 to its final position in a controlled manner requires that the attraction be resisted, requiring an applied force which is sufficiently opposed to the motion of the charge as to render the net amount of external work negative.

> ASIDE: In bringing q_2 from infinity to its ultimate destination, at distance r from q_1, it must first attain distances r_a and r_b, with
>
> $$r < r_a < r_b \,.$$
>
> If instead q_2 were released from rest at r_b, then the attractive electric force would draw it in to r_a at a later time, and there it would be moving with speed v_a. As the charged particle would have gained kinetic energy as it fell inward, its potential energy must have decreased.

THE ELECTROSTATIC POTENTIAL ENERGY OF A THREE-CHARGE SYSTEM

The analysis of two charges, performed just above, can be self-consistently extended to accommodate systems with more than two charges.

1 Start with just the charge q_1 in empty space. It is free to move anywhere as there are no other charges extant to produce a potential. So, nail it down at point $\mathcal{O}$.

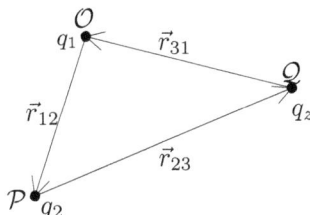

FIGURE 9.3   Electrostatic Potential Energy of a Three-Point-Charge System

2   Bring q_2 in from infinity to the point $\mathcal{P}$ in the potential produced by q_1. There is a certain amount of work performed by the electric force of interaction acting between q_1 and q_2. [Countervailing work must be done by an external agent.] The electrostatic potential energy of the q_1 plus q_2 system is

$$U_{12} = \frac{k\,q_1\,q_2}{|\vec{r}_{12}|}\,.$$

3   Bring q_3 in from infinity to the point $\mathcal{Q}$ in the net potential produced by q_1 and q_2. The two electrostatic potential energy contributions can be computed independently:

$$U_{13} = \frac{k\,q_1\,q_3}{|\vec{r}_{13}|}\,, \qquad \text{and} \qquad U_{23} = \frac{k\,q_2\,q_3}{|\vec{r}_{23}|}\,.$$

The net electrostatic potential energy associated with the three-charge system is:

$$U_{123} = U_{12} + U_{13} + U_{23} = \frac{k\,q_1\,q_2}{|\vec{r}_{12}|} + \frac{k\,q_1\,q_3}{|\vec{r}_{13}|} + \frac{k\,q_2\,q_3}{|\vec{r}_{23}|}\,.$$

IF there were four charges, THEN another step would occur in which the fourth charge is brought from infinity to a point $\mathcal{R}$ in the net potential produced by the set of three charges, $\{q_1, q_2, q_3\}$, at the points $\{\mathcal{O}, \mathcal{P}, \mathcal{Q}\}$, respectively. Abbreviating, we can discern the pattern more easily as we compute:

$$U_{1234} = U_{12} + U_{13} + U_{14} + U_{23} + U_{24} + U_{34}\,,$$

using the two-point charge expression for each of the U_{ij}.

The generalisation to collections of five or more point charges is straightforward: the potential energies associated with all pairwise combinations must be summed. The extension of these results to the case of continuous distributions of electric charge is investigated next.

$$V(\mathcal{P}) = \sum_{i=1}^{\mathcal{N}} V_i(\mathcal{P}) = \sum_{i=1}^{\mathcal{N}} \frac{k \, \Delta q_i}{|\vec{r}_i|}$$

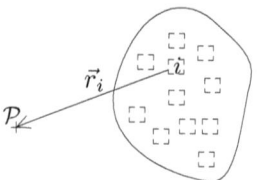

FIGURE 9.4   Obtaining the Electrostatic Potential outside a Distribution of Charge

Electrostatic Potential Energy and Distributions of Charge

As has become customary, the PARTITION, COMPUTE, and SUM method is employed to deal with continuous distributions of charge. The distributed charge is partitioned into $\mathcal{N}$ small charge elements, Δq_i, as shown in Figure 9.4. The contribution to the potential at the field point $\mathcal{P}$ due to each charge element, treated as a point charge, is computed. These are summed to yield the net potential at $\mathcal{P}$.

Infinitely refining the partition, via the simultaneous limits

$$\lim_{\mathcal{N} \to \infty} \qquad \text{AND} \qquad \lim_{\Delta q_i \to 0} \quad \forall \, i \in [1, \ldots, \mathcal{N}],$$

transforms the discrete sum into an integral. In which case,

$$V_{\text{net}}(\mathcal{P}) = \int dV = \int \frac{k \, dq}{r},$$

where r is the distance from the particular charge element dq in the partition to the field point $\mathcal{P}$. As is always the case, dq depends on geometrical factors describing the size and shape of the distribution.

To bring a point charge, q_0, from infinite distance to the field point $\mathcal{P}$ in the presence of the electric field due to the charge distribution requires work. This amount of work becomes the potential energy of the system comprised of the charge distribution plus the point charge q_0, *i.e.*,

$$U_{\text{net}, q_0} = q_0 \, V_{\text{net}}(\mathcal{P}) = k \, q_0 \int \frac{dq}{r}.$$

Our final word is that there exist two ways of computing the electric potential produced by a charge distribution.

$\boldsymbol{E}$ Knowledge of the electric field enables computation of potential differences via line integration along paths in space:

$$V_{\text{dist}}(\mathcal{P}) = -\int_{\vec{r}_0}^{\vec{r}} \vec{\boldsymbol{E}}_{\text{dist}} \cdot d\vec{s}.$$

ρ, σ, λ Knowing the charge distribution permits computation of the potential with respect to the point at infinity, via

$$V_{\text{dist}}(\mathcal{P}) = \int \frac{k \, dq}{r}.$$

Chapter 10

Rife with Potential

Electric potential at a point in space described by the position vector $\vec{r_P}$ was first defined by integrating the electric field along a path from some reference point [which may be taken to be the point at infinity] to the field point:

$$V(\vec{r_P}) - V(\vec{r_0}) = -\int_{\vec{r_0}}^{\vec{r_P}} \vec{E} \cdot d\vec{s}.$$

This expression for the change in potential is valid irrespective of how the field is produced.

In the last chapter, we saw that electric potential could be calculated directly from knowledge of a charge distribution by adding together the potential contributions from all of the charge elements. Concisely,

$$V(\vec{r}) = k \int \frac{dq}{|\vec{r}|},$$

where $|\vec{r}|$ is the distance from the charge element, dq, to the field point $\mathcal{P}$. [$\mathcal{P}$ is at position $\vec{r}$ with respect to dq.] In this formulation, it is implicitly assumed that $V(\infty) = 0$.

EXAMPLE [*Electrostatic Potential on the Axis of a Uniformly Charged Thin Ring*]

Consider a thin ring of radius a lying in the yz-plane, centred on the origin, and bearing constant lineal charge density λ_0. Also consider a field point $\mathcal{P}$ located a distance x along the x-axis.

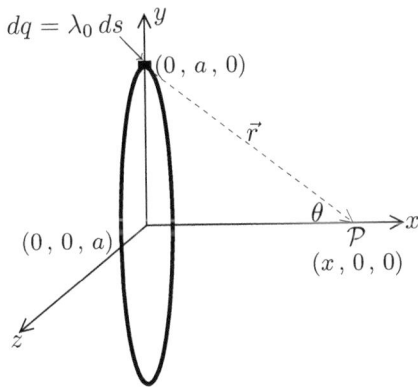

FIGURE 10.1   Computing the Electric Potential of a Thin Uniformly Charged Ring

ASIDE: A thin ring has a unique specific value a for its radius. A finite width [annular] ring possesses a range of radii. This feature complicates the analysis, as we shall see when we consider the thin disk. A ring with finite thickness may have the geometry of a thin cylindrical shell. Other rings might have toroidal geometry. The analysis of these cases can be quite involved.

The total charge on the ring is computed by integrating the charge density over the ring's extent. All of the charge elements on the ring lie at the same radial distance, a, from the centre. Each dq $[dq = \lambda_0\,ds = \lambda_0\,a\,d\theta]$ subtends a particular angle, $d\theta$. Hence,

$$Q_{\text{Total}} = \int dq = \int \lambda_0\,ds = \lambda_0 \int_0^{2\pi} a\,d\theta = \lambda_0\,a\,\left[\theta\Big|_0^{2\pi}\right] = \lambda_0\,2\,\pi\,a\,.$$

This result, while tautological, serves as a warm-up for the computation of potential. The distance from each charge element to the field point, $\mathcal{P}$, is

$$r = |\vec{r}| = \sqrt{a^2 + x^2}\,.$$

This distance is precisely the same for all of the charge elements comprising the ring. Thus,

$$V(\vec{r_p}) = k \int \frac{dq}{r} = k \int \frac{dq}{\sqrt{a^2 + x^2}} = \frac{k}{\sqrt{a^2 + x^2}} \int dq = \frac{k\,Q_{\text{Total}}}{\sqrt{a^2 + x^2}}\,.$$

A sketch of the potential function, $V(\vec{r_x}) = V((x\,,0\,,0))$, *vs.* x appears in Figure 10.2.

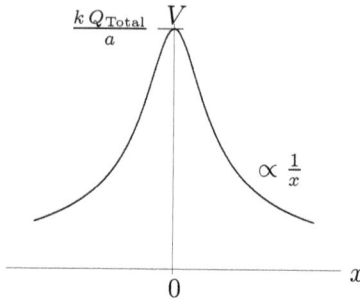

FIGURE 10.2   The Electric Potential along the Ring Axis

Two brief comments propel us into our investigation of the disk.

SYMMETRY

+   Under reflection in the yz-plane $[x \leftrightarrow -x]$, the electric potential is symmetric: $V((-x\,,0\,,0)) = V((x\,,0\,,0))$.

+   The electric potential is also unchanged under rotation about the x-axis.

DIFFERENTIABILITY   The potential function is continuous and differentiable everywhere along the axis of symmetry, *i.e.*, $\forall\,(x\,,0\,,0)$.

EXAMPLE [*Electrostatic Potential on the Axis of a Uniformly Charged Disk*]

A thin disk of radius a bears a uniform surface charge density σ_0. The electric potential at any point in space may be obtained by linear superposition of the potential contributions of the set of thin rings of charge which together comprise the disk. We partition the disk into thin rings of radius r', employ the result derived just previously for the potential of such a ring at a field point on the axis, and add these contributions to obtain the potential produced by the entire disk at the field point.

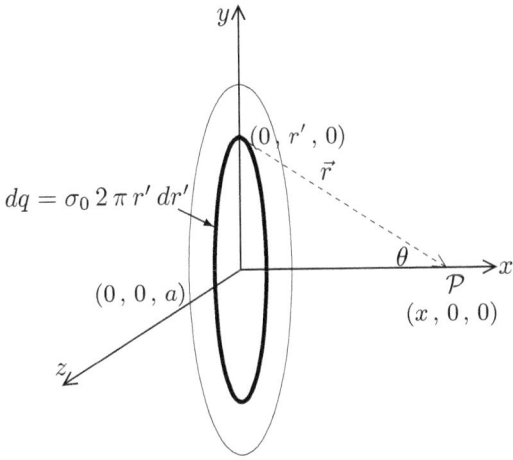

FIGURE 10.3   Computing the Electric Potential of a Thin Uniformly Charged Disk

First, for grins, let's compute the total charge on the disk by integrating the charge density over its extent.

$$Q_{\text{Total}} = \int dq = \int \sigma_0 \, dA = \int \sigma_0 \, r' \, dr \, d\theta = \sigma_0 \int_0^a r' \, dr' \int_0^{2\pi} d\theta$$
$$= \sigma_0 \left[\tfrac{1}{2} a^2 - 0 \right] \left[2\pi - 0 \right] = \sigma_0 \, \pi \, a^2 .$$

That the total charge on the disk is precisely equal to the product of the [constant] charge per unit area and the total area is unsurprising.

Although all parts of each circular ring are at fixed distance from the field point $\mathcal{P}$, this distance depends on the radius of the ring, a, *i.e.*,

$$r = \sqrt{r'^2 + x^2} \ .$$

Incorporation of this Pythagorean result into the general expression for the potential function leads to an elementary integral:

$$V_{\text{disk}}(\mathcal{P}) = k \int \frac{dq}{r} = k \int_0^a \frac{\sigma_0 \, 2\pi \, r' \, dr'}{\sqrt{r'^2 + x^2}} = k \, \sigma_0 \, 2\pi \int_0^a \frac{r' \, dr'}{\sqrt{r'^2 + x^2}}$$
$$= k \, \sigma_0 \, 2\pi \, (r'^2 + x^2)^{1/2} \Big|_0^a = k \, \sigma_0 \, 2\pi \left(\sqrt{a^2 + x^2} - |x| \right) .$$

Recognition that $\sqrt{x^2} = |x|$ has informed the last equality. A sketch of $V_{\text{disk}}((x,0,0))$ appears in Figure 10.4.

Two brief comments follow [*cf.* the thin ring example].

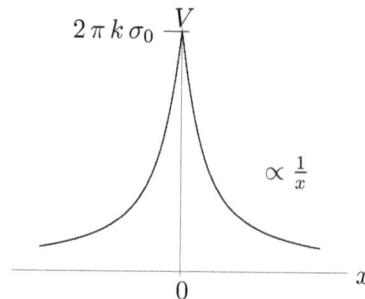

FIGURE 10.4   The Electric Potential along the Axis of a Uniformly Charged Disk

SYMMETRY   The electric potential of the disk is invariant under reflection in the yz-plane and rotation about the x-axis.

DIFFERENTIABILITY   The potential function is continuous everywhere along the axis of symmetry. The potential is differentiable everywhere except at the origin, the intersection of the axis with the disk.

ASIDE: **Q:** What about the electric potential above an infinite plane of charge?

A: Previously [in Chapter 3], the electric field associated with the infinite plane bearing uniformly distributed electric charge was determined by taking the improper limit, $a \to \infty$, of the finite disk result.

This relatively simple procedure fails in this instance, as the value so-obtained is infinite:

$$\lim_{a \to \infty} V(\mathcal{P}) = \lim_{a \to \infty} k\sigma_0\, 2\pi \left(\sqrt{a^2 + x^2} - |x| \right) \to \infty \,.$$

Things go spectacularly wrong here, because we are attempting to commit a mathematical/physical *faux pas*. One cannot simultaneously insist that the potential vanish at the point at infinity [implicit in the PARTITION, COMPUTE, and SUM method] and have an infinite [unbounded] extension of charge. Recall that the electrostatic potential energy stored in a charge configuration is equal to the work done in assembling the system from charges brought from infinite distance. It is not altogether surprising that there is infinite energy associated with unbounded distributions containing infinite total charge.

Differences in potential above an infinite plane of uniform charge density may be computed employing the definition couched in terms of the electric field.

EXAMPLE   [*Electrostatic Potential Energy Due to a Spherical Shell of Charge*]

A thin spherical shell of radius R possesses uniform surface charge density σ_0.

ASIDE: The radius of the thin shell has a unique value, R. If the shell were thick, then there might be a range of radii associated with the distribution of charge. This complicates the analysis, as we'll see in the next section.

The total charge on the spherical shell is obtained by integrating the charge density over the surface. As the charge density is assumed to be uniform, and the area of a sphere of radius R is $4\pi R^2$, the total charge is

$$Q_{\text{Total}} = \sigma_0\, 4\pi R^2 \,.$$

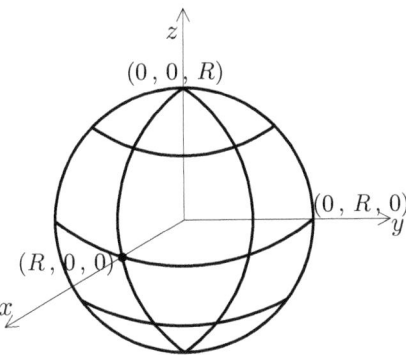

FIGURE 10.5   A Thin Spherical Shell Bearing Uniform Areal Charge Density

A field point, $\mathcal{P}$, is located at distance r from the centre of the spherical shell. It matters crucially whether the field point is outside or inside the shell.

EXTERNAL   Outside the spherical shell, $r > R$.

Let's compute the potential function by integrating the electric field produced by the charged shell along a path from the reference point, located infinitely far away, to $\mathcal{P}$ at $\vec{r_{\mathcal{P}}}$:

$$V_{\text{shell},o}(\vec{r_{\mathcal{P}}}) = -\int_{\infty}^{\vec{r_{\mathcal{P}}}} \vec{E}_{\text{shell},o} \cdot d\vec{s} .$$

Recall that, outside the shell, the electric field is indistinguishable from that produced by a point charge, Q_{Total}, located at the centre of the shell. Hence,

$$\vec{E}_{\text{shell},o}(\vec{r}) = \frac{k\, Q_{\text{Total}}}{|\vec{r}|^2}\, \hat{r} .$$

Taking into account the spherical symmetry of the system simplifies the computation of the potential outside the shell. At distance $r_{\mathcal{P}} > R$ from its centre,

$$V_{\text{shell},o}(\vec{r_{\mathcal{P}}}) = \int_{\infty}^{r_{\mathcal{P}}} -k\, Q_{\text{Total}} \frac{dr}{r^2} = k\, Q_{\text{Total}} \left[\frac{1}{r}\Big|_{\infty}^{r_{\mathcal{P}}}\right] = \frac{k\, Q_{\text{Total}}}{r_{\mathcal{P}}} .$$

Furthermore, as the surface of the shell is approached from without, *i.e.*, $r \to R^{+}$, the potential approaches the limiting value

$$V_{\text{shell},o}(R) = \frac{k\, Q_{\text{Total}}}{R} .$$

INTERNAL   Inside the spherical shell, $r < R$.

Gauss's Law assures us that the electric field is zero everywhere inside the shell because there is no net charge enclosed within any Gaussian sphere lying entirely within the spherical shell. [No enclosed charge means no electric flux for all instances and geometries of Gaussian surfaces completely within the shell, which in turn requires that the electric field itself vanish everywhere within the shell.]

CLAIM: The potential within the shell is not zero, despite the vanishing field, as the reference point lies at infinity.

ASIDE: PROOF of the Claim (by *reductio ad absurdum*)

Suppose that the potential function were zero both inside the shell and at the point at infinity. Accordingly, it would require zero net energy to move a point test charge, q_0, from infinity to the region inside the shell, *i.e.*,

$$\Delta U_{\mathrm{E}} = q_0 \, \Delta V = 0 \,.$$

However, it is only once the charged particle passes inside the shell that the local electric field [and force] that it experiences vanishes. While it is outside the shell, the field, force, potential, and electrostatic potential energy are non-vanishing.

IF q_0 and Q_{Total} are $\begin{bmatrix} \text{LIKE} \\ \text{UNLIKE} \end{bmatrix}$, THEN

the electrostatic potential energy of the system is $\begin{bmatrix} \text{POSITIVE} \\ \text{NEGATIVE} \end{bmatrix}$.

Here is a direct contradiction of the initial claim that the change in electrostatic potential energy is ZERO, and thus the premise, that the potential vanishes within the shell, is shown to be incorrect.

Correct reasoning proceeds from the realisation that the reference value for the potential function is fixed once and for all, and that a "bootstrap" technique must be used in regions of space cut off from the reference point. Simply put, in this case, the reference point for the potential inside the shell must be chosen to lie on the shell.

[Isotropy and spherical symmetry ensure that the shell is an equipotential surface.]

Thus,

$$V(r) - V(R) = -\int_{R^+}^{r} \vec{\boldsymbol{E}}_{\mathrm{shell},i} \cdot d\vec{s}$$

$$V(r) - \frac{k\,Q_{\mathrm{Total}}}{R} = 0 \qquad \Longrightarrow \qquad V(r) = \frac{k\,Q_{\mathrm{Total}}}{R} \,,$$

everywhere inside the shell.

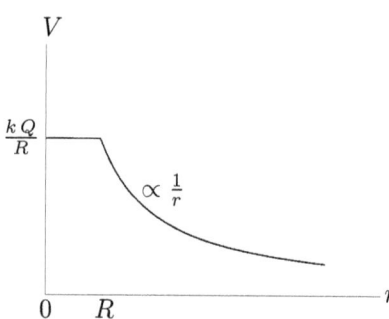

FIGURE 10.6   Electric Potential of a Uniformly Charged Spherical Shell

This potential function is continuous and differentiable everywhere, except at the shell, $r = R$, where it is only continuous.

EXAMPLE [*Spherical Volume of Uniform Charge Density*]

A solid spherical volume has radius R and constant volume charge density ρ_0.

> ASIDE: The material comprising such a solid charged sphere must be insulating. The presence of net electric charge density inside the body requires that regions of non-vanishing electric field must be found within the sphere. Therefore, at least some of the charge-bearing particles are immersed in a non-zero electric field and subjected to an electric force.
>
> If the source were composed of conducting material, then these charged particles would move freely and easily in response to the net field. Two consequences would then ensue.
>
> ○ The system would not be in electrostatic equilibrium.
> ○ The distribution of charge would rapidly become non-uniform.

Once again, we first compute the total charge borne by the source and compare it to the expected value, ρ_0 (Volume). The integral over the charge elements yields

$$Q_{\text{Total}} = \int dq = \int \rho_0 \, dV = \rho_0 \int_0^R \int_0^\pi \int_0^{2\pi} r^2 \sin(\theta) \, dr \, d\theta \, d\phi$$

$$= \rho_0 \left[\frac{1}{3} r^3 \Big|_0^R \right] \left[-\cos(\theta) \Big|_0^\pi \right] \left[\phi \Big|_0^{2\pi} \right] = \rho_0 \frac{R^3}{3} \times 2 \times 2\pi$$

$$= \rho_0 \frac{4\pi}{3} R^3 \,.$$

The computed value of the total electric charge comports precisely with our expectations.

The field point at which the potential is to be determined is at radial distance r from the centre of the source. As in the shell case, our analysis depends on whether the field point is within, or outside of, the extended source.

EXTERNAL   Outside the spherical distribution, $r > R$.

As the electric field, external to the sphere $\vec{E}_{\text{sphere},o}$, is indistinguishable from that of a point charge, Q_{Total}, located at the centre of the sphere, so too must be the potential. Hence,

$$V(r) = \frac{k\,Q_{\text{Total}}}{r} \,, \quad \forall \, r > R \,.$$

Approaching the surface of the sphere from the outside, *i.e.*, $r \to R^+$,

$$\lim_{r \to R^+} V(r) = \lim_{r \to R^+} \frac{k\,Q_{\text{Total}}}{r} = \frac{k\,Q_{\text{Total}}}{R} \,.$$

INTERNAL   Inside the spherical distribution, $r < R$.

In Chapter 6, it was argued that the electric field inside a uniformly charged sphere varies linearly with distance from the centre:

$$\vec{E}_{\text{sphere},i}(r) = \frac{k\,Q_{\text{Total}}}{R^3} r \,\hat{r} \,, \quad \forall \, r < R \,.$$

That the field depends only on the magnitude of the distance to the field point, and that its direction is purely radial, are manifestations of the spherical symmetry and isotropy of the source distribution.

The exterior edge of the sphere acts as the reference point for the determination of values of the potential within, so

$$V(r) - V(R) = -\int_R^r \vec{E}_{\text{sphere},i} \cdot d\vec{s}$$

$$V(r) - \frac{k\,Q_{\text{Total}}}{R} = -\frac{k\,Q_{\text{Total}}}{R^3}\left[-\tfrac{1}{2}r^2\Big|_R^r\right] = \frac{k\,Q_{\text{Total}}}{R^3}\frac{1}{2}\left[R^2 - r^2\right]$$

$$\implies \quad V(r) = \frac{k\,Q_{\text{Total}}}{2\,R^3}\left[3\,R^2 - r^2\right], \quad \forall\, r < R.$$

Written more compactly, this reads

$$V(r) = \frac{k\,Q_{\text{Total}}}{2\,R}\left[3 - \frac{r^2}{R^2}\right].$$

A sketch of the dependence of the potential function on the radial distance to the field point appears in Figure 10.7.

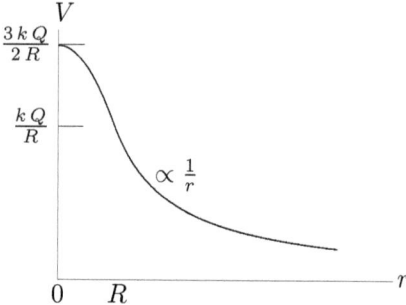

FIGURE 10.7 The Potential Produced by a Uniform Spherical Distribution of Charge

This potential function is continuous and differentiable everywhere. In this instance, owing to the cusp in the field at the surface, the second derivative of the potential is discontinuous at $r = R$.

Q: While this has been lots of fun, why should one concern oneself with the electric potential when the electric field seems more fundamental?

A: As the electrostatic force is CONSERVATIVE, the electric field can be derived from knowledge of the potential. In many instances, it proves less onerous to ascertain the scalar electric potential field than it would be to directly determine the vector electric field.

Chapter 11

Potentials, Fields, and All That

The claim that it was possible to compute the electric field associated with a given potential function was made at the end of Chapter 10.

Q: How does one go about doing this?

A: In short, the electric field is obtained by taking "minus the gradient of the potential," i.e., $\vec{E} = -\vec{\nabla} V$.

To explicate this abbreviated answer, we present a brief synopsis of the analysis of the relation between conservative forces and potential energy functions found in VOLUME I, Chapter 26 [adapted to accommodate our present concerns].

$\vec{\nabla}$ The potential difference between two points, both of which are situated in an arbitrarily small neighbourhood of a field point, $\mathcal{P}$, is

$$dV = -\vec{E}(\mathcal{P}) \cdot d\vec{s},$$

where the infinitesimal line element, $d\vec{s}$, "steps" from one point to the other.

$\vec{\nabla}$ Try as one might, it proves impossible to assert that

$$\frac{dV}{d\vec{s}} = -\vec{E}(\mathcal{P}),$$

on account of (at least) two conundrums:

♮   How does one divide by a vector?

♮   How might one undo [a.k.a. invert] the dot product?

$\vec{\nabla}$ A resolution is to choose coordinates and expand the RHS of the expression for dV. When Cartesian coordinates are employed,

$$dV = -\vec{E}(\mathcal{P}) \cdot d\vec{s} = -\big[E_x \, dx + E_y \, dy + E_z \, dz \big].$$

As the initial and final points are unrestricted, the expansion is formally valid for particular steps taken in each of the coordinate directions:

$$d\vec{s}_x = (dx, 0, 0), \qquad d\vec{s}_y = (0, dy, 0), \quad \text{and} \quad d\vec{s}_z = (0, 0, dz).$$

Separate consideration of each of these yields expressions for the components of the electric field in terms of partial derivatives of the potential function, i.e.,

$$E_x = -\frac{\partial V}{\partial x}, \qquad E_y = -\frac{\partial V}{\partial y}, \quad \text{and} \quad E_z = -\frac{\partial V}{\partial z}.$$

[Partial derivatives are evaluated holding the other variables fixed.]

$\vec{\nabla}$ We codify the above process of determining the field in terms of the **gradient** operator, $\vec{\nabla}$, as follows:

$$\vec{E}(\mathcal{P}) = -\vec{\nabla}V \, .$$

This expression has the virtue of being independent of the choice of coordinates.

ASIDE: For example, as seen above, in Cartesian coordinates, $(x\,,\,y\,,\,z)$,

$$\vec{\nabla} \equiv \hat{\imath}\,\frac{\partial}{\partial x} + \hat{\jmath}\,\frac{\partial}{\partial y} + \hat{k}\,\frac{\partial}{\partial z}\,,$$

whereas in polar coordinates, $(r\,,\,\theta\,,\,\phi)$,

$$\vec{\nabla} \equiv \hat{r}\,\frac{\partial}{\partial r} + \hat{\theta}\,\frac{1}{r}\frac{\partial}{\partial \theta} + \hat{\phi}\,\frac{1}{r\sin(\theta)}\frac{\partial}{\partial \phi}\,,$$

while in cylindrical coordinates, $(r\,,\,\theta\,,\,z)$,

$$\vec{\nabla} \equiv \hat{r}\,\frac{\partial}{\partial r} + \hat{\theta}\,\frac{1}{r}\frac{\partial}{\partial \theta} + \hat{k}\,\frac{\partial}{\partial z}\,.$$

In cases where the physical situation evinces spherical symmetry, the polar form quoted above simplifies to

$$\vec{\nabla} = \hat{r}\,\frac{\partial}{\partial r}\,,$$

acting on functions which solely depend on r. The same formal result is obtained for cylindrical systems possessing axial symmetry.

Let's re-examine several situations encountered in previous chapters.

....

EXAMPLE [*Electric Field and Potential along the Axis of a Uniformly Charged Ring*]

A thin ring of radius a lying centred in the yz-plane, as shown in Figure 11.1, possesses uniform lineal charge density λ_0. The total charge borne by the ring is $Q_{\text{Total}} = \lambda_0\,2\,\pi\,a$.

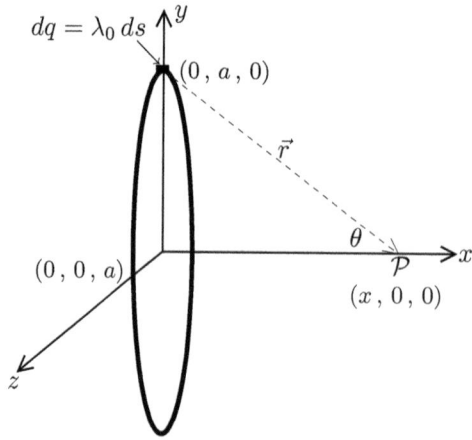

FIGURE 11.1   The Uniformly Charged Thin Ring, *Encore*

The electric field at points located along the symmetry axis of the ring,

$$\vec{E}\big((x\,,0\,,0)\big) = k\,Q_{\text{Total}}\,\frac{x}{\left(a^2 + x^2\right)^{3/2}}\,\hat{\imath}\,,$$

was computed in Chapter 3. The symmetry of the situation dictates that the field at points along the x-axis be directed along the x-axis. The axial component of the electric field is plotted, as a function of position along the axis, in the left panel of Figure 11.2.

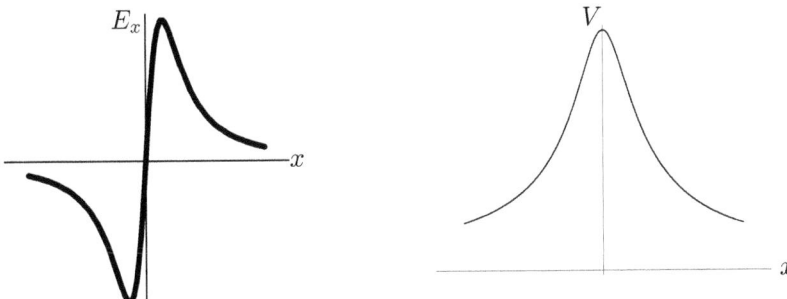

FIGURE 11.2   On-Axis Field and Potential of a Uniformly Charged Thin Ring

The electric potential function along the symmetry axis for the distribution of charge was computed in Chapter 10, and there we obtained

$$V\big((x\,,0\,,0)\big) = \frac{k\,Q_{\text{Total}}}{\left(a^2 + x^2\right)^{1/2}}\,.$$

A sketch of $V\big((x\,,0\,,0)\big)$ *vs.* x is found in the right panel in Figure 11.2.

Examination of the potential and field in Figure 11.2 corroborates the formal relation:

$$-\frac{\partial V}{\partial x} = k\,Q_{\text{Total}}\,\frac{x}{\left(a^2 + x^2\right)^{3/2}} = E_x\,.$$

EXAMPLE [*Electric Field and Potential along the Axis of a Uniformly Charged Disk*]

A disk of radius a, lying in the yz-plane, with uniform surface charge density σ_0, bears a total charge of $Q_{\text{Total}} = \sigma_0\,\pi\,a^2$.

In Chapter 3, the electric field for this distribution was determined to be

$$\vec{E}\big((x\,,0\,,0)\big) = 2\,\pi\,k\,\sigma_0\left[1 - \frac{x}{\sqrt{a^2 + x^2}}\right]\hat{\imath}$$

along the positive x-axis. Symmetry ensures that only E_x is non-zero at field points along the axis.

In Chapter 10, the electric potential function along the symmetry axis was determined to be

$$V\big((x\,,0\,,0)\big) = 2\,\pi\,k\,\sigma_0\left[\sqrt{a^2 + x^2} - |x|\right]\,.$$

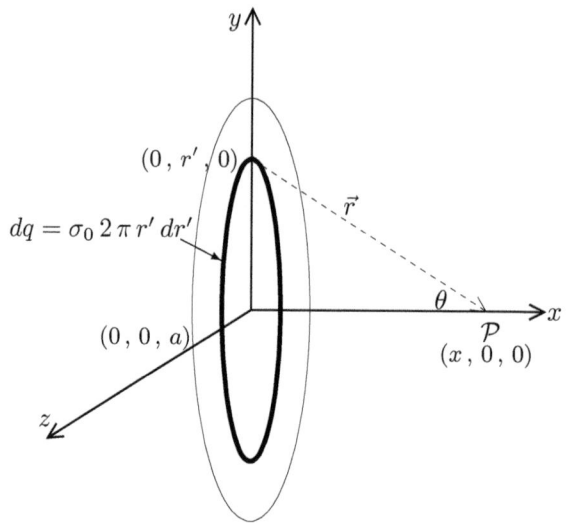

FIGURE 11.3   The Uniformly Charged Thin Disk, *Encore*

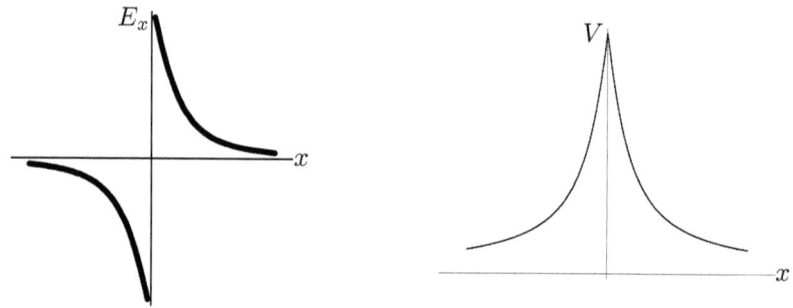

FIGURE 11.4   On-Axis Field and Potential of a Uniformly Charged Thin Disk

Sketches of the field and the potential along the axis appear in Figure 11.4.

The relation between field and potential is observed to hold here as well, for when $x > 0$,

$$-\frac{\partial V}{\partial x}\Bigg|_{x>0} = -\frac{\partial}{\partial x}\left(2\pi k\,\sigma_0\left[\sqrt{a^2 + x^2} - x\right]\right) = -2\pi k\,\sigma_0\left[\frac{x}{\sqrt{a^2 + x^2}} - 1\right] = E_x\Bigg|_{x>0},$$

and when $x < 0$,

$$-\frac{\partial V}{\partial x}\Bigg|_{x<0} = -\frac{\partial}{\partial x}\left(2\pi k\,\sigma_0\left[\sqrt{a^2 + x^2} + x\right]\right) = -2\pi k\,\sigma_0\left[\frac{x}{\sqrt{a^2 + x^2}} + 1\right] = E_x\Bigg|_{x<0}.$$

DISCONTINUOUS   The electric field is DISCONTINUOUS at the surface of the disk.

CUSP   Concomitant with the field discontinuity is the cusp [sharp peak] at $x = 0$ in the sketch of the potential along the x-axis.

EXAMPLE [*Field and Potential of a Uniformly Charged Spherical Shell*]

A thin spherical shell of radius R, with uniform surface charge density σ_0, bears net charge $Q_{\text{Total}} = \sigma_0\, 4\,\pi\, R^2$. The symmetry of this situation demands that the electric field be everywhere radially directed and spherically symmetric [isotropic].

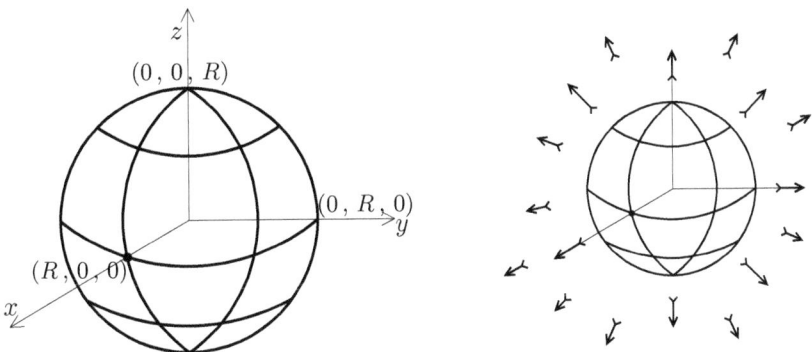

FIGURE 11.5   The Uniformly Charged Spherical Shell

Outside the shell, the electric field is indistinguishable from that of a point charge, Q, placed at its centre, while inside, the field vanishes. That is,

$$\vec{E}_o(r) = \frac{k\,Q_{\text{Total}}}{r^2}\,\hat{r} \qquad \text{and} \qquad \vec{E}_i(r) = \vec{0}\,.$$

[The field is discontinuous at the shell radius, R.]

Plotting the magnitude of the radial electric field as a function of distance from the centre of the shell yields the curve found in the left panel of Figure 11.6.

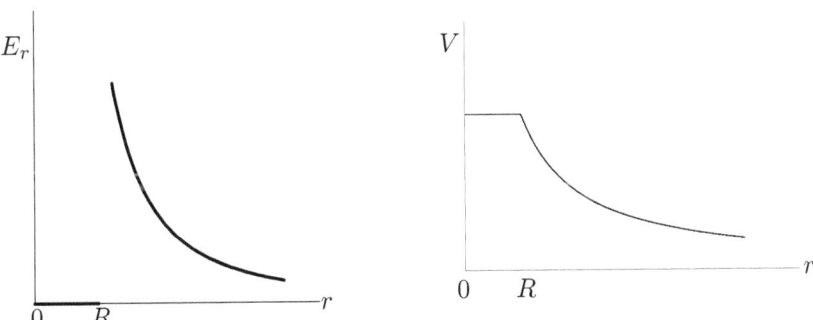

FIGURE 11.6   Field and Potential of a Uniformly Charged Spherical Shell

The electric potential associated with the uniformly charged spherical shell is

$$V_o(r) = \frac{k\,Q_{\text{Total}}}{r} \qquad \text{and} \qquad V_i(r) = \frac{k\,Q_{\text{Total}}}{R}\,.$$

The fundamental relation between potential and field holds on the outside,

$$-\frac{\partial V_o}{\partial r} = \frac{k\,Q_{\text{Total}}}{r^2} = E_r\Big|_{r>R},$$

and within the shell,

$$-\frac{\partial V_i}{\partial r} = 0 = E_r\Big|_{r<R}.$$

EXAMPLE [*Potential and Field of a Uniform Spherical Distribution of Charge*]

A sphere of radius R and uniform volume charge density ρ_0 bears net charge

$$Q_{\text{Total}} = \rho_0\,\frac{4}{3}\,\pi\,R^3\,.$$

Outside, the situation is just as it was for the shell. The field and potential are indistinguishable from those produced by a point charge, Q_{Total}, lying at the centre of the sphere:

$$\vec{E}_o(r) = \frac{k\,Q_{\text{Total}}}{r^2}\,\hat{r} \qquad \text{and} \qquad V_o(r) = \frac{k\,Q_{\text{Total}}}{r}\,.$$

Note that

$$-\frac{\partial V_o}{\partial r} = \frac{k\,Q_{\text{Total}}}{r^2} = E_r\Big|_{r>R}$$

demonstrates explicitly that the potential–field relation holds external to the sphere.

It was shown in Chapter 6 that the electric field is directed radially inside the sphere, and its magnitude grows from zero at the centre to its value, $k\,Q_{\text{Total}}/R^2$, at the surface. In Chapter 10, the potential inside the sphere was determined. It has the requisite quadratic behaviour to comport with the linearly increasing field strength. The potential function is continuous at the surface of the sphere and consistent with the established reference value: $V = 0$ at $R \to \infty$. For the uniformly charged sphere,

$$\vec{E}_i(r) = \frac{k\,Q_{\text{Total}}\,r}{R^3}\,\hat{r} \qquad \text{and} \qquad V_i(r) = \frac{k\,Q_{\text{Total}}}{2\,R}\left[3 - \frac{r^2}{R^2}\right].$$

The gradient of this spherically symmetric potential is

$$-\vec{\nabla}V_i(r) = -\frac{\partial V_i}{\partial r}\,\hat{r} = +\frac{k\,Q_{\text{Total}}\,r}{R^3}\,\hat{r} = \vec{E}_i(r)\,.$$

Let's investigate the electric potential of a charged conductor in electrostatic equilibrium. Three conditions must be met.

- The electric field vanishes within the body of the conductor.
- All excess charge resides on the surface of the conductor.
- The surface electric field is directed outward along the perpendicular.

Pick any two field points on the surface of the conductor, $\mathcal{A}$ and $\mathcal{B}$, along with a path joining them which runs along the surface.[1] According to its definition, the potential difference existing between the two points is

$$V(\mathcal{B}) - V(\mathcal{A}) = -\int_{\mathcal{A}}^{\mathcal{B}} \vec{E} \cdot d\vec{s}.$$

The electric field is orthogonal to the surface, and thus the integrand vanishes everywhere along the path. Thus the potential difference between all such points lying on the surface is identically zero; **the surface of a conductor in electrostatic equilibrium is an equipotential surface.**

Furthermore, all points inside the conductor have the same potential as those on the surface, because the field vanishes everywhere[2] inside. Thus it takes zero net work to move a test charge, q_0, about in the body of the conductor, say from point $\mathcal{C}$ to $\mathcal{D}$, since the resultant potential energy difference,

$$\Delta U_{\mathcal{C}\mathcal{D}} = q_0 \, \Delta V_{\mathcal{C}\mathcal{D}} = 0 \,,$$

vanishes for all interior points. Recall that conductors which are subjected to externally applied fields polarise so as to quench the field within the conductor. Even these conductors and their surfaces are equipotentials, provided that quenching has occurred and the charge distribution on the surface is static.

Q: What happens within a cavity completely inside a conductor in equilibrium?

A: Irrespective of the external conditions [*e.g.*, charge distribution, ambient applied electric field, *etc.*], the electric field in the cavity is ZERO![3] While this can be argued by invoking Gauss's Law, we shall instead employ a potential-based argument.

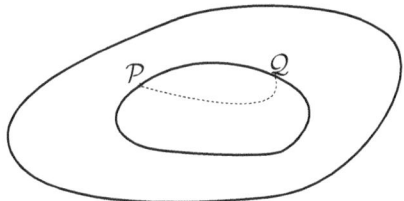

FIGURE 11.7   A Geode-Like Cavity within a Conductor in Electrostatic Equilibrium

Figure 11.7 shows a cavity within a lump of conductor. It is empty insofar as there are no isolated sources of charge residing inside the cavity. The distribution of charge, whatever its form, is assumed to be unchanging. Accordingly, all points on the walls of the cavity are at the same potential. In particular, for two such points, $\mathcal{P}$ and $\mathcal{Q}$,

$$0 = \Delta V_{\mathcal{P}\mathcal{Q}} = V(\mathcal{Q}) - V(\mathcal{P}) = -\int_{\mathcal{P}}^{\mathcal{Q}} \vec{E} \cdot d\vec{s}.$$

The only way that this integral relation can be true for **all** paths leading from $\mathcal{P}$ to $\mathcal{Q}$ is for the electric field to vanish throughout the cavity.

---

[1] The restriction that the path be confined to the surface, while aiding us in visualising the situation, is not necessary and easily dropped.

[2] This feature makes for an easier way of establishing that the potentials at the surface points $\mathcal{A}$ and $\mathcal{B}$ are the same – take a *subterranean* route!

[3] The lawyers insist that we disclaim, "Provided that the cavity is empty of charge."

ASIDE: PROOF (by *reductio ad absurdum*)

Suppose that there is a region completely inside the cavity within which the local electric field is non-zero. Consider three paths, $\{\gamma_1, \gamma_2, \gamma_3\}$, joining the points $\mathcal{P}$ and $\mathcal{Q}$. These points lie on the periphery of the cavity [outside of the non-zero field region]. The first path skirts the region of non-zero electric field. The second path passes through the region once, while the third is coincident with the second, except where it doubles back so as to twice overlap the second in the non-zero field region. Representative paths are sketched in the figure below.

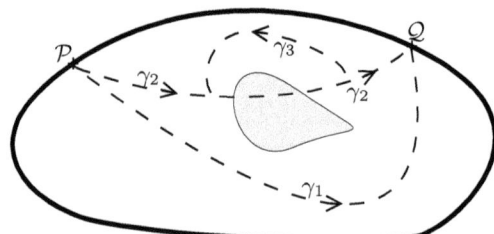

Let's evaluate the potential difference between $\mathcal{P}$ and $\mathcal{Q}$ along each of the three paths in turn.

1.   The electric field vanishes everywhere on the locus of points comprising γ_1. Thus, $\Delta V_{\mathcal{P}\mathcal{Q}, \gamma_1} = 0$. This result is completely consistent with the bulk of the conductor, on whose surface $\mathcal{P}$ and $\mathcal{Q}$ reside, being an equipotential region.

2.   A portion of γ_2 passes through the region of non-zero electric field. It is almost inevitable that $\Delta V_{\mathcal{P}\mathcal{Q}, \gamma_2} \neq 0$.

3.   Along the third path, the portions outside the region contribute nothing to the net potential difference, while the part within doubles that computed for γ_2, viz., $\Delta V_{\mathcal{P}\mathcal{Q}, \gamma_3} = 2\,\Delta V_{\mathcal{P}\mathcal{Q}, \gamma_2}$.

The electric potential function must be single-valued, and hence its differences ought to be unique. Agreement among the three paths requires $\Delta V_{\mathcal{P}\mathcal{Q}, \gamma_2} = 0$. However, this is generally impossible to enforce unless $\boldsymbol{E} = \vec{0}$. Therefore, we are forced to concede that the electric field must vanish everywhere throughout the empty cavity lying within the conductor [in electrostatic equilibrium].

Chapter 12

Capacitance

Two isolated masses of conducting material, in electrostatic equilibrium, bear equal magnitude, opposite sign charges $-Q$ and $+Q$, respectively. While each of the lumps comprises an equipotential region, there must be a potential difference between them, as a [non-zero] electric field must exist somewhere in the surrounding space. The field lines in Figure 12.1 refer to the net electric field produced by the charges distributed on each of the [surfaces of the] lumps.

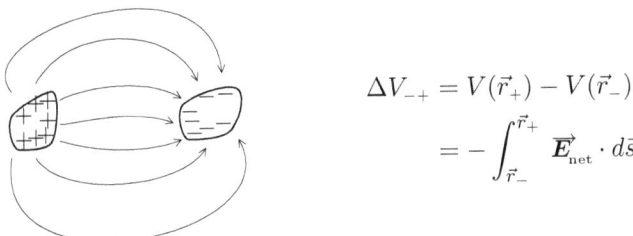

$$\Delta V_{-+} = V(\vec{r}_+) - V(\vec{r}_-)$$
$$= -\int_{\vec{r}_-}^{\vec{r}_+} \vec{E}_{\text{net}} \cdot d\vec{s}$$

FIGURE 12.1   Two Lumps of Oppositely Charged Conductor Form a Capacitor

Any path through space from one lump to the other must encounter the field. Thus, there must be a non-zero potential difference between the two lumps. As POSITIVE charges are sources of field lines, while NEGATIVE charges are sinks,

$$V(\vec{r}_+) > V(\vec{r}_-) \qquad \text{or, equivalently,} \qquad \Delta V_{-+} > 0.$$

That is, any path from the negatively charged lump to the positive one is necessarily "uphill" energetically.

CAPACITANCE   The capacitance of a system of isolated conductors bearing charges $+Q$ and $-Q$ is the ratio of the magnitude of the charge to the potential difference, V, extant between the conductors,

$$C = \frac{Q}{V}.$$

- In many simple cases, the partitioning/distribution of the charges is unique. In more complicated cases, additional assumptions are necessary.

- Capacitance is, by definition, positive.

- The SI unit of capacitance is the farad, $[\text{F}] = \left[\frac{\text{C}}{\text{V}}\right]$. In practice, typical **capacitors** found in common electrical and electronic devices have capacitances of picofarads, pF, or nanofarads, nF, or microfarads, μF.

- Surprisingly, **C depends** NEITHER **on Q**, NOR **on V**. The capacitance is entirely determined by the geometry of the arrangement of conductors.

EXAMPLE [*Capacitance of an Isolated Spherical Conductor*]

As our first example, we shall consider an isolated spherical conductor of radius R and calculate its capacitance with respect to infinity. [Electric field lines emanating from the conductor, when it bears charge $+Q$, terminate at infinity[1] in the absence of other charged bodies].

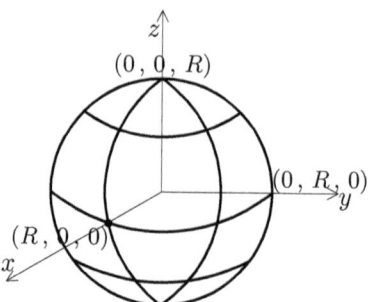

FIGURE 12.2   An Isolated Spherical Conductor Bearing Charge $+Q$

We proceed by placing $+Q$ on the spherical lump.

[The observed net neutrality of matter requires that $-Q$ accrue to the point at infinity.]

Absent any external perturbations, electrostatic equilibrium arises with the charge distributed uniformly on the surface of the lump. According to Gauss's Law, the electric field and potential outside of the sphere are indistinguishable from those of a point charge, $+Q$, located at the centre of the sphere. Retaining the customary convention that $V \to 0$ at infinite distance, the potential difference between the collections of positive and negative charge is simply equal to

$$V_{\infty+} = \frac{kQ}{R}.$$

Thus, the capacitance is

$$C_{\text{sphere}} = \frac{Q}{V_{\infty+}} = \frac{Q}{\frac{kQ}{R}} = \frac{R}{k} = 4\pi\epsilon_0 R.$$

The most important thing to note about this result is that its sole dependence, aside from physical and mathematical constants, is on the quantity that characterises the geometry of the sphere, *i.e.*, its radius.

Four steps are needed to compute the capacitance of an arrangement of conductors.

1.   Characterise the geometry and configuration of the conducting materials.

2.   Assume an equilibrium charge distribution (partition $\pm Q$, as needed).

3.   Compute ΔV [via the electric field, if necessary] for this distribution.

4.   Take $C = \frac{Q}{\Delta V}$.

---

[1] This option was explicitly included when electric field lines were introduced in Chapter 2.

EXAMPLE [*A Parallel Plate Capacitor*]

In this example, we shall slavishly follow the four-step recipe for finding capacitances.

Geometry and Configuration   Two identical thin plates with face area A are separated by a small distance d. The plates may be any shape, except long and thin [*i.e.*, filamentous], and they must completely overlap.[2]

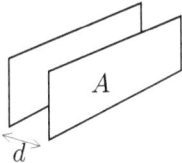

FIGURE 12.3   Two Parallel Conducting Plates

Charge Distribution   It seems reasonable to propose that the total charges $\pm Q$ are distributed uniformly across their respective plates.

> ASIDE: Three related comments are in order. Were the plates of infinite size, in-plane isotropy would militate for uniformity. For finite plates, **fringing** occurs, whereby the field at the edges is stronger and extends beyond the volume enclosed between the completely overlapping plates.
>
> [Think of how charge accumulates at points of high curvature!]
>
> This effect occurs primarily in a narrow strip along the periphery. In situations where the ratio of boundary area to total area is relatively small, the effects of fringing can be safely ignored. Should the plates be long and narrow, or otherwise strip-like, fringing effects could become appreciable. Finally, fringing is also affected by the plate separation, d. The narrower the gap, the less fringing occurs, and the field is better confined within the region between the plates.

IF the charge is uniformly distributed, THEN the charge densities are $\pm \sigma_0 = \pm \frac{Q}{A}$.

Computing ΔV   As established in Chapter 6, the field between infinite parallel plates is uniform.

The electric field is

$$\vec{E} = \frac{\sigma_0}{\epsilon_0} \left[+ \rightarrow - \right].$$

A path leading directly from the negatively charged plate to the positively charged plate is shown in the sketch.

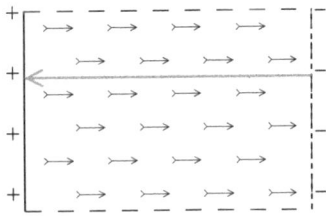

FIGURE 12.4   Electric Field (Model) and Path between the Charged Parallel Plates

---

[2] "Overlap" is a less cumbersome way of saying that the orthogonal projections of the two plates coincide.

In the region of space between the plates, the electric field [in the model describing the system] is uniform, with magnitude $\sigma_0/\epsilon_0 = (Q/A)/\epsilon_0$. Knowledge of the electric field in the space between the parallel plates enables computation of the potential difference across the gap:

$$V = \Delta V_{-+} = V(+) - V(-) = -\int_{\vec{r}_-}^{\vec{r}_+} \vec{E} \cdot d\vec{s} = -\frac{Q}{\epsilon_0 A} \int_{\vec{r}_-}^{\vec{r}_+} [\rightarrow] \cdot d\vec{s} = \frac{Q}{\epsilon_0 A} [d] = \frac{Q\,d}{\epsilon_0 A}.$$

In the penultimate equality, just above, the path chosen for evaluation of the integral went directly across the gap, $[\leftarrow]$, from negative to positive. As potential differences are path independent, and each plate is an equipotential, this is the potential difference between the plates.

Take the ratio of total charge to potential difference   The capacitance of the system of two identical, parallel, overlapping, and nearby plates is

$$C = \frac{Q}{V} = \frac{Q}{\frac{Q\,d}{\epsilon_0 A}} = \frac{\epsilon_0 A}{d}.$$

The capacitance of a pair of parallel identical plate conductors is proportional to the plate area and inversely proportional to their separation.

Two brief comments show how these dependences are not altogether unexpected.

A   Greater area provides additional *capacity* to store charge at the same density.

d   For fixed charge density and plate face area, the electric field in the region between the plates is determined and uniform in the interstitial region. Thus, the potential difference existing between the plates must be directly proportional to the size of the gap.

Conversely, to keep the potential difference between the two plates while narrowing the gap requires additional charge on each plate.

EXAMPLE [*A Cylindrical Capacitor*]

Here, we again follow the recipe.

Geometry and Configuration   Two coaxial cylindrical shells of radii a and b, where $b > a$, have common length L and completely overlap. We insist that the length of the cylinders be much greater than their radii, $L \gg b$, so as to make negligible the effects of fringing at the end caps.

[The cylindrical capacitor is topologically different from the parallel plate capacitor.]

Charge Distribution   Place $+Q$ on the inner cylinder and $-Q$ on the outer. The end cap effects are presumed to be small, because the strips of area along the ends make insubstantial contributions to the total areas of the sides of the cylinders. Thus, we assume that the charges are distributed uniformly on each of the cylinders.

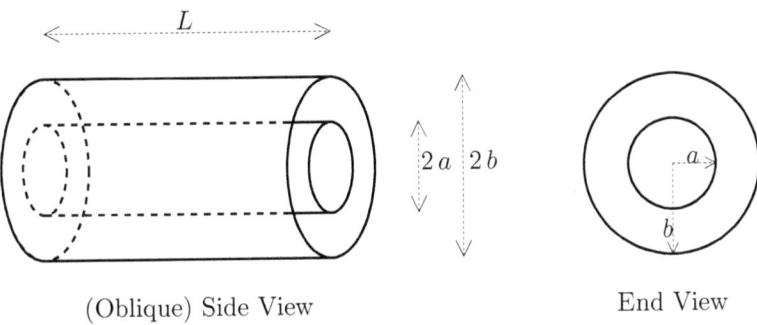

L

$2\,a$ $2\,b$

a

b

(Oblique) Side View   End View

FIGURE 12.5 A Cylindrical Arrangement of Conductors

Computing ΔV   To compute the potential difference, we require the electric field in the interstitial region. For grins, we now set about determining the electric field everywhere in the thick slab of space pierced orthogonally by the concentric cylinders.

GAUSS'S LAW   The axial symmetry of the system demands that the electric field point purely in the radial direction and be independent of angle and position along the axis, *i.e.*,

$$\vec{E}_{\text{net}} = \left[E_r\,\hat{r} + E_\theta\,\hat{\theta} + E_z\,\hat{k} \right]\Big|_{\substack{E_\theta=0 \\ E_z=0}} = E_r(r)\,\hat{r}.$$

The LHS of Gauss's Law, for the [closed] Gaussian surface consisting of the two end caps, $\{S_{c1}, S_{c2}\}$, each of face area $A_{c1,c2} = \pi R^2$, and the side surface, S_s, with area $A_s = 2\pi R L$, reads

$$\Phi_{E,S} = \oint_S \vec{E}_{\text{net}} \cdot d\vec{A} = \int_{S_{c1}+S_{c2}+S_s} \vec{E}_{\text{net}} \cdot d\vec{A}.$$

Given the restriction on the direction of the field, the net flux through the entire Gaussian surface simplifies to

$$\Phi_{E,S} = \int_{S_{c1}} E_r\,\hat{r} \cdot \hat{k}\, dA_t + \int_{S_{c2}} E_r\,\hat{r} \cdot (-\hat{k})\, dA_b + \int_{S_s} E_r\,\hat{r} \cdot \hat{r}\, dA_s = \int_{S_s} E_r\, dA_s.$$

All points on the side surface lie at radius R, fixing the value of the integrand everywhere in the domain of integration. Hence,

$$\Phi_{E,S} = E_r(R) \int_{S_s} dA_s = E_r(R)\, 2\pi R L.$$

The RHS of Gauss's Law reads: $q_{\text{encl.}}/\epsilon_0$.

$R < a$ Within the inner cylinder, there is no charge present to be enclosed within a Gaussian surface. Thus, $E_r(R)\, 2\pi R L = 0$, for all $R < a$, and one must conclude that the electric field vanishes throughout this region.

$a < R < b$ In the region between the conducting shells, all of the charge deposited on the inner surface, $+Q$, is enclosed within the Gaussian cylinder. Hence, $E_r(R)\, 2\pi R L = Q/\epsilon_0$, and thus

$$E_r(R) = \frac{Q}{2\pi \epsilon_0 R L} = \frac{2 k Q}{R L}.$$

The physical situation of a very long and narrow system of conductors motivates re-expression of this result in terms of the effective lineal charge density, $\lambda_0 = Q/L$. Doing so, one obtains

$$E_r(R) = \frac{2\,k\,\lambda_0}{R}\,,$$

which is identical to the field produced by a thin rod bearing uniformly distributed charge, *cf.* Chapter 3.

$b < R$ Outside of the system of conductors, there is no net charge enclosed within the Gaussian cylinder owing to the equality in magnitude of the [OPPOSITE] charges assigned to each shell. Thus, $E_r(R)\,2\,\pi\,R\,L = 0$, for all $R > b$, and one must conclude that the electric field vanishes everywhere throughout this region of space.

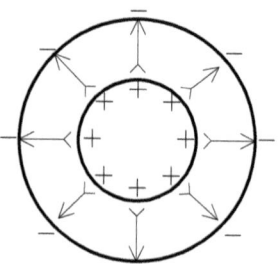

FIGURE 12.6 The Electric Field within a Cylindrical Capacitor

With the field now determined, it is possible to ascertain the potential difference existing between the charged cylinders. Integrating along a purely radial path from the negatively charged cylinder at $R = b$ to the positive one at $R = a$,

$$V = \Delta V_{-+} = \Delta V_{ba} = V(a) - V(b) = -\int_b^a \vec{E}_{\text{net}} \cdot d\vec{R} = +\int_a^b \frac{2\,k\,\lambda_0}{R}\,dR$$

$$= 2\,k\,\lambda_0\,\ln(R)\Big|_a^b = 2\,k\,\lambda_0\,\ln\left(\frac{b}{a}\right) = \frac{2\,k\,Q}{L}\,\ln\left(\frac{b}{a}\right)\,.$$

In the final equality, the effective lineal charge density was re-expressed in terms of the charge placed on each conductor. The potential difference is positive for $b > a$, as expected.

Take the ratio of total charge to potential difference The capacitance of this system of concentric cylindrical conductors is

$$C = \frac{Q}{V} = \frac{Q}{\frac{2\,k\,Q}{L}\,\ln(b/a)} = \frac{L}{2\,k\,\ln(b/a)}\,.$$

Two comments finish off this example.

L The capacitance is proportional to L, the length of the system, because longer cylinders can carry more charge at the same surface charge density.

a, b The capacitance increases as the separation between the plates is reduced, because the potential difference is diminished.

Chapter 13

Capacitors in Series and Parallel

In Chapter 12, we computed the capacitances associated with certain configurations of paired conductors. This was done by distributing charges $\pm Q$ on the conductors, determining the electric field throughout the nearby region(s), computing the potential difference between the pair of conductors, and finally taking the ratio of the charge distributed to the potential difference.

ASIDE: That this sounds complicated is no surprise, as it *is* complicated.

We've also verified in our examples that capacitance, $C = Q/V$, depends neither on Q nor on V, but rather on the physical geometry of the conductors. In this chapter, we shall not concern ourselves with the computation of capacitances, but instead study how capacitances may be combined.

A particular parallel plate capacitor with plate area A and separation distance d has capacitance $C = \epsilon_0 A/d$. Should the need arise for a capacitor with twice the capacitance, several options present themselves. One could fabricate another system of plates with the same area and half the original separation, since $\epsilon_0 A/(d/2) = 2 \epsilon_0 A/d = 2\,C$; or another system of plates with twice the area, held at the same separation. Another alternative, not unlike the last, is to take two copies of the existing capacitor and cleverly link them so as to effect a system with twice the area.

Q: How might one link the pairs of conducting planes so as to accomplish this goal?

A: The idea of doubling the area of each plate suggests connecting the positive plate on one capacitor to the positive plate on the other; the negative to the negative. In order to ensure that each pair of joined plates is at exactly the same potential, they should be linked by a small[1] amount of conducting material, a **thin wire**.

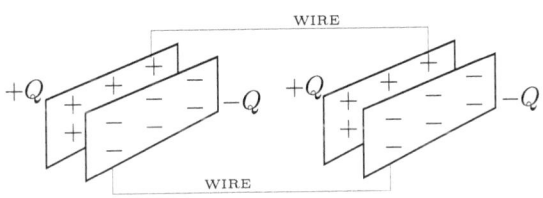

FIGURE 13.1   Connecting Two Identical Capacitors to Double the Capacitance

ASIDE: In Chapter 11, it was shown that all points on the surface and within the bulk of a conductor in electrostatic equilibrium are at the same potential. If two points in the conductor were not at the same potential, then an electric field would exist in the space between the points, and charge would flow until the field was quenched and the potentials

---

[1] A large amount of conductor would contribute materially to the capacitance of the system.

equalised. Two features of such a scenario bear mentioning: (1) the flow of charge falsifies any claim that the system is in electrostatic equilibrium, and (2) the spontaneous response of the system to a discrepancy in potential at contiguous points is to approach equilibrium. Non-equilibrium steady state potential differences will be investigated anon.

Henceforth, we shall portray the joining of the respective parallel plates with conducting wires **schematically**, as in Figure 13.2. When the individual capacitances are C_0, the effective capacitance of the parallel combination is $2\,C_0$.

FIGURE 13.2   Two Identical Capacitors Joined in Parallel

Three ways to generalise and extend this result are to be investigated. The first and second link two parallel plate systems which are **NOT** identical. The third subjects the system of capacitors, joined in parallel, to a constant electric potential maintained by a **battery**.

Q: What if we combine two non-identical plate systems [first making sure that the potentials match]?

[The connection is effected by simultaneously throwing switches in each of the two joining wires.]

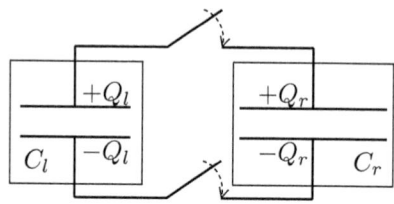

FIGURE 13.3   Two Generic Capacitors Joined in Parallel

A: Each capacitor has a well-defined capacitance, say C_l for the leftmost and C_r for the rightmost, where

$$C_l = \frac{Q_l}{V_l} \qquad \text{and} \qquad C_r = \frac{Q_r}{V_r}$$

hold initially [before the connection is made]. **IF** the potentials of the respective positive and negative plates are matched, $V_l = V_r = V$, **THEN** no charge will flow when the plates are connected. That is, the electrostatic equilibrium of the original disjoint system is undisturbed, and $\{Q_l, Q_r\}$ are invariant. After the connections are made, the effective capacitor bears a net charge of $Q = Q_l + Q_r$, held at a potential difference of V. The effective capacitance is

$$C_{\text{Eff}} = \frac{Q_l + Q_r}{V} = \frac{Q_l}{V} + \frac{Q_r}{V} = \frac{Q_l}{V_l} + \frac{Q_r}{V_r} = C_l + C_r\,.$$

Q: What if we connect two plate systems whose potentials do not match?

A: Each capacitor has a well-defined capacitance: C_l for the left and C_r for the right, where

$$C_l = \frac{Q_l}{V_l} \qquad \text{and} \qquad C_r = \frac{Q_r}{V_r}$$

hold initially. IF the potentials of the respective positive and negative plates are mismatched, $V_l \neq V_r$, THEN upon making the connections, charge will migrate[2] until the potentials are equalised at some intermediate value V'. At such time, the charges on the capacitors will be Q'_l and Q'_r respectively. Charge conservation[3] dictates that the total charge be invariant: $Q'_l + Q'_r \equiv Q_l + Q_r$. In addition, the capacitances are not affected by charge and voltage, so

$$C_l = \frac{Q'_l}{V'} \qquad \text{and} \qquad C_r = \frac{Q'_r}{V'}.$$

Thus, after the onset of electrostatic equilibrium, the effective capacitor holds the total charge, $Q_l + Q_r$, at common potential V'. Therefore, we find ourselves back in the first case (investigated above), and the effective capacitance is

$$C_{\text{Eff}} = \frac{Q_l + Q_r}{V'} = \frac{Q'_l + Q'_r}{V'} = \frac{Q'_l}{V'} + \frac{Q'_r}{V'} = C_l + C_r \,.$$

The result is not so surprising when we recall that the capacitance is a property dependent on the geometry of the conductors and not on the particulars of charge and potential.

Q: What if a constant source of electric potential [*a.k.a.* a BATTERY] is connected to a capacitor which initially bears no charge, *i.e.*, $Q_i = 0$ and $V_i = 0$?

A: The battery in Figure 13.4 is represented as a box with two posts [terminals], between which a constant potential difference, V_0, is maintained.

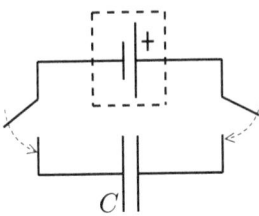

FIGURE 13.4 A Battery and a Capacitor Connected via Two Switches

The two switches are thrown simultaneously at the instant $t = 0$. Consequently, positive charge migrates outward from the higher-potential post of the battery and accrues on one plate of the capacitor, while a precisely equal amount of negative charge flows from the lower-potential side of the battery to the other plate.

> ASIDE: At each instant, CHARGE CONSERVATION dictates that the charges on the capacitor plates must be equal and opposite.

---

[2]For the moment we shall not concern ourselves with how this migration transpires, except to recall that we argued in Chapter 11 that it took no energy to move charge around within a conductor in electrostatic equilibrium.

[3]Charge conservation is one of the two most important properties of electric charge, listed in Chapter 1!

The flow of charge from the battery to the plates stops once the potential difference across the capacitor matches the potential difference across the battery, and electrostatic equilibrium is attained. The capacitor is then "charged up" to potential V_0, and, concomitantly, it bears a charge, $Q_0 = C V_0$, on its plates.

Our established custom in other contexts has been to name physical objects or systems by reference to their relevant features. Continuing in this mode, arrangements of conductors shall be henceforth labelled by their capacitances.

EXAMPLE [*Two Capacitors in Parallel*]

Two initially uncharged[4] capacitors, C_1 and C_2, are combined in parallel and subsequently connected, by means of switches, to a source of constant electric potential V. The switches were thrown and sufficient time elapsed for the system to come to electrostatic equilibrium.

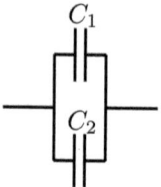

FIGURE 13.5   Two Capacitors in Parallel

Qc: What is the effective capacitance of the parallel system of two capacitors?

Qq: How much charge is borne by the system once electrostatic equilibrium is attained?

Aq: The net flow of charge ceases once the applied potential difference is **mirrored across**[5] both capacitors, *viz.*, $V_1 = V = V_2$. The charges on the capacitors are

$$Q_1 = C_1 V \qquad \text{and} \qquad Q_2 = C_2 V \,,$$

respectively. Thus, the total charge is $Q_{[12]} = Q_1 + Q_2 = (C_1 + C_2) V$.

Ac: The effective capacitance of the parallel combination of C_1 and C_2 is

$$C_{[12]} = \frac{Q_{[12]}}{V_{[12]}} = \frac{Q_1 + Q_2}{V} = C_1 + C_2 \,.$$

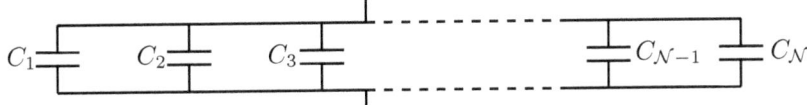

FIGURE 13.6 $\mathcal{N}$ Capacitors in Parallel

EXAMPLE [*Parallel Arrangement of $\mathcal{N}$ Capacitors*]

Suppose that $\mathcal{N}$ capacitors, $\{C_i,\ i \in 1, 2, \ldots, \mathcal{N}\}$, are connected strictly in parallel, as illustrated in Figure 13.6.

Qc: What is the effective capacitance of $\mathcal{N}$ capacitors in parallel?

Qq: How much charge is borne by the system when it is subjected to voltage V ?

Aq: Any applied potential difference V is mirrored across each of the capacitors, *viz.*, $V = V_1 = V_2 = \ldots = V_{\mathcal{N}}$. The charges on the respective capacitors are

$$Q_1 = C_1 V, \qquad Q_2 = C_2 V, \qquad \ldots, \qquad Q_{\mathcal{N}-1} = C_{\mathcal{N}-1} V, \quad \text{and} \quad Q_{\mathcal{N}} = C_{\mathcal{N}} V.$$

Thus, the total charge is

$$Q_{[12\ldots\mathcal{N}]} = Q_1 + Q_2 + \ldots + Q_{\mathcal{N}} = \left(C_1 + C_2 + \ldots + C_{\mathcal{N}} \right) V = \left(\sum_{i=1}^{\mathcal{N}} C_i \right) V.$$

Ac: The effective capacitance of the purely parallel combination of N capacitors is

$$C_{[12\ldots\mathcal{N}]} = \frac{Q_{[12\ldots\mathcal{N}]}}{V_{[12\ldots\mathcal{N}]}} = \frac{Q_1 + Q_2 + \ldots + Q_{\mathcal{N}}}{V} = C_1 + C_2 + \ldots + C_{\mathcal{N}} = \sum_{i=1}^{\mathcal{N}} C_i.$$

Four comments pertinent to this result follow.

$C_{[\ldots]} > C_i$ The effective capacitance of the system is greater than any constituent capacitance. One way to apprehend this is that any addition of conductor increases the capacity to hold charge at fixed potential difference.

$C_i = C_0$ IF all $\mathcal{N}$ capacitors are identical, THEN $C_{[12\ldots\mathcal{N}]} = \mathcal{N} C_0$.

SCALE   The largest constituent capacitors set the scale for the effective capacitance.

SUM/SAME   The analysis of the parallel case was predicated on the existence of two combinatorial constraints,[6] which we shall whimsically dub SUM and SAME.

SUM   Charges residing on the paired conductors in the system SUM to yield the total charge held on the system.

SAME   The SAME potential difference occurs across each of the capacitors.

---

[4] This is not a critical assumption. It is made so as to explicitly avoid any circumstance in which the initial potential difference across either capacitor exceeds that of the battery.

[5] This turn of phrase is commonly employed to describe situations in which one (or more) electrical component(s) is energetically "opposed to" another (set of) component(s). The term *mirrored* has a nice kinesthetic aspect.

[6] Precisely these constraints arose repeatedly, in various guises, in the analyses of elastic materials, springs, thermal conductors, *etc.*, found in VOLUME II. Their ubiquity is not accidental.

Q: Do any other combinations of two capacitors exist?

A: The capacitors could be connected in **series**.

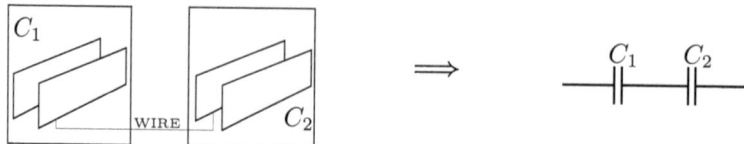

FIGURE 13.7   Two Capacitors Joined in Series

EXAMPLE [*Two Capacitors in Series*]

Two initially uncharged capacitors are joined in series and then hooked up to a constant source of potential V. The switches shown in Figure 13.8 were thrown sufficiently long ago that the system has reached electrostatic equilibrium.

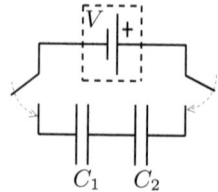

FIGURE 13.8   Two Capacitors in Series and a Battery

Qc: What is the effective capacitance of two capacitors arranged in series?

Qq: How much charge is borne by the system in electrostatic equilibrium?

Ac: The positive charge, $Q_{(12)}$, which flows to the rightmost plate of C_2, is precisely matched by negative charge, $-Q_{(12)}$, deposited on the left plate of C_1. The presence of these charges on the EXTERIOR plates produces a net electric field which bathes the INTERIOR plates. The conducting material comprising the interior plates and the wire joining them polarises so as to **quench** the imposed field.

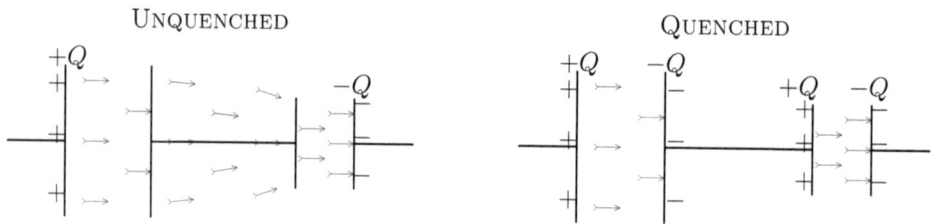

FIGURE 13.9   Two Capacitors Arranged in Series and Subjected to a Potential

Complete quenching occurs when the induced charges residing on the interior plates are $\mp Q_{(12)}$ [mirroring the charges on the exterior plates]. The physical situation is depicted in Figure 13.9.

[The net charge neutrality of the system is scrupulously maintained.]

The potential difference across each capacitor is

$$V_1 = \frac{Q_1}{C_1} = \frac{Q_{(12)}}{C_1}, \qquad \text{and} \qquad V_2 = \frac{Q_2}{C_2} = \frac{Q_{(12)}}{C_2}.$$

The entirety of the interior conductor is at an intermediate potential, V_i, between the potentials found on the exterior faces of the respective capacitors, V_- and V_+. Electric potential *vs.* position in the two-capacitor circuit is plotted in Figure 13.10.

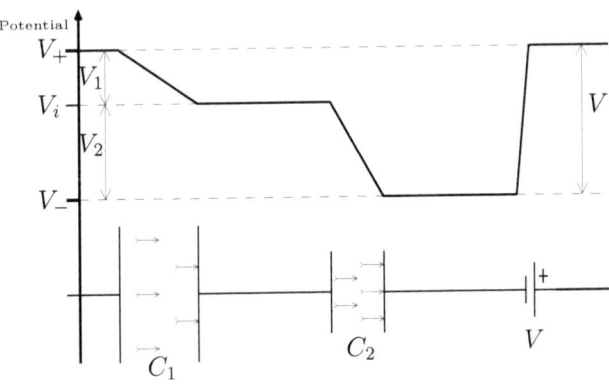

FIGURE 13.10   Energetics of Two Capacitors in Series

Irrespective of the value of the intermediate potential, energy conservation requires that the sum of the potential differences across the two capacitors precisely mirror the potential difference across the terminals of the battery, *viz.,*

$$V = V_1 + V_2 = \frac{Q_{(12)}}{C_1} + \frac{Q_{(12)}}{C_2} = Q_{(12)} \left[\frac{1}{C_1} + \frac{1}{C_2} \right].$$

The definition of capacitance [as the ratio of the magnitude of the opposing charges held on the system of conductors to the potential difference between them] yields the expression

$$C_{(12)} = \frac{Q_{(12)}}{V_{(12)}} = \frac{Q_{(12)}}{V}.$$

Combining the above relations leads to

$$C_{(12)} = \left[\frac{1}{C_1} + \frac{1}{C_2} \right]^{-1} = \frac{C_1 C_2}{C_1 + C_2} \qquad \Longrightarrow \qquad \frac{1}{C_{(12)}} = \frac{1}{C_1} + \frac{1}{C_2}.$$

Aq: The charge held on the series arrangement of capacitors under investigation is

$$Q_{(12)} = C_{(12)} V_{(12)} = \frac{V}{\frac{1}{C_1} + \frac{1}{C_2}}.$$

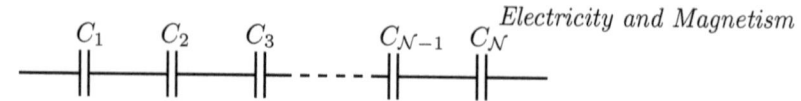

FIGURE 13.11 $\mathcal{N}$ Capacitors in Series

EXAMPLE [*Series Arrangement of $\mathcal{N}$ Capacitors*]

Suppose that $\mathcal{N}$ capacitors, $\{C_i, \; i \in 1, 2, \ldots, \mathcal{N}\}$, are connected strictly in series, as illustrated in Figure 13.11. The switches connecting this network to a battery with potential difference V were thrown long ago, and the system has come to electrostatic equilibrium.

Qc: What is the effective capacitance of $\mathcal{N}$ capacitors in series?

Qq: How much charge is borne by the system when it is held at potential V?

Ac: Any applied potential difference is mirrored across the totality of the set of capacitors, *viz.*, $V = V_1 + V_2 + \ldots + V_{\mathcal{N}}$. The charges on the respective capacitors are all precisely equal:

$$Q_{(12\ldots\mathcal{N})} = Q_1 = Q_2 = \ldots = Q_{\mathcal{N}}\,.$$

Thus,

$$\frac{1}{C_{(12\ldots\mathcal{N})}} = \frac{V_{(12\ldots\mathcal{N})}}{Q_{(12\ldots\mathcal{N})}} = \frac{\sum_{i=1}^{\mathcal{N}} V_i}{Q_{(12\ldots\mathcal{N})}} = \sum_{i=1}^{\mathcal{N}} \frac{V_i}{Q_{(12\ldots\mathcal{N})}} = \sum_{i=1}^{\mathcal{N}} \frac{V_i}{Q_i} = \sum_{i=1}^{\mathcal{N}} \frac{1}{C_i}\,.$$

Aq: The charge borne on the system is the same as that on each of the individual capacitors. In terms of the capacitances and the applied potential difference,

$$Q_{(12\ldots\mathcal{N})} = C_{(12\ldots\mathcal{N})}\, V_{(12\ldots\mathcal{N})}\,.$$

Four comments pertinent to this result follow below.

$C_{(\ldots)} < C_i$ The effective capacitance of the system is less than any constituent capacitance.

$C_i = C_0$ IF all $\mathcal{N}$ capacitors are identical, THEN $C_{(12\ldots\mathcal{N})} = \frac{C_0}{\mathcal{N}}$.

SCALE The smallest constituent capacitors set the scale for the effective capacitance.

SUM/SAME The archetypical constraints for purely series arrangements of capacitors are:

SUM the potential differences across the paired conductors in the system SUM to yield the overall potential difference, and

SAME each of the capacitors bears the SAME amount of charge.

ADDENDUM to Chapter 13

Consider the [daunting] network of 15 capacitors, labelled using **hexadecimal** [base-16] notation, $\{C_i, i \in 1, 2, \ldots, 9, A, \ldots, E, F\}$, illustrated in the figure below. We shall compute the effective capacitance of the network by iterative application of the combinatorial rules for purely parallel and series arrangements. Two considerations merit notice.

- IF the network is attached to an overall source of voltage, OR charge resides on some of the capacitors prior to their being assembled into the network, THEN we assume that the connections were made sufficiently far in the past that [approximate] electrostatic equilibrium is in effect.

- The thin wires joining the capacitors may be stretched, shrunk, bent, or straightened without changing the overall capacitance of the network by any appreciable amount.

For the sake of definiteness, we shall suppose that each of the odd-numbered capacitors has capacitance C, while each even-numbered capacitor has capacitance $2C$.

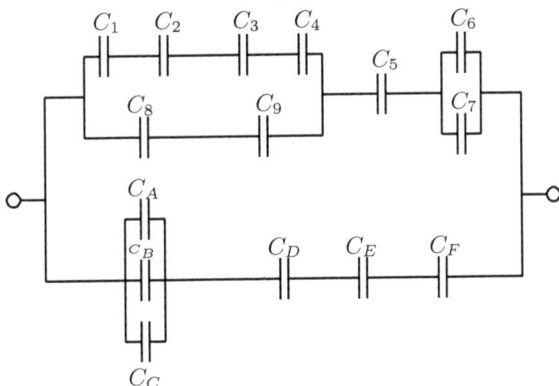

Looking at the smallest scales, one may identify three separate collections of capacitors which are arranged strictly in series, and two which are purely in parallel. Effective capacitances for these five subnetworks may be straightforwardly obtained.

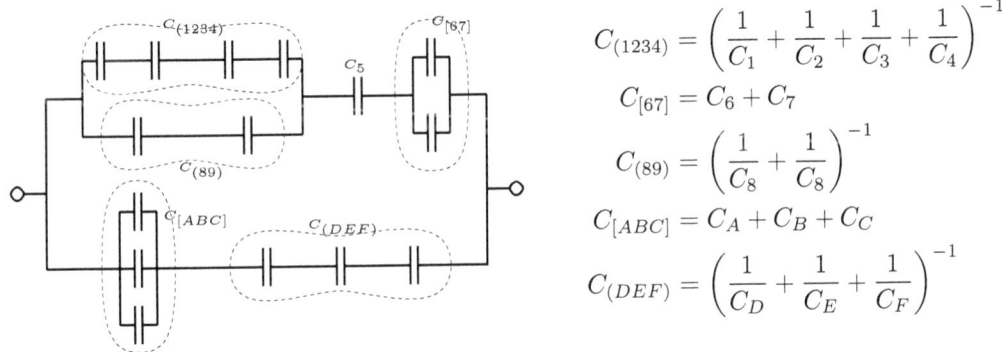

$$C_{(1234)} = \left(\frac{1}{C_1} + \frac{1}{C_2} + \frac{1}{C_3} + \frac{1}{C_4}\right)^{-1}$$

$$C_{[67]} = C_6 + C_7$$

$$C_{(89)} = \left(\frac{1}{C_8} + \frac{1}{C_8}\right)^{-1}$$

$$C_{[ABC]} = C_A + C_B + C_C$$

$$C_{(DEF)} = \left(\frac{1}{C_D} + \frac{1}{C_E} + \frac{1}{C_F}\right)^{-1}$$

In this particular instance, $C_{\text{odd}} = C$ and $C_{\text{even}} = 2C$, these combinations work out to be:

$$C_{(1234)} = \frac{C}{3}, \quad C_{(89)} = \frac{2C}{3}, \quad C_{(DEF)} = \frac{2C}{5}, \quad C_{[67]} = 3C, \quad C_{[ABC]} = 5C.$$

Replacing these smallest-scale subnetworks with equivalent effective capacitances reveals another set of higher-level parallel and series combinations.

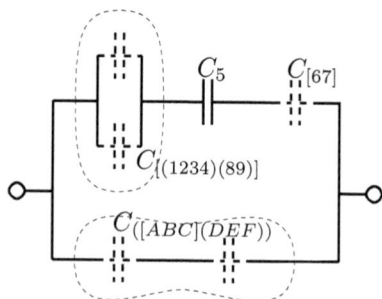

The aggregrations displayed in the figure above have effective capacitances

$$C_{[(1234)(89)]} = C_{(1234)} + C_{(89)} \, ,$$

$$C_{([ABC](DEF))} = \left(\frac{1}{C_{[ABC]}} + \frac{1}{C_{(DEF)}} \right)^{-1} = \left(\frac{1}{C_{[ABC]}} + \frac{1}{C_D} + \frac{1}{C_E} + \frac{1}{C_F} \right)^{-1} \, .$$

When the odd capacitances are C and the evens are $2\,C$, these turn out to be:

$$C_{[(1234)(89)]} = C \quad \text{and} \quad C_{([ABC](DEF))} = \frac{10\,C}{27} \, .$$

With this further aggregation, higher-level parallel and series structures, amenable to analysis, are revealed.

$$C_{([(1234)(89)]5[67])} = \left(\frac{1}{C_{[(1234)(89)]}} + \frac{1}{C_5} + \frac{1}{C_{[67]}} \right)^{-1}$$

The effective network is finally seen to be a parallel arrangement of $C_{([(1234)(89)]5[67])}$ and $C_{([ABC](DEF))}$, as illustrated in the penultimate schematic diagram below.

$\Longrightarrow$ C_{Eff}

$$C_{\text{Eff}} = C_{[([(1234)(89)]5[67])([ABC](DEF))]}$$
$$= C_{([(1234)(89)]5[67])} + C_{([ABC](DEF))}$$

In the particular instance under consideration,

$$C_{([(1234)(89)]5[67])} = \left(\frac{1}{C} + \frac{1}{C} + \frac{1}{3\,C} \right)^{-1} = \frac{3\,C}{7} \, ,$$

and hence

$$C_{\text{Eff}} = C_{[([(1234)(89)]5[67])([ABC](DEF))]} = \frac{3\,C}{7} + \frac{10\,C}{27} = \frac{151}{189} C \, .$$

Chapter 14

Energetics of Capacitance

Recently, we loosely argued that upon connecting an uncharged capacitor to something that had a voltage difference, some charge would flow to the plates. Any question of energetics was cursorily dismissed by saying,

> "Within a conductor in electrostatic equilibrium, charge can move around without requiring net energy input (or dissipation), because the entire conductor is at constant electric potential, and besides, the electric field vanishes, too."

At the time, we merely needed to establish the plausibility of charge migration and the system's evolution toward electrostatic equilibrium.

> [These arguments break down for non-equilibrium situations.]

In this chapter, we shall compute the amount of energy required to charge up a capacitor. Then, we shall posit that this energy is stored in the electric field! Finally, we shall consider a novel arrangement of conducting plates to generalise the parallel plate capacitor results of Chapter 12.

A parallel plate capacitor, C, bearing charge q [$+q$ on one plate and $-q$ on the other], is shown in the leftmost part of the sketch in Figure 14.1.

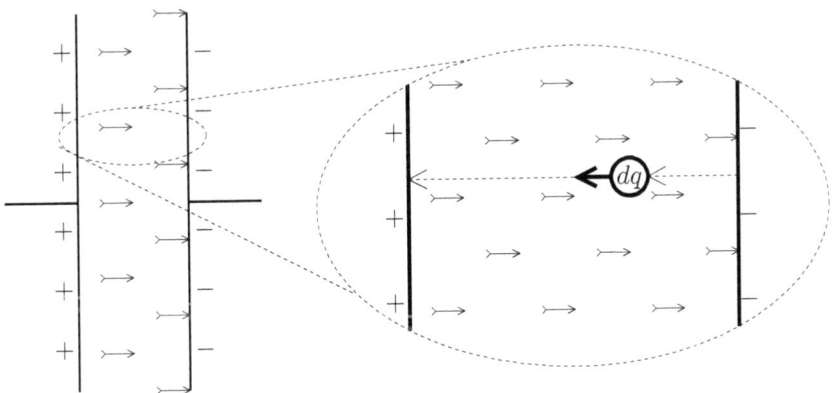

FIGURE 14.1   Increasing the Charge Borne on a Parallel Plate Capacitor

Suppose that, in some time interval, the charge on the plates is increased, $q \rightarrow q + dq$. According to the models studied in previous chapters, this change is accomplished by providing equal and opposite charge bundles, $\pm dq$, to the respective plates. And yet, one also can increase the charge held on the capacitor by sending a dq-bundle of positive charge **from** the negative plate **to** the positive plate.

[Charge conservation is upheld, and the results of the two processes are indistinguishable.]

The potential difference and electric field lines are oriented in such a manner that the bundle of positive charge moves "uphill" across the gap from the negative plate to the positive plate. In crossing over, the electrostatic potential energy of the bundle of charge is increased by the amount of work needed to push it uphill, *viz.*, $dW = V\,dq$. The potential difference, prior to the addition of the charge bundle, is $V = q/C$, and thus,

$$dW = \frac{q\,dq}{C}\,.$$

To charge a capacitor from $q = 0$ [zero initial charge] to $q = Q$ [a final charge] requires an amount of external work given by:

$$W = \int dW = \int_0^Q \frac{q}{C}\,dq = \frac{1}{C}\left[\frac{1}{2}q^2\Big|_0^Q\right] = \frac{Q^2}{2\,C}\,.$$

Recall from earlier discussions that the electrostatic potential energy of a collection of point-like charges is equal to the net mechanical work expended by an external agent in the course of assembling the system. Hence, the electrostatic potential energy of the capacitor with capacitance C, bearing charge Q, held at potential difference V, is

$$U_E = \frac{Q^2}{2\,C} = \frac{1}{2}\,Q\,V = \frac{1}{2}\,C\,V^2\,.$$

The first of the above expressions is quoted from further above, while the latter two are obtained by successive substitutions of $Q = C\,V$. Although equivalent, they are mentioned separately because in the context of specific problems, it is usually the case that one of these forms is more directly applicable.

Consider the logical implications of these energetic results.

- IF Q is fixed, THEN U decreases as C increases.   [Higher capacitance makes it easier to store a given amount of charge; it takes less energy.]

- IF V is fixed, THEN U increases as C increases.   [Increasing capacitance at fixed potential requires additional charge and therefore additional energy.]

> ASIDE: There is always a limit to how much energy can be stored in a capacitive circuit element. Trying to maintain too high a potential difference between nearby plates can result in **electrical breakdown** [*a.k.a.* electrical discharge, arcing, *etc.*] of the material[1] found between the plates.

CLAIM:[2] **The energy stored in a capacitor is held in the electric field.**
What follows is a very cute and deceptively simple derivation of the electrostatic potential energy per unit volume stored in the electric field within a parallel plate capacitor.

A parallel plate capacitor, with face area A and plate separation d, bears charge Q. Our model for these systems assumes that the charge is distributed evenly upon on the surfaces of the plates,

$$\sigma_0 = \frac{Q}{A}\,.$$

---

[1] The "stuff" that resides between the plates is discussed in Chapter 15.
[2] We shall not prove this claim in general; its plausibility in a particular case is illustrated.

This, in turn, leads to uniformity of the electric field between the plates,

$$\vec{E} = \frac{\sigma_0}{\epsilon_0} \, [\, + \to - \,],$$

as well as its vanishing outside the capacitor [no fringing].

> ASIDE: Recall from Chapter 6 that each plate contributes a field with magnitude $\sigma_0/(2\,\epsilon_0)$ in the perpendicular direction with consistent orientation either directly away from or toward the source plate. Between the plates the superposition is purely constructive, leading to a doubling of the one-plate magnitude. Everywhere outside the paired plates, the superposition is entirely destructive, and hence the net field vanishes there.

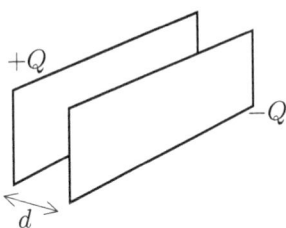

FIGURE 14.2 A Parallel Plate Capacitor

The spatial volume between the plates, throughout which the field is non-vanishing, is $A\,d$. In Chapter 12, the capacitance of a pair of parallel conducting plates was determined to be $C = \epsilon_0\,A/d$. Just above, the amount of electrostatic potential energy stored on the capacitor was ascertained to be $U_E = Q^2/(2\,C)$. Substitution yields

$$U_E = \frac{Q^2}{2\,\frac{\epsilon_0 A}{d}} = \frac{Q^2\,d}{2\,\epsilon_0\,A} = \frac{Q^2}{2\,\epsilon_0\,A^2}\,A\,d = \frac{\sigma_0^2}{2\epsilon_0}\,A\,d\,.$$

The fraction in the last equality above may be re-expressed in terms of the [constant] electric field strength, $E_0 = \frac{\sigma_0}{\epsilon_0}$,

$$\frac{\sigma_0^2}{2\epsilon_0} = \frac{1}{2}\,\epsilon_0\,E_0^2\,.$$

The factor multiplying the fraction is the interstitial volume [the domain of the non-zero electric field]. Hence, the **electrostatic potential energy density** is

$$u_E = \frac{U_E}{\text{Volume}} = \frac{U_E}{A\,d} = \frac{1}{2}\,\epsilon_0\,E_0^2\,.$$

ELECTROSTATIC ENERGY DENSITY   The electrostatic potential energy density [a.k.a. electrostatic potential energy per unit volume] at a field point, $\mathcal{P}$, is locally determined by the electric field strength and the permittivity of free space, via

$$u_E(\mathcal{P}) = \frac{1}{2}\,\epsilon_0\,E^2(\mathcal{P})\,.$$

The SI units work out consistently:

$$[\,u_E\,] = \left[\frac{\text{C}^2}{\text{N}\cdot\text{m}^2}\right]\left[\frac{\text{N}}{\text{C}}\right]^2 = \left[\frac{\text{N}}{\text{m}^2}\right] = \left[\frac{\text{N}\cdot\text{m}}{\text{m}^3}\right] = \left[\frac{\text{J}}{\text{m}^3}\right]\,.$$

EXAMPLE [*A Novel Conducting Plate Geometry*]

Geometry and Configuration   Three parallel plates, each with area A, are stacked, with separation d, in such a manner that they completely overlap. The length and width of the plates are much greater than the separation distance, so fringing may be neglected. The three-plate system is connected to a battery, as shown in Figure 14.3. Switches were thrown, and the system has attained electrostatic equilibrium.

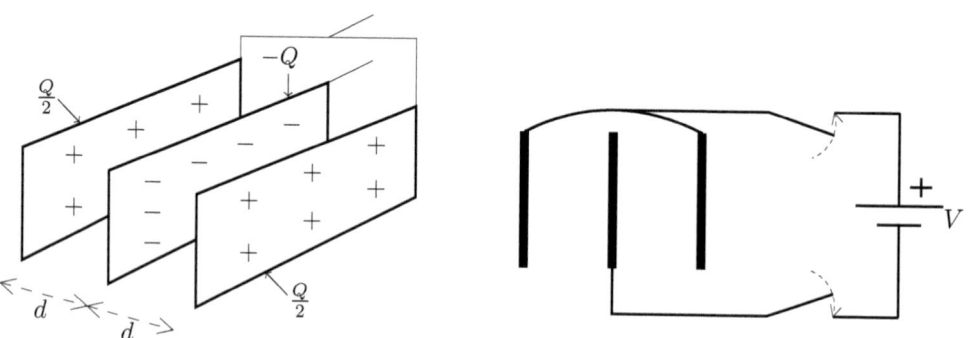

FIGURE 14.3   Three Stacked Conducting Plates: $+ - +$

Charge Distribution   Distributing $\pm Q$ on this system of plates is consistently accomplished with $+\frac{Q}{2}$ on each of the outer pair of positive plates and $-Q$ on the centre plate. The charge densities on each of the outer plates, and on the centre plate, are

$$\sigma_+ = \frac{Q}{2A} \qquad \text{and} \qquad \sigma_- = -\frac{Q}{A},$$

respectively.

Computing ΔV   With the charges assigned to each of the plates, we are equipped to determine the associated electric fields and their superposition.

LEFT	CENTRE	RIGHT
$\left|\vec{E}_l\right| = E_+ = \frac{\sigma_+}{2\,\epsilon_0}$	$\left|\vec{E}_c\right| = E_- = \frac{\sigma_-}{2\,\epsilon_0}$	$\left|\vec{E}_r\right| = E_+ = \frac{\sigma_+}{2\,\epsilon_0}$

The magnitudes of the electric fields produced by the plates of charge are:

$$E_+ = \frac{\sigma_+}{2\,\epsilon_0} = \frac{Q}{4\,\epsilon_0\,A} \qquad \text{and} \qquad E_- = \frac{\sigma_-}{2\,\epsilon_0} = \frac{Q}{2\,\epsilon_0\,A}\,.$$

The net field throughout space is the linear superposition of the fields from each of the three parallel plates.

OUTSIDE   To the left of the three plates,

$$\vec{E}_{\text{net},ol} = E_+\,[\ \leftarrow\ ] + E_-\,[\ \rightarrow\ ] + E_+\,[\ \leftarrow\ ]$$

$$= \left\{ -\frac{Q}{4\,\epsilon_0\,A} + \frac{Q}{2\,\epsilon_0\,A} - \frac{Q}{4\,\epsilon_0\,A} \right\}\,[\rightarrow] = \vec{0}.$$

The same cancellation occurs to the right (with the directions reversed),

$$\vec{E}_{\text{net},or} = \vec{0}.$$

Thus, the net electric field produced by the charges distributed evenly on the set of plates vanishes everywhere outside of the system.

INSIDE   Within the left interstitial space,

$$\vec{E}_{\text{net},il} = E_+\,[\ \rightarrow\ ] + E_-\,[\ \rightarrow\ ] + E_+\,[\ \leftarrow\ ]$$

$$= \left\{ +\frac{Q}{4\,\epsilon_0\,A} + \frac{Q}{2\,\epsilon_0\,A} - \frac{Q}{4\,\epsilon_0\,A} \right\}\,[\rightarrow] = \frac{Q}{2\,\epsilon_0\,A}\,[\rightarrow].$$

The field contributions from the left and right plates cancel, and the net field is that produced by the centre plate alone. The same effect occurs in the right interstitial region,

$$\vec{E}_{\text{net},ir} = \frac{Q}{2\,\epsilon_0\,A}\,[\leftarrow].$$

Within the system of plates, the field is as though only the centre plate [bearing charge $-Q$] exists.

Now that the electric field throughout space has been inferred, the potential difference between the centre and either outer plate may be computed.

$$\Delta V = V(\text{outer}) - V(\text{centre}) = -\int_{\vec{r}_c}^{\vec{r}_o} \vec{E}_{\text{net}} \cdot d\vec{s}\,.$$

Path independence of the potential difference allows us to choose a straight-line path perpendicular to each surface, in which case the integrand simplifies to a negative constant. The line integral yields the plate separation distance, d, and thus,

$$V = \Delta V = +E_-\,[d] = \frac{Q\,d}{2\,\epsilon_0\,A}\,.$$

Ratio of total charge to potential difference   Finally, the capacitance of this particular configuration of three conducting plates is computed to be

$$C = \frac{Q}{V} = \frac{Q}{\frac{Q\,d}{2\,\epsilon_0\,A}} = \frac{2\,\epsilon_0\,A}{d}\,.$$

The net capacitance is equal to that of two identical parallel plate capacitors connected in parallel. The novelty is that this was accomplished with three pieces of conductor rather than four.

Chapter 15

Dielectrics

Until now, we have investigated the capacitive properties of sundry conductor geometries, but have not considered what material, if any, lies between the plates.

ASIDE: Actually, we've assumed a **vacuum**. Fortunately, the physical difference between air and vacuum in this regard is slight.

Suppose that a given arrangement of conductors [with vacuum filling[1] the space(s) between them] is found to have capacitance C. IF an **insulating** material[2] is interposed between the conductors, THEN the capacitance of the new system is greater than C. This increase arises from the dielectric properties of the insulating material and is characterised by its **dielectric constant**, k. The value $k = 1$ is assigned to the vacuum, and thus $k > 1$ for all materials. Some representative values[3] of k are presented in the table below.

MATERIAL	Vacuum	Air	Paper	Silicon	Particular Alloys & Doped Ceramics
k	1	1.00054	3.5	12	100+

HOW DIELECTRICS WORK: PHENOMENOLOGY

A parallel plate capacitor, with vacuum between the plates, bears charges $\pm Q_0$, held at potential difference V_0. Accordingly, the capacitance of this system is

$$C_0 = \frac{Q_0}{V_0},$$

and it should be remembered that C_0 is independent of both Q_0 and V_0.

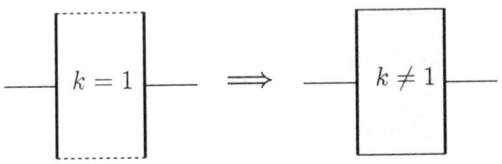

FIGURE 15.1   Replacing the Dielectric Material in a Parallel Plate Capacitor

IF a material with dielectric constant k is inserted between the plates so it fully occupies the interstitial space, AND the charge on the plates is held constant, THEN the potential difference across the capacitor is reduced, *i.e.*,

$$V_0 \quad \rightarrow \quad V_k = \frac{V_0}{k},$$

---

[1]This sounds oxymoronic, eh?

[2]Should one try to fill the interstitial space with a conducting material, the [former] capacitor would be reduced to a single solid lump of conductor.

[3]These particular values occur at Standard Temperature, 20 C, and Pressure, 1 atm $= 101.3\,\text{kPa}$.

AND consequently the capacitance is increased:

$$C_0 \quad \rightarrow \quad C_k = \frac{Q_0}{V_k} = \frac{Q_0}{\frac{V_0}{k}} = k \frac{Q_0}{V_0} = k C_0 > C_0 .$$

Three comments are necessary before we continue.

|| This is a description of HOW changing the dielectric medium between the conductors changes the capacitance property of the system. A more fundamental explication of WHY dielectrics behave in this manner is forthcoming.

|| The dielectric substance must be an electrically neutral insulating material, as the amount of charge on the plates is [assumed to be] not affected by interposition of the dielectric.

> **NEUTRAL**   Net charge on the dielectric would induce polarisation on the conducting plates, changing their charges, disrupting the electric field, and shifting the values of the (equi-)potential on each plate. These changes in turn affect the capacitance of the system.

> **CONDUCTING**   If electric charge is able to move freely through the dielectric material, then the prospect of electrostatic equilibrium at any plate charge other than $Q = 0$ is untenable.

|| The presence of the dielectric reduces the store of electrostatic potential energy held on the capacitor by a factor of the dielectric constant:

$$U_{E,0} = \frac{Q_0^2}{2\,C_0} \quad \rightarrow \quad U_{E,k} = \frac{Q_k^2}{2\,C_k} = \frac{Q_0^2}{2\,(k\,C_0)} = \frac{\frac{Q_0^2}{2\,C_0}}{k} = \frac{U_{E,0}}{k} .$$

We obtain the same result, diminishment by a factor of k, irrespective of the means by which the potential energy is evaluated:

$$U_{E,k} = \frac{1}{2}\,C_k\,V_k^2 = \frac{1}{2}\,(k\,C_0)\left(\frac{V_0}{k}\right)^2 = \left(\frac{1}{2}\,C_0\,V_0^2\right)\Big/ k ,$$

$$U_{E,k} = \frac{1}{2}\,Q_k\,V_k = \frac{1}{2}\,Q_0\left(\frac{V_0}{k}\right) = \left(\frac{1}{2}\,Q_0\,V_0\right)\Big/ k .$$

**Increasing capacitance (holding charge fixed)
reduces the amount of stored energy.**

ASIDE: The **dielectric strength** of a material is an estimate of the maximum internal electric field that it can withstand. When V_0/d exceeds the dielectric strength, cascading ionisation within the material causes it to break down and briefly become conducting. The dielectric material sustains damage whenever this occurs. The capacitor may even *blow up* and a fire may start.

HOW DIELECTRICS WORK: MICROSCOPICALLY

Our choice of microscopic model for dielectrics depends on whether the molecules comprising the material are **polarised** or not.

POLARISATION   A molecule is said to be polarised IFF it possesses a net non-zero dipole moment in the absence of any externally applied electric field.

The atoms comprising the molecule are neutral [possessing equal numbers of protons and electrons]. The negative charge is distributed throughout a region of space corresponding to the extent of the atom, while the positive charge is concentrated within the very much smaller nucleus. The precise details of molecular formation are fearsomely complicated, and yet the polarisation effect is describable in simple terms.

> ASIDE: The notion that **the description of physical systems on a certain scale can be made effectively independent of the much smaller-scale models which describe the system's constituents** is a VERY powerful idea.

For our present purposes we require only that the neutral molecule possess positively and negatively charged constituents, and that the distributions of these constituents be somewhat independent and not completely fixed.

UNPOLARISED MEDIA   Although the dielectric material is unpolarised, it acquires a net polarisation on account of its being bathed in the ambient electric field generated by the charges in the conductor lumps. Figure 15.2 illustrates this idea.

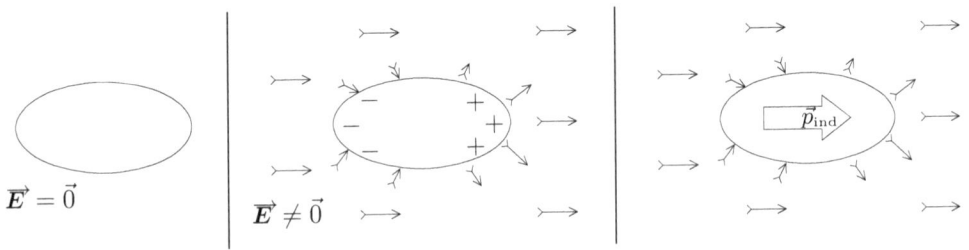

FIGURE 15.2   A Non-Polar Molecule in Zero and Non-Zero Electric Field

> **LEFT**   In the leftmost panel, a neutral molecule, residing in a region free of any ambient electric field, is depicted as an ellipsoidal blob. The spatial averages of the overlapping distributions of positive and negative charge coincide in this case and the molecule remains unpolarised.

> **CENTRE**   The centre panel portrays the same neutral molecule, immersed in an externally applied electric field. The electric field bathing the molecule has pushed the distribution of positive charge downhill in the field and pulled the negative charge uphill.

>> The magnitudes of the shifts may differ for the two species of charge.
>> Depletion of one species is equivalent to enhancement of the other.

> The molecule now has a downfield region of net positive charge and an upfield region of net negative charge. It has polarised.

RIGHT   In the rightmost panel, the perturbed molecule is represented by its **induced dipole moment**. In Chapter 2, dipoles were constructed from monopolar sources, and it was determined that $\vec{p} = 2\,a\,q$ [from $-$ to $+$]. Here, $\vec{p}_{ind}$ is meaningfully ascribed to the molecule, even when the quantities a and q are ill-defined.

Thus, **under the influence of the applied electric field, the molecules in the material each act as tiny electric dipoles** with the following properties.

$\vec{p}_{ind}$ The induced dipole is

–   aligned with the external field responsible for its creation, and

–   proportional to the strength of the applied field. [The precise manner of the proportionality is not specified. However, the constant term is ZERO, since polarisation is absent when the applied field vanishes. Also, the linear and higher coefficients may be expected to depend on the composition of the dielectric material.]

$\vec{E}_{ind}$ Each induced dipole contributes to a net induced electric field. Every induced dipole is aligned in opposition to the external field which produced the polarisation in the first place. This has implications.

−+−⋯+−+   The dipoles are oriented head-to-tail, and thus neutrality [charge cancellation] is maintained within the bulk of the material. A net buildup of positive [negative] charge occurs on the surface of the dielectric facing the negative [positive] plate.

− ∘ ∘ ⋯ ∘ +   This net separation of charge on the surfaces of the dielectric material gives rise to an induced electric field opposed to the applied external field due to the charges on the plates.

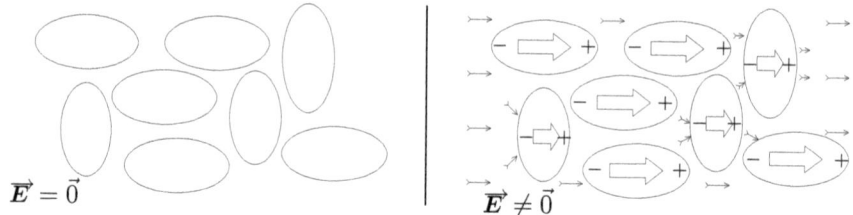

$\vec{E} = \vec{0}$ $\vec{E} \neq \vec{0}$

FIGURE 15.3   Non-Polar Molecules in Zero and Non-Zero Electric Field

The net field within the dielectric is weaker than the applied field.

POLARISED MEDIA   Polar molecules possess a non-zero electric dipole moment, even in the absence of an external electric field.

ASIDE: **Q:** Is this really possible? Are there examples?
   A: Yep. Water [H_2O] is a polar molecule. The polar covalent chemical bonds which join hydrogen atoms to the oxygen atom give rise to an overall net dipole moment for the molecule, because, contrary to naive expectations,[4] H_2O is NOT a linear molecule.[5] Salt [$Na\,Cl$] is polar, because of its strongly ionic bond. There are many other examples.

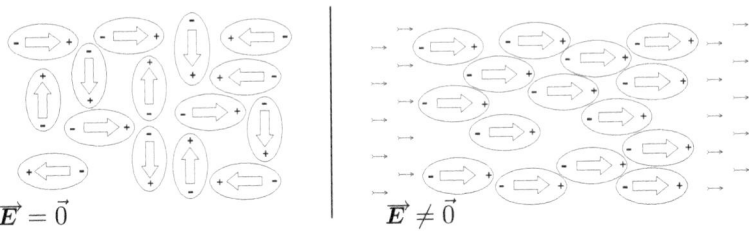

$$\vec{E} = \vec{0} \qquad\qquad\qquad \vec{E} \neq \vec{0}$$

FIGURE 15.4 Polar Molecules in Zero and Non-Zero Electric Field

LEFT In the left-hand panel of Figure 15.4, a collection of neutral polar molecules occupies a region free of any ambient electric field. The dipole moments of the molecules are randomly oriented, and hence the net dipole moment of the collection vanishes.

RIGHT The right panel portrays the collection of polar molecules subjected to an externally applied electric field. The free-standing dipoles attempt to align with the field in two ways. One is the polarisation of the slightly perturbed molecular charge distributions, as in the non-polar case described just above. The other, more significant, contribution comes from rotation[6] of the dipoles into alignment with the external field [conformational change].

ASIDE: In Figure 15.4, the degree of alignment is exaggerated for emphasis.

Under the influence of an externally applied field, the neutral polar medium undergoes conformational change which, while preserving overall and interior neutrality, shifts equal and opposite charges to the surfaces adjacent to the charged conductor plates.

$\vec{p}_{\mathrm{perm}}$ The degree of alignment of the permanent dipoles is contingent on the strength of the field, the composition of the dielectric, and the temperature.[7]

$-\mathrm{oo}\cdots\mathrm{o}+$ The net separation of charge on opposing surfaces gives rise to an induced electric field within the bulk of the dielectric material which is opposed to the external field applied by the charged plates.

The net field within the dielectric is weaker than the applied field.

Thus, for both polarised and unpolarised dielectric media, the electric field within the material is less than the applied external field. Put another way, the net field is the linear superposition of the applied and induced electric fields:

$$\vec{E}_{\mathrm{net}} = \vec{E}_{\mathrm{applied}} + \vec{E}_{\mathrm{induced}}\,.$$

ASIDE: If the symmetry of the situation allows reduction to 1-d, we elide the vector symbols and write
$$E_{\mathrm{net}} = E_{\mathrm{applied}} - E_{\mathrm{induced}} = E_0 - E_i\,.$$

---

[4] For instance, CO_2 is a linear molecule.
[5] Quantum mechanical and group-theoretic principles explain why water is not linear.
[6] The dynamics and energetics of this rotation are the subject of the upcoming Chapter 16.
[7] At higher temperatures, the dipoles are more vigorously jostled, making them slightly freer to rotate. At lower temperatures, the dipoles are more firmly held in place by neighbouring molecules.

In the following two-part *Gedanken* experiment, slabs of neutral dielectric material are immersed in the electric field between two parallel charged plates. In the first part, the slab occupies only a portion of the volume between the plates, as illustrated in Figure 15.5. In the second part, the slab completely fills the gap, as shown in Figure 15.6.

The [equal and opposite] electric charge residing upon the parallel plates is externally prescribed and independent of details pertaining to the dielectric. WHEN the dielectric block is neutral, AND has face area A, precisely matching that of the plates, AND the usual parallel plate assumptions[8] hold, THEN it is reasonable to assume that the charge is distributed uniformly across the plates:

$$\sigma_0 = \frac{Q_{\text{Total}}}{A} = \text{constant}.$$

In Chapter 6, where the parallel plate capacitor was first studied, we determined that a uniform electric field[9] with magnitude

$$\left|\vec{E}_o\right| = E_o = \frac{\sigma_0}{\epsilon_0}$$

occupies the interstitial space. Consider what happens when a dielectric substance partially or completely impinges upon this space.

PARTIAL   Locality and the neutrality of the dielectric together militate that the electric field in the non-dielectric region within the plates is precisely the same as if the dielectric weren't there. Hence, the strength of the electric field in the region between the dielectric and the plates is simply E_o.

The dielectric material, bathed in the field produced by the charge on the plates, responds by inducing a field within its bulk. This induced field is uniform [under our assumptions], with magnitude

$$E_i = \frac{\sigma_i}{\epsilon_0},$$

and is directed in opposition to the externally applied field. The symbol σ_i denotes the induced [residual] surface charge densities on the faces of the dielectric.

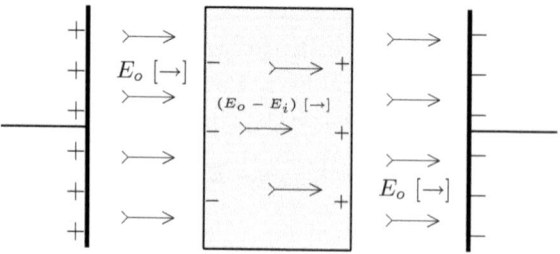

FIGURE 15.5 A Dielectric Substance Partially Filling a Region of Uniform Applied Electric Field

---

[8]The dimensions of the plates (and dielectric) are much larger than their separations, the plates (and dielectric) overlap exactly, and fringing effects are negligible.

[9]We call this field $\vec{E}_o$ because, from the perspective of the dielectric, it is imposed from outside.

ASIDE: While all of this sounds quite reasonable, one ought not to take it too seriously, since σ_i is NOT an observable quantity.

Thus, the net field inside the dielectric material is

$$E_{\text{net}} = E_o - E_i = \frac{\sigma_0 - \sigma_i}{\epsilon_0}.$$

COMPLETE   The situation in which the dielectric completely fills the gap may be modelled by taking the limit of the partially filled case.

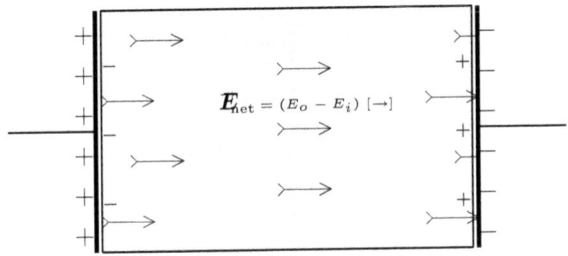

FIGURE 15.6   A Dielectric Substance Completely Filling a Region of Uniform Applied Electric Field

Here, the net field within the dielectric material *is* the uniform electric field everywhere within the capacitor. The potential difference across the gap is readily computed to be

$$\Delta V = V_+ - V_- = -\int_-^+ \vec{E}_{\text{net}} \cdot d\vec{s} = +\frac{\sigma_0 - \sigma_i}{\epsilon_0} \int_-^+ ds = +\frac{\sigma_0 - \sigma_i}{\epsilon_0} \, d.$$

Recall that the dielectric strength of a material is the factor by which the potential difference across it is reduced, *viz.*,

$$\Delta V = \frac{V_0}{k} = \frac{E_o \, d}{k} = \frac{(\sigma_0/\epsilon_0) \, d}{k} = \frac{\sigma_0 \, d}{k \, \epsilon_0}.$$

Equating these two expressions for the potential difference yields

$$\frac{\sigma_0 \, d}{k \, \epsilon_0} = \frac{\sigma_0 - \sigma_i}{\epsilon_0} \, d \quad \Longrightarrow \quad \frac{\sigma_0}{k} = \sigma_0 - \sigma_i \quad \Longrightarrow \quad \frac{\sigma_i}{\sigma_0} = \frac{k-1}{k} = 1 - \frac{1}{k}.$$

Several comments are in order.

- The neutral dielectric material polarises when subjected to an applied external electric field and thereby partially screens the applied field.

- Coherent polarisation throughout the dielectric substance leads to net separation of charge while maintaining bulk neutrality within interior regions.

- The induced charge density depends on σ_0 and k, since σ_0 is proportional to the strength of the applied uniform electric field which drives the polarisation, and larger values of k correspond to greater dielectric effect.

- There are two limiting cases.

  k = 1 The vacuum has $k = 1$, because, however strong the applied field, there is no material present between the plates to polarise, and hence the induced charge density, $\sigma_{i,\text{vacuum}}$, is always exactly zero.

  > ASIDE: Strictly speaking, this is only true classically. In Quantum Field Theory, the vacuum *froths* and *roils* with virtual particle-antiparticle pairs which *can* and *do* interact with external fields.

  k = ∞ A conducting material has a divergent dielectric constant, since a conductor in electrostatic equilibrium is necessarily an equipotential region. That is,

  $$0 \equiv \Delta V = \frac{V_0}{k_{\text{cond}}} \qquad \Longleftrightarrow \qquad k_{\text{cond}} \to \infty \,.$$

  Also, the induced charge density cancels the applied charge density:

  $$\left. \frac{\sigma_i}{\sigma_o} \right|_{\text{cond.}} = \lim_{k \to \infty} \frac{k-1}{k} = 1 \,.$$

  These results are fully consistent with, and generalise, the notion that conductors in electrostatic equilibrium fully quench any applied electric field.

- The reader might note a seeming incongruity in the analysis, for when computing the induced electric field we blithely employed the vacuum form,

  $$E_i = \frac{\sigma_i}{\epsilon_0} \,,$$

  despite the fact that we were working inside the dielectric. Now, before attempting to do something rash, like setting $E_i = \sigma_i/(k\,\epsilon_0)$, one ought to remember that the induced charge density is not directly observed, whereas the dielectric constant has directly measurable energetic implications.

The *quick and dirty* way of dealing with dielectric materials is to:

1st Obtain the VACUUM formula for a physical electric quantity of interest.

2nd Replace the permittivity of free space, ϵ_0, with $k\,\epsilon_0$.

Chapter 16

Energetics of Dipoles

In Chapter 15, we learned that dielectric materials have diminished net electric fields within their bulk because the external applied field induces net [coherent] alignment of molecular dipoles. Let's now consider the energetics favouring this process of alignment.

The primitive model of an electric dipole, consisting of two equal and opposite point charges $\pm q$, separated by distance $2\,a$, was developed in Chapter 2. The electric dipole moment, $\vec{p} = 2\,a\,q\,[-\rightarrow+]$, was defined in terms of the quantities shown in Figure 16.1.

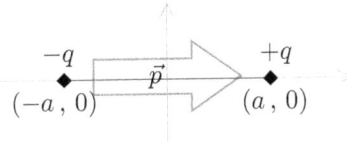

FIGURE 16.1   An Elementary Electric Dipole Composed of Two Monopoles

Now, imagine such an electric dipole immersed in a uniform electric field, as illustrated in Figure 16.2.

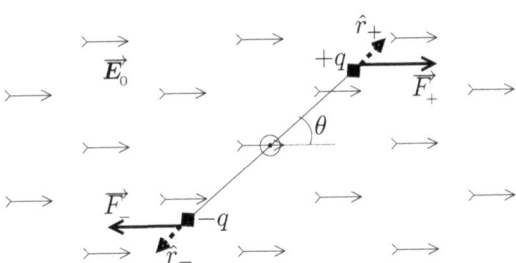

FIGURE 16.2   An Elementary Electric Dipole Immersed in a Uniform Electric Field

Evidently, the net electric force acting on the dipole vanishes:

$$\vec{F}_{\text{NET}} = \vec{F}_+ + \vec{F}_- = (+q)\,\vec{E}_0 + (-q)\,\vec{E}_0 = \vec{0}.$$

The net torque $\vec{\tau}_{\text{net}}$, associated with the electric forces, need not vanish. Choosing to compute the torques about the axis perpendicular to the page and through the centre of the dipole [indicated by the $\odot$ in the figure], one may write

$$\vec{\tau}_{\text{net}} = \vec{\tau}_+ + \vec{\tau}_-, \qquad \text{where} \qquad \vec{\tau}_+ = \vec{r}_+ \times \vec{F}_+ \qquad \text{and} \qquad \vec{\tau}_- = \vec{r}_- \times \vec{F}_-.$$

Both the forces and the vectors from the axis to the point are equal and opposite, *i.e.*,

$$\vec{F}_- = -\vec{F}_+ \qquad \text{and} \qquad \vec{r}_- = -\vec{r}_+,$$

and therefore the sum of the two torques is twice that of each individually:

$$\vec{\tau}_{net} = 2\,\vec{r}_+ \times \vec{F}_+ = 2\,|\vec{r}_+|\,|\vec{F}_+|\,\sin(\theta)\,[\otimes] = 2\,a\,q\,E\,\sin(\theta)\,[\otimes]\,.$$

Re-expressed in a coordinate-independent manner, this reads

$$\vec{\tau}_{net} = \vec{p} \times \vec{E}\,.$$

The ambient electric field attempts to rotate the dipole moment vector into alignment. Complete alignment of a uniform electric field and dipole moment, with $\vec{\tau}_{net} = \vec{p} \times \vec{E}_0 = \vec{0}$, is illustrated in the sketch to the left in Figure 16.3.

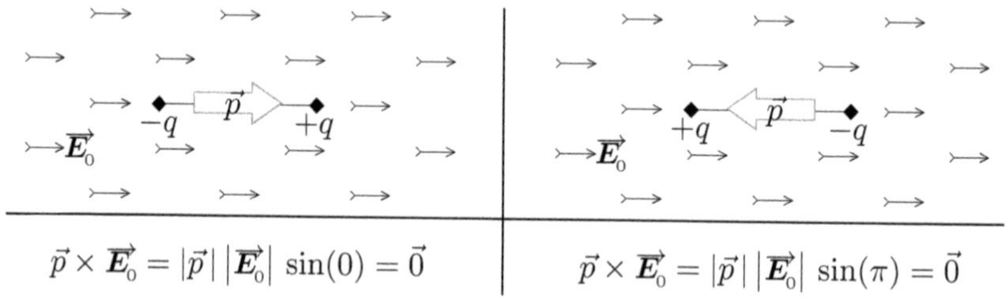

$$\vec{p} \times \vec{E}_0 = |\vec{p}|\,|\vec{E}_0|\,\sin(0) = \vec{0} \qquad\qquad \vec{p} \times \vec{E}_0 = |\vec{p}|\,|\vec{E}_0|\,\sin(\pi) = \vec{0}$$

FIGURE 16.3   Equilibrium Configurations of Dipoles in Ambient Electric Fields

Remarkably, the net torque also vanishes when the dipole moment is anti-aligned with the [local] electric field, as illustrated in the right panel of Figure 16.3.

The behaviour of the dipole interacting with the uniform external electric field may be re-examined in light of the energetics of the system. As we'll see, the aligned case corresponds to **stable** equilibrium, while the anti-aligned situation is **unstable**.

The work done by the torque[1] acting on the electric dipole as it rotates from initial angle θ_i, to final angle θ_f, is

$$W_{if}[\vec{\tau}] = \int_{\theta_i}^{\theta_f} \vec{\tau} \cdot d\vec{\theta}\,.$$

The torque on the dipole in Figure 16.2, $\vec{p} \times \vec{E} = 2\,a\,q\,E_0\,\sin(\theta)\,[\otimes]$, is anti-parallel to the axis along which the angle is measured, and thus the integrand becomes

$$\vec{\tau} \cdot d\vec{\theta} = 2\,a\,q\,E_0\,\sin(\theta)\,d\theta\,(\otimes \cdot \odot) = -2\,a\,q\,E_0\,\sin(\theta)\,.$$

The work is computed to be

$$W_{if}[\vec{\tau}] = -2\,a\,q\,E_0 \int_{\theta_i}^{\theta_f} \sin(\theta)\,d\theta = -2\,a\,q\,E_0 \left[-\cos(\theta) \Big|_{\theta_i}^{\theta_f} \right]$$

$$= 2\,a\,q\,E_0 \left[\cos(\theta_f) - \cos(\theta_i) \right]\,.$$

The expression obtained for the work depends solely on the initial and final angles, indicating that the dipole–field interaction is **conservative**. Hence, an electric dipole residing in a uniform electric field possesses a **potential energy function**, $U_p(\theta)$, satisfying

$$\Delta U_p = U_p(\theta_f) - U_p(\theta_i) = -W_{if}[\vec{\tau}] = -2\,a\,q\,E_0 \left[\cos(\theta_f) - \cos(\theta_i) \right]\,.$$

---

[1] Rotational work was formulated in VOLUME I, Chapter 40.

The simplest suitable potential energy function is $-U_0 \cos(\theta)$, for U_0, a positive constant bearing units of energy. Implicit in this choice is the association of the ZERO of potential energy with the angle $\theta_0 = \frac{\pi}{2}$, corresponding to a dipole lying at right angles to the [local] electric field. The particular potential energy function, written explicitly [and in its most general form], is

$$U_p(\theta) = -2\,a\,q\,E_0 \cos(\theta) = -\vec{p} \cdot \vec{E}.$$

This potential energy function is plotted in Figure 16.4.

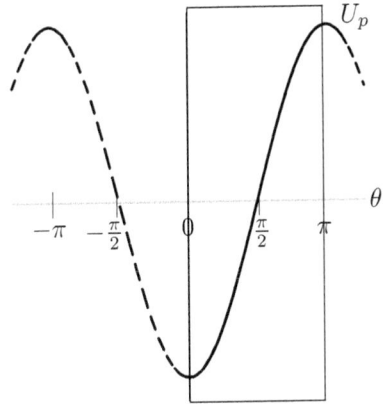

FIGURE 16.4   Potential Energy of an Electric Dipole in a Uniform Electric Field

The physical situation exhibits axial symmetry. That is, the potential energy depends on the [polar] angle between the dipole vector and the electric field, and not on the angle with which the dipole projects onto the plane perpendicular to the field. Hence, the physical domain of θ is $[0, \pi]$. The potential function on this domain is drawn within the box in Figure 16.4. Notwithstanding the importance of this axial symmetry, it proves useful to extend the domain of the potential energy function to include fictitious negative angles and angles greater than π. These extensions are shown by the dashed lines.

The extended potential energy function has several advantages. Foremost of these is that it exhibits the 2π periodicity which is superficially [and misleadingly] evident in Figure 16.2.

> ASIDE: For clarity's sake, it is best to consider a constrained system in which the dipole is allowed only to swivel about the axis pointing directly out of the page in Figure 16.2. The extended potential energy function accurately describes this case.

Secondly, while the simplest potential energy function relegates its maximum and its minimum to endpoints, the extended function is centred on a global minimum.[2] This feature makes more obvious the existence and properties of the equilibrium points. In accord with the figure and the formula, the extended potential energy function, $U_p(\theta)$, is

$$\begin{bmatrix} \text{MINIMISED} \\ \text{ZERO} \\ \text{MAXIMISED} \end{bmatrix} \text{ when the dipole and the local electric field are } \begin{bmatrix} \text{ALIGNED} \\ \text{PERPENDICULAR} \\ \text{ANTI-ALIGNED} \end{bmatrix}.$$

Therefore, $\theta = 0$ is a point of stable equilibrium, while $\theta = \pm\pi$ are unstable.

[These observations follow from the direct computation of torques earlier in this chapter.]

---

[2]This is but one of an infinite number of equivalent minima.

The dynamical behaviour exhibited by a physical system whose present state is in the vicinity of an energetic minimum is almost[3] always well-described by an associated SHO system.[4] We shall verify that this is the case here by forming a truncated Maclaurin series[5] associated with $U_p(\theta)$. Generally speaking, the Maclaurin series corresponding to a function $f(x)$ is a power series in x, *viz.*,

$$S_f(x) = \sum_{n=0}^{\infty} \frac{f^{(n)}(0)}{n!}\, x^n \, ,$$

where $f^{(n)}(0)$ represents the nth derivative of f, evaluated at $x = 0$.

Thus, for the potential energy function, $U_p(\theta) = -p\, E_0\, \cos(\theta)$,

$$S_{U_p}(\theta) = \sum_{n=0}^{\infty} \frac{1}{n!} \left.\frac{d^n U_p(\theta)}{d\theta^n}\right|_{\theta=0} \theta^n \, .$$

The derivatives of $U_p(\theta)$ at the origin are:

$$\left.\frac{d^0 U_p(\theta)}{d\theta^0}\right|_{\theta=0} = \cos(\theta)\Big|_{\theta=0} = 1 \qquad\qquad \left.\frac{dU_p(\theta)}{d\theta}\right|_{\theta=0} = -\sin(\theta)\Big|_{\theta=0} = 0$$

$$\left.\frac{d^2 U_p(\theta)}{d\theta^2}\right|_{\theta=0} = -\cos(\theta)\Big|_{\theta=0} = -1 \qquad\qquad \left.\frac{d^3 U_p(\theta)}{d\theta^3}\right|_{\theta=0} = +\sin(\theta)\Big|_{\theta=0} = 0$$

$$\left.\frac{d^4 U_p(\theta)}{d\theta^4}\right|_{\theta=0} = \cos(\theta)\Big|_{\theta=0} = 1 \qquad\qquad \left.\frac{d^5 U_p(\theta)}{d\theta^5}\right|_{\theta=0} = -\sin(\theta)\Big|_{\theta=0} = 0$$

$$\vdots \qquad\qquad\qquad\qquad\qquad\qquad\qquad \vdots$$

This series includes only even terms, which is entirely consistent with the symmetry of the potential under reflection $\theta \to -\theta$. Collecting these results in order to explicitly render the lowest-order terms in the series yields

$$S_{U_p}(\theta) = -p\, E_0\left(1 - \frac{1}{2}\theta^2 + \frac{1}{24}\theta^4 - \dots\right) \simeq p\, E_0\left(-1 + \frac{1}{2}\theta^2\right) \, ,$$

where the polynomial approximation has been truncated at second order. The first term is a constant and hence has no effect on the dynamics. The next term has precisely the SHO, quadratic, form. In the vicinity of $\theta = 0$, the higher-order terms may be neglected [in most cases] and the SHO approximation works extremely well.

Q: What if this approximation isn't "good enough?"

A: Then we are stuck modelling the system as an **anharmonic** oscillator, and the analysis is considerably harder. Incidently, the qualification "almost" [all minima], made earlier, is to accommodate potential energy functions for which the quadratic term vanishes at the minimum.

---

[3] We'll see momentarily why this qualifier is necessary.

[4] Simple harmonic oscillator systems are discussed briefly in VOLUME I and extensively in VOLUME II of these notes.

[5] The Maclaurin series is a Taylor series computed about the origin.

Chapter 17

Electric Current

Thus far, we've focussed on **electrostatics**, studying only those situations in which charge distributions are constant in time. Now we're ready to consider **charge currents**.

MICROSCOPIC APPROACH TO ELECTRIC CURRENT

Were one to look [ultra-] microscopically at a portion of an isolated metallic conductor, one would observe a lattice of metallic ions and some number of "free" electrons. The ions have essentially no mobility; they may be considered fixed in space. The number of free electrons per atom, *a.k.a.* the degree of ionisation, is typically fractional (since it is averaged throughout the lattice).

[Charge conservation is rigorously enforced on both the atomic and the bulk material scales.]

The epithet "free" is qualified because, while these electrons are not directly associated with any specific ion in the lattice, they are ordinarily confined within the bulk of the metallic substance.

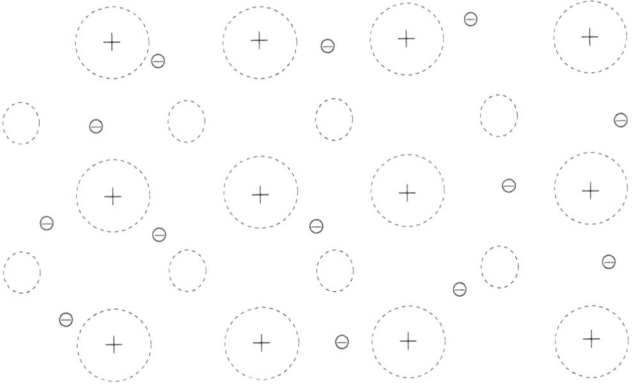

FIGURE 17.1 An Ultra-Microscopic Slice through a Metallic Solid

The simplest metallic lattice is a [near-]regular 3-d arrangement of atomic "balls." Figure 17.1 shows a cross-sectional slice through such a lattice. The larger disks represent the [positively charged] lattice sites that lie in the plane of the slice. The smaller disks represent lattice ions whose centres lie above and below the slice. There are a variety of ways in which the atoms may be arranged within the planes, and different manners in which the planes may be stacked.

[Figure 17.1 displays a face-centred cubic configuration.]

The tiny circles represent free electrons, each one bearing one fundamental unit of negative charge. While free electrons are not tied to any specific lattice site, mesoscopic volumes of the material are approximately neutral.

ASIDE: **Q:** What does it mean to be "approximately" neutral?

A1: The net charge in any region containing a large number of lattice sites, at a particular instant, is small compared to the number of sites.

A2: The time average of the net charge in any region containing a vast number of lattice sites vanishes.

Q: What scales are operative here?

A: There are several separate and overlapping scales.

? At today's limit of spatial resolution, $\lesssim 10^{-17}$ m, electrons are point-like.

fm Atomic nuclei have radii on the order of a few fm $[1\,\mathrm{fm} = 10^{-15}\ \mathrm{m}]$. The **femtometre** is dubbed the *fermi*, in honour of Enrico Fermi[1]

Å A diaphanous cloud[2] of evanescent [bound] electrons extends from the nucleus at the centre of each metallic ion to distances of 1 or 2 angstroms $[1\,\text{Å} = 10^{-10}\ \mathrm{m}]$, or so.

#Å The lattice sites are typically spaced several angstroms apart.

While the lattice ions occupying distinct sites are packed rather densely, the electrons are of such exceedingly small size that they move about the lattice in much the manner of the molecular constituents of a highly rarefied gas. Unlike gas molecules, however, the electrons interact strongly with one another, with the lattice ions, and with the coherent vibrational modes of the lattice [phonons]. Although the existence of these strong and long-range interactions makes it very difficult to determine specific dynamical behaviours, the rarefied gas model is a useful guide to our intuition.

Now that we've obtained a slightly better idea of the microscopic details of the lump of conductor pictured in Figure 17.1, let's consider the FLUX and INTENSITY of free electrons penetrating a plane surface passing through the bulk of the conducting material.

CLAIM: IF the conductor is in electrostatic equilibrium, THEN the net flux of electrons through any embedded surface must be zero.[3]

PROOF: Were this not the case, bulk material would spontaneously polarise. Since this behaviour is not observed,[4] one must conclude that the net flux vanishes.

ELECTROMOTIVE FORCE An electromotive force, [EMF], is, as its name suggests, a CAUSE which has the EFFECT of producing an **electric current**. For our present and near-future purposes, a battery is a source of EMF.

IF the neutral conductor is subjected to an EMF [perhaps this is accomplished by direct connection to a battery] THEN an electric field is induced within the bulk of the conductor! This, in turn, incites the free electrons to move coherently and thereby affords the possibility of non-zero charge flux through imaginary surfaces embedded within the conductor.[5]

---

[1] Fermi (1901–1954, 1938 Nobel Prize) made significant advances in theoretical *and* experimental nuclear, quantum, and statistical physics.

[2] The terms cloud and evanescent gain fuller meaning in quantum mechanics.

[3] The flux actually is expected to exhibit statistical fluctuations about zero which average to zero over long time scales.

[4] The lawyers insist that we qualify this by saying that spontaneous polarisation within the bulk is not observed to persist on long time scales.

[5] Zero flux is still obtained when the moving electrons flow alongside the imaginary surface without passing through it.

Q: Haven't we argued strenuously and convincingly that an external field applied to a conductor is quenched within its bulk?

A: Yep. These arguments are pertinent to situations of electrostatic equilibrium, and the presence of the EMF makes this a non-equilibrium situation.

We must make a necessary distinction between **equilibrium states** on the one hand, and **steady states** on the other. [Examples from hydrodynamics may prove *apropos.*]

Electrostatic Equilibrium   For a charged conductor in electrostatic equilibrium, the [time-averaged] charge distribution remains constant in time.

> In a fluid which is at rest, the pressure at a given point in space is constant in time. According to Pascal's Formula, this pressure supports the base of a thin column of fluid extending upwards from the point in space to the nearest fluid boundary.

Steady State   The state of a conductor bearing a constant electric current, *i.e.*, a steady flux of electric charge, is also not changing in time.

> The pressure is also constant at a particular point in space through which ideal fluid is moving in laminar, non-turbulent, fashion. In this case, the constancy of the pressure follows from Bernoulli's Equation, which contains a term dependent on the fluid-velocity.

The distinction between these two cases is important. In the former, equilibrium is established and maintained in the absence of interference from outside the system. In the latter, a steady state requires persevering effort on the part of an external agent [in the electric case, the EMF; in the hydrodynamic instance, some sort of pump] to maintain the time-independence of the observed physical state.

While all equilibrium states are steady states, not all steady states are equilibria.

EXAMPLE [*EMF in a Conducting Wire*]

A five metre long segment of uniform wire [conductor] is subjected to a 10 V EMF.

Q: What electric field is established within the wire?

A: The potential difference between the two ends of the wire owes its existence to a local electric field at points lying between the endpoints. The field is everywhere identified with "minus the gradient of the potential,"

$$E_x = -\frac{dV}{dx}.$$

The average magnitude of the electric field, $\langle |\vec{E}| \rangle$, throughout the length of wire, is $\frac{\Delta V}{\Delta x} = 2\,\mathrm{V/m}$. Since the wire is uniform and homogeneous, the local electric field does not differ from its average value, and hence $|\vec{E}| = 2\,\mathrm{V/m}$ everywhere within the wire.

> ASIDE: In common parlance, electric field magnitudes are quoted in volts per metre, rather than in newtons per coulomb: $\mathrm{V/m} = \frac{\mathrm{J/C}}{\mathrm{m}} = \frac{\mathrm{J/m}}{\mathrm{C}} = \mathrm{N/C}$.

While an applied EMF has little effect upon the lattice ions, as these are *frozen in place*, the "free" electrons *drift* in the direction opposite to that of the induced electric field.

[This net drift of charged particles produces an **electric current**.]

(AVERAGE) **ELECTRIC CURRENT**   The average electric current in a solid conductor is the average charge flux passing through an imaginary surface spanning the solid.

$$I_{av} = \frac{\Delta Q}{\Delta t},$$

where ΔQ is the total amount of charge passing through the imaginary surface throughout the time interval [of duration Δt] under consideration.

(INSTANTANEOUS) **ELECTRIC CURRENT**   The electric current in a conductor is the instantaneous charge flux through an imaginary surface spanning the conductor.

$$I = \lim_{\Delta t \to 0} I_{av} = \frac{dq}{dt}.$$

The SI unit of current, the ampere [A], is considered to be a fundamental physical unit, like the metre, the second, and the kilogram. Long ago [in Chapter 1], it was remarked that the coulomb is actually a derived unit:

$$1\,\mathrm{C} \equiv 1\,\mathrm{A}\cdot\mathrm{s}.$$

Electric current is defined to be the flux of POSITIVE electric charge, even though, in all but exotic instances, it is produced by the coherent drift of negatively charged electrons.

ASIDE: The convention for labelling POSITIVE and NEGATIVE electric charge was proposed by Ben Franklin and accepted by the broader scientific community (*circa* 1750). The realisation that typical currents result from the coherent flow of negatively charged microscopic constituents [see Chapter 28] did not occur until 1879. The notion of an electron [a quantum of electric charge] was bruited about in the nineteenth century as an aid to modelling the arrangement and chemical properties of elements in the periodic table. The electron was identified as a point-like particulate subatomic constituent of matter by J.J. Thomson in 1897. Thomson was awarded the Nobel Prize in 1906, primarily for this discovery.

Let's consider carefully how positive electric current can result from the coherent motion of particles bearing electric charge.

NEGATIVE   IF the particles each carry some amount of negative electric charge, THEN the electric current is in the direction opposite to that in which the electrons drift, as is illustrated in Figure 17.2.

POSITIVE   IF the charge carriers bear positive charge [*e.g.*, positrons, protons, nuclei, positive ions, *etc.*, or some combination of these], THEN the electric current flows in the same direction as that in which the charge carriers drift.

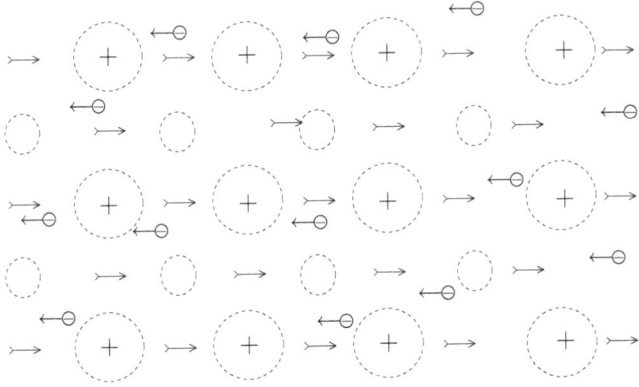

FIGURE 17.2 Electric Current Arising from the Drift of Negatively Charged Particles

While the intuitive analysis is well and good, let's now construct a rigorous and quantitative microscopic model of electric current cast in terms of the coherent drift of charge carriers under the influence of an applied EMF.

Consider a long, uniform wire with narrow cross-sectional area A made of some conducting material [*e.g.*, metal] that possesses a particular number density of charge carriers, η. The units of number density are

$$[\eta] = \frac{\#}{\text{m}^3} = \text{Volume}^{-1}.$$

Suppose that each of the charge carriers bears charge q, and drifts forward with speed v_d, on average. [Other cases are easily imagined, but shall not be considered here.]

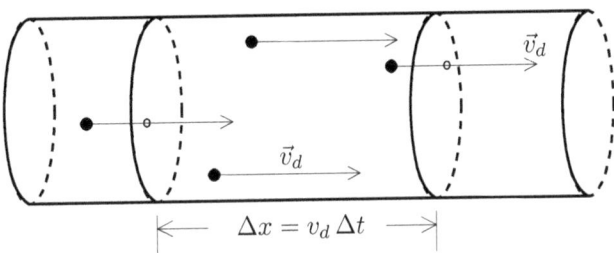

FIGURE 17.3 A Microscopic Charge Flux Model for Electric Current

Hydrodynamic flux arguments suggest that in the course of a short time interval Δt, one can expect, on average, that all of the charge carriers that are located at a distance $\Delta x = v_d \, \Delta t$ up the wire will have drifted past the shaded plane.

Q: How much charge flows through an [imaginary] plane surface lying perpendicular to and spanning the cylindrical wire with cross-sectional area A, throughout the time interval Δt?

A: The amount of charge expected to pass through the imaginary surface is

$$\Delta Q = q \, \Delta N \,,$$

where ΔN is equal to the number of charge carriers passing through the imaginary surface. This in turn will be equal to the number of drifting charge carriers in a cylindrical segment of the wire with volume $A \, \Delta x$.

To summarise, ΔN depends on the density of carriers available in the metal and the "upstream" volume drifting past the imaginary plane in time Δt.

$$\Delta N = \eta \, A \, \Delta x = \eta \, A \, v_d \, \Delta t .$$

Thus,

$$\Delta Q = q \, \eta \, A \, v_d \, \Delta t , \qquad \text{and hence} \qquad I_{\text{av}} = \frac{\Delta Q}{\Delta t} = q \, \eta \, v_d \, A .$$

In a steady state, the instantaneous current is constant and equal to its average value. Thus, the current in the wire, I, is proportional to q, η, A, and v_d.

Two comments finish off this chapter.

I   Foreshadowing Chapter 18, we observe that the electric current [charge flux], like all other fluxes, admits expression as (intensity) · (Area).

I   Complicating the picture somewhat, and calling into question our mental image of electrons flowing sedately up the wire in response to the applied EMF, is the fact that the drift velocity of the free electrons, v_d, is very much **smaller** than the expected thermal velocity, v_{thermal}, of the same electrons!

$$v_{\text{drift}} \ll v_{\text{thermal}}.$$

What saves the model is the **coherence** of the drift effect. The thermal effects are **random** and therefore [on average] cancel out completely.

Chapter 18

Electric Current Density, Ohm's Law, and Resistance

In the last chapter, a formula for the electric current in a wire,

$$I = q \eta v_d A,$$

was derived, and much ado was made about its interpretation as **charge flux**. In the above expression, $q \eta$ is the free electric charge density, v_d is the drift speed [assumed to be uniform], and A is the cross-sectional area. Strictly speaking, a proper expression of flux accounts for the vector nature of two of the above quantities, and so a more general formula is

$$I = q \eta \vec{v}_d \cdot \vec{A}.$$

This expression for the current suffices for uniform drift velocities and planar surfaces. IF the drift velocity field is not uniform OR the [imaginary] area is not planar, THEN an integral formulation is required.

ASIDE: *Déjà vu!* The same concerns arose when we considered electric fluxes in Chapter 5.

Instead of pursuing these generalisations now, let's define the intensity counterpart to the electric flux.

ELECTRIC CURRENT DENSITY   The electric current density, J, is the electric current per unit cross-sectional area of the imaginary surface through which the charge flux passes. The dimensions of the electric current density are

$$[J] = \frac{\text{current}}{\text{unit area}} = \frac{\text{charge}}{(\text{area} \cdot \text{time})},$$

so its SI units are $[J] = \frac{A}{m^2} =$ amperes per square metre.

In the simplest case [that of a uniform straight wire], $J = I/A = q \eta v_d$. More generally, $\vec{J} = q \eta \vec{v}_d$ accommodates those situations in which the drift velocity is not uniform.

• The vector current density everywhere points in the direction of the local flow of positive charge.

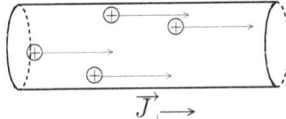   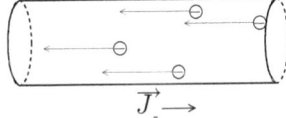

FIGURE 18.1   Current Density for Positive and Negative Charge Carriers

• The CHARGE FLUX is consistently obtained as the integral of the INTENSITY over the area of the imaginary surface: $I = \int \vec{J} \cdot d\vec{A}$.

• IF the current density is uniform AND the area planar, THEN the integration of the current density over the surface becomes $I = \vec{J} \cdot \vec{A}$.

A Prelude to Resistance

As was alluded to in Chapter 17, the charge carriers [usually electrons] do not flow in an unimpeded manner through a normal wire. In fact, they scatter from [*i.e.*, collide with] lattice defects, impurities, or even the lattice ions themselves.

[By analogy, think of the flow of water around boulders or islands in a stream.]

Furthermore, the motions of the conduction electrons are also affected by coherent vibrations of the entire lattice [*a.k.a.* phonons].

Ohm's Law[1]

By construction, the current density is proportional to the drift velocity, $\vec{J} = \eta\, q\, \vec{v}_d$, and the drift velocity arises in dynamical response to the electric field induced throughout the conducting material. These facts inspire the claim that, for a large class of materials, the electric current density is [linearly] proportional to the applied electric field, *i.e.*,

$$\vec{J} = \sigma\, \vec{E},$$

where the proportionality constant, σ, is called the [electrical] **conductivity**. Several comments follow immediately below.

- In a uniform [homogeneous] and straight wire, one expects the electric field and the steady state electric current density to be uniform vector fields.

- The conductivity, σ, is a phenomenological property characteristic of the particular material comprising the wire. We'll defer discussion of the units of σ for the time being.

- The relation $\vec{J} = \sigma\, \vec{E}$ is called **Ohm's Law**.

Materials for which Ohm's Law is operative are called **Ohmic**, whereas those for which Ohm's Law does not hold are termed **non-Ohmic**.

Let's apply Ohm's Law to a straight segment of uniform wire with length L and cross-sectional area $\mathcal{A}$.

[As an added bonus, we shall reveal the "$V = I\,R$" version of Ohm's Law.]

Impose an EMF on the wire segment, in such a manner that a constant potential difference, $V = V_b - V_a$, is maintained between its farthest ends. The presence of the [constant] EMF requires that there be a steady state electric field throughout the length of the wire. The electric field promotes a coherent drift of charge carriers within the wire, which becomes manifest as the electric current.

In Figure 18.2, $V_b > V_a$, and so the electric field points to the right. The positive current flows to the right also. The uniformity and homogeneity of the wire ensures that the electric

[1]Georg Ohm (1789–1854) was a German physicist and mathematician.

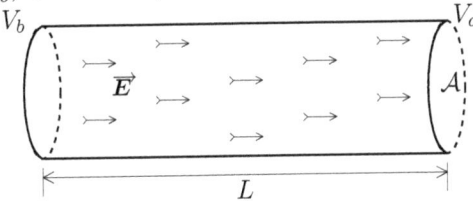

FIGURE 18.2 A Straight Segment of Homogeneous Conductor, a Potential Difference, and an Electric Field

field is constant through the bulk of the conducting material comprising the wire. From the potential difference then, the [average] strength of the electric field may be inferred:

$$V = V_b - V_a = -\int_{x_a}^{x_b} \vec{E} \cdot d\vec{s} = -\int_{x_a}^{x_b} E_0 \cos(\pi)\, ds = E_0 \int_{x_a}^{x_b} ds = E_0 \left(x_b - x_a \right) = E_0\, L \,,$$

and therefore $E_0 = V/L$.

> ASIDE: We occasionally remind ourselves that the free electrons which constitute the current in the wire are coherently drifting in the direction opposite to the phenomenological current.

Considering magnitudes only, on account of the essentially one-dimensional character of the wire, one realises that $J = \sigma\, E_0 = \sigma\,(V/L)$ and $I = J\,\mathcal{A}$ together imply that

$$I = \sigma\, V \frac{\mathcal{A}}{L} \,, \qquad \text{and hence} \qquad V = \frac{L}{\sigma\,\mathcal{A}}\, I \,.$$

By defining **electric resistance**, R, in the manner described next, the above equation simplifies to yield the familiar version of Ohm's Law.

RESISTANCE   A block of homogeneous material with uniform geometry has electrical resistance R, determined by the electrical conductivity of the substance, σ, and the length, L, and cross-sectional area, A, of the block.

$$R = \frac{L}{\sigma\,\mathcal{A}} \,.$$

Electrical resistance is the parameter relating the current response of the block to an applied electric potential difference.

The SI unit of resistance is the **ohm** $[\Omega]$, where

$$1\,\Omega = \frac{\text{V}}{\text{A}} \,.$$

That is, the resistance is inferred to be $1\,\Omega$ when an applied potential difference of one volt causes one ampere of current to flow.

RESISTIVITY   The converse[2] of conductivity, σ, is **resistivity**, ρ.

$$\rho = \frac{1}{\sigma} \,.$$

The SI units of resistivity are $[\rho] = \Omega \cdot \text{m}$. Conductivity is whimsically measured in terms of "mhos," $[\mho]$, where $1\,\mho = 1\,(\Omega \cdot \text{m})^{-1}$.

---

[2]In fact, they are reciprocals.

The resistance of a particular piece of material can be expressed in terms of the resistivity characteristic of the substance of which it is comprised, along with geometric factors describing its shape. In cases where the shape has uniform cross-section,

$$R = \rho \frac{L}{\mathcal{A}},$$

where L is the length of the wire and $\mathcal{A}$ is its cross-sectional area.

⇑   Resistance increases as the length increases.

⇓   Resistance decreases as the cross-sectional area increases.

OHM'S LAW   The general expression of Ohm's Law (quoted earlier) is $\vec{J} = \sigma \vec{E}$, where $\vec{J}$ is the local electric current density, σ is the conductivity, and $\vec{E}$ is the local electric field. Nevertheless, the analysis of the straight segment of uniform wire yields: $V = IR$, which is the oft-quoted **familiar form of Ohm's Law!**

EXAMPLE [*Resistivity Scales*]

Copper is a good conductor and rubber is a good insulator. At $20\,c$ the respective resistivities[3] of copper and rubber are approximately

$$\rho_{\mathrm{Cu}} = 1.7 \times 10^{-8}\ \Omega \cdot \mathrm{m} \qquad \text{and} \qquad \rho_{\mathrm{rubber}} \simeq 10^{10}\ \Omega \cdot \mathrm{m}.$$

WOW! These materials have 18 orders of magnitude difference in resistivity! This enormous range enables fine technological control of flows of electricity.

For an Ohmic material, a plot of current response *vs.* applied EMF yields a straight line, as illustrated on the left in Figure 18.3. A **diode**, a particular type of non-Ohmic device, has the current response to applied voltage curve presented on the right.

FIGURE 18.3   Current Response to Applied EMF for Ohmic and Non-Ohmic Materials

Q: Why frame this as the current response to an applied voltage?

A: It so happens that voltage sources are much easier to produce and to calibrate than are current sources, and so, experimentally, voltage better serves as the independent variable.

---

[3]The resistivity of a material is temperature dependent. In Chapter 19, a simple [linear] effective model for the temperature dependence of resistivity shall be presented.

Chapter 19

Resistance Is Not Futile

Over limited ranges of temperature, resistivities are very well modelled by a linear approximation,

$$\rho(T) = \rho_0 \left(1 + \alpha \left(T - T_0 \right) \right).$$

An alternate expression is obtained by writing

$$\Delta \rho = \alpha \, \rho_0 \, \Delta T.$$

The various terms in the above formulae are explained below.

T_0 The reference temperature, T_0, is the temperature about which the linear expansion is performed. [It is conventional to fix $T_0 = 20\,\mathrm{C}$.]

ρ_0 The base resistivity, $\rho_0 = \rho(T_0)$, is the value of the resistivity at T_0.

α The [LINEAR] **temperature coefficient of resistivity**, α, determines the change in resistivity effected by a given change in ambient temperature. It has units $[\alpha] = [\mathrm{C}^{-1}]$.

ASIDE: There is nothing particularly mysterious or fundamental about this linear approximation (except perhaps that it is efficacious). Rather, it is none other than the linear approximation to a differentiable function arising from a first-order Taylor expansion in the neighbourhood of a particular point.

[Some refer to this as the "tangent line approximation."]

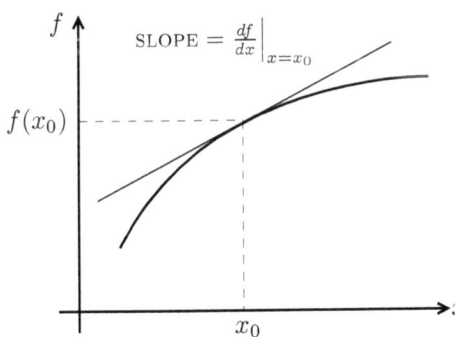

In a neighbourhood of the point x_0, the Taylor expansion of $f(x)$ yields

$$f(x_0 + \delta) \simeq f(x_0) + \left. \frac{df}{dx} \right|_{x=x_0} \delta$$
$$+ \frac{1}{2} \left. \frac{d^2 f}{dx^2} \right|_{x=x_0} \delta^2 + \ldots$$
$$\sim f(x_0) + \left. \frac{df}{dx} \right|_{x=x_0} \delta$$

to first-order.

FIGURE 19.1 Linear Approximation to a Function in a Neighbourhood of x_0

The [linear] temperature coefficient of resistivity, α, may be POSITIVE, as one might expect, or NEGATIVE, which is rather a surprise.[1] The behaviour at low temperatures is widely divergent in these two cases.

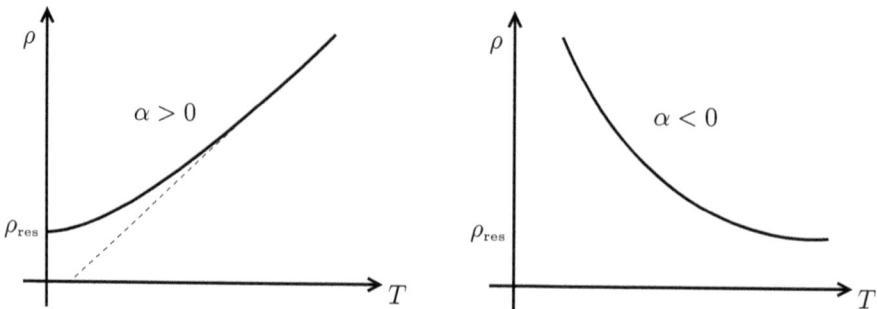

FIGURE 19.2   Positive- and Negative-Alpha Materials

The precise shapes of the resistivity curves shown in Figure 19.2 are material-dependent and dictated by the degree to which the free electrons scatter with the lattice, phonons, and each other. In the $\alpha > 0$ case, naive extrapolation [indicated by the dashed line] would suggest that the resistivity declines to zero at some value of temperature. That this does not occur[2] is apparent from the convergence of the low temperature part of the curve to a residual resistivity, ρ_{res}. In the $\alpha < 0$ case, the resistivity tends toward a positive residual value as the temperature of the material increases. In either case, the residual resistivity prevents one from rendering resistances arbitrarily small, $R \to 0$, by cooling or heating.

Q: Why should one expect higher temperatures to lead to increased resistivity?

A: Higher temperature means greater thermal agitation of lattice ions and free (conduction) electrons. Thus, we may expect greater disruption of electric charge flux.

[Imagine the stream of charge flowing amidst boulders which are bouncing about.]

Q: How might the resistivity increase as the temperature decreases?

A: Perhaps one could employ the analogy of a frozen stream, in which the cold causes the flow of fluid to [nearly] cease. At higher temperatures, the stream begins to melt somewhat and flow occurs more readily.

More curiouser than $\alpha < 0$ is the remarkable property of **superconductivity**. For certain $\alpha > 0$ materials, there exists a substance-specific CRITICAL TEMPERATURE, T_C. At temperatures below T_C, the resistivity of the material is EXACTLY ZERO. A consequence of zero resistivity is that current can flow even when no voltage is applied. These so-called persistent currents have been experimentally observed!

What is now regarded as low-temperature superconductivity [$T_C < 25\,\text{K}$] was serendipitously discovered in 1911 by Heike Kamerlingh-Onnes (1853 - 1926), who led what was then the world's premier laboratory for low-temperature physics at the University of Leiden.

[Kamerlingh-Onnes was awarded the 1913 Nobel Prize for his sensational discoveries.]

---

[1] Materials with $\alpha < 0$ are typically of a class called **semiconductors**. Silicon and germanium are the most well-known examples of semiconducting materials.

[2] "For most materials," the lawyers insist that we hasten to add.

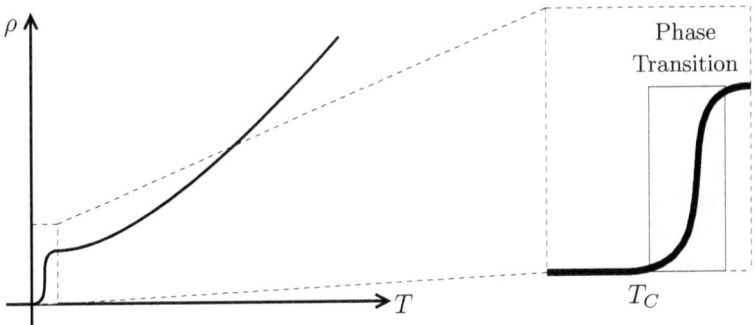

FIGURE 19.3 The Superconducting Phase Transition

ASIDE: Although it is unusual for this prize to follow so closely upon the work for which it is awarded, two factors made his selection compelling. First, the liquification of helium, making these low-temperature investigations possible, was an experimental *tour de force*. Second, superconductivity defies explanation in terms of classical physics, and thus was a profoundly important discovery.

No convincing explanation of superconductivity was advanced until the Bardeen-Cooper-Schrieffer [BCS] Theory was proposed in the late 1950s. In the BCS model, superconductivity arises from a quantum mechanical effect involving pairing of electrons.

[Bardeen, Cooper, and Schrieffer shared the 1972 Nobel Prize.]

In 1986, certain ceramic metal oxides were found to exhibit High-T_C superconductivity.

[J.G. Bednorz and K.A. Müller were awarded the 1987 Nobel Prize for this discovery.]

The critical temperatures of these materials are on the order of 100 K.[3] Despite there having been many tens of thousands of scientific papers published on the subject of High-T_C since 1986, the underlying mechanism is still unknown. Three additional comments ought to be made.

o   These critical temperatures are much warmer than those associated with low-temperature superconductivity.

o   Nitrogen gas, N_2, liquifies at $T \sim 77$ K [under 1 atm pressure]. There are well-developed industrial-scale cooling methods for attaining these temperatures.

o   One eagerly awaits the discovery and engineering of room-temperature super-conducting materials. This breakthrough will earn some lucky bloke(s) a Nobel Prize, and will generate oodles of money for those who commercialise the resulting technology.

---

[3] Even by Canadian standards, this is rather cold!

THE ENERGETICS OF RESISTANCE

In microscopic terms, resistance arises when the flow of electrons[4] is impeded by collisions or interactions with the metallic ions in the lattice [along with crystal defects, phonons, *etc.*]. When the electrons scatter, they transfer energy and momentum to the lattice, and this energy eventually manifests itself as HEAT.

> ASIDE: Friction acts dissipatively on macroscopic mechanical systems. For example, when a block slides down a frictional inclined plane, some of its gravitational potential energy is converted into heat. In a similar manner, electrical potential energy is converted into thermal energy when current passes through a resistor. This phenomenon is termed **Joule heating**.

Recall that the time rate at which energy is transferred or exchanged is the **power**. To facilitate our investigation of electric power into and out of a resistor, we construct the simplest possible electric circuit, as illustrated in the following [schematic] **circuit diagram**.

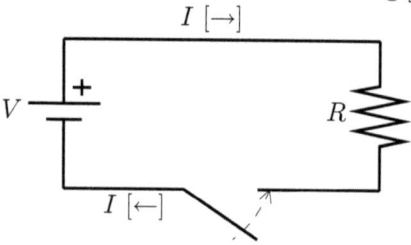

FIGURE 19.4   A Simple Electric Circuit

Although we have already employed schematic circuit diagrams in the discussion of capacitor networks in Chapter 13, we shall here formalise the [idealised] interpretation of each of the symbols appearing in Figure 19.4.

BATTERY

$V \overset{\mathbf{+}}{\underset{}{\rule{1cm}{0.4pt}}}$	An **ideal battery** is a device which maintains a constant potential difference between two points in space, represented by the terminals in the symbol.
	The positive [+] side is at potential V_+, while the negative [−] side is at V_-. The positive side is at higher [relative] potential $$ V_+ - V_- = V > 0 \,. $$
	Often, $V_- = 0$ is chosen for simplicity.

WIRE

$\rule{2cm}{0.4pt}$ $I\ [\rightarrow]$ Direct(ed) Current	The **ideal wires** joining circuit elements have zero resistance. This is a valid approximation in most circumstances.
	[If this approximation is NOT valid, then one models the physical wire as two ideal wires joined by a resistor of the appropriate magnitude.]
	The arrow accompanying the I indicates the direction or sense in which current in the wire is deemed to flow.

---

[4]In virtually all cases of interest, the fundamental charge carriers are electrons.

RESISTOR

R

An **ideal resistor** is an electrical device [*a.k.a.* component] possessing a particular resistance, R. Physical resistors may be fabricated from lengths of thin wire or blocks of insulating material.

SWITCH

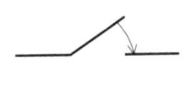

An **ideal switch** is thrown at a specified time so as to establish an electric circuit and thus to allow the flow of current. The switch itself has no resistance.

A charge element, ΔQ, flowing as part of the current in the simplest electric circuit, passes through the battery from low- to high-potential. As a consequence, the electrostatic potential energy of the charge element increases by the amount

$$\Delta U = \Delta Q\,V\,.$$

This increase in electrical energy of the charge element is brought about by the conversion of chemical energy stored in the battery. As the charge element moves through the ideal wire joining the high-potential terminus of the battery to the resistor, it encounters no resistance, and therefore no electrical energy is dissipated (in the wire). **En route to the resistor, each charge element retains all of the electrostatic potential energy that it gained while traversing the battery.** An analogous statement holds for the return of the charge elements to the low-potential end of the battery. The PRINCIPLE OF ENERGY CONSERVATION then requires that:

> **As charge elements flow through the resistor, each must lose exactly the amount of potential energy, ΔU, that it gained in traversing the battery.**

ASIDE: If this were not the case, then the potential difference, $V_+ - V_-$, would not be sustained, contradicting the assumption that the battery is ideal, or the energy conservation principle would be violated.

The energy deposited into the resistor is dissipated as HEAT.

ASIDE: In thermodynamics, the **quantity of heat** is usually denoted by Q [or ΔQ]. Here we shall call the quantity of heat produced in the resistor "ΔH," since Q is used to represent an amount of electric charge.

Thus,

$$\Delta H = \Delta U = \Delta Q\,V\,.$$

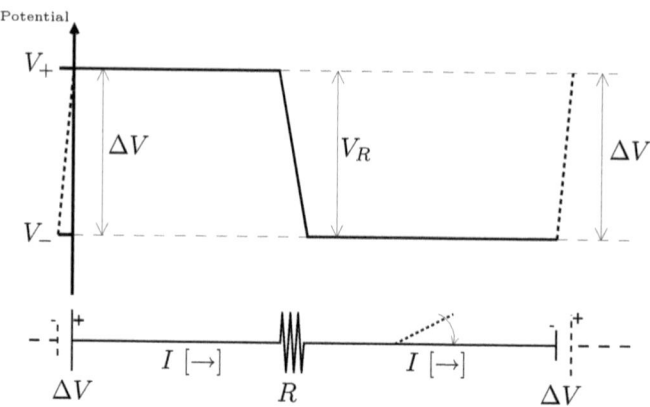

FIGURE 19.5   Potential Energy Diagram for the Simplest Circuit

The energetic aspect may be somewhat obfuscated by the closedness of the circuit. In the energy diagram presented in Figure 19.5, we cleave this *Gordian Knot* by formally displacing the high- and low-potential terminals of the battery to the left and right extremities of a finite line. This particular scheme for representing the potential ensures that there is a net downward energy slope in the central portion of the figure. Within the battery [considered to reside at one end or the other of the figure], the potential rises steeply. More precisely, the potential is constant within contiguous regions of ideal wire, punctuated by abrupt changes within the [isolated and compact] battery and resistor.

The circuit structure in the model is retained in the figure through the imposition of PE-RIODIC BOUNDARY CONDITIONS. Following the flow of current in the diagram, one returns to the left edge once one reaches the right. The energetics of the circuit are independent of the details of the scheme employed to illustrate them. Starting from any other point in the circuit amounts to a finite shift along the horizontal axis shown in Figure 19.5 and does not affect the periodicity. The inescapable conclusion is that, for this simple circuit, it is the battery's potential difference, $\Delta V = V_+ - V_-$, mirrored across the resistor, that provides the current-driving EMF, $V_R = I\,R$.

The [instantaneous] power produced in the resistor is the rate at which heat is generated:

$$P = \lim_{\Delta t \to 0} \frac{\Delta H}{\Delta t} = \lim_{\Delta t \to 0} \frac{\Delta Q}{\Delta t} V = I\,V\,,$$

where the instantaneous current, I, is the $\Delta t \to 0$ limit of $\frac{\Delta Q}{\Delta t}$. This result for the power produced in the resistor readily admits three modes of expression,

$$P = I\,V = I^2\,R = \frac{V^2}{R}\,,$$

where application of Ohm's Law, $V = I\,R$, is sufficient to convert any one of these forms into another. The unit for power is the watt [equal to one joule per second, *i.e.*, $1\,\text{W} = 1\,\text{J/s}$]. Here, it is also evident that

$$\left[\text{W}\right] = \left[\text{A} \cdot \text{V}\right] = \text{ampere} \cdot \text{volt}\,.$$

Chapter 20

Resistors in Series and Parallel

In Chapter 18, we computed the resistance of a uniform segment of conductor [relating its current response to an applied EMF]. It was also shown that the resistance depends neither on the magnitude or orientation of the applied EMF, nor on the current response. Instead, it is determined by the resistivity of the material and the geometry of the conductor. In this chapter, we shall study how resistances combine.

In light of our previous work with combinations of capacitors, we are inspired to ask:

Q: What arrangements of two resistors are possible?

A: Two resistors may be combined in **series** or in **parallel**. We'll describe in detail these, and their generalisations to $\mathcal{N}$ resistors, in the forthcoming suite of examples.

> **Our analyses will consist in the imposition of constraints arising from the conservation of charge and energy.**

In all situations, Ohm's Law, $V = IR$, relating the current THROUGH a resistive electric component to the potential difference ACROSS the same component, is assumed to be operative. Additionally, we shall assume that the connections between resistive elements are effected ideally, *viz.*, without contributing to the overall resistance of the system. This latter assumption is easily relaxed once we generalise the two-resistor results.

EXAMPLE [*Two Resistors in Series*]

Two resistors, R_1 and R_2, connected in series, are shown in Figure 20.1.

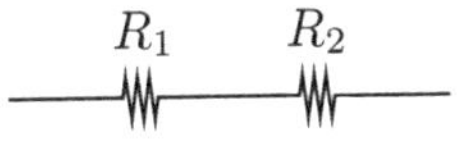

FIGURE 20.1   Two Resistors Joined in Series

Q: What is the effective resistance of the series system of two resistors?

A: Let's assume that the pair of resistors is subjected to an applied EMF, V, [and that a sufficient time interval has elapsed for the current to have reached its steady state value].

Conservation of electric energy and of electric charge imposes SUM and SAME constraints on the potential difference across and the current through each of the constituent resistors.

SUM   The total potential difference is precisely equal to the sum of the potential differences across the two resistors,
$$V = V_1 + V_2 .$$

A pictorial representation[1] of this relation appears in Figure 20.2.

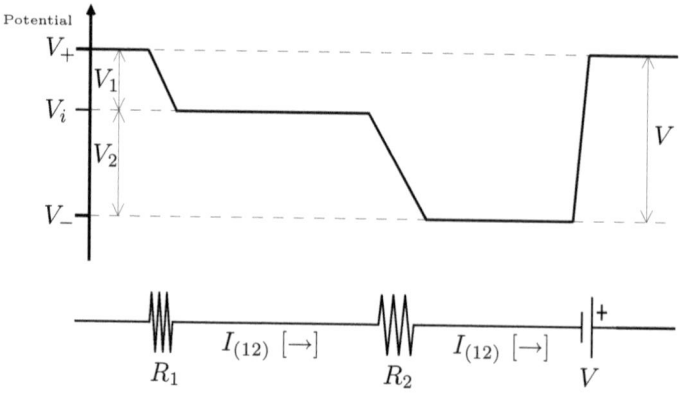

FIGURE 20.2   The Potential Throughout the Battery–Two-Resistor Circuit

SAME   The same current flows through each of the resistors,

$$I_1 = I_2 = I_{(12)} ,$$

and so we speak of the current, $I_{(12)}$.

The constant of proportionality relating the potential difference across the pair to the unique current through them is the effective resistance, $R_{(12)}$.

$$V = I_{(12)} \, R_{(12)} .$$

Also,

$$V_1 = I_1 \, R_1 = I_{(12)} \, R_1 \qquad \text{and} \qquad V_2 = I_2 \, R_2 = I_{(12)} \, R_2 .$$

The energy constraint, $V = V_1 + V_2$, implies that

$$I_{(12)} \, R_{(12)} = I_{(12)} \, R_1 + I_{(12)} \, R_2 = I_{(12)} \left(R_1 + R_2 \right) ,$$

and thus

$$R_{(12)} = R_1 + R_2 .$$

**The effective resistance of two resistors in series is
the sum of the individual resistances.**

. . . .

EXAMPLE [*Two Resistors in Parallel*]

Two resistors, R_1 and R_2, are connected in parallel as pictured in Figure 20.3.

Q: What is the effective resistance of this combination?

A: Let's assume that the pair of resistors is connected to a battery with EMF V, and that sufficient time has elapsed for the total current to have reached its steady state value, $I_{[12]}$. Conservation of charge and of energy imposes SUM and SAME constraints on the current through and the potential difference across each of the constituent resistors.

---

[1]This is a natural and intuitive construction, since electric potential [potential energy per unit charge] is necessarily position dependent.

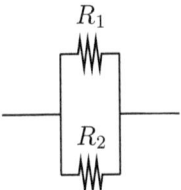

FIGURE 20.3 Two Resistors in Parallel

SUM There is an electric current passing through each of the resistors. The total electric current through the parallel pair of resistors is the sum of these currents,

$$I_{[12]} = I_1 + I_2 .$$

SAME The same potential difference,

$$V = V_1 = V_2 = V_{[12]} ,$$

is applied across each of the resistors.

The effective resistance, $R_{[12]}$, relates the potential difference across the parallel combination to the total current through the system.

$$V_{[12]} = I_{[12]} R_{[12]} \quad \Longrightarrow \quad I_{[12]} = \frac{V_{[12]}}{R_{[12]}} .$$

Also,

$$V_{[12]} = V_1 = I_1 R_1 \quad \Longrightarrow \quad I_1 = \frac{V_{[12]}}{R_1} , \quad \text{and} \quad V_{[12]} = V_2 = I_2 R_2 \quad \Longrightarrow \quad I_2 = \frac{V_{[12]}}{R_2} .$$

The current conservation constraint, $I_{[12]} = I_1 + I_2$, implies that

$$\frac{V_{[12]}}{R_{[12]}} = \frac{V_{[12]}}{R_1} + \frac{V_{[12]}}{R_2} = V_{[12]} \left(\frac{1}{R_1} + \frac{1}{R_2} \right) ,$$

and thus

$$\frac{1}{R_{[12]}} = \frac{1}{R_1} + \frac{1}{R_2} .$$

The reciprocal of the effective resistance of two resistors in parallel is the sum of the reciprocals of the individual resistances.

Having completed the analyses of both manners in which two resistors can be combined, let's undertake the generalisations to strictly series and purely parallel combinations of $\mathcal{N}$ resistors. In the Addendum, we'll tackle a particularly thorny **mixed** case.

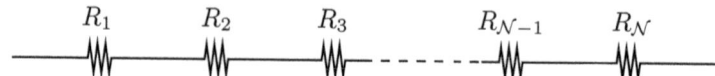

FIGURE 20.4 $\mathcal{N}$ Resistors in Series

EXAMPLE [*Series Arrangement of $\mathcal{N}$ Resistors*]

Suppose that $\mathcal{N}$ resistors, $\{R_n\, ,\ n \in 1, 2, \ldots, \mathcal{N}\}$, are connected strictly in series, as illustrated in Figure 20.4.

Q: What is the effective resistance of $\mathcal{N}$ resistors in series?

A: For each resistor, $n \in \{1, 2, \ldots, \mathcal{N}\}$, Ohm's Law applies locally, *i.e.*, $V_n = I_n R_n$. Globally, the potential differences across all of the resistors sum to yield the overall EMF,

$$V_{(12\ldots\mathcal{N})} = \sum_{n=1}^{\mathcal{N}} V_n \, ,$$

and the same current, $I_n = I_{(12\ldots\mathcal{N})}$, flows through each.

For the system as a whole, $V_{(12\ldots\mathcal{N})} = I_{(12\ldots\mathcal{N})}\, R_{(12\ldots\mathcal{N})}$, and

$$V_{(12\ldots\mathcal{N})} = \sum_{n=1}^{\mathcal{N}} I_n R_n = I_{(12\ldots\mathcal{N})} \sum_{n=1}^{\mathcal{N}} R_n \, .$$

Therefore,

$$R_{(12\ldots\mathcal{N})} = \sum_{n=1}^{\mathcal{N}} R_n \, .$$

The effective resistance of $\mathcal{N}$ resistors arranged in series is the sum of their respective resistances.

Four comments pertinent to this result follow below.

$R_{(\ldots)} > R_n$ The effective resistance is greater than any constituent resistance.

$R_n = R_0$ IF all $\mathcal{N}$ resistors are identical, THEN $R_{(12\ldots\mathcal{N})} = \mathcal{N}\, R_0$.

SCALE   The largest constituent resistors set the scale for the effective resistance.

BLOW UP   IF the nth resistor should fail [colloquially, blow up], $R_n \to \infty$, THEN the effective resistance also diverges to infinity and the current through the network ceases.

> ASIDE: When electric Christmas lights were first introduced, they were wired in series, and the entire string would go dark whenever a single bulb failed. Determining which of the bulbs was defunct was a difficult and tedious task.

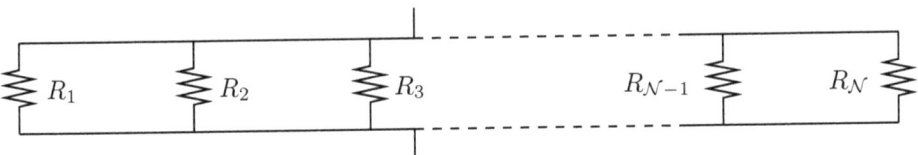

FIGURE 20.5 $\mathcal{N}$ Resistors in Parallel

EXAMPLE [*Parallel Arrangement of $\mathcal{N}$ Resistors*]

Suppose that $\mathcal{N}$ resistors, $\{R_n, \ n \in 1, 2, \ldots, \mathcal{N}\}$, are connected perfectly in parallel, as illustrated in Figure 20.5.

Q: What is the effective resistance of $\mathcal{N}$ resistors in parallel?

A: For each resistor, $n \in \{1, 2, \ldots, \mathcal{N}\}$, Ohm's Law, $V_n = I_n R_n$, is locally in force. Globally, the potential differences across all of the resistors are the same, $V_n = V_{[12\ldots\mathcal{N}]}$, while the total current is the sum of the currents passing through the resistors,

$$I_{[12\ldots\mathcal{N}]} = \sum_{n=1}^{\mathcal{N}} I_n \,.$$

The current through the nth resistor is

$$I_n = \frac{V_n}{R_n} = \frac{V_{[12\ldots\mathcal{N}]}}{R_n} \,.$$

The total current satisfies both

$$I_{[12\ldots\mathcal{N}]} = \frac{V_{[12\ldots\mathcal{N}]}}{R_{[12\ldots\mathcal{N}]}} \qquad \text{and} \qquad I_{[12\ldots\mathcal{N}]} = \sum_{n=1}^{\mathcal{N}} \frac{V_{[12\ldots\mathcal{N}]}}{R_n} = V_{[12\ldots\mathcal{N}]} \sum_{n=1}^{\mathcal{N}} \frac{1}{R_n} \,.$$

Thus, for $\mathcal{N}$ resistors arranged in parallel,

$$\frac{1}{R_{[12\ldots\mathcal{N}]}} = \sum_{n=1}^{\mathcal{N}} \frac{1}{R_n} \,.$$

The reciprocal of the effective resistance of $\mathcal{N}$ resistors purely in parallel is the sum of the reciprocals of the individual resistors.

Four comments pertinent to this result follow.

$R_{[\ldots]} < R_n$ The effective resistance of the system is less than any constituent resistance.

$R_n = R_0$ IF all $\mathcal{N}$ resistors are identical, THEN $R_{[12\ldots\mathcal{N}]} = \frac{R_0}{\mathcal{N}}$.

SCALE   The smallest constituent resistors set the scale for the effective resistance.

BLOW UP   If the nth resistor blows up, then the effective resistance does increase somewhat, but an electric current continues to flow through the network.

> ASIDE: The second, and subsequent, generations of Christmas lights have employed parallel wiring. Should an individual bulb fail, only it goes dark. The remaining lights continue to spread their cheerful glow.

ADDENDUM to Chapter 20

Consider the [daunting] network of 15 resistors, labelled using **hexadecimal** [base-16] notation, $\{R_i, i \in 1, 2, \ldots, 9, A, \ldots, E, F\}$, illustrated in the figure below. We shall compute the effective resistance of the network by iterative application of the combinatorial rules for purely parallel and series arrangements. Two considerations merit notice.

- IF the network is attached to an overall source of voltage, THEN we assume that the connections were made sufficiently far in the past that [approximate] steady state current flow has been attained.

- The thin wires joining the resistors may be stretched, shrunk, bent, or straightened without appreciably changing the overall resistance of the network.

For the sake of definiteness, we shall suppose that each of the odd-numbered resistors has resistance R, while each even-numbered resistor has resistance $2R$.

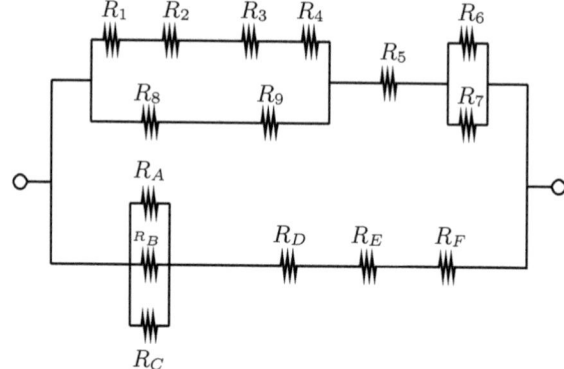

Looking at the smallest scales, one may identify three separate collections of resistors which are arranged strictly in series, and two which are purely in parallel. Effective resistances for these five subnetworks may be straightforwardly obtained.

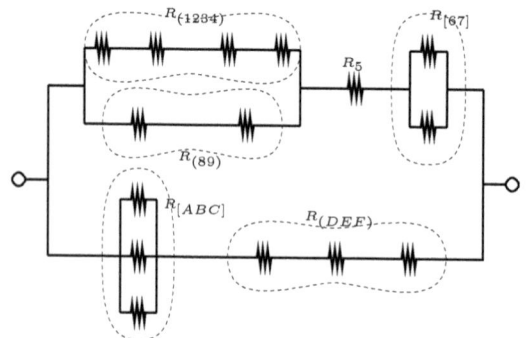

$$R_{(1234)} = R_1 + R_2 + R_3 + R_4$$

$$R_{[67]} = \left(\frac{1}{R_6} + \frac{1}{R_7}\right)^{-1}$$

$$R_{(89)} = R_8 + R_9$$

$$R_{[ABC]} = \left(\frac{1}{R_A} + \frac{1}{R_B} + \frac{1}{R_C}\right)^{-1}$$

$$R_{(DEF)} = R_D + R_E + R_F$$

In this particular instance, $R_{\text{odd}} = R$ and $R_{\text{even}} = 2R$, these combinations work out to be:

$$R_{(1234)} = 6R\,, \qquad R_{(89)} = 3R\,, \qquad R_{(DEF)} = 4R\,, \qquad R_{[67]} = \frac{2R}{3}\,, \qquad R_{[ABC]} = \frac{R}{2}\,.$$

Replacing these smallest-scale subnetworks with equivalent effective resistances reveals another set of higher-level parallel and series combinations.

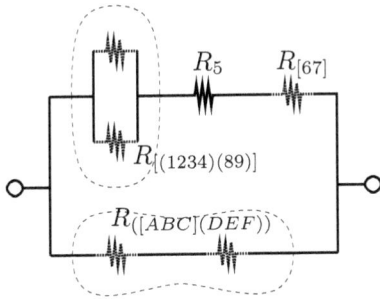

The aggregrations displayed in the figure above have effective resistances

$$R_{[(1234)(89)]} = \left(\frac{1}{R_{(1234)}} + \frac{1}{R_{(89)}}\right)^{-1},$$

$$R_{([ABC](DEF))} = R_{[ABC]} + R_{(DEF)} = R_{[ABC]} + R_D + R_E + R_F.$$

When the odd resistors are R and the evens are $2R$, these turn out to be

$$R_{[(1234)(89)]} = 2R \quad \text{and} \quad R_{([ABC](DEF))} = \frac{9C}{2}.$$

With this further aggregation additional higher-level structure, amenable to analysis, is revealed.

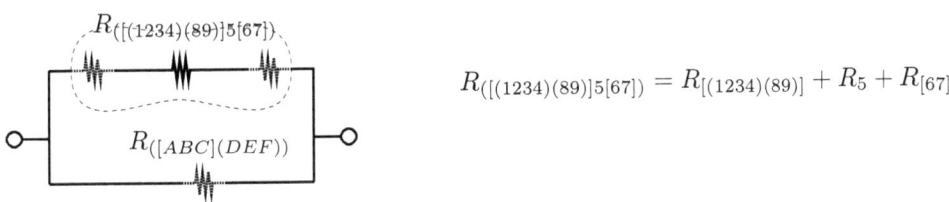

$$R_{([(1234)(89)]5[67])} = R_{[(1234)(89)]} + R_5 + R_{[67]}$$

The effective network is finally seen to be a parallel arrangement of $R_{([(1234)(89)]5[67])}$ and $R_{([ABC](DEF))}$, as illustrated in the penultimate schematic diagram below.

$$R_{\text{Eff}} = R_{[([(1234)(89)]5[67])([ABC](DEF))]}$$

$$= \left(\frac{1}{R_{([(1234)(89)]5[67])}} + \frac{1}{R_{([ABC](DEF))}}\right)^{-1}$$

In the particular instance under consideration,

$$R_{([(1234)(89)]5[67])} = 2R + R + \frac{2}{3}R = \frac{11}{3}R,$$

and hence

$$R_{\text{Eff}} = R_{[([(1234)(89)]5[67])([ABC](DEF))]} = \left(\frac{3}{11R} + \frac{2}{9R}\right)^{-1} = \left(\frac{49}{99R}\right)^{-1} = \frac{99}{49}R.$$

Chapter 21

DC Circuits Mélange

In this chapter, we shall discuss diverse aspects of **direct current** [DC] electric circuits. First, we consider how the connectedness of the components of a network dictates its overall properties. Earlier chapters strongly suggested that this was the case for networks comprised entirely of capacitors [Chapter 13] and resistors [Chapter 20]. Second, we develop a more realistic model of a battery, accounting for the observation that its supply voltage tends to decrease as its current drain increases. Third, we expose a serendipitous offshoot of this development, the notion of IMPEDANCE MATCHING, when we employ the improved battery in a simple resistive circuit. Finally, we quote **Kirchoff's Rules**. These provide convenient expression of the energy and charge constraints in a form best suited for the determination of current responses.

CIRCUITS AND TOPOLOGY

The geometry of a circuit is secondary to its TOPOLOGY.

> **GEOMETRY**   The circuit geometry is determined by how the components are positioned in space. In point of fact, since we are modelling these systems, the geometry boils down to how we choose to draw the circuit.

> **TOPOLOGY**   The circuit topology is established by the manner in which the circuit elements are connected to one another. In practice, choices are made in drawing circuit diagrams to emphasise and clarify the topology.

The science of topology is the study of connectedness in its pure mathematical form. Fortunately, we need not be overly concerned with the technical details to employ topological reasoning in the analysis of circuits. The essential idea is that a circuit exhibiting an evidently complex geometric structure may possess an underlying simplicity. This simplicity can be made manifest by iterative redrawing of the circuit according to the following rules.

> **SHAPE**   The wires joining electric components may be stretched, contracted, bent, or slid along other wires. [A wire may slide through a junction[1] provided that due care is taken to not change the potential difference across any of the circuit components.]

> **MOVE**   Circuit elements may be moved, provided that the wires which connect them are not cut or otherwise reconfigured.

> **PASS**   Individual wires may be carefully cut and their paired ends rejoined so as to effect the passage of the wire through some sort of obstacle (often another wire). [The reconnection may take place across a junction[2] provided that the potential difference experienced by each circuit element remains unaffected by this action.]

---

[1]This alters the formal—mathematical—topology, but preserves the Kirchoffian circuit topology.
[2]Ditto.

EXAMPLE [*Resistor Topology: Looks May Deceive*]

The arrangement of five resistors illustrated on the left in Figure 21.1 seems complicated and appears to be non-planar. The first step toward computation of the effective resistance of the network is to determine its local series and parallel substructures. Attempting to divine the effective resistance without first clarifying the topological structure amounts to [wild] guesswork.

FIGURE 21.1   Exposing the Underlying Structure of a Circuit

Careful inspection of both the circuit on the left and the simpler circuit on the right shows that they are topologically equivalent. Resistors R_b and R_c are in parallel with the series combination of R_e and [R_a and R_d in parallel]. *I.e.*,

$$R_{\text{Eff}} = R_{[((ad]e)bc]} \,,$$

and hence

$$R_{\text{Eff}} = \left[\frac{1}{\frac{R_a\,R_d}{R_a+R_d} + R_e} + \frac{1}{R_b} + \frac{1}{R_c} \right]^{-1} .$$

The effective resistance of this network is now fully determined, for any given values of $R_a \ldots R_e$. The essential step in this analysis was the determination of the simpler equivalent circuit.

BATTERIES AND INTERNAL RESISTANCE

Recall that an ideal battery is a device which maintains a constant potential difference between two posts [points in space] irrespective of the current flowing through it. For real batteries, the voltage tends to diminish as the current draw increases.

> ASIDE: This is seen to happen when an automobile is started while its headlights are on. Provided that the battery is in good working order, the headlights are at their normal operational brightness prior to engaging the starter. While the starter is cranking the engine, the lights dim appreciably. [Once the engine catches and the starter is turned off, the lights return to their normal brightness.] The starter draws a large current from the battery and this produces a significant decline in its effective voltage. Thus, the EMF across the light bulb filaments [resistors] is reduced, which in turn leads to diminished power transfer,
>
> $$P_R = \frac{V^2}{R} \,,$$
>
> and less Joule heating, and therefore also diminished incandescent light intensity.

By ascribing the voltage/current behaviour to a so-called **internal resistance**, one is able to model a real battery. In other words, take an ideal battery [a source of constant EMF] and a resistor, connect them in series, place them in a *box*, and *voilà*, you have an improved model of a battery.

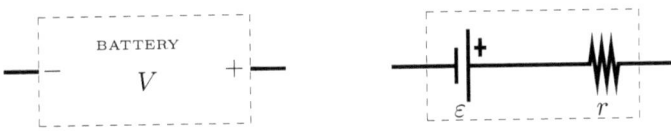

FIGURE 21.2   A More Realistic Battery

If the source of constant EMF has strength ε, the internal resistance is r, and the current flowing through the battery is I, then the effective EMF of the battery is

$$V = \varepsilon - I\,r.$$

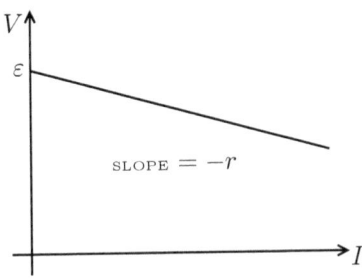

FIGURE 21.3   Voltage Diminishes with Current Draw in a Real Battery

EXAMPLE [*Power Deposited into a Load Resistor by a Real Battery*]

The circuit represented twice in Figure 21.4 is comprised of a real battery, V, and a single load resistor, R.

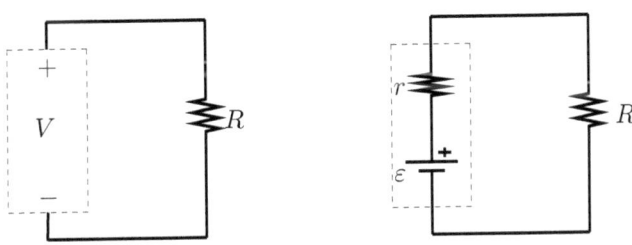

FIGURE 21.4   A Real Battery in Series with a Load Resistor

Let's investigate the changes in electric potential as charge elements [comprising the current] traverse the circuit. We start in the lower left-hand corner of Figure 21.4 and proceed clockwise. WLOG, we shall set the zero of potential at the battery's negative post [recognising that the assumption of ideal wire places the entire wire in the lower half of the figure at zero potential]. Upon passing through the ideal battery, each charge element enjoys an increase in its potential from 0 to ε. Subsequently the current passes through internal resistor r, and there suffers a loss of potential equal to $I\,r$. The charge elements emerging from the battery are at an intermediate potential of $V_i = \varepsilon - I\,r$ with respect to the negative terminal of the battery. All of this potential must be given up in passing through the load resistor, since the charge elements emerging from it find themselves in a region of zero potential.

ASIDE: Were not all of the [excess] potential consumed in traversing the load, the electrostatic potential energy of the charge elements would become ambiguous.

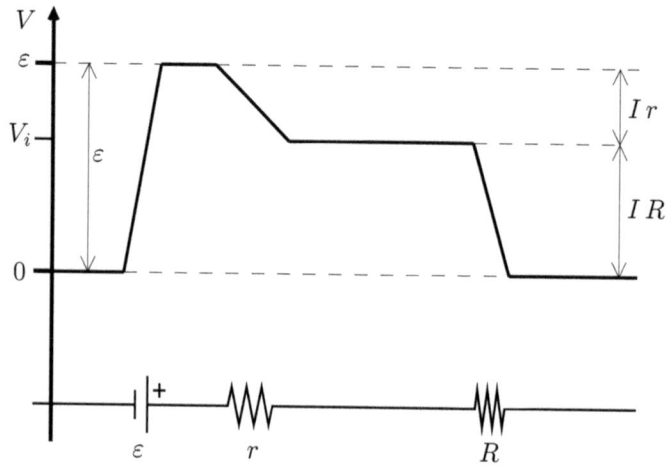

FIGURE 21.5 A Real Battery and a Resistive Load: Potentials

The voltage output of the real battery, $V_i = \varepsilon - I\,r$, is mirrored across the resistive load R.

The real battery's potential is equal to that of its ideal counterpart if and only if no current flows. *I.e.,*

$$V = \varepsilon \qquad \mathsf{IFF} \qquad I = 0\,.$$

Resuming the analysis of the circuit, we realise that the ideal source of EMF, ε, lies parallel to the series combination of the internal and [external] load resistors.[3] Accordingly,

$$\varepsilon = I\,(r + R)\,, \qquad \text{and hence} \qquad I = \frac{\varepsilon}{r + R}\,.$$

In this more realistic model of a battery, the current in the circuit depends both on the resistive load and on the internal resistance.

Thinking practically,[4] it is usually safe to presume that the electric circuit is designed to accomplish some useful good with the energy dissipated in the load resistor. Also, it is generally desired that electric circuits perform their functions in an efficient manner.

---

[3]Rearranging the resistors [in conformity with the SHAPE, MOVE, and PASS rules] makes this evident.
[4]Some might snark that this is the better alternative to "practically thinking."

Q: How might the power delivered to the load resistance be maximised?

A: The power input to the circuit from the ideal source of EMF contained within the real battery is

$$P_{\text{input}} = \varepsilon\,I\,.$$

Meanwhile, the amounts of power deposited into the internal and load resistances are

$$P_r = I^2\,r \qquad \text{and} \qquad P_R = I^2\,R\,,$$

respectively.

> ASIDE: Corroborative evidence for the two channels of power dissipation is that both the load resistor and the battery experience Joule Heating.

Conservation of energy requires[5] that the rate at which power is input to the circuit be equal to the rate at which it is dissipated in the resistors. Therefore, the power deposited in the load resistor is

$$P_R = P_{\text{input}} - P_r = I^2\,r = \varepsilon\,I - I^2\,r = \varepsilon\left(\frac{\varepsilon}{R+r}\right) - \left(\frac{\varepsilon}{R+r}\right)^2 r = \varepsilon^2\,\frac{R}{(R+r)^2}\,.$$

The previously derived expression for the current was used in the above chain of equalities. A plot of the power input to the load resistor as a function of the load resistance is sketched in Figure 21.6.

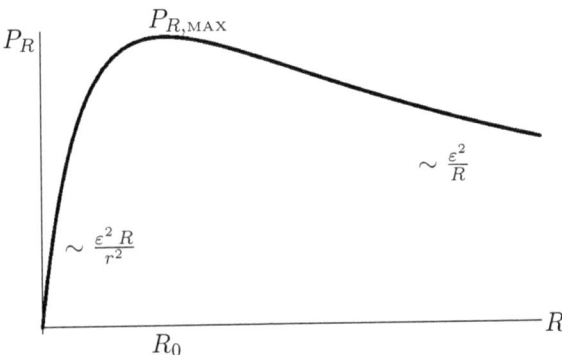

FIGURE 21.6 Power Dissipated in the Load Resistor as a Function of Its Resistance

The power dissipated by the resistor, P_R, tends to zero for both small and large load resistances R. The result, while a surprise to some, is readily seen by fixing r and taking the $R \to 0$ and $R \to \infty$ limits of P_R. Rolle's Theorem [a particular instance of the Mean-Value Theorem] then ensures that there exist at least one value of the load resistance which maximises the power input to the load. The sketch shows that there is only one such critical value of the resistance, R_0. One sets

$$\left.\frac{dP}{dR}\right|_{R=R_0} = 0\,,$$

---

[5]This constraint might hold only in a time-averaged sense, if there were a way in which the circuit could store energy. In the case of a purely resistive load, the system has no capacity to store energy and thus the energy fluxes in and out must balance at every instant.

and solves for R_0 in terms of the circuit parameters ε and r. The R derivative of P_R is

$$\frac{\partial P_R}{\partial R} = \frac{\varepsilon^2}{(R+r)^2} - 2\frac{\varepsilon^2 R}{(R+r)^3} = \frac{\varepsilon^2}{(R+r)^3}[R+r-2R] = \frac{\varepsilon^2}{(R+r)^3}[r-R].$$

Setting this last expression for the derivative equal to zero and solving for the critical value of the load resistance yield the non-trivial solution

$$R_0 = r$$

[along with the extraneous/spurious solutions $\varepsilon = 0$ and $R \to \infty$].

**The power output to the load is maximised
when the load resistance is matched to the internal resistance!**

We conclude this chapter with a concise presentation of **Kirchoff's Rules**.

KIRCHOFF'S RULES   Kirchoff's Rules [*a.k.a.* Kirchoff's Laws] provide practical expression of two foundational principles to be employed in circuit analysis.

K1   CONSERVATION OF CHARGE/CURRENT

**The sum of currents entering a junction
equals the sum of currents leaving the same junction.**

Another formulation is: **The oriented sum of currents vanishes everywhere.**

K2   CONSERVATION OF ENERGY

**The sum of the changes in potential around any closed loop
in the circuit must equal zero.**

Physically, these two rules make perfect sense. Charge is neither created nor destroyed; and in the course of traversing a closed circuit one must return to one's starting position with one's original potential energy!

ASIDE: Notwithstanding the inherent rationality of these rules, care must be exercised to ensure that K1 and K2 are consistently applied.

[*Not to worry! We shall have plenty of opportunities to practice.*]

Chapter 22

Timely Applications of Kirchoff's Rules

In this chapter we shall practice applying Kirchoff's Rules in both time-independent[1] and time-dependent situations.

EXAMPLE [*Two Resistors and an Ideal Battery Arranged All in Series*]

 ASIDE: This case has already been discussed in Chapters 20 and 21. It is revisited here to provide us with practice applying Kirchoff's Rules.

Two resistors, $R_1 = 6\,\Omega$ and $R_2 = 3\,\Omega$, are connected in series with an ideal battery producing an EMF, $\varepsilon = 6\,\mathrm{V}$.

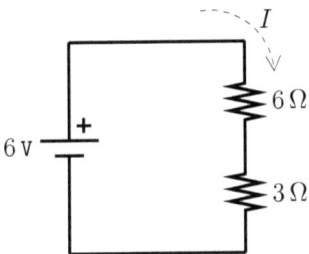

FIGURE 22.1 Two Resistors in Series across an Ideal Battery

Application of K1 and K2 begins with two observations.

1 There are no junctions, so K1 applies trivially.

 There is a unique, *a priori* unknown, steady state current flowing in the circuit. It passes through both resistors, and hence $I_1 = I = I_2$. A clockwise orientation has been assigned to the current [anticipating that this is the sense in which the actual current flows].

2 There is one closed-loop path in this circuit along which we apply K2.

 Traversing the loop in a clockwise sense [starting in the lower left corner] and summing the changes in potential as they occur *en route* yields

$$0 = +\varepsilon - I_1\,R_1 - I_2\,R_2 = \varepsilon - I\,(R_1 + R_2)\,.$$

 The signs of the $I\,R$ terms are a consequence of the assigned orientation of the current[2] and the decision to sum the potential differences in a clockwise sense.

---

[1]The abiding assumption is that the circuit being analysed has been in operation for a sufficiently long time for its current to have attained its asymptotic steady state [constant] value. The relevant time scales on which the steady state develops will be discussed here and in Chapter 38.

[2]Positive current passing through a resistor flows from HIGHER to LOWER potential.

The dynamics of the current response in the present case are governed by the expression of K2 quoted above. The solution for the current is straightforwardly obtained:

$$I = \frac{\varepsilon}{R_1 + R_2}.$$

For the resistors and battery in this example, the steady state current is $I = 6\,\text{v}/9\,\Omega = 2/3\,\text{A}$.

EXAMPLE [*Two Resistors and Two Ideal Batteries All in Series*]

Two resistors, R_1 and R_2, and two ideal batteries possessing EMFs ε_1 and ε_2, are arranged purely in series as shown in Figure 22.2. Further suppose, for definiteness, that

$$R_1 = 2\,\Omega\,, \qquad R_2 = 4\,\Omega\,, \qquad \varepsilon_1 = 4\,\text{v}, \quad \text{and} \quad \varepsilon_2 = 3\,\text{v}\,.$$

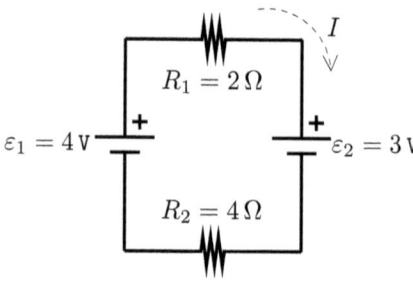

FIGURE 22.2   Two Resistors and Two Batteries All in Series

Application of Kirchoff's Rules begins with two observations.

1   There are no junctions in this circuit, so K1 is trivial.

Let I denote the unique steady state current flowing clockwise in the circuit.

2   There is one closed loop for which K2 applies.

Summing the changes in potential as the loop is traversed in a clockwise sense [starting in the lower left corner] yields

$$0 = +\varepsilon_1 - I\,R_1 - \varepsilon_2 - I\,R_2\,.$$

Solving this equation for the current, I, yields

$$I = \frac{\varepsilon_1 - \varepsilon_2}{R_1 + R_2}\,.$$

With the resistors and batteries specified above, the steady state current is

$$I = \frac{4 - 3}{2 + 4}\,\frac{\text{v}}{\Omega} = \frac{1}{6}\,\text{A}\,.$$

EXAMPLE [*Four Resistor plus Two Battery Network*]

Two sources of EMF, ε_1 and ε_2, and four resistors, $\{R_1, \ldots, R_4\}$, are joined by wires to produce the formidable-looking circuit illustrated below in Figure 22.3. We shall content ourselves with analysing this one algebraically and shall forego the final step of plugging in the specific numerical values pertinent to the problem.

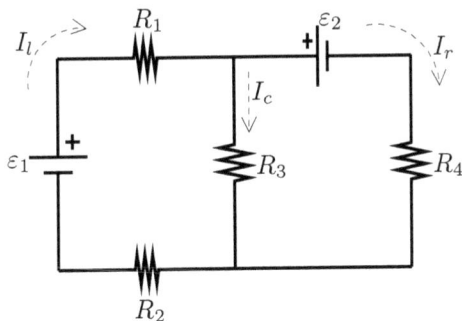

FIGURE 22.3   A Network with Two Batteries and Four Resistors

Application of Kirchoff's Rules begins with various observations.

1   There are two junctions serving to unite the **left**, **central**, and **right** branches of the circuit. We may consistently assign labels to the unique currents in the three branches: I_l, I_c, and I_r, respectively. In addition, we must ascribe a direction, or sense, to each current. Let's choose to have I_l enter the top junction, and both I_c and I_r leave.

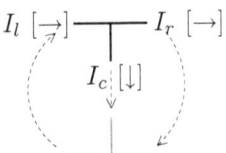

In effect, the current through the leftmost branch of the circuit is SPLIT at the upper junction and REJOINED at the lower junction. Thus, having two junctions gives rise to one constraint upon the currents in the branches.

Having $\mathcal{N}$ junctions provides, at most, $\mathcal{N} - 1$ constraints.

> ASIDE: The labels and orientations assigned to the [unknown] currents are chosen freely by us and are, in this sense, arbitrary. The precise form of the algebraic relations among the currents [following from application of Kirchoff's rules] depends on these choices. However, consistent physical results may be obtained by solving this set of algebraic equations for the unknown currents. CAVEAT: Obtaining a negative value for a particular current simply means that its initial orientation assignment was incorrect. A positive current of the determined magnitude flows in the opposite sense along the specified wire.

**Insofar as circuit properties are concerned,
negative current entering a junction is indistinguishable from
positive current leaving, and *vice versa*.**

Application of K1 to the circuit under consideration enforces current conservation
in terms of a mathematical constraint on the labelled and oriented currents:

$$I_l = I_c + I_r .$$

2   There are three closed loops in the circuit.

    l   The left-side loop is comprised of the left and central circuit branches.

    r   The right-side loop consists of the right and central branches.

    o   The outer loop is formed by the left and right branches.

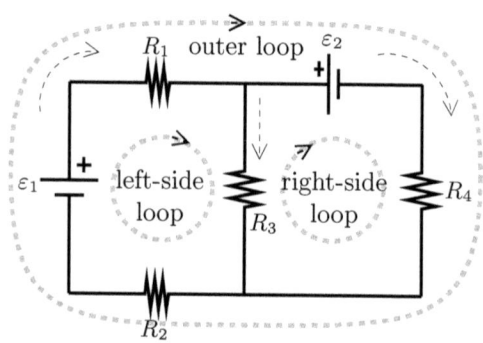

FIGURE 22.4   K2 Loops in the Network

For each of these three loops, the net change in potential must be zero.

[It will be shown that there are only two independent loop equations here.]

Summing potential changes while traversing the left-side loop in a clockwise direction, beginning at the lower left corner, yields the following relation:

$$0 = +\varepsilon_1 - I_l\, R_1 - I_c\, R_3 - I_l\, R_2 .$$

Summing potential differences clockwise around the right-side loop produces

$$0 = -\varepsilon_2 - I_r\, R_4 + I_c\, R_3 .$$

The relation obtained by traversing the outer loop in a clockwise direction beginning at the lower left and summing the voltage increments is

$$0 = +\varepsilon_1 - I_l\, R_1 - \varepsilon_2 - I_r\, R_4 - I_l\, R_2 = \varepsilon_1 - \varepsilon_2 - I_l\big(R_1 + R_2\big) - I_r\, R_4 .$$

This outer equation is redundant, as it is the sum of the two equations corresponding to the left- and right-side loops, quoted directly above. In more technical
terms, this outer-loop relation is not linearly independent of the two inner-loop
equations.

There are at most $N - 1$ independent loop equations in an N-loop circuit.
CAVEAT: Counting loops in non-planar circuits may be a difficult task.

Careful application of Kirchoff's Rules has yielded a system of three independent linear equations for the three unknown currents:

$$0 = I_l - I_c - I_r$$
$$0 = +\varepsilon_1 - I_l\left(R_1 + R_2\right) - I_c\,R_3\,.$$
$$0 = -\varepsilon_2 + I_c\,R_3 - I_r\,R_4$$

The solution to this set of coupled linear algebraic equations, $\{I_l, I_c, I_r\}$, may be obtained by one's own favourite method of tackling such systems.

ASIDE: While the lawyers insist that the standard disclaimer "provided that solutions exist" must be mentioned here, we need not worry too much, since solutions do exist for all physical [read this as "reasonable"] EMFs and resistances.

TIME-DEPENDENT DIRECT CURRENT (DC) ELECTRIC CIRCUITS

It is not enough to only know the steady state behaviour of DC circuits; it is desirable to understand and describe each circuit's approach to steady state behaviour from various possible initial conditions, as well. Let's explore, in greater detail, the manner in which a current flow causes the buildup of charge on a capacitor [*cf.* Chapters 13 and 14].

RC Circuit with a Switch

An ideal battery with EMF ε, a capacitor, C, a resistor, R, and a switch, are connected together in series, as shown in Figure 22.5.

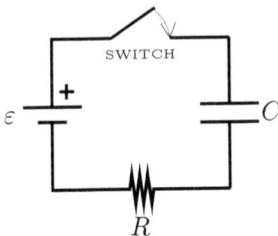

FIGURE 22.5 An RC Circuit

Until time $t = 0$ the switch is open and no current flows. Furthermore, with only slight loss of generality, we shall presume that there is zero charge on the capacitor prior to $t = 0$.

ASIDE: This seems very reasonable, since the open switch prevents the flow of current from the battery which would deliver charge to the capacitor. However, one must consider the possibility that the capacitor may have been given a net charge before it was connected to the other components in the circuit.

Let's throw the switch at $t = 0$. From thence onward, **current flows in the circuit and charge builds up on the capacitor**. Furthermore, the behaviour of this electric circuit changes in time.

QUALITATIVE ANALYSIS OF THE TIME-DEPENDENCE OF THE RC CIRCUIT

$t = 0$ At the closing of the switch, the entire potential of the battery is mirrored in the resistor [given the assumed zero-charge initial state of the capacitor]. Thus the initial current must be

$$I_0 = \frac{\varepsilon}{R} \, .$$

$t > 0$ As the capacitor accrues charge, a potential difference [oriented in opposition to that of the battery] is built up across it. CONSERVATION OF ENERGY demands that the potential difference across the resistor decline, and thus so too does the current in the circuit.

$t \to \infty$ After a long time [where the time scale is determined by the characteristics of the circuit itself, as we'll see presently], the potential of the battery will be almost entirely mirrored across the capacitor, and thus the current will approach zero. The charge which will have accrued to the capacitor after a long time is computable by the following argument.

$$\text{As} \quad t \to \infty, \qquad V_{\text{capacitor}} = \frac{Q}{C} \to \varepsilon, \qquad \text{and thus} \qquad Q \to \varepsilon C \, .$$

This is an accurate description of the essential physics.

QUANTITATIVE ANALYSIS OF THE TIME-DEPENDENCE OF THE RC CIRCUIT

We shall now employ Kirchoff's Rules to model the buildup of charge on the capacitor, $q(t)$, by explicit analysis of the time-dependent dynamics of the circuit.

1 There is one unique time-dependent current, *viz.*, $I(t)$.

2 There is one loop. Clockwise summation of potential differences yields

$$0 = \varepsilon - \frac{q}{C} - I R \, .$$

The qualitative predictions for early and very late times are confirmed in this relation.

- As $t \to 0^+$, $q \to 0$. The initial current is $I_0 = \varepsilon/R$.
- Subsequent to $t = 0$, the charge on the capacitor $q(t)$ increases, and the current, $I(t)$, diminishes.
- As $t \to \infty$, $q \to \varepsilon C$ and $I \to 0$.

Continuing with the analysis, one notes that the electric current, I, is the charge flux to the capacitor, and hence

$$I = \frac{dq}{dt} \, .$$

The flux of charge deposited on the capacitor is provided solely by the current in the circuit.

The K2 relation, quoted above, is a differential equation for the charge as a function of time. We shall ascertain the solution, $q(t)$, in two major steps. The first step is to reformulate K2 in terms of the current response, and to develop qualitative and quantitative solutions for $I(t)$. The second step is to integrate the current to obtain the charge, $q(t)$, accruing to the capacitor at time t.

  ASIDE: This method of computing the current in the circuit, and then the charge on the capacitor, is somewhat indirect. We do this because, as we will see in many future instances, the circuit dynamics is better described in terms of current.

The realisation that derivatives of a dynamical equation are *bona fide* relations in their own right inspires us to take the time derivative of the above expression of K2. When taking the derivative, it is assumed that the circuit parameters ε, C, and R are all constant in time, and thus

$$0 = -\frac{1}{C}\frac{dq}{dt} - R\frac{dI}{dt}$$

is obtained. By identifying the current with the time rate of change of charge on the capacitor, and rearranging, one obtains a differential equation solely involving the current,

$$\frac{dI}{I} = -\frac{1}{RC}\,dt\,.$$

This equation is of the form characteristic of exponential behaviour and admits the general solution

$$I(t) = I_0\,e^{-\frac{t}{RC}}\,.$$

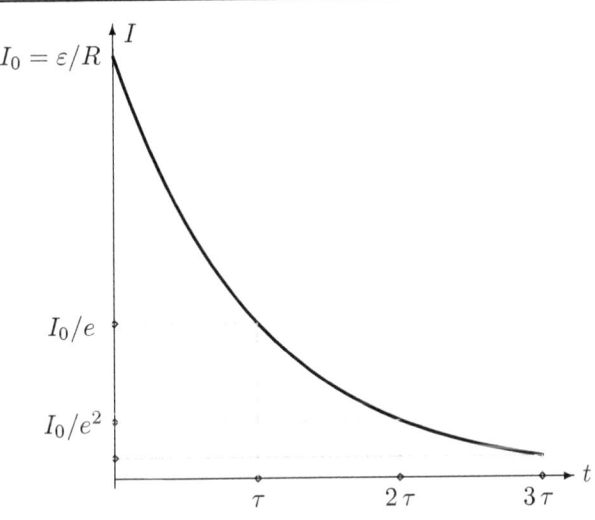

FIGURE 22.6 Current *vs.* Time for an RC Circuit

To summarise: the current response within the RC Circuit varies with time according to

$$I(t) = I_0\,e^{-t/RC}\,.$$

In order that the argument appearing in the exponent be dimensionless, the combination RC must have units of time. That the dynamical properties of the circuit unfold on the time scale fixed by τ is revealed in Figure 22.6.

RC CIRCUIT TIME CONSTANT   For a circuit with effective resistance R in series with effective capacitance C, the time scale characteristic of the circuit dynamics is set by the RC time constant,

$$\tau = RC.$$

The symbol denoting the time constant is the Greek letter τ, pronounced "tau." In SI, $1\ \Omega \cdot F = 1\ s$, showing the consistency of the unit scheme.

The current response of the circuit as a function of time is precisely the time rate at which the charge accumulates on the capacitor [after the switch is thrown]. Thus,

$$\frac{dq}{dt} = I(t) \qquad \Longrightarrow \qquad dq = I_0\, e^{-\frac{t}{RC}}\, dt = \frac{\varepsilon}{R}\, e^{-\frac{t}{RC}}\, dt.$$

Self-consistent integration of this expression

$$\text{from initial values} \left\{ \begin{array}{c} q_0 = 0 \\ t = 0 \end{array} \right\} \text{ to final values} \left\{ \begin{array}{c} q(t) \\ t\ [\text{UNSPECIFIED}] \end{array} \right\}$$

results in

$$q(t) = \varepsilon\, C \left(1 - e^{-\frac{t}{RC}} \right) = q_\infty \left(1 - e^{-\frac{t}{RC}} \right).$$

For the last equality we have represented the asymptotic value of the charge appearing on the capacitor by $q_\infty = \varepsilon\, C$.

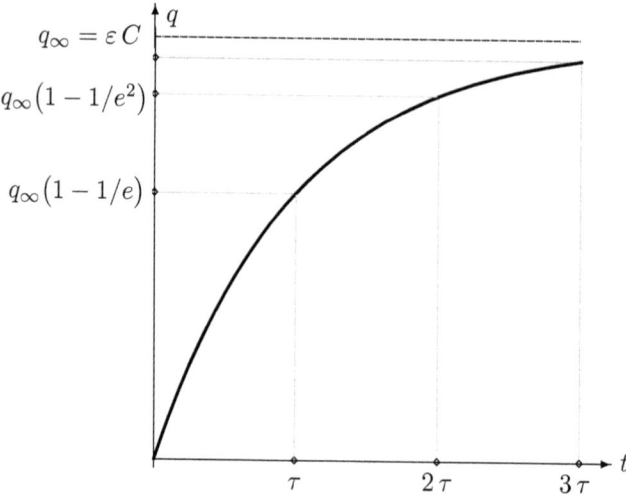

FIGURE 22.7   Capacitor Charge Accumulation in an RC Circuit

Chapter 23

More RC Circuits and Segue to Magnetism

A capacitor, C, resistor, R, and switch are joined by ideal wires as shown in Figure 23.1. The switch is open for all times $t < 0\,\text{s}$, and the capacitor has charges $\pm Q_0$ on its plates prior to $t = 0$.

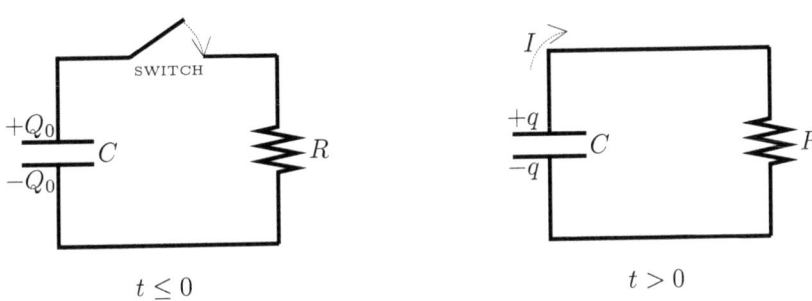

FIGURE 23.1 Discharging a Capacitor through a Resistor

At time $t = 0$, the switch is closed and the capacitor begins to discharge as current flows through the wires leading to and from the resistor.

QUALITATIVE ANALYSIS

The current will taper off as the charge on the plates is reduced because it is the potential difference across the capacitor, $V_C = \frac{Q}{C}$, which drives the current.

QUANTITATIVE ANALYSIS

As in the charging case, examined in Chapter 22, the behaviour of this circuit is amenable to mathematical modelling via careful application of Kirchoff's Rules.

1   There are no junctions, so K1 is trivial. There is a single unique current, I, flowing clockwise in the circuit.

2   There is one loop in the circuit. A thorough [clockwise] accounting of potential differences [K2] yields
$$0 = +\frac{q}{C} - I\,R\,.$$

The current in this expression arises from the DISCHARGING of the capacitor, and thus
$$I = -\frac{dq}{dt}\,.$$

In light of this relation, the differential equation governing the amount of charge on the capacitor is

$$\frac{dq}{q} = \frac{-1}{RC}\,dt\,.$$

This equation for $q(t)$ is identical to that obtained for $I(t)$ in Chapter 22.

[Except, of course, that this one is for charge whilst the other was for current.]

Let's self-consistently integrate this differential equation, from time $t = 0$, when the switch is thrown and the charge on each plate of the capacitor is $\pm Q_0$, to time t and charge $\pm q(t)$. The solution obtained for the magnitude of the charge borne on each plate is

$$q(t) = Q_0\,e^{-\frac{t}{RC}}\,.$$

This function is sketched in the left panel of Figure 23.2. For grins, we may determine its time rate of change. This flux of charge [the current in the circuit] is plotted in the right panel.

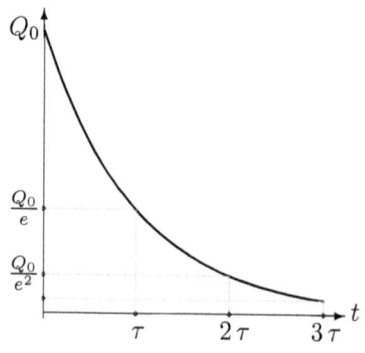
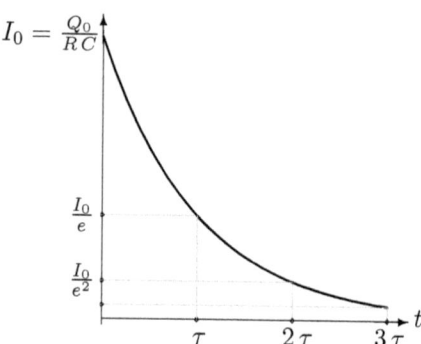

FIGURE 23.2 Capacitor Charge and Current When Discharging through a Resistor

EXAMPLE [*The Time Scale of RC Circuit Dynamics*]

The time scale on which RC circuits approach steady state behaviour is given by the time constant. When one pops open an electrical device, one typically finds resistors and capacitors with values of R and C on the order of $100\,\Omega$ and $1\,\mu F$, respectively. In such cases,

$$\tau = RC \sim 10^{-4}\,\mathbf{s}\,.$$

Of course, it is almost always possible to select resistors and capacitors with values far from these. This does not hinder our concluding that:

RC circuits typically relax on relatively short time scales.

In Chapters 19–21, we undertook analyses of resistive circuits with little or no concern for the manner in which the current evolved from its initial value to its steady state value, other than to comment that one has to wait for this to occur. The *a posteriori* justification for this cavalier attitude is that all circuits have an effective capacitance owing to the presence of conducting material in the wires and the circuit components. These capacitances are ordinarily very small [*i.e.*, nF or pF scales], and thus, for all but the largest-scale resistors, the transition to steady state behaviour occurs within the blink of an eye.

   ASIDE: We will re-encounter this need for an *a posteriori* justification when **inductance** is introduced!

EXAMPLE [*RC Circuit Behaviour*]

A capacitor, $C = 50\,\mu\text{F}$, a resistor, $R = 120\,\text{k}\Omega$, a battery with EMF $\varepsilon = 9\,\text{V}$, and a switch are all arranged in series. Prior to time $t = 0$, there was no charge on the capacitor. At $t = 0$, the switch is thrown.

Q: What is the current in the circuit at $t = 12\,\text{s}$?

A: In general,

$$I(t) = I_0\,e^{-t/\tau}\,, \qquad \text{where} \qquad I_0 = \frac{\varepsilon}{R} \qquad \text{and} \qquad \tau = RC\,.$$

In this instance, $I_0 = 9\,\text{V}/120\,\text{k}\Omega = 75\,\mu\text{A}$, and the time constant characterising the dynamical response of this circuit is $\tau = RC = 50 \times 10^{-6} \times 120 \times 10^3 = 6\,\text{s}$. At the instant $t = 12\,\text{s}$, two time constant intervals have elapsed since the switch was thrown. The current in the circuit at that time is

$$I(12) = \frac{75}{e^2} \times 10^{-6} \simeq 10.15\,\mu\text{A}\,.$$

Thus far, we have investigated **electric fields**: first in the context of ELECTROSTATICS, and more recently in that of electric current. At this juncture, two additional—interlinked—phenomena must be considered. These are that

Magnetic Fields Act on Moving Charges,

and

Moving Charges Produce Magnetic Fields.

Throughout the remainder of these notes, we shall be concerned with the intertwining of electrical and magnetic phenomena.

[It's **electromagnetism** rather than **electricity and magnetism**.]

A hint of the entanglement of electricity and magnetism is revealed when a charged particle accelerated from rest by an ambient electric field [*à la* Chapter 2] produces a magnetic field which can then influence other moving charged particles. Rather than risk becoming lost in circular reasoning, let's first investigate how a magnetic field acts on charged particles. Later we'll worry about how magnetic fields are produced, and later still, how everything fits nicely together.

LORENTZ FORCE LAW When a point-like particle bearing charge q and moving with velocity $\vec{v}$ is immersed in a magnetic field of strength $\vec{B}$, it experiences a **magnetic force** [*a.k.a.* **Lorentz force**],

$$\vec{F_{\text{B}}} = q\,\vec{v} \times \vec{B}\,.$$

The **magnetic field, $\vec{B}$,** is variously named the "magnetic induction field," the "magnetic flux density," or simply the "magnetic field."

The vector cross-product appears in the expression for the Lorentz force. Recall that it is a bilinear, associative, ANTI-commutative vector-valued operator acting on pairs of input vectors. Here we devote a few words to its reintroduction.

The cross product of two given vectors, $\vec{A}$ and $\vec{B}$, may be consistently viewed from either a geometric or a Cartesian-component perspective. These equivalent formulations lend themselves to effective use in different circumstances.

GEOMETRIC   With the vectors represented as directed magnitudes:

$$\vec{A} \times \vec{B} = |\vec{A}|\,|\vec{B}|\,\sin\left(\theta\right)\,[\text{RHR}]\,.$$

The magnitude on the RHS is the product of the magnitudes of the two vectors and the **sine** of the smaller angle lying between the vectors when they are placed tail-to-tail. The direction of the resultant is perpendicular to the plane containing both vectors, with orientation specified by a RIGHT HAND RULE [RHR].

PK's favourite version of this RHR is: Align the metacarpals on your right hand with the first vector, sweep your digits/phalanges toward the second vector, and *presto*, your thumb signifies the direction, perpendicular to the plane generated by the first and second vectors, in which the cross-product vector points.

ALGEBRAIC   With the vectors represented as ordered N-tuples of Cartesian components:

$$\vec{A} \times \vec{B} = (A_x,\,A_y,\,A_z) \times (B_x,\,B_y,\,B_z)$$
$$= (A_y B_z - A_z B_y,\,A_z B_x - A_x B_z,\,A_x B_y - A_y B_x)\,.$$

In terms of unit vectors, one may write

$$\vec{A} \times \vec{B} = \left(A_x\,\hat{\imath} + A_y\,\hat{\jmath} + A_z\,\hat{k}\right) \times \left(B_x\,\hat{\imath} + B_y\,\hat{\jmath} + B_z\,\hat{k}\right)$$
$$= (A_y B_z - A_z B_y)\,\hat{\imath} + (A_z B_x - A_x B_z)\,\hat{\jmath} + (A_x B_y - A_y B_x)\,\hat{k}\,.$$

The mathematical structure of matrix determinants affords a shorthand means of expression of the cross product in Cartesian bases:

$$\vec{A} \times \vec{B} = \begin{vmatrix} \hat{\imath} & \hat{\jmath} & \hat{k} \\ A_x & A_y & A_z \\ B_x & B_y & B_z \end{vmatrix}\,.$$

In both the geometric and algebraic points of view, the cross-product vanishes **IFF EITHER** the vectors are parallel, **OR** either is equal to the zero vector.

ASIDE: It is noteworthy that the cross-product acts on the Cartesian basis vectors in the manner of a PERMUTATION OPERATOR.

$\hat{\imath} \times \hat{\imath} = \vec{0}$	$\hat{\imath} \times \hat{\jmath} = \hat{k}$	$\hat{\imath} \times \hat{k} = -\hat{\jmath}$
$\hat{\jmath} \times \hat{\imath} = -\hat{k}$	$\hat{\jmath} \times \hat{\jmath} = \vec{0}$	$\hat{\jmath} \times \hat{k} = \hat{\imath}$
$\hat{k} \times \hat{\imath} = \hat{\jmath}$	$\hat{k} \times \hat{\jmath} = -\hat{\imath}$	$\hat{k} \times \hat{k} = \vec{0}$

Chapter 24

The Lorentz Force

A particle bearing electric charge q and moving with velocity $\vec{v}$ through a region of space containing a magnetic field, $\vec{B}$, experiences a magnetic force, $\vec{F}_{M,q} = q\,\vec{v} \times \vec{B}$, called the Lorentz force. Noteworthy aspects of the Lorentz force include the following.

- The magnitude of the force, $|\vec{F}_{M,q}|$, depends on q, $|\vec{v}|$, and the **sine** of the angle between $\vec{v}$ and $\vec{B}$.

  o   IF the velocity of the particle is perpendicular to the magnetic field, THEN the Lorentz force is maximised.

  o   IF the velocity of the particle is [anti-]parallel to the magnetic field, THEN the magnitude of the Lorentz force is zero.

- The Lorentz force is directed perpendicular to the plane generated by $\vec{v}$ and $\vec{B}$, with its orientation given by the RIGHT HAND RULE [RHR].

- The orientation of the Lorentz force depends on the sign of the charge borne by the particle, *i.e.*, $q \longrightarrow -q \quad \implies \quad \vec{F}_{M,q} \longrightarrow -\vec{F}_{M,q}$.

- The SI unit of magnetic field strength is the tesla [T], named in honour of the prolific inventor Nicola Tesla (1856–1943). Consistency of the units in the definition of the Lorentz force requires that

$$ \mathtt{T} = \frac{\mathtt{N}}{\mathtt{C} \cdot (\mathtt{m/s})} = \frac{\mathtt{N}}{\mathtt{A} \cdot \mathtt{m}} \, . $$

The CGS-unit for the magnetic field, the gauss [G] refuses to fall into desuetude because it is nicely scaled.[1] The conversion factor from gauss to tesla is $1\,\mathtt{G} = 1 \times 10^{-4}$ T.

- The magnetic field, $\vec{B}$, has a number of aliases. One of these is **magnetic flux density** [the intensity of magnetic flux]. Thus, introducing the weber[2] [Wb] as the unit of magnetic flux, the tesla may also be expressed as a weber per square metre: $\mathtt{T} = \frac{\mathtt{Wb}}{\mathtt{m}^2}$.

- Most importantly, **the magnetic force does no work on a particle.**

Q: How can this be true?

A: As the Lorentz force is perpendicular to the velocity of the particle, it follows that the magnetic power input to the particle must vanish at all instants:

$$ P_M = \vec{F}_{M,q} \cdot \vec{v} = q\,(\vec{v} \times \vec{B}) \cdot \vec{v} \equiv 0 \, . $$

Power is the rate at which work is done, so **no mechanical work is performed by the Lorentz force acting on a charged particle.**

---

[1]The Earth's magnetic field is about $\frac{1}{2}$ G.

[2]Wilhelm Weber (1804–1891), a German physicist, was co-inventor, along with Gauss, of the electromagnetic telegraph.

ASIDE: Equivalently, one may reason that

$$dW = \vec{F}_{M,q} \cdot d\vec{s} = \vec{F}_{M,q} \cdot (\vec{v}\,dt) = \left(\vec{F}_{M,q} \cdot \vec{v}\right)\,dt = 0 \,,$$

and hence that $\int dW = \int 0\,dt = 0$.

As the magnetic force does no work on the particle, its kinetic energy remains constant[3] [according to the WORK–ENERGY THEOREM]. Constancy of kinetic energy implies that its speed remains constant, and hence only the direction of its velocity may change. *Déjà vu!* This is precisely what occurs in uniform circular motion.

[Practical applications which rely on Lorentz force properties are considered in future chapters.]

Q: Now that the magnetic force acting on a single charged particle moving through space has been considered, what can be said about the magnetic force exerted upon a wire which carries a current?

A: Lots, as we'll see anon.

MAGNETIC FORCE ON A SEGMENT OF CURRENT CARRYING WIRE

A straight segment of current-carrying wire, with length L and cross-sectional area $|\vec{A}|$, is immersed in a uniform magnetic field, $\vec{B}$, as illustrated in the sketch in Figure 24.1. We assume that the geometry of the wire is uniform and that it is homogeneous in composition.

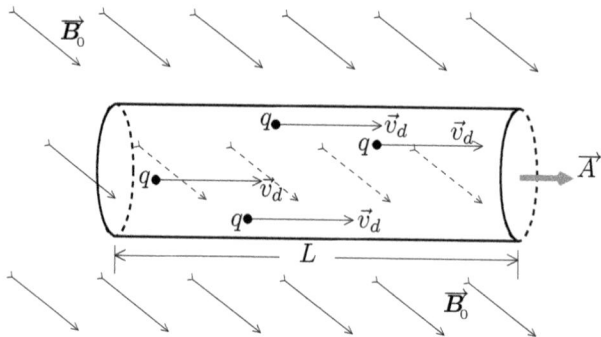

FIGURE 24.1   A Current-Carrying Wire Immersed in a Uniform Magnetic Field

In the microscopic model, the electric current in the wire arises through the coherent drift of a collection of charge carriers. We shall assume that there is but one species of charge carrier [with all representatives indistinguishable], labelled by q, the electric charge that each bears. All of the charge carriers may be thought to drift along the wire with the same characteristic speed, $\vec{v}_d$.

[It may help to think of the current as a uniformly flowing (charged) fluid.]

As there is no *a priori* reason to suppose that the ambient uniform magnetic field, $\vec{B}$, does not penetrate into the wire, we must assume that it does. Consequently, the charge carriers

---

[3]Assuming the absence, or cancellation, of other external forces acting on the particle.

are drifting through a region permeated by a magnetic field, and one might be tempted to simply claim that the Lorentz force acts on each and every drifting charge carrier thus:

$$\vec{F}_{\text{M},q} = \text{magnetic force per charge carrier} = q\,\vec{v}_d \times \vec{B}_0\,.$$

While such temptations are often to be avoided, this one warrants serious examination.

> ASIDE: On the one hand, this claim is so very obviously false. Were one to compute the Lorentz force acting on a particular charge carrier within the conducting wire, one would employ its actual velocity rather than the drift velocity associated with the current. Unfortunately, in almost all conceivable circumstances, the **thermal velocity**[4] is many times greater than the drift velocity.
>
> On the other hand, the thermal velocities of the collection of charge carriers moving in the wire are expected to be uncorrelated and isotropically distributed, and thus their unweighted sum ought to vanish. Meanwhile, the drift velocity is a small [coherent] effect which in aggregate yields a non-zero net contribution to the Lorentz force acting on [the collection of charge carriers found in] this segment of the wire.

The segment of conducting wire has many charge carriers. Recalling the analysis of electric current [in Chapter 17], suppose that the material from which the wire is made has η charge carriers per unit volume. The volume of the segment of wire under consideration is $V = |\vec{A}|\,L = A\,L$. Therefore, the total number of charge carriers in the wire segment is $\mathcal{N} = \eta\,A\,L$. Each of these moves with [average] drift velocity $\vec{v}_d$. Thus, the total Lorentz force acting on the wire [the coherent superposition of the Lorentz forces acting on each charge carrier] is

$$\vec{F}_{\text{M,net}} = \mathcal{N}\,\vec{F}_{\text{M},q} = \mathcal{N}\,q\,\vec{v}_d \times \vec{B}_0 = \eta\,A\,L\,q\,\vec{v}_d \times \vec{B}_0\,.$$

In the microscopic analysis of Chapter 18, the electric current was expressed in terms of the [local] current density, $\vec{J}$, via

$$I = \int \vec{J} \cdot d\vec{A}\,.$$

In the present case involving a straight wire segment, the assumptions of uniformity and homogeneity allow one to write:

$$\vec{J} = \eta\,q\,\vec{v}_d\,.$$

In light of this last relation, we define a directed length vector, $\vec{L}$, having magnitude equal to the length of the straight-line wire segment and direction parallel to that of the electric current. To preclude any ambiguity arising from the orientation of the cross-sectional area vector, which directly affects the sign of the current, it is safest to write

$$\vec{F}_{\text{M},I} = |I|\,\vec{L} \times \vec{B}_0\,.$$

In practice, one often chooses to *live dangerously* and omit the absolute value symbol.

Q: What does one do when the segment of wire is not straight?

A: IF the wire segment is curved or bent, THEN we partition it into subsegments over which it is [approximately] straight, analyse each subsegment in turn, and superpose the results to determine the net force.

Q: What does one do when the magnetic field is not uniform?

---

[4] The heat in a substance is a manifestation of the disordered kinetic energy of its microscopic constituents. The small masses of the charge carriers imply that they must have correspondingly large thermal speeds.

A: IF the field is not uniform over the extent of the wire, THEN we partition the wire into subsegments over which the field is [approximately] uniform, compute the force on each subsegment, and sum to obtain the net force.

In summary: the total magnetic force on a wire segment with initial and final endpoints $\{\vec{r}_i, \vec{r}_f\}$, carrying current I through a region of space containing a magnetic field, $\vec{B}$, is

$$\vec{F}_{\mathrm{M}, I, i \to f} = \int_{\vec{r}_i}^{\vec{r}_f} |I| \, d\vec{s} \times \vec{B} .$$

Three comments must be made.

$\quad$ **I** $\quad$ The same current, I, passes through each and every point in the wire. [Electric charge and current are conserved. By assumption, there are no bifurcations or junctions in the wire, *i.e.*, there is a unique path from the initial point to the final point.]

$i \to f$ $\quad$ The sense in which the integral is performed, *i.e.*, from $\vec{r}_i$ to $\vec{r}_f$, corresponds to the direction in which the current flows.

$\quad$ **$\vec{F}$** $\quad$ The symbol "$\vec{F}_{\mathrm{M}, I, i \to f}$" is deceptively compact. As in the explication of mechanical work in VOLUME I, Chapter 21, one must possess vector-valued expressions for the path through space corresponding to the wire and the field, choose a 1-d parameterisation of the entire path, pull back the integrand to the parameter space, AND then effect the final integration.

$\qquad\qquad$ [In many circumstances, one or more of these hurdles is insurmountable.]

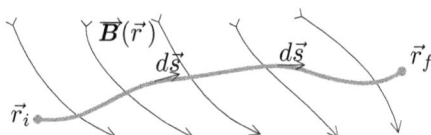

FIGURE 24.2 $\quad$ A Current-Carrying Wire Immersed in a Magnetic Field

Do note that in Figure 24.2, the magnetic field is indicated by the drawing of several **magnetic field lines**. Just as an electric field gives rise to electric field lines [*cf.* Chapter 2], the magnetic field generates magnetic field lines. The local field is dense within the region of space it occupies [a technical CAVEAT is "unless it vanishes"], and it is tangentially directed along the field lines which are shown. The magnetic field strength at a point in space is proportional to the local density of the field lines.

In the remainder of these notes, our attentions shall be confined to two special cases in which the analyses are more tractable. The ambient magnetic field is assumed to be **uniform**, while the wire may adopt either of two possible topologies:

$\qquad\qquad$ $\vec{r}_i \neq \vec{r}_f$ $\qquad$ distinct endpoints to the segment, or

$\qquad\qquad$ $\vec{r}_i = \vec{r}_f$ $\qquad$ identical endpoints, *i.e.*, a closed loop.

Chapter 25

Current, Lorentz Force, and Torque

In the previous chapter, it was determined that the magnetic force acting on a straight segment of wire with length $|\vec{L}|$, carrying current I in the direction of $\vec{L}$ through a region of space permeated by a uniform magnetic field, $\vec{B}_0$, is

$$\vec{F}_{M,I} = |I|\,\vec{L} \times \vec{B}_0.$$

In situations in which the wire is not straight, or the magnetic field is not uniform, one employs the PARTITION, COMPUTE, SUM [and REFINE] strategy.

P   Partition the wire into [approximately] straight and uniform segments.

C   Compute [or otherwise determine] the force acting on each segment.

S   Sum to get the total force acting on the wire.

One may choose to refine the partition into an infinite number of infinitesimal segments, $d\vec{s}$, in which case the differential contribution to the net force coming from each segment is of the form:

$$d\vec{F}_{M,I} = I\,d\vec{s} \times \vec{B}_0.$$

It has been assumed that the entire electric current, I, passes through each (sub-)segment of the wire.

> ASIDE: If this is not the case, there must be a junction present [since current is conserved]. In this event, one computes the forces acting on each individual wire separately and adds them [LINEAR SUPERPOSITION] to get the grand total force.

The differential force contributions are summed [a.k.a. integrated] to obtain the total force:

$$\vec{F}_{M,I,\,i \to f} = \int_{\vec{r}_i}^{\vec{r}_f} |I|\,d\vec{s} \times \vec{B}.$$

The generic case is illustrated in the figure from Chapter 24 which is reproduced here.

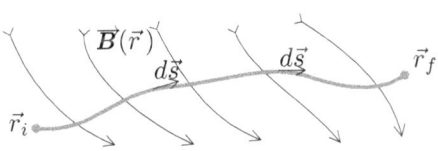

FIGURE 25.1   A Current-Carrying Wire Immersed in a Magnetic Field

Consider, in turn, two special cases for which the analysis is more straightforward.

First, IF the magnetic field is uniform over the entire length of the wire, AND the wire has distinct endpoints, THEN [owing to the linearity of both integration and the cross product] the integral simplifies to

$$\vec{F}_{\mathrm{M},I,\,i\rightarrow f} = I\left[\int_{\vec{r}_i}^{\vec{r}_f} d\vec{s}\right] \times \vec{B}_0\,.$$

Q: What is the factor $\left[\int_{\vec{r}_i}^{\vec{r}_f} d\vec{s}\right]$? Can it be simply ascertained?

A: The factor appearing in the brackets is the sum of all of the infinitesimal displacements along the wire from the initial point to the final point. It is the vector which starts at the initial point of the wire segment and ends at its final point, *i.e.*, precisely the net displacement,

$$\vec{L}_{if} = \vec{r}_f - \vec{r}_i\,.$$

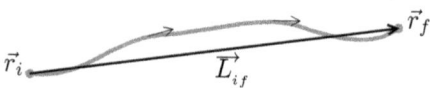

FIGURE 25.2 The Current Flowing through a Curved Wire is Effectively Displaced

Thus,

$$\vec{F}_{\mathrm{M},I} = I\,\vec{L}_{if} \times \vec{B}_0\,.$$

The curved wire segment with distinct endpoints experiences exactly the same Lorentzian magnetic force as would a straight wire segment joining the same endpoints.

Second, IF the magnetic field is uniform over the entire length of the wire, AND the current loop is closed [*i.e.*, it starts and terminates at the same point in space], THEN the same simplifications as in the preceding case ensue:

$$\vec{F}_{\mathrm{M},I,\,i\rightarrow f=i} = I\left[\int_{\vec{r}_i}^{\vec{r}_f=\vec{r}_i} d\vec{s}\right] \times \vec{B}_0 = I\left[\oint d\vec{s}\right] \times \vec{B}_0\,.$$

The bracketed factor amounts to the net displacement associated with the wire, and is thus identically ZERO. **A closed current-carrying loop which is completely immersed in a uniform magnetic field experiences zero net magnetic force.** *I.e.*,

$$\vec{F}_{\mathrm{M},I,\,a\rightarrow a} = \vec{0}\,.$$

[The net force need not vanish if the magnetic field is not uniform over the entire loop of wire.]

Q: Since the net magnetic force vanishes, aren't such configurations devoid of interest?

A: Nope! Although the net magnetic force assuredly vanishes, the net magnetic torque will generally be non-zero.

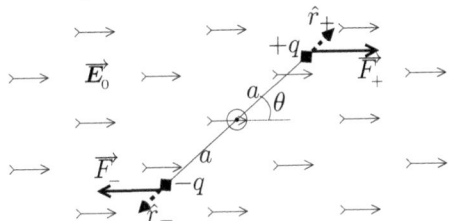

FIGURE 25.3 An Electric Dipole Immersed in a Uniform Electric Field, *Encore*

MAGNETIC TORQUE EXERTED ON A CURRENT-CARRYING LOOP OF WIRE

Before we get to the magnetic torque, let's warm up with a review of the torque experienced by an electric dipole residing in a uniform electric field, *cf.* Chapter 16.

The electric forces exerted on the positively and negatively charged constituents of the dipole are:

$$\vec{F}_{\mathrm{E},+q} = q\,\vec{E}_0 = q\,E_0\,[\ \rightarrow\ ] \qquad \text{and} \qquad \vec{F}_{\mathrm{E},-q} = -q\,\vec{E}_0 = -q\,E_0\,[\ \rightarrow\ ] = q\,E_0\,[\ \leftarrow\ ].$$

Clearly, the net force vanishes:

$$\vec{F}_{\mathrm{E,net}} = \vec{F}_{\mathrm{E},+q} + \vec{F}_{\mathrm{E},-q} = \vec{0}.$$

Choosing the centre of the dipole to be the location of the torque point, the torques associated with the electric forces are

$$\vec{\tau}_{\mathrm{E},+q} = (a\,\hat{r}_+) \times \vec{F}_{\mathrm{E},+q} = a\,q\,\hat{r}_+ \times \vec{E}_0 \qquad \text{and}$$
$$\vec{\tau}_{\mathrm{E},-q} = (a\,\hat{r}_-) \times \vec{F}_{\mathrm{E},-q} = -a\,q\,\hat{r}_- \times \vec{E}_0 = +a\,q\,\hat{r}_+ \times \vec{E}_0.$$

Therefore the net electric torque exerted about the midpoint of the dipole is

$$\vec{\tau}_{\mathrm{E,net}} = \vec{\tau}_{\mathrm{E},+q} + \vec{\tau}_{\mathrm{E},-q} = 2\,a\,q\,\hat{r}_+ \times \vec{E}_0 = \vec{p} \times \vec{E}_0.$$

Recall that $\vec{p}$ is the **electric dipole moment vector**.

ASIDE: Just for grins, one might compute the torque about some other, less symmetric torque point. So, for example, the electric torques exerted about the point occupied by the negative charge in the dipole are

$$\vec{\tau}_{\mathrm{E},+q} = (2\,a\,\hat{r}_+) \times \vec{F}_{\mathrm{E},+q} = 2\,a\,q\,\hat{r}_+ \times \vec{E}_0 \qquad \text{and} \qquad \vec{\tau}_{\mathrm{E},-q} = -q\,(\vec{0}) \times \vec{F}_{\mathrm{E},-q} = 0,$$

and hence the net electric torque is

$$\vec{\tau}_{\mathrm{E,net}} = \vec{\tau}_{\mathrm{E},+q} + \vec{\tau}_{\mathrm{E},-q} = 2\,a\,q\,\hat{r}_+ \times \vec{E}_0 = \vec{p} \times \vec{E}_0.$$

We knew that this had to be the case, because the net force on the dipole vanishes, and thus the TORQUE UNIQUENESS THEOREM [VOLUME I, Chapter 44] applies.

Let's now investigate the torque exerted upon a rectangular loop of current by a uniform magnetic field oriented so as to graze[1] the surface formed by its edges.

ASIDE: **Q:** Why rectangular? Why grazing?

A: This is to facilitate the computation. Grazing incidence happens to maximise the effect that we seek to exhibit, while also reducing the complexity of the calculation. Emboldened by the experience gained here, we shall find the somewhat more generic case, to be examined in Chapter 26, less daunting.

A region of space is suffused with a uniform magnetic field. Coordinates may be chosen such that $\vec{B}_0 = B_0\,\hat{\imath}$, or, equivalently, $\vec{B}_0 = (B_x\,,\,B_y\,,\,B_z) = (B_0\,,\,0\,,\,0)$.

Suppose that a rectangular loop of wire, with sides of length a and b, lies in the xy-plane and carries constant current I. Further suppose that the a-sides are parallel to the x-axis, while the b-sides lie in the y-direction. In accord with Figure 25.4, we label the edges $\{1, 2, 3, 4\}$, proceeding anti-clockwise from the lower right corner.

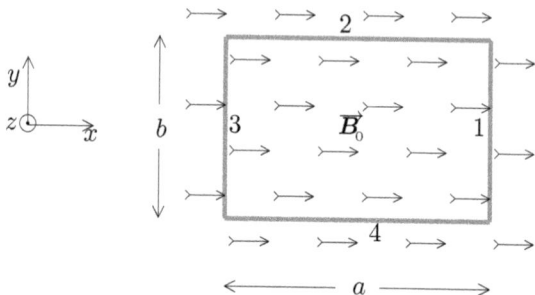

FIGURE 25.4   A Current-Carrying Loop Immersed in a Grazing Magnetic Field

ASIDE: Of course, any real wire loop is never quite rectangular. The corners are always at least slightly rounded, to spare the charge carriers the infinite acceleration that accompanies a ninety-degree shift in [drift] velocity. We choose to ignore the rounding, instead pretending that each of the four segments comprising the loop is perfectly straight.

The recently derived result for the magnetic force acting upon a straight current-carrying segment of wire,

$$\vec{F}_i = I\,\vec{L}_i \times \vec{B}_0\,,$$

applies to each of the four segments. The same current passes through each segment and the field is uniform throughout the entire region of space occupied by the loop.

Application of the generic formula to each segment in turn yields the following.

1   The current in the first segment is directed toward the top of the page, and thus $\vec{L}_1 = b\,[\uparrow]$, leading to

$$\vec{F}_1 = I\,(b\,[\uparrow]) \times B_0\,[\rightarrow] = I\,b\,B_0\,[\otimes] = -I\,b\,B_0\,\hat{k}\,.$$

---

[1] The field that grazes the loop is everywhere non-zero, and yet no flux passes through the loop. Magnetic flux will be discussed extensively in Chapter 33 *et seq.*

2   The second segment's current is to the left in the page, hence $\vec{L}_2 = a \ [\leftarrow]$, and

$$\vec{F}_2 = I \left(a \ [\leftarrow] \right) \times B_0 \ [\rightarrow] = I \, a \, B_0 \ [\leftarrow \times \rightarrow] = \vec{0} \,.$$

3   The current in the third segment is directed toward the bottom of the page. Thus $\vec{L}_3 = b \ [\downarrow]$, and

$$\vec{F}_3 = I \left(b \ [\downarrow] \right) \times B_0 \ [\rightarrow] = I \, b \, B_0 \ [\odot] = I \, b \, B_0 \ \hat{k} \,.$$

4   The fourth segment's current flows to the right. Therefore $\vec{L}_4 = a \ [\rightarrow]$, and

$$\vec{F}_4 = I \, a \, B_0 \ [\rightarrow \times \rightarrow] = \vec{0} \,.$$

Evidently, the net Lorentz force acting on the rectangular loop vanishes, *i.e.*,

$$\vec{F}_{\text{net}} = \vec{F}_1 + \vec{F}_2 + \vec{F}_3 + \vec{F}_4 = -I \, b \, B_0 \ \hat{k} + \vec{0} + I \, b \, B_0 \ \hat{k} + \vec{0} = \vec{0} \,.$$

While it is edifying to see that net magnetic force acting on the current loop is zero, we recall that our goal has been to compute the net torque.[2]

The net magnetic torque acting on the rectangular loop is the sum of the torques exerted by the four magnetic forces. Rather than compute the torque about a single point [which is the most general case], we simplify somewhat and compute the torque about the axis passing through the midpoints of sides 2 and 4, parallel to the y-axis. Three separate views of the loop and the axis are presented in Figures 25.5 and 25.6, accompanying the torque computations.

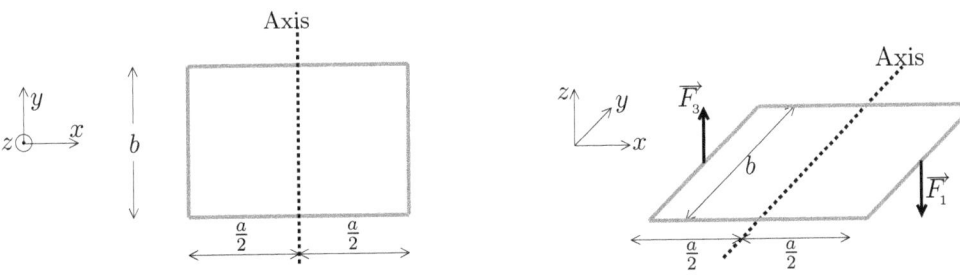

FIGURE 25.5   Dynamics of the Current-Carrying Loop in a Grazing Magnetic Field

1   The torque exerted by the magnetic force acting on the first segment, about the axis, is $\vec{\tau} = \vec{r}_1 \times \vec{F}_1$. In this case, $\vec{r}_1 = \frac{a}{2} \, \hat{\imath}$, $\vec{F}_1 = I \, b \, B_0 \ [-\hat{k}]$, and thus

$$\vec{\tau}_1 = \vec{r}_1 \times \vec{F}_1 = \frac{1}{2} I \, a \, b \, B_0 \ [\hat{\imath} \times (-\hat{k})] = \frac{1}{2} I \, a \, b \, B_0 \ \hat{\jmath} \,.$$

---

[2]Significant progress has been made. The vanishing of the net magnetic force ensures [via the TORQUE UNIQUENESS THEOREM] that the result we obtain for the torque is independent of the point about which we compute its value.

2   There is no torque exerted by this force, on account of its vanishing.

$$[\text{Zero force} \implies \text{zero torque!}]$$

One might also deem the magnetic force to be acting at the midpoint of the straight segment [which is bisected by the axis]. The moment arm of the force would thus vanish, and so too would the torque contribution.

3   The torque about the axis exerted by the force acting on the third segment is $\vec{\tau}_3 = \vec{r}_3 \times \vec{F}_3$. In this instance, $\vec{r}_3 = \frac{a}{2}\left[-\hat{\imath}\right]$, $\vec{F}_3 = I\,b\,B_0\,[\hat{k}]$, so

$$\vec{\tau}_3 = \vec{r}_3 \times \vec{F}_3 = \frac{1}{2}\,I\,a\,b\,B_0\left[-\hat{\imath}\times\hat{k}\right] = \frac{1}{2}\,I\,a\,b\,B_0\,\hat{\jmath}.$$

4   This torque vanishes for the same reasons as mentioned above for segment 2.

Stripped of all extraneous detail, the situation is as sketched in Figure 25.6.

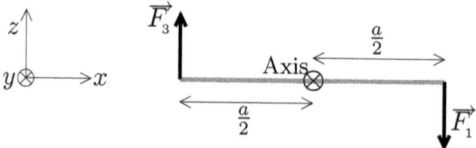

FIGURE 25.6   A Rectangular Current Loop Situated in a Uniform Magnetic Field

The net torque exerted on the rectangular loop by the magnetic forces acting on the current is

$$\vec{\tau}_{M,\text{net}} = \sum_{i=1}^{4} \vec{\tau}_i = \frac{1}{2}\,I\,a\,b\,B_0\,\hat{\jmath} + \vec{0} + \frac{1}{2}\,I\,a\,b\,B_0\,\hat{\jmath} + \vec{0} = I\,a\,b\,B_0\,\hat{\jmath}.$$

Here are a few comments.

DIRECTION   The net torque is directed along the axis, which makes sense when $\vec{F}_1$ and $\vec{F}_3$ are identified as a "couple," and justifies our computing the torque about the axis, rather than a particular point in space.

COEFFICIENT   The magnitude of the torque depends on a combination of factors.

I	The current circulating in the loop	EXPECTED
B_0	The strength of the ambient magnetic field	EXPECTED
a b	The area of the loop	NOVEL

Chapter 26

Magnetic Torque on Current Loops

In Chapter 25, we obtained an expression for the net torque exerted on a rectangular loop of current by a uniform magnetic field barely grazing the loop. We found that, in addition to the expected dependences on the magnitudes of the current and the field, there appeared a factor identifiable as the area of the loop. A more general case situates the rectangular current loop obliquely in the uniform magnetic field. Let's tackle this case now.

A region of space is suffused with a uniform magnetic field. Once again, coordinates may be chosen such that $\vec{B_0} = B_0\,\hat{\imath}$, or, equivalently, $\vec{B_0} = (B_x\,,\,B_y\,,\,B_z) = (B_0\,,\,0\,,\,0)$.

A rectangular loop of wire, with sides of length a and b, lies with its b edges parallel to the y-axis and its a edges confined to planes parallel to the xz-plane. The situation is illustrated in Figure 26.1, where we again employ the edge-numbering scheme used in the previous chapter.

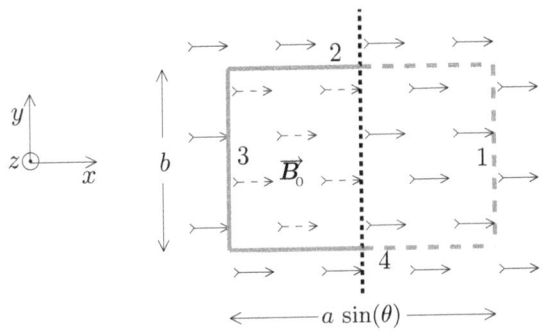

FIGURE 26.1 A Current-Carrying Loop Immersed Obliquely in a Magnetic Field

To quantify the direction, we specify the angle that the normal to the rectangular loop makes with respect to the x-axis, θ.

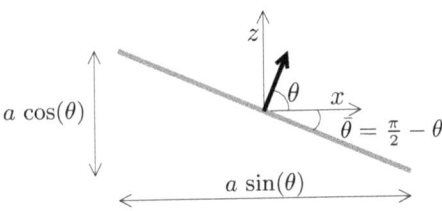

FIGURE 26.2 The Rectangular Planar Loop Lying Obliquely in the Uniform Field

Each of the straight segments of wire, bearing current I through the uniform field, experiences a magnetic force, $\vec{F}_i = I\,\vec{L}_i \times \vec{B}_0$.

1   The current in the first segment flows in the $\hat{j}$ direction, and thus $\vec{L}_1 = b\,\hat{j}$. Hence,

$$\vec{F}_1 = I\left(b\,[\hat{j}]\right) \times B_0\,[\hat{\imath}] = I\,b\,B_0\,[\hat{j} \times \hat{\imath}] = I\,b\,B_0\,[-\hat{k}] = -I\,b\,B_0\,\hat{k}.$$

2   The second segment's current is in the direction $[-\sin(\theta)\,\hat{\imath} + \cos(\theta)\,\hat{k}]$, and therefore $\vec{L}_2 = a\,[-\sin(\theta)\,\hat{\imath} + \cos(\theta)\,\hat{k}]$. The force on this segment does not vanish:

$$\vec{F}_2 = I\left(a\,[-\sin(\theta)\,\hat{\imath} + \cos(\theta)\,\hat{k}]\right) \times B_0\,[\hat{\imath}] = I\,a\,B_0\,\cos(\theta)\,[\hat{k} \times \hat{\imath}] = I\,a\,B_0\,\cos(\theta)\,\hat{j}.$$

3   The current in the third segment is directed anti-parallel to the y-axis. Thus, $\vec{L}_3 = b\,[-\hat{j}]$, and

$$\vec{F}_3 = I\left(b\,[-\hat{j}]\right) \times B_0\,[\hat{\imath}] = I\,b\,B_0\,[-\hat{j} \times \hat{\imath}] = I\,b\,B_0\,\hat{k}.$$

4   The fourth segment's current flows oppositely to that in the second segment. Therefore, $\vec{L}_4 = a\,[\sin(\theta)\,\hat{\imath} - \cos(\theta)\,\hat{k}]$, and

$$\vec{F}_4 = I\,a\,B_0\,\left[-\cos(\theta)\,\hat{k} \times \hat{\imath}\right] = -I\,a\,B_0\,\cos(\theta)\,\hat{j}.$$

The net Lorentz force acting on the oblique rectangular loop vanishes:

$$\vec{F}_{\text{net}} = \vec{F}_1 + \vec{F}_2 + \vec{F}_3 + \vec{F}_4 = -I\,b\,B_0\,\hat{k} + I\,a\,B_0\,\cos(\theta)\,\hat{j} + I\,b\,B_0\,\hat{k} - I\,a\,B_0\,\cos(\theta)\,\hat{j} = \vec{0}.$$

The net magnetic torque acting on the rectangular loop is the sum of the torques exerted by the four magnetic forces. We shall compute these torques about the axis parallel to the y-axis and passing through the midpoint of sides 2 and 4. Three separate views of the loop and the axis are presented in Figures 26.3 and 26.4, alongside the torque computations.

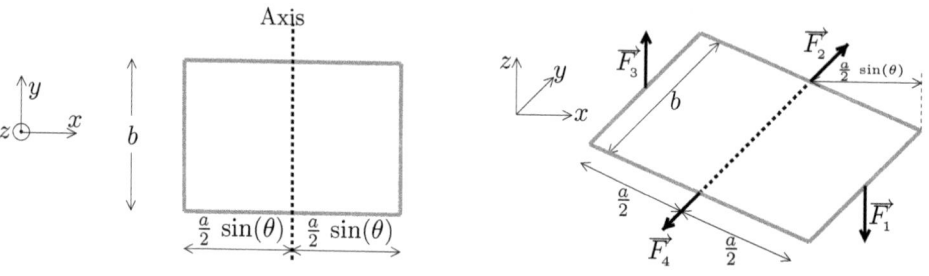

FIGURE 26.3 Dynamics of the Rectangular Current Loop

1   The torque about the axis exerted by the magnetic force acting on the first segment is $\vec{\tau}_1 = \vec{r}_1 \times \vec{F}_1$. In this [tilted] case, $\vec{r}_1 = \frac{a}{2}\left[\sin(\theta)\,\hat{\imath} - \cos(\theta)\,\hat{k}\right]$. The force, $\vec{F}_1 = I\,b\,B_0\,[-\hat{k}]$, was determined just above. Hence,

$$\vec{\tau}_1 = \vec{r}_1 \times \vec{F}_1 = \frac{1}{2}\,I\,a\,b\,B_0\,\sin(\theta)\left[\hat{\imath} \times (-\hat{k})\right] = \frac{1}{2}\,I\,a\,b\,B_0\,\sin(\theta)\,\hat{\jmath}.$$

2   Although the force on the second segment is non-zero, the torque that this force produces about the y-axis vanishes since it has vanishing moment arm.

[Should you have reservations about this argument, another (incontrovertible) line of reasoning is advanced further below.]

3   The torque about the axis exerted by the force acting on the third segment is $\vec{\tau}_3 = \vec{r}_3 \times \vec{F}_3$, where $\vec{r}_3 = \frac{a}{2}\left[-\sin(\theta)\,\hat{\imath} + \cos(\theta)\,\hat{k}\right]$ and $\vec{F}_3 = I\,b\,B_0\,\hat{k}$. Thus,

$$\vec{\tau}_3 = \vec{r}_3 \times \vec{F}_3 = \frac{1}{2}\,I\,a\,b\,B_0\,\sin(\theta)\left[-\hat{\imath} \times \hat{k}\right] = \frac{1}{2}\,I\,a\,b\,B_0\,\sin(\theta)\,\hat{\jmath}.$$

4   This torque vanishes for the same reasons as quoted above for segment 2.

Close inspection of the figure showing the forces acting on the loop reveals that the force pair $\vec{F}_2$ and $\vec{F}_4$ forms a **couple** sharing the same line of action. Together, they cannot possibly produce a net non-zero torque.

Stripped of all extraneous detail, the situation is as sketched in Figure 26.4. The figure shows clearly that the moment arms associated with the forces acting on the first and third segments are foreshortened.

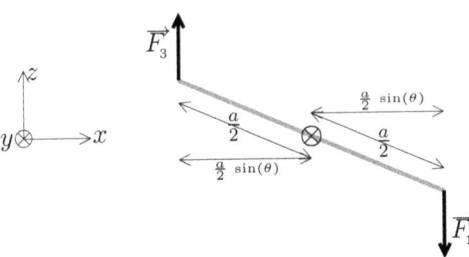

FIGURE 26.4   An Oblique Current Loop Situated in a Uniform Magnetic Field

The net torque exerted on the rectangular loop by the magnetic forces acting on the current that it carries is

$$\vec{\tau}_{M,\text{net}} = \sum_{i=1,3}\vec{\tau}_i = \frac{1}{2}\,I\,a\,b\,B_0\,\sin(\theta)\,\hat{\jmath} + \frac{1}{2}\,I\,a\,b\,B_0\,\sin(\theta)\,\hat{\jmath} = I\,a\,b\,B_0\,\sin(\theta)\,\hat{\jmath}.$$

The direction and the magnitude of the [vector] torque merit comment, as does the question of whether this result is consistent with the grazing field case studied in Chapter 25.

DIRECTION   The net torque is directed along the y-axis.

[This validates our choice to evaluate the torque about this axis.]

COEFFICIENT   The magnitude of the torque depends on a combination of factors.

I	The current circulating in the loop	EXPECTED
B_0	The strength of the ambient magnetic field	EXPECTED
a b	The area of the loop	AS PREVIOUSLY
$\sin(\theta)$	sine(angle between normal and field)	NOVEL

CONSISTENCY   The result for the oblique case is completely consistent with that obtained when the field was grazing, $\theta = \pi/2$, in Chapter 25. It is also evident, from this more general expression and Figure 26.4, that the torque is maximised when the field is grazing.

────────────

Careful reflection upon the expression for the magnetic torque acting on a [planar] current loop suggests the general form:

$$\vec{\tau}_{M,I} = I\,\vec{A} \times \vec{B}_0\,,$$

where the **area vector** associated with the loop, $\vec{A}$, has the following properties.

$|\vec{A}|$ Its magnitude is equal to the [planar] surface area enclosed by the loop.

$\hat{A}$ It is directed perpendicular to the plane containing the loop, with its orientation assigned by a RIGHT HAND RULE [RHR].

RHR   (1) Align the digits of your right hand with the direction of the electric current flowing in the wire loop.

(2) Rotate your arm so that your right palm faces inwards [*i.e.*, toward the wire on the other side of the loop].

[This is always possible and unambiguous (even for very convoluted loops).]

(3) Your right thumb indicates the direction along the perpendicular axis in which the area vector points.

The structure of the expression for the torque cries out for a deeper interpretation.

────────────

MAGENTIC DIPOLE MOMENT   The magnetic dipole moment of a planar current loop is

$$\vec{m} = I\,\vec{A}\,.$$

The magnetic moment is a vector directed perpendicular to the plane of the loop, with its orientation dictated by the direction of flow of the electric current. It is directly proportional to the magnitudes of the current circulating in, and the area enclosed by, the loop. The SI units associated with $\vec{m}$ are $\text{A} \cdot \text{m}^2$. With the substitution $I\,\vec{A} = \vec{m}$, the expression for the torque on the current loop reads:

$$\vec{\tau}_{M,I} = \vec{m} \times \vec{B}_0\,.$$

It is precisely this compact form which inspires the identification of $\vec{m}$ as the magnetic dipole moment, in light of the response of an electric dipole situated in a uniform electric field [*cf.* Chapter 25, most recently, as well as Chapter 16].

$$\begin{array}{c|c}
\vec{\tau}_{E,dipole} = \vec{p} \times \vec{E}_0 & \vec{p} = \text{electric dipole moment} \\
\hline
\vec{\tau}_{M,I-loop} = \vec{m} \times \vec{B}_0 & \vec{m} = \text{magnetic dipole moment}
\end{array}$$

EXAMPLE [*Magnetic Moments*]

Three loops of wire are illustrated in Figure 26.5. Each loop lies entirely in the plane of the page. The area of the ith loop is A_i, while the current in the associated wire is I_i. The magnitudes and directions of the magnetic dipole moments are quoted alongside the loops in the figure.

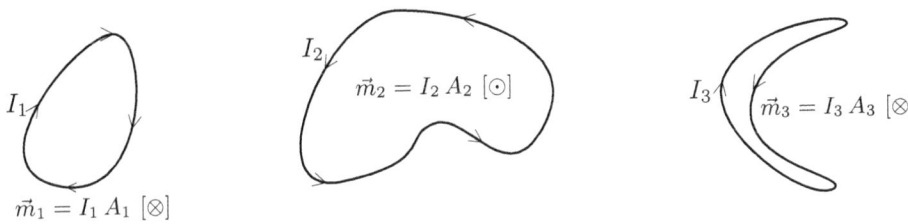

FIGURE 26.5   Magnetic Moments Associated with Three Planar Current Loops

Q: What about non-planar loops?

A: Suppose that a current, I, flows in the Salvador Dali-esque[1] loop shown in the left panel of Figure 26.6.

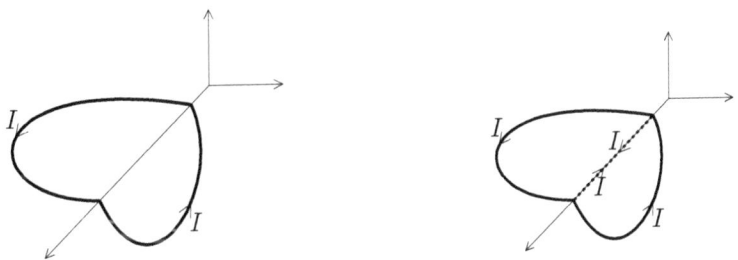

FIGURE 26.6   A Salvador Dali-esque Current Loop [Impediment AND Resolution]

The beauty of this figure is that it both illustrates the obstacle presented by the non-planar loop, and suggests a stratagem for overcoming it.

IMPEDIMENT   Unique normal vectors to non-planar loops do not exist.

RESOLUTION   The solution is to TILE the area bounded by the non-planar loop.

---

[1]This loop is evocative of a painting by Dali, *The Persistence of Memory* (1931).

[Each planar tile is deemed to have a conducting edge.]

In the example at hand, the natural tiling has two elements: one in the xy-plane, the other in the xz-plane [as shown]. The current in the loop coincident with the outside edge of the xy-tile passes along the internal tile edge, and thus the first tile [formally] represents a planar current loop. The magnetic moment associated with the current, I, encircling the xy-tile is straightforwardly computed to be

$$\vec{m}_{xy} = I\,\vec{A}_{xy} = I\,A_{xy}\,\hat{k}\,.$$

Also, the current residing in the portion of the original loop bordering the xz-tile is deemed to be directed along the interior edge, thus forming a second planar loop with

$$\vec{m}_{xz} = I\,\vec{A}_{xz} = I\,A_{xz}\,\hat{\jmath}\,.$$

Three comments follow.

I_{net} The common edge of the two tiles accommodates oppositely directed flows of current. Thus, the net current flowing in the interior of the overall loop vanishes, while the current everywhere on the perimeter of the original loop is I. In this way, the tiling and redirection of currents have not affected the overall characteristics of the original current–loop system.

$\vec{m}_{\text{net}}$ The net magnetic moment is obtained by linear superposition of the magnetic moments of the tiled regions, *i.e.*,

$$\vec{m} = \vec{m}_{xy} + \vec{m}_{xz}\,.$$

P, C, S   This has been yet another successful application of the PARTITION, COMPUTE, and SUM strategy. Here, it was the area enclosed by the loop that was subjected to partitioning.

> ASIDE: Enquiring minds wanting to know WHETHER and HOW the REFINE step might be carried out are encouraged to go look up **Green's** and **Stokes'** theorems in their favourite multivariable calculus reference.

ENERGETICS OF CURRENT LOOPS IN UNIFORM MAGNETIC FIELDS

A current loop, with magnetic moment $\vec{m}$, is situated in an ambient uniform magnetic field, $\vec{B}_0$. The net magnetic force acting on the loop vanishes, but not the net magnetic torque, $\left|\vec{m} \times \vec{B}_0\right| = m\,B_0\,\sin(\theta_{m,B})$. Suppose that the loop undergoes a rotation about an axis through its centre, perpendicular to both $\vec{m}$ and $\vec{B}_0$.

> ASIDE: The implicit assumption here is that the magnetic dipole moment and the ambient uniform field are NOT [anti-]aligned. The restriction on the motion of the loop enables us to focus exclusively on the most interesting case without distraction.

To visualise, we appeal again to Figure 26.1 and subsequent diagrams, in which the axis of rotation is parallel to the y-axis. The angle between the magnetic moment vector and the direction of the local constant field is precisely the θ employed in the figures and earlier formulae. [See Figure 26.2.] The rotation experienced by the loop is quantified by the difference between the initial and final angles, θ_i and θ_f.

The amount of mechanical work performed by the magnetic torque[2] acting on this rotating current loop is

$$W_{if}\left[\vec{\tau}\right] = \int_{\vec{\theta}_i}^{\vec{\theta}_f} \vec{\tau} \cdot d\vec{\theta} = \int_{\theta_i}^{\theta_f} -m\,B_0\,\sin(\theta)\,d\theta = m\,B_0\left[\cos(\theta)|_{\theta_i}^{\theta_f}\right]$$
$$= m\,B_0\,\cos(\theta_f) - m\,B_0\,\cos(\theta_i)\,.$$

The minus sign in the second equality arose because the directions of the torque, $\hat{\jmath}$, and increasing angle, $-\hat{\jmath}$, are anti-parallel, and thus their dot product is negative. Note that the work done depends only on the initial and final angles. This is the necessary and sufficient condition for there to exist an associated magnetic dipole potential energy function:

$$\Delta U_m = U_{m,f} - U_{m,i} = U_m(\theta_f) - U_m(\theta_i) = -W_{if}\left[\vec{\tau}\right] = -m\,B_0\,\cos(\theta_f) + m\,B_0\,\cos(\theta_i)\,.$$

The simplest potential energy function [able to reproduce the ΔU_m] is

$$U_m(\theta) = -m\,B_0\,\cos(\theta)\,,$$

where the reference zero of potential energy occurs when $\theta = \pi/2$. This corresponds to the grazing case investigated in Chapter 25. The table and the sketch in Figure 26.7 display the salient properties of this potential energy function.

Field	Moment	Angle	Potential	Energy
$\rightarrow$	$\leftarrow$	π	$+m\,B_0$	MAX
$\rightarrow$	$\uparrow$	$\pi/2$	0	ZERO
$\rightarrow$	$\rightarrow$	0	$-m\,B_0$	MIN

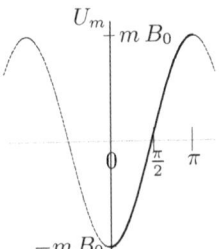

FIGURE 26.7   Magnetic Dipole Potential Energy

The subscripted "m" on the potential energy function does double duty. It reminds us that this energy is associated with a magnetic interaction, and that the magnetic dipole moment of the loop [its magnitude and relative direction with respect to the local field] is the controlling parameter. Reflecting upon the form of U_m reveals its natural [geometric] expression:

$$U_m = -\vec{m} \cdot \vec{B}\,.$$

ASIDE: Recall from Chapter 16 that the electrostatic potential energy associated with an electric dipole immersed in an electric field is given by

$$U_p = -\vec{p} \cdot \vec{E}\,.$$

Déjà vu, eh?

Our analyses of forces, torques, and energetics compel us to conclude that **a loop of electric current, in the presence of a magnetic field, acts like a magnetic dipole.**

---

[2]The mechanical work associated with a torque acting on a rigidly rotating object was discussed in Chapter 40 of VOLUME I.

ADDENDUM: A Practical Electro-Magnetic Device—the Galvanometer

For all intents and purposes, a Galvanometer, named in honour of Luigi Galvani[3] (1737–1798), is just a current loop in a [nearly] uniform magnetic field, with mechanical connections to four things, listed below.

⊥   a needle, or some other device, pointing to a calibrated scale

⊥   a [spring or torsion fibre] mechanism which provides a restoring torque opposed to the displacement of the loop from its reference or "null" position

⊥ ˙   a damping mechanism, in order that the loop and needle eventually come to rest   [*Cf.* the discussion of the DAMPED HARMONIC OSCILLATOR in VOLUME II]

⊥   input and output terminals for directing an external electric current around the loop

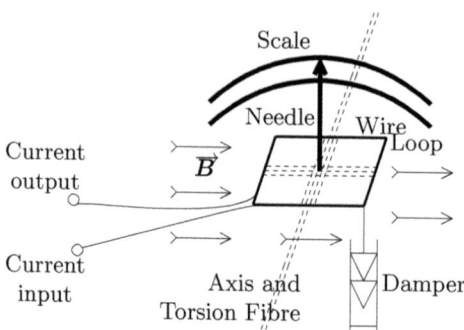

FIGURE 26.8   A Galvanometer Detects and Measures Electric Currents

IF one increases the current in the loop [say from zero], THEN the magnetic moment associated with the loop increases, thus increasing the magnetic torque on the loop. The increase in magnetic torque then requires a correspondingly larger oppositional torque [provided by the torsion fibre] to keep the needle stationary. The angle at which the needle comes to rest (after damping) can be recorded on the scale. In this way, electric currents in wires may be detected and measured!

---

[3]Galvani determined that the workings of the nervous system are essentially electrical, rather than mechanical or thermal.

Chapter 27

Back to Moving Charged Particles

Consider a particle with mass m, charge q, and velocity $\vec{v}$, moving through a region of space containing a uniform magnetic field, $\vec{B_0}$. The Lorentz force acting on the particle is

$$\vec{F}_{\text{M},q} = q\,\vec{v} \times \vec{B_0}\,.$$

Recall that the Lorentz force does no work on the charged particle.

[The force is always orthogonal to the particle's velocity.]

Provided that there are no other forces acting [or the other forces all precisely cancel], the kinetic energy of the particle remains constant.[1]

Without any loss of generality, let's choose coordinates such that the z-axis lines up with the magnetic field, so $\vec{B_0} = B_0\,\hat{k}$. Further suppose that at an initial time, t_i, the velocity of the particle is perpendicular to the magnetic field, i.e., $\vec{v} = (v_x\,, v_y\,, 0)$. The Lorentz force acting on the particle is

$$\vec{F}_{\text{M},q} = q\,(v_x\,, v_y\,, 0) \times (0\,, 0\,, B_0) = q\,B_0\left[v_y\,\hat{i} - v_x\,\hat{j}\right]\,.$$

Thus, the acceleration of the particle is

$$\vec{a} = \frac{q}{m}\,B_0\left[v_y\,\hat{i} - v_x\,\hat{j}\right] = \frac{q}{m}\,B_0\,(v_y\,, -v_x\,, 0)\,.$$

Several aspects of this result are noteworthy.

$\propto$ The acceleration of the particle is proportional to its **charge-to-mass ratio**,[2] $\frac{q}{m}$, the strength of the magnetic field, B_0, and the speed with which the particle is moving, $\left|(v_y\,, -v_x\,, 0)\right| = \sqrt{v_x^2 + v_y^2}$.

$\perp$ Although the acceleration and velocity reside in different vector spaces, it is fair to say that they are orthogonal, since $\vec{v} \cdot \vec{a} \propto (v_x\,, v_y\,, 0) \cdot (v_y\,, -v_x\,, 0) = v_x\,v_y - v_y\,v_x = 0$.

This orthogonality of velocity and acceleration has kinematic consequences. Recalling the discussions of archetypical accelerations and circular motion [VOLUME I, Chapters 6 and 8], one realises that, subsequent to t_i, the particle moves along a segment of circular arc.[3] From the definition of centripetal acceleration [found in Lecture 8 of VOLUME I],

$$\vec{a}_c = -\frac{v^2}{R} \text{ [radially inwards]}\,,$$

---

[1] The WORK–ENERGY THEOREM provides assurance of this.

[2] The charge-to-mass ratio is a characteristic, or defining, property of certain types of particles, especially those deemed fundamental.

[3] If the region of space permeated by the uniform field is sufficiently large, then the particle trajectory will wrap around a complete circle multiple times. Otherwise, the acceleration regime lasts for only a portion of one period and only a portion of a circle is traversed. Going forward, we shall assume that the motion is circular, all the while realising that the regime may end before a complete period elapses.

and the dependencies of the Lorentz [net] force, it follows that

$$\frac{v^2}{R} = \frac{q}{m} \, v \, B_0 \, .$$

Therefore,

$$R = \frac{v}{\frac{q}{m} \, B_0} = \frac{m \, v}{q \, B_0} \, .$$

Other features of the circular motion are predictable, too. For instance, the angular velocity of the particle is

$$\omega = \frac{v_{\text{tangential}}}{R} = \frac{v}{R} = \frac{q}{m} \, B_0 \, .$$

The frequency [denoted by the Greek letter ν, pronounced "nu"] of revolution[4] of the particle is

$$\nu = \frac{\omega}{2 \, \pi} = \frac{1}{2 \, \pi} \frac{q}{m} \, B_0 \, ,$$

while the associated period, T, is

$$T = \frac{1}{\nu} = \frac{2 \, \pi}{\omega} = 2 \, \pi \, \frac{m}{q \, B_0} \, .$$

All of the above quantities bear their proper SI units: m, rad/s, Hz, and s, respectively. The chief significance of these results is that the period and frequency depend exclusively on the particle's charge-to-mass ratio and the strength of the [externally prescribed] uniform magnetic field. The radius also depends on the momentum of the particle.

....

EXAMPLE [*Charged Particle Moving Orthogonally to a Uniform Magnetic Field*]

A large region of otherwise empty space [vacuum] is filled with a uniform magnetic field, $\vec{B} = 0.0625 \, \text{T} \, [\hat{k}]$. A proton, with elementary charge $+e = 1.6 \times 10^{-19}$ C and mass $m_p = 1.67 \times 10^{-27}$ kg, is injected into the middle of this expansive region with initial velocity $\vec{v}_0 = 1000 \, \text{m/s}$.

Q: Granted that the proton's subsequent trajectory is a circular path, what are the properties of this orbit?

A: The radius and the cyclotron frequency/period fix the essential properties.

Radius In this instance, the radius is

$$R = \frac{m \, v}{q \, B_0} = \frac{1.67 \times 10^{-27} \times 1000}{1.6 \times 10^{-19} \times 0.0625} \, \frac{\text{kg} \cdot \text{m/s}}{\text{C} \cdot \text{T}} = 1.67 \times 10^{-4} \, \frac{\text{kg} \cdot \text{m/s}}{\text{N}/(\text{m/s})} = 1.67 \times 10^{-4} \, \text{m} \, .$$

This fast-moving proton's orbital radius is about one-sixth of one millimetre. There should be no practical obstacle to ensuring[5] that the magnetic field is approximately uniform throughout regions of this size.

To increase the orbital radius, one might

+   increase the mass of the particle [perhaps by using a positive ion, rather than a proton],
+   increase the speed of the proton, or
+   weaken the magnetic field.

---

[4] For historical reasons, this is often called the "cyclotron frequency."
[5] How magnetic fields are produced will be discussed starting in the next chapter.

Period, Frequency   The period is

$$T = 2\pi\,\frac{m}{q\,B_0} = 2\pi\,\frac{1.67\times10^{-27}}{1.6\times10^{-19}\times0.0625}\,\frac{\text{kg}}{\text{C·T}} = 1.05\times10^{-6}\,\frac{\text{kg}}{\text{N}/(\text{m}/\text{s})}$$
$$= 1.05\times10^{-6}\text{ s} = 1.05\ \mu s\,.$$

ASIDE: The value computed for the period is kinematically commensurate with the radius determined above, as, under uniform circular motion,

$$T = \frac{\text{circumference}}{\text{speed}} = \frac{2\pi\times1.67\times10^{-4}}{1000}\,\frac{\text{m}}{\text{m}/\text{s}} = 1.05\times10^{-6}\text{ s}\,.$$

The cyclotron frequency is the reciprocal of the period, $\nu = \frac{1}{T} = 9.53\times10^5$ Hz $= 953$ kHz.

In the cases that we have examined, the initial [and subsequent] velocity of the particle was orthogonal to the field. More generally, the initial velocity may be decomposed into its parallel part [one component aligned with the uniform magnetic field] and its remaining parts [$N-1$ components perpendicular to the field]. That is,

$$\vec{v}_0 = \vec{v}_{0,\parallel} + \vec{v}_{0,\perp}\,.$$

The cross-product is bilinear in its inputs. This may be exploited to simplify the expression for the Lorentz force initially exerted upon the particle:

$$\vec{v}_0 \times \vec{B} = (\vec{v}_{0,\parallel} + \vec{v}_{0,\perp}) \times \vec{B} = \vec{v}_{0,\parallel} \times \vec{B} + \vec{v}_{0,\perp} \times \vec{B} = \vec{v}_{0,\perp} \times \vec{B}\,.$$

The component of the initial velocity parallel to the local field neither contributes to the Lorentz force, nor does it directly feel its effects.

[Zero acceleration occurs in the direction of the field.]

For the perpendicular components of the velocity, the situation is identical to the restricted case discussed at the start of this chapter.

The parallel and perpendicular motions decouple. Thus, the trajectory of the particle is helical: a superposition of inertial motion in the parallel direction and uniform circular motion in the [comoving] perpendicular plane.

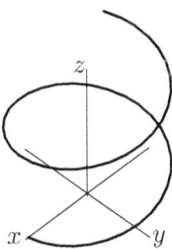

FIGURE 27.1   A Portion of the Helical Trajectory of a Charged Particle Moving in a Uniform Magnetic Field

Hey, wait a minute! We saw much earlier that a charged particle residing in an electric field feels an electric force, and more recently that a charged particle moving through a magnetic field experiences a magnetic force. As there seems to be no *a priori* reason why electric and magnetic fields might not co-exist, the general expression for the force acting on a [point-like] particle bearing charge q is[6]

$$\vec{F}_{\text{EM}} = q\,\vec{E} + q\,\vec{v} \times \vec{B} = q\,[\,\vec{E} + \vec{v} \times \vec{B}\,]\,.$$

A particular arrangement of electric and magnetic fields, termed **crossed fields**, possesses attractive features and merits explicit analysis.

CROSSED FIELDS The epithet, crossed, vividly describes the property

$$\vec{E} \perp \vec{B}$$

everywhere throughout a region of space.

Suppose that [approximately uniform] crossed electric and magnetic fields are found within a region of space. WLOG we employ Cartesian coordinates, $(x,\,y,\,z)$, and write these fields as:

$$\vec{E}_0 = E_0\,[-\hat{k}\,] = -E_0\,\hat{k} \qquad \text{and} \qquad \vec{B}_0 = B_0\,\hat{j}\,.$$

A charged particle is injected into this region, with an initial velocity perpendicular to both the electric and magnetic fields, *i.e.*,

$$\vec{v}_0 = v_0\,\hat{\imath}\,.$$

The physical situation is displayed in Figure 27.2.

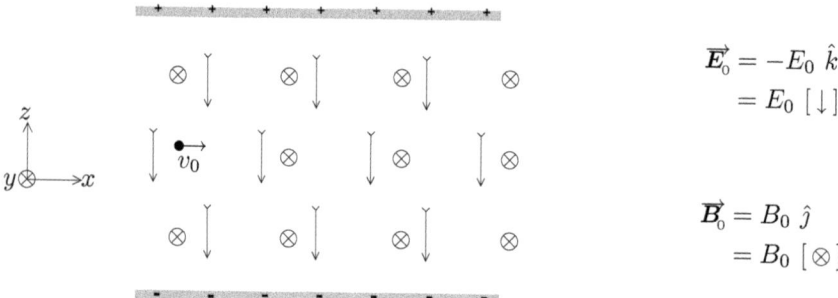

FIGURE 27.2 A Charged Particle Moving Orthogonally through Crossed Fields

The electric and magnetic forces acting on the charged particle are $\vec{F}_{\text{E}} = -q\,E_0\,\hat{k}$ and $\vec{F}_{\text{M}} = q\,v_0\,B_0\,\hat{k}$, respectively. Thus, the net [electromagnetic] force is

$$\vec{F}_{\text{EM}} = q\,[v_0\,B_0 - E_0]\,\hat{k}\,.$$

A positively charged particle must then behave in one of three distinct possible manners, depending on its speed.

---

[6]Some authors call this the Lorentz force. We demur, restricting the appellation "Lorentz" to the magnetic part only.

$-\hat{k}$ IF $v_0 < \frac{E_0}{B_0}$, *i.e.*, the particle is slow-moving, **THEN** the electric force dominates, **AND** the particle accelerates in the $-\hat{k}$ direction.

$\vec{0}$ IF $v_0 = \frac{E_0}{B_0}$, the *Goldilocks* speed, **THEN** the electric and magnetic forces cancel exactly, **AND** the particle moves inertially through the region.

$+\hat{k}$ IF $v_0 > \frac{E_0}{B_0}$, *i.e.*, the particle is fast-moving, **THEN** the magnetic force is the greater, **AND** the particle accelerates in the $\hat{k}$ direction.

IF the particle bears negative charge, **THEN** the above analysis is qualitatively correct, except that all of the accelerations are reversed.

Crossed fields are usefully employed to discriminate between, or filter, charged particles based upon their speed.[7]

THE VELOCITY SELECTOR

Imagine that a collimated beam of identical[8] charged particles has been prepared. By "collimated" we mean that the particles are made to move in [approximately] the same direction. No specification or restriction is made on the speeds of the particles.

ASIDE: The typical, if wasteful, way to prepare such a beam is to pass particles emanating from a source through two, or more, collimating slits. Two conditions must be met for a given particle to appear in the output beam: it must pass through the first slit, and its velocity must lie within a small cone of allowed angles. All other particles from the source are rejected.

Without loss of generality, we align the x-axis with the central direction of the collimated beam, and assume that all of the particles have velocities parallel to $\hat{i}$. Let the index n label the particles which comprise the beam during some specified time interval. The velocity of the nth particle is $\vec{v}_n = v_n\,\hat{i}$.

The individual particles are expected to have differing speeds. And yet, meaningful information is found in the population-aggregated data. The **velocity distribution** gives the fraction of particles in the beam with a particular speed, as illustrated in Figure 27.3. The **kinetic energy distribution** provides the fraction of particles with a given kinetic energy.

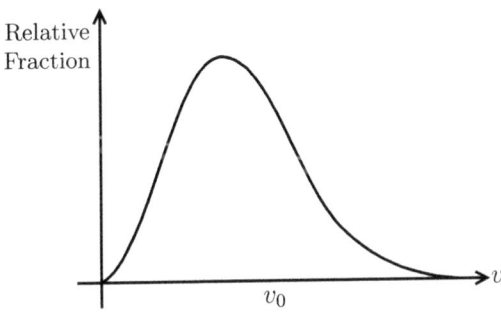

FIGURE 27.3   A Distribution of Speeds in a Collimated Beam of Charged Particles

---

[7] Such speed discrimination is tantamount to kinetic energy selection, too.

[8] Okay, this is overly restrictive. All that we really require is that all particles in the beam bear the same electric charge. Demanding that they are identical obviates [minor] complications.

Let's make a few incisive comments about the velocity distribution.

o   Figure 27.3 is meant to be heuristic. The particular shape of the distribution of speeds is strongly influenced by the circumstances under which the beam of particles was produced.

o   The **spectral radiance** [the energy intensity of electromagnetic radiation as a function of wavelength or frequency, *cf.* VOLUME II, Chapter 41] is analogous to the kinetic energy distribution.

o   The continuous distribution is an approximation to the discrete, binned histograms that arise for finite numbers of particles in the beam and coarse-grained velocity discrimination.

o   The fraction of particles drops to zero in the limit as $v \to 0$. Particles at rest will certainly not appear in the beam.

o   The velocity distribution function also tends toward zero in the high-speed limit. Rapidly increasing amounts of kinetic energy must be provided to accelerate particles to ever-higher speeds.

Suppose now that the collimated beam of charged particles proceeds into a region of crossed electric and magnetic fields as portrayed in Figure 27.2. Beyond the crossed field zone, there lies another collimating slit, aligned with the first.

The strengths of the electric and magnetic fields are [at least to a some extent] externally controllable, and thence their ratio, $v_0 = E_0/B_0$, may be prescribed. Particles with speed[9] v_0 move inertially, while all others are deflected and fail to negotiate the second slit.

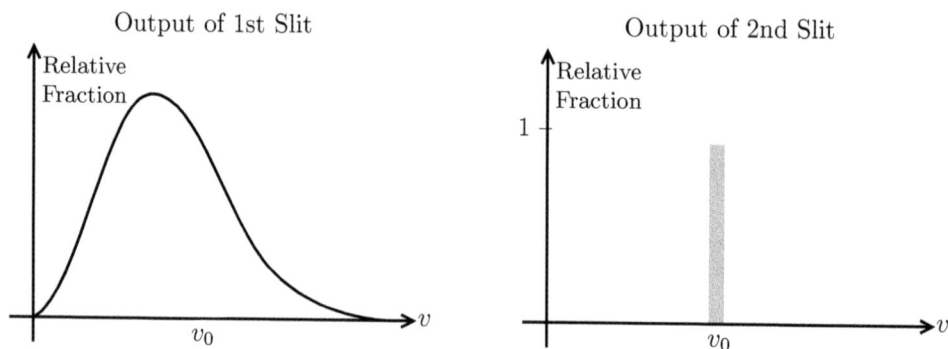

FIGURE 27.4 The Action of a Velocity Selector

By tuning the ratio of strengths of the crossed electric and magnetic fields, one is able to select the speed of the charged particles emerging from the second collimating slit.

This device is called a velocity selector!

---

[9]One really ought to say "within a narrow range of speeds, centred on v_0."

Chapter 28

The Hall Effect

Thus far, in all cases where we have encountered electric currents, our analyses did not depend on, nor differentiate between, the possible signs of the charge borne by the carriers. In other words, a given phenomenological [macroscopic] current could be achieved [microscopically] by either positive charge carriers drifting in the forward direction, or negative charge carriers drifting in the reverse direction. Nothing that has been considered until now allows one to distinguish between the two possible microscopic scenarios.

> ASIDE: These scenarios may be extended to allow for a variety of particles bearing distinct charges and drifting at different rates.

It is through the Hall Effect, in which the sign of the charge carriers matters crucially, that we know that the carriers are almost always electrons.

HALL EFFECT: POSITIVE CHARGE CARRIERS

The microscopic model of a current, I, flowing through a straight segment of conducting wire, was developed in Chapter 17 and employed to good effect in Chapter 24. Two novel augmentations are introduced here. The first is that the current flows uniformly through a wire of rectangular cross-section. The second is that a uniform magnetic field, $\vec{B}_0$, directed perpendicular both to the wire [and hence to the current] and to one of its planar faces, bathes the conductor. Figure 28.1 presents, in cross-section, the assumptions about the wire, the current, and the field, without additional preconceptions.

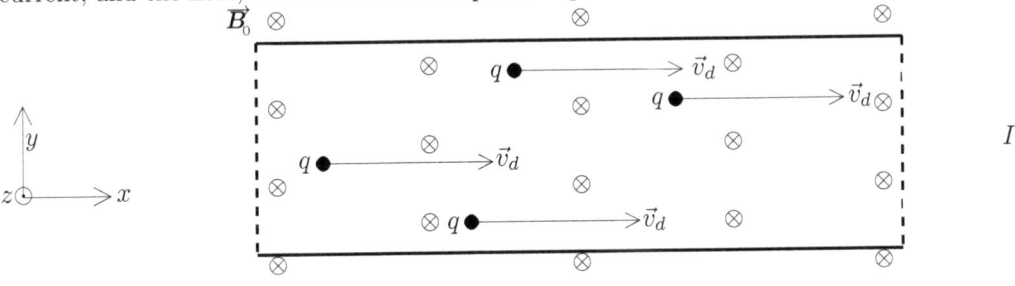

FIGURE 28.1   Flowing Positive Charge Carriers Bathed in a Uniform Magnetic Field

As the charge carriers bear positive charge, the magnetic force acting on each of them is[1]

$$\vec{F}_{\mathrm{M}} = q\,\vec{v}_d \times \vec{B}_0 = q\,v_d\,[\hat{\imath}] \times B_0\,[-\hat{k}] = q\,v_d\,B_0\,\hat{\jmath}.$$

Thus, there is [in an aggregated sense] a magnetic force acting in the $+\hat{\jmath}$ direction on the [positive by assumption] charge carriers. In response to this force, the carriers deflect toward

---

[1] Recall the CAVEATS expressed in Chapter 24 regarding the use of the drift velocity in the computation of magnetic forces.

the top of the wire. Inexorably, with the passage of time, **a surplus of positive charge accrues on top, while a deficit of positive** [equivalent to a surplus of negative] **charge accumulates on the bottom of the wire.**

±   Separation of charge **induces** a transverse[2] electric field inside the wire.

±   Even when steady state conditions [*e.g.*, a uniform current] are established, the wire is NOT in electrostatic equilibrium.

The induced electric field is directed from the surplus of positive charge on the top of the wire, to the excess of negative charge on the bottom of the wire. Homogeneity of the wire and uniformity of its rectangular geometry together determine the induced electric field to be uniform throughout the wire. Although the induced field acts in opposition, charge separation continues, albeit at an increasingly slower rate. Eventually, enough charge separates to produce an internal electric field of strength sufficient to cancel the magnetic force acting on the drifting charge carriers. That is,

$$\vec{0} = \vec{F}_{\text{EM}} = q\,(v_d\,B_0 - E_i)\,\hat{j}\,,$$

where E_i is the asymptotic strength of the electric field induced in the wire. Once this happens, the charge carriers resume drifting uniformly [as they would in the absence of the applied magnetic field].

ASIDE: Upon reflection, one realises that the wire polarises just enough to act as a velocity selector tuned to the drift speed of the charge carriers.

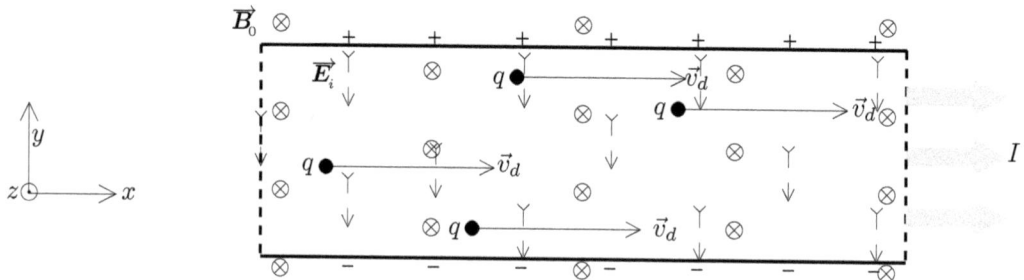

FIGURE 28.2 Positive Charge Carriers Bathed in an Applied Magnetic Field and Induced Electric Field

Application of a magnetic field has led to the formation of a countervailing induced electric field oriented transversely within the wire. Hence, there exists a potential difference, the **Hall Potential**, across the wire [orthogonal to both the current/wire and the applied magnetic field]. With the charge carriers assumed to be positive, the Hall Potential,

$$V_{\text{Hall}} = V_{\text{top}} - V_{\text{bottom}} > 0\,,$$

is certain to be POSITIVE.

ASIDE: In the model system we are considering, a rectangular wire bathed in a uniform perpendicular magnetic field, the Hall Potential may be straightforwardly computed:

$$V_{\text{Hall}} = E_i\,\Delta Y = v_d\,B_0\,\Delta Y.$$

---

[2]This induced field is distinct from the electric field within the wire, acting in the $\hat{i}$ direction, which drives the steady state electric current.

While correct, this expression depends on the drift speed, which is not directly observable. Recognising that $I = q \eta v_d A$ for uniform and homogeneous wires and $A = \Delta Y \Delta Z$ [ΔY and ΔZ are the rectangular edge lengths of the wire] allows re-expression of the Hall Potential entirely in terms of experimentally accessible quantities:

$$V_{\text{Hall}} = \frac{I B_0 \Delta Z}{\eta q} \,.$$

Hall Effect: Negative Charge Carriers

Let's repeat the analysis, changing only the assumed sign of the charge borne by the carriers. Again, starting without preconceptions, the wire, current, and applied magnetic field are illustrated in Figure 28.3. The magnetic force exerted on these $q < 0$ carriers is:

$$\vec{F}_{\text{M}} = q\,\vec{v}_d \times \vec{B}_0 = q\,v_d\,B_0 \left[-\hat{\imath} \times (-\hat{k}) \right] = q\,v_d\,B_0\,[-\hat{\jmath}] = |q|\,v_d\,B_0\,\hat{\jmath}\,.$$

Surprisingly, the magnetic force acting on the negative charge carriers also acts to deflect them in the $+\hat{\jmath}$ direction, *i.e.*, toward the top of the wire.

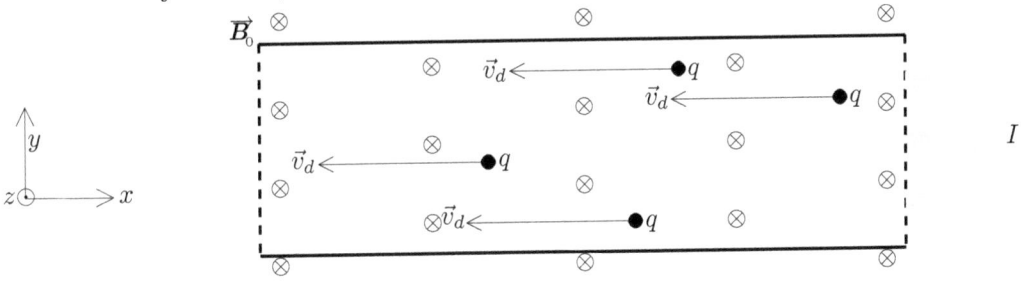

FIGURE 28.3 Flowing Negative Charge Carriers in an Applied Magnetic Field

In contrast to our earlier analysis of the positive case, the deflection of the negative charge carriers to the top of the wire gives rise to a deficit of negative charge near the bottom surface. This **polarisation** within the wire leads inevitably to the establishment of an induced electric field. This induced field opposes additional charge separation, and thus a steady state is approached in which the upward magnetic force is precisely cancelled by the downward electric force.

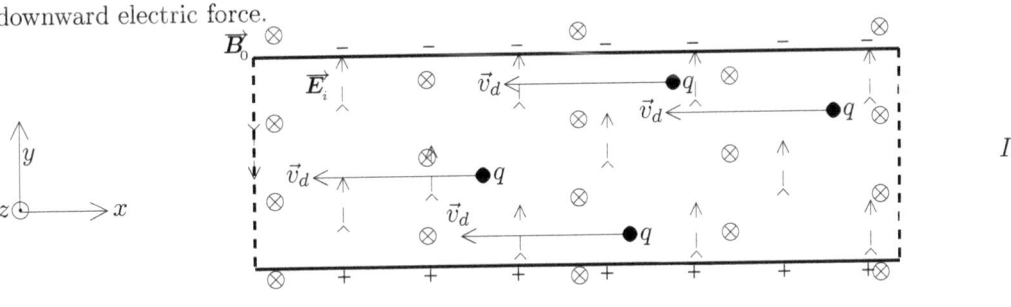

FIGURE 28.4 Flowing Negative Charge Carriers in an Applied Magnetic Field and Induced Electric Field

Here too, the wire acts as a velocity selector tuned to the drift velocity.

A more accurate representation of the steady state physical situation appears in Figure 28.4. The induced electric field within the wire produces a Hall Potential difference between the upper and lower faces of the rectangular wire,

$$V_{\text{Hall}} = V_{\text{top}} - V_{\text{bottom}} < 0 \,,$$

which is certain to be NEGATIVE, owing to the orientation of the field.

———————

Edwin Hall (1855–1938) realised that this provides an unambiguous means of determining [within a single species charge carrier model] the sign of the charge borne by the charge carriers. *Circa* 1879, Hall measured his eponymous potential to be NEGATIVE and thus concluded that (in ordinary circumstances) **positive [forward-directed] current arises from the microscopic drift of negatively charged particles in the reverse direction.**

> ASIDE: Afterward, *circa* 1895, J.J. Thomson (1858–1940), a British physicist and Nobel Laureate, immersed himself in the study of electric currents passing through samples of rarefied gases. In the context of these **cathode ray tube**[3] **experiments**, Thomson discovered a tiny [point-like] negatively charged particle which was later dubbed the "**electron**." Physicists of the day *put two and two together*, identifying Hall's negative bearers of electric charge with Thomson's electrons.
>
> Thomson's experiments yielded the first direct evidence of subatomic structure. At the end of the nineteenth century [*la fin de siècle*], the atomic hypothesis had gained widespread, if grudging, acceptance. Thomson showed that chemical atoms were not "indivisible," and hence not quite "atomic" in the Greek sense.

———————

MAGNETIC SOURCES

In recent chapters, we have studied the effects of magnetic fields on moving charged particles and currents. At this juncture we are prepared to tackle the question of HOW magnetic fields are produced.

The short answer is:

Magnetic Fields are Produced by Moving Charges!

We'll elaborate on this in the remainder of this chapter and in several more to come.

> ASIDE: Do not be unduly troubled by the prospect of [unphysical] self-interaction, *viz.*, a moving charged particle responding to the magnetic field that its own motion produces.[4] Rest assured that, as long as the proper distinction is made between the source of the field and the particle(s) upon which the field acts, no logical conundra arise.

———————

[3]These were progenitors of the CRTs mentioned in Chapter 2.

[4]This same pitfall was avoided earlier in our investigations by the tacit understanding that an electrically charged particle is not acted upon by its own electric field.

LAW of BIOT and SAVART   Current I, flowing through differential directed segment of wire $d\vec{s}$, is the source of a differential contribution to the net magnetic field at a point, $\mathcal{P}$, located at position $\vec{r}$ with respect to the segment of wire. This differential field has magnitude and direction specified by

$$d\boldsymbol{B}(\mathcal{P}) = \frac{\mu_0}{4\pi}\frac{I\,d\vec{s}\times\hat{r}}{r^2}\,, \qquad \text{for} \quad \vec{r} = r\,\hat{r}\,.$$

An illustrative example involving a current-bearing wire, two current elements, and a common field point appears in Figure 28.5. The various factors in the expression for $d\boldsymbol{B}(\mathcal{P})$ are elaborated upon further below. The net magnetic field at a specific field point is obtained by integrating over the current elements:

$$\boldsymbol{B}(\mathcal{P}) = \int d\boldsymbol{B}(\mathcal{P}) = \int \frac{\mu_0}{4\pi}\frac{I\,d\vec{s}\times\hat{r}}{r^2}\,.$$

This integral expression for the field is called the Biot–Savart Law.[5] Performing this integration can be rather daunting, as both $d\vec{s}$ and $\vec{r}$ may vary in manners which are hard to characterise.

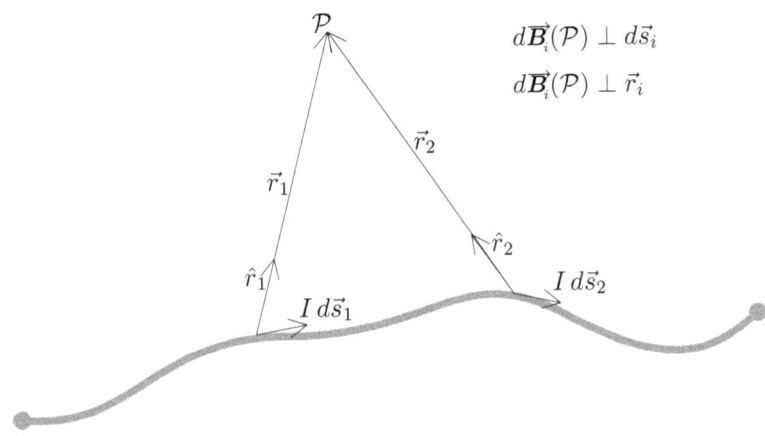

FIGURE 28.5   A Current-Carrying Wire Generates a Magnetic Field

The various constitutive parts of the Biot–Savart Law are described below.

$\boldsymbol{\mu_0}$ The constant μ_0 is the **permeability of free space**. Its value is defined to be

$$\mu_0 = 4\pi\times10^{-7}\ \frac{\text{Wb}}{\text{A}\cdot\text{m}} = 4\pi\times10^{-7}\ \frac{\text{T}\cdot\text{m}}{\text{A}}$$

in the SI scheme. Recall that the weber [Wb] is the unit of magnetic flux: $1\,\text{Wb} = 1\,\text{T}\cdot\text{m}^2$.

> ASIDE: This curious assignment of an exact [irrational] value to a physical constant follows from an **electromagnetic** relation involving ϵ_0 and μ_0. We shall make explicit this relation in Chapter 48. Nevertheless, we shall continue to write $\frac{\mu_0}{4\pi}$, wherever it is needed, in order to accommodate other unit systems.

---

[5] Jean-Baptiste Biot (1774–1862) was a prolific physicist, mathematician, astronomer, and geologist. Félix Savart (1791–1841) was a physicist with abiding interests in oscillations, vibrations, and acoustics. These French scientists conducted experiments (*circa* 1820) by which it was revealed that electric current in a wire produces a magnetic field throughout the space nearby.

$I\,d\vec{s}$ The product $I\,d\vec{s}$ is the **current element**. Strictly speaking, I is the magnitude of the [positive] current. Its direction is incorporated into $d\vec{s}$.

$\vec{r} = r\,\hat{r}$ Both the magnitude, r, and the direction, $\hat{r}$, of the relative position of the field point with respect to the current element affect the magnitude and direction of the differential magnetic field. In particular, **the magnitude falls off in proportion to the squared distance**.

$I\,d\vec{s} \times \hat{r}$ The differential contribution to the magnetic field at $\mathcal{P}$ lies orthogonal to the plane formed by the current element $d\vec{s}_i$ and the displacement vector leading from the current element to the field point, $\vec{r}_i$.

EXAMPLE [*Magnetic Field along the Bisector of a Straight Segment of Wire*]

A thin straight wire of length L carries current I. A field point, $\mathcal{P}$, lies at distance a along a perpendicular bisector of the wire, as illustrated in Figure 28.6.

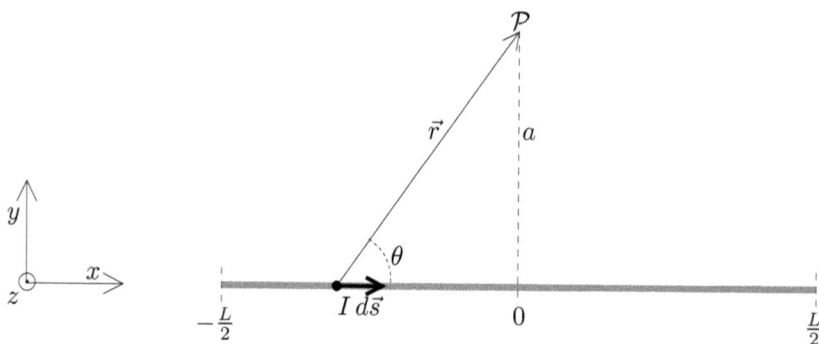

FIGURE 28.6   A Current-Carrying Thin Straight Wire

With this geometry,

$$d\vec{s} \times \hat{r} = ds\,\hat{\imath} \times (\cos(\theta)\,\hat{\imath} + \sin(\theta)\,\hat{\jmath}) = \sin(\theta)\,ds\,\hat{k}$$

for each current element, and thus all of the differential magnetic field contributions precisely align.

ASIDE: The axial symmetry of the current in the wire requires that the net magnetic field at $\mathcal{P}$ be directed orthogonally to the page.

The vector integration in the Biot–Savart Law is thus greatly simplified:

$$d\vec{B}(\mathcal{P}) = \frac{\mu_0\,I}{4\,\pi}\,\frac{\sin(\theta)\,ds}{r^2}\,\hat{k} = dB\,\hat{k}\,, \qquad \text{for} \quad dB = \frac{\mu_0\,I}{4\,\pi}\,\frac{\sin(\theta)\,ds}{r^2}\,.$$

Although the evident complexity of the integrand has been drastically reduced, it cannot yet be evaluated, because both the distance to the field point and the angle θ vary with the location of the current source. To proceed, we shall transform the integration from ds, the line element along the wire, to $d\theta$, the angle at which the field point appears with respect to the current-directed line element.

A right triangle with height a, base $-x$, and hypotenuse r is pictured in Figure 28.7. This parameterisation of the triangle places the origin of coordinates at the rightmost extent of its base.

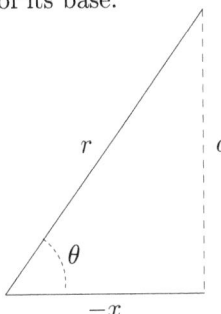

From the sketch,

$$\tan(\theta) = \frac{a}{-x} \quad \text{and} \quad \sin(\theta) = \frac{a}{r},$$

which together imply that

$$\left\{ \begin{array}{l} x = -a\cot(\theta) \\ dx = a\csc^2(\theta) \end{array} \right\} \quad \text{and} \quad r = a\csc(\theta).$$

FIGURE 28.7 Right Triangle Geometry for the Current-Carrying Straight Wire

Re-expression of dB entirely in terms of the angle θ is accomplished via

$$dB = \frac{\mu_0 I}{4\pi} \frac{\sin(\theta)\,ds}{r^2} = \frac{\mu_0 I}{4\pi} \frac{\sin(\theta)\,a\csc^2(\theta)\,d\theta}{\left(a\csc(\theta)\right)^2} = \frac{\mu_0 I}{4\pi a} \sin(\theta)\,d\theta.$$

The limits of integration are the angles between $d\vec{s}$ and $\vec{r}$ at the two ends of the wire. Let's denote the angle at the left edge by "θ_L" and the one on the right by "θ_R."

ASIDE: The analysis so far has been general enough to apply to all field points, not just those which happen to lie on the perpendicular bisector.

Integrating dB from θ_L to θ_R yields

$$B(\mathcal{P}) = \frac{\mu_0 I}{4\pi a} \left[-\cos(\theta_R) + \cos(\theta_L)\right].$$

WHEN the field point lies along the perpendicular bisector, as illustrated in Figure 28.8, THEN the left and right angles are SUPPLEMENTARY, *i.e.*, $\theta_R = \pi - \theta_L$, and as a consequence $-\cos(\theta_R) = \cos(\theta_L)$.

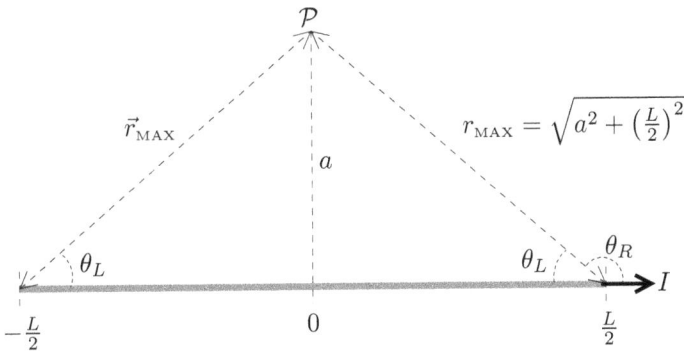

FIGURE 28.8 Determining the Field outside a Long Straight Wire

Thus, in this instance,

$$B(\mathcal{P}) = \frac{\mu_0 I}{4\pi a} \left[2\cos(\theta_L)\right].$$

The cosine of θ_L is readily expressed in terms of the [half-]length of the straight segment and the [perpendicular] distance to the field point, *i.e.*,

$$\cos\left(\theta_L\right) = \frac{L/2}{r_{\text{MAX}}} = \frac{L/2}{\sqrt{a^2 + \left(\frac{L}{2}\right)^2}} = \frac{1}{\sqrt{\left(\frac{a}{L/2}\right)^2 + 1}} = \frac{1}{\sqrt{1 + \left(\frac{2a}{L}\right)^2}} = \left[1 + \left(\frac{2a}{L}\right)^2\right]^{-\frac{1}{2}}.$$

Hence, the magnitude of the magnetic field at distance a along the perpendicular bisector of a thin wire has been determined to be

$$B = \frac{\mu_0 I}{4\pi a}\, 2\left[1 + \left(\frac{2a}{L}\right)^2\right]^{-\frac{1}{2}} = \frac{\mu_0 I}{2\pi a}\, \frac{1}{\sqrt{1 + \left(\frac{2a}{L}\right)^2}}.$$

o   It is worth reiterating that the magnetic field is cylindrically [axially] symmetric about the straight wire bearing the current.

o   In the limit where $\frac{a}{L} \to 0$, the magnitude of the field approaches

$$B \to \frac{\mu_0 I}{2\pi a}.$$

This limit is attained whenever the wire is very long, or when the field point is near to the thin wire.

Chapter 29

M-M-More Magnetic Sources

Recall the example from the previous chapter, in which a thin straight wire of length L carries current I. At a field point, $\mathcal{P}$, located distance a away from the wire, along its perpendicular bisector, the magnitude of the magnetic field is

$$B(\mathcal{P}) = \frac{\mu_0\,I}{2\,\pi\,a}\left[1 + \left(\frac{2\,a}{L}\right)^2\right]^{-\frac{1}{2}}.$$

The direction of the field is perpendicular to the plane containing the field point and the wire. The magnetic field wraps around the wire in a manner respecting the cylindrical symmetry of the space outside the wire.

    ASIDE: Each differential contribution to the field is perpendicular to its current element **source** [and hence to the straight wire], as well as to the vector leading from the current element to the field point.

Two lengths appear in this expression: a, the distance from the source to the field point, and L, the extent of the source. The two natural limits are of phenomenological interest.

    0   IF $\frac{a}{L} \to 0$ [as occurs when a long straight wire is observed from a nearby field point], THEN the strength of the field is well-approximated by

$$B(\mathcal{P}) \simeq \frac{\mu_0\,I}{2\,\pi\,a}.$$

This limit is in effect whenever the angle subtended by the current source, as viewed from the perspective of the field point, approaches π.
[E.g., think of train tracks traversing a vast plain.]

    ∞   IF $\frac{a}{L} \to \infty$ [as when a short wire is viewed from a large distance], THEN

$$B(\mathcal{P}) \simeq \lim_{\frac{a}{L}\to\infty} \frac{\mu_0\,I}{2\,\pi\,a}\left[1 + \left(\frac{2\,a}{L}\right)^2\right]^{-\frac{1}{2}}$$

$$\simeq \lim_{\frac{L}{a}\to 0} \frac{\mu_0\,I}{4\,\pi\,a}\left[\frac{2}{\frac{2\,a}{L}}\right] \simeq \lim_{\frac{L}{a}\to 0} \frac{\mu_0\,I}{4\,\pi\,a}\frac{L}{a} \to 0 \quad [\text{as } a^{-2}].$$

In this case, the angle subtended by the source current—as viewed from the field point—vanishes.

[Biot–Savart superposes a collection of infinitesimal current elements to yield finite results.]

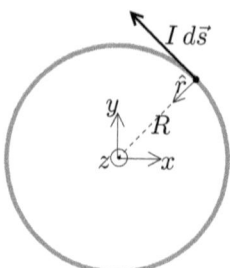

FIGURE 29.1 A Current Element in a Circular Loop

EXAMPLE [*Magnetic Field at the Centre of a Circular Loop of Current*]

Let us compute the magnetic field produced at the centre of a circular loop of current. For ease of visualisation and simplicity of analysis, we shall choose coordinates such that the circular loop lies centred in the xy-plane, as illustrated in Figure 29.1.

In applying the Biot–Savart Law, one must form the cross product of the current element and the unit vector in the direction of the field point. For the field point at the centre of the thin circular current-carrying loop,

$$I\, d\vec{s} \times \hat{r} = I\, ds\, \hat{k}\,,$$

where ds is the differential arc length along the wire. Meanwhile, the distance from each current element to the common field point is equal to the radius of the circle, *i.e.*, $r = R$. Thus, the expression for the magnetic field simplifies considerably to

$$\vec{B}(\vec{0}) = \frac{\mu_0}{4\pi} \int \frac{I\, d\vec{s} \times \hat{r}}{r^2} = \frac{\mu_0\, I}{4\pi\, R^2} \oint ds\, \hat{k}\,.$$

The integral which remains is precisely equal to the circumference of the circular path: $\oint ds = 2\pi R$. Hence, the magnitude of the field at the centre of the loop is

$$B(\vec{0}) = \frac{\mu_0\, I}{2\, R}\,,$$

while its direction, $\hat{k}$, is perpendicular to the plane of the loop.

A few comments are appropriate.

OFF-CENTRE   The Biot–Savart integrand does not simplify to the degree noted above when the field point is located elsewhere in the xy-plane. When the field point is inside the loop, all of the field contributions from the current elements align in the same direction, whereas when the field point is outside the loop, the contributions from widely spaced current elements may be anti-parallel. In the panels in the lower right portion of Figure 29.2, the differential magnetic fields produced by opposing current elements are illustrated.[1] One must bear in mind two things: that the figure shows only the two current elements penetrating a cross-section of the ring [and thus they nicely superpose], and that the entire system is axially symmetric, complicating the general off-centre analysis.

---

[1] The [horizontal] density of arrows represents the relative strength of the local field.

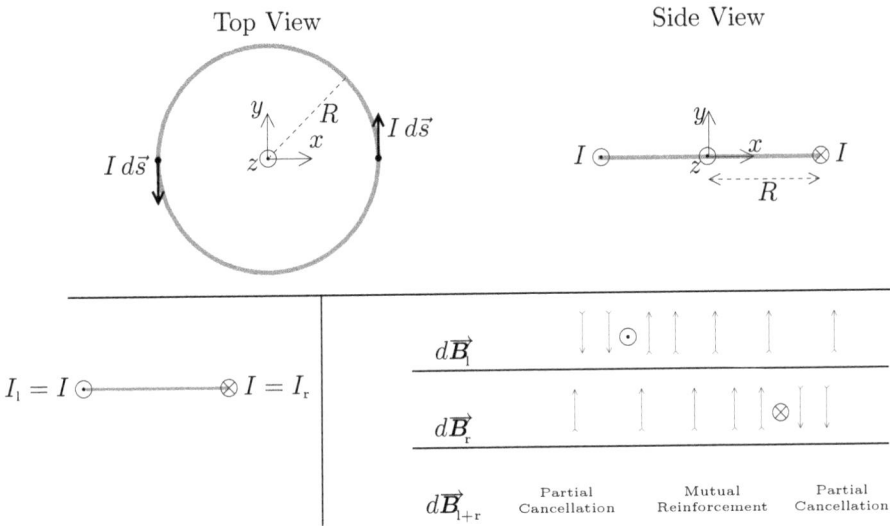

FIGURE 29.2 Superposition of Magnetic Field Contributions from Opposite Current Elements in a Circular Ring of Current

To perform the integration determining the magnetic field everywhere in the xy-plane is a bit too daunting for us now. Instead, let us employ our "little grey cells"[2] to examine the net field far outside the loop.

Consider a pair of opposite current elements, $\{I\,d\vec{s}_n, I\,d\vec{s}_f\}$, labelled "near" and "far," and a field point at a large distance, $r \gg R$, from the centre of the loop. The field contributions are written

$$dB_n = \frac{\mu_0\, I}{4\,\pi}\,\frac{ds}{r_n^2} \qquad \text{and} \qquad dB_f = \frac{\mu_0\, I}{4\,\pi}\,\frac{ds}{r_f^2},$$

in terms of the near and far distances,

$$r_n = r - R \qquad \text{and} \qquad r_f = r - R.$$

The two field contributions are anti-aligned, and thus

$$dB_{n+f} \propto \frac{1}{(r-R)^2} - \frac{1}{(r+R)^2} \simeq \frac{1}{r^2}\left[\left(1 + \frac{R}{r} + \dots\right) - \left(1 - \frac{R}{r} + \dots\right)\right] \simeq \frac{2\,R}{r^3}.$$

Q: Where have we seen this sort of thing before?

A: During the discussion of the electric dipole, way back in Chapter 2!

$\qquad\qquad$ **The far-field regime in the plane of the loop looks "dipole-ish."**

COIL $\quad$ IF the circular ring in Figures 29.1 and 29.2 represents a **coil** comprised of $\mathcal{N}$ tightly wrapped loops of wire,[3] THEN the magnitude of the magnetic field everywhere in space

---

[2] Taking a page from M. Hercule Poirot, Agatha Christie's fictional Belgian detective.

[3] In our model, the extent of the collection of looped wires both out of and within the plane of the coil is deemed insignificant on the scale set by the radius of the coil.

is $\mathcal{N}$ times that produced by a single loop. In particular, at the centre of the coil, the magnitude of the field is

$$B_{\text{coil}}(\vec{0}) = \mathcal{N}\,\frac{\mu_0\,I}{2\,R}\,.$$

EXAMPLE [*Magnetic Field along the Axis of a Circular Loop of Current*]

Let us extend the computation in the previous example to field points lying along the z-axis of the loop centred in the xy-plane.

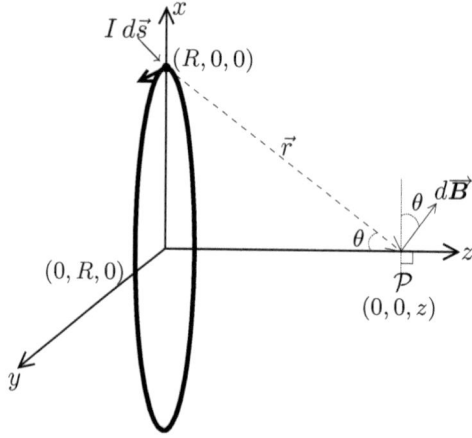

FIGURE 29.3 Magnetic Field Produced along the Axis of a Current Loop

Integrating over all of the current elements in the loop to determine the net field at $\mathcal{P}$ is tantamount to summing a set of differential field contributions all lying on the surface of a cone. By symmetry, the off-axis components will exactly cancel in the sum, while the on-axis contributions are all of the same value. The z-component of the infinitesimal field produced by the each current element is

$$dB_z(\mathcal{P}) = \left| d\vec{B} \right| \sin(\theta)\,,$$

where $\sin(\theta)$ is fixed by the radius of the loop and the position of $\mathcal{P}$ to be

$$\sin(\theta) = \frac{R}{r} = \frac{R}{\sqrt{R^2 + z^2}}\,.$$

It follows straightforwardly, in this case, that

$$\left| d\vec{B} \right| = \frac{\mu_0}{4\,\pi}\left| \frac{I\,d\vec{s} \times \hat{r}}{r^2} \right| = \frac{\mu_0\,I}{4\,\pi\,r^2}\,ds\,,$$

since $d\vec{s} \perp \hat{r}$ and the distance from each part of the loop to the point on the axis is constant. Thus,

$$B_z = \int dB_z = \int \left| d\vec{B} \right| \sin(\theta) = \int \frac{\mu_0\,I}{4\,\pi\,r^2}\,\frac{R}{\sqrt{R^2 + z^2}}\,ds = \frac{\mu_0\,I\,R}{4\,\pi\left(R^2 + z^2\right)^{\frac{3}{2}}}\oint ds\,.$$

As in the first example in this chapter, $\oint ds = 2\pi R$, and hence

$$B_z(\mathcal{P}) = \frac{\mu_0\, I\, R^2}{2\left(R^2 + z^2\right)^{\frac{3}{2}}}\,.$$

Since symmetry demands the vanishing of both B_x and B_y everywhere along the axis, the entire field at $\mathcal{P}$ has been determined.

DIRECTION   The magnetic field along the z-axis [the axis of the loop] points in the $\hat{k}$ direction, irrespective of whether the field point is at positive or negative z.

DIPOLE MOMENT   The magnitude of the loop's on-axis field may be re-expressed in favour of the magnetic dipole moment, $m = |\vec{m}| = I\,A = I\,\pi\,R^2$, introduced [in the context of our discussion of the response of current loops to ambient magnetic fields] in Chapter 26. Doing so yields

$$B_z(\mathcal{P}) = \frac{\mu_0\, m}{2\pi\left(R^2 + z^2\right)^{\frac{3}{2}}}\,.$$

$z \gg R$ Far out along the axis, the field is well-approximated by

$$B_z(\mathcal{P}) \simeq \frac{\mu_0\, m}{2\pi\, z^3}\,.$$

Here, the magnetic field is proportional to the dipole moment and falls off as the cube of the distance from the centre of the loop.

The far-field regime along the axis of symmetry is evidently dipolar, too.

Extending this analysis to OFF-AXIS field points reveals that the magnetic field lines for a circular current loop possess precisely the dipole form everywhere in space.

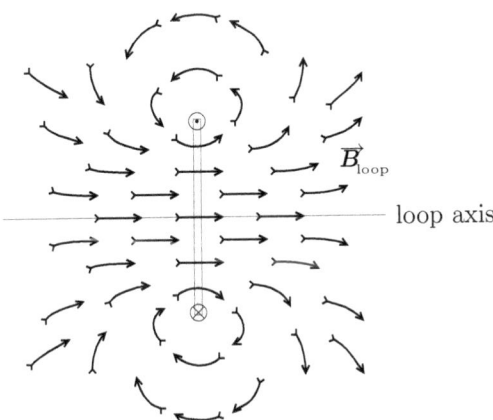

FIGURE 29.4   The Dipole Magnetic Field Associated with a Current Loop

Chapter 30

Interacting Wires and Ampère's Law

Knowing, as we now do, that a magnetic field is produced by the flow of current in a wire and how a current-carrying wire responds to an ambient magnetic field, we are in position to compute the magnetic force between two current-carrying wires.

THE MAGNETIC FORCE ACTING BETWEEN TWO CURRENT-CARRYING WIRES

In the model for this analysis, four assumptions are made about the wires.

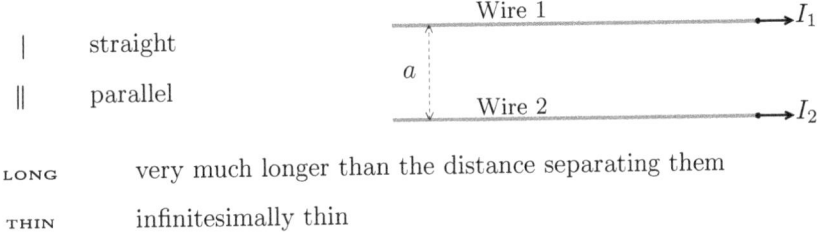

	straight
‖	parallel

LONG   very much longer than the distance separating them

THIN   infinitesimally thin

FIGURE 30.1   Two Parallel Current-Carrying Wires

The separation distance sets a physical scale, and thus the assumption that the wires are long amounts to a declaration that both are effectively infinite in length. Let's adopt the point of view in which the current in wire 2, I_2, establishes a magnetic field, $\vec{B}_2$, throughout space. The current in wire 1, I_1, responds to this field. It is in this [field-mediated] manner that the currents **interact**.

While $\vec{B}_2$ is not uniform within any 3-d spatial region, it is [approximately[1]] constant everywhere along wire 1.

MAGNITUDE $B_{21} = \left| \vec{B}_2(\mathcal{P}_1) \right| = \dfrac{\mu_0\, I_2}{2\,\pi\,a}$, for any point $\mathcal{P}_1$ coincident with wire 1.

DIRECTION $\vec{B}_2(\mathcal{P}_1)$ is directed straight out from the page.

The Lorentz force acting on wire 1 is identical to that which it would experience if it were bathed in a uniform field, $\vec{B}_{21} = B_{21}\,[\odot]$. Thus, $\vec{F}_{21} = \vec{F}_{M2,I1} = I_1\, \vec{L}_1 \times \vec{B}_{21}$, according to the formulae developed in Chapter 25. In this expression, the current displacement vector points in the direction in which I_1 flows and has magnitude ... OOOOOOPS!
The force acting on wire 1 is ill-defined, as its magnitude is formally infinite.

---

[1]The assumptions that the wires are parallel and thin validate this approximation.

Thwarted in our attempt to compute the actual force acting on wire 1 due to its unbounded length, we shall instead compute the force per unit length that it experiences:

$$\vec{f}_{21} = \frac{\vec{F}_{21}}{L_1} = I_1 \, \widehat{L}_1 \times \vec{B}_{21} = I_1 \frac{\mu_0 \, I_2}{2 \, \pi \, a} \, [\rightarrow] \times [\odot] = \frac{\mu_0 \, I_1 \, I_2}{2 \, \pi \, a} \, [\downarrow].$$

Thus, wire 1 experiences a constant force per unit length acting toward wire 2, owing to the interaction between the currents. This force

I_1 , I_2 is proportional to each of the currents

a^{-1} diminishes with increasing separation of the wires

μ_0 has its scale set by the coupling constant μ_0

Three nettlesome aspects of current–current interactions, SYMMETRY, ORIENTATION, and SCALE, must be discussed before we tackle Ampère's Law.

Q: What if one were to compute the response of the current in wire 2 to the magnetic field produced by the current flowing in wire 1?

A: Wonder no more! Let's redo the analysis from this perspective.

The magnetic field produced by the current in wire 1 at any point in space occupied by wire 2,

$$\vec{B}_1(\mathcal{P}_2) = \frac{\mu_0 \, I_1}{2 \, \pi \, a} \, [\otimes],$$

is constant. Thus, the magnetic force per unit length acting on wire 2 is

$$\vec{f}_{12} = \frac{\vec{F}_{12}}{L_2} = I_2 \, \widehat{L}_2 \times \vec{B}_{12} = I_2 \frac{\mu_0 \, I_1}{2 \, \pi \, a} \, [\rightarrow] \times [\otimes] = \frac{\mu_0 \, I_1 \, I_2}{2 \, \pi \, a} \, [\uparrow].$$

Hence, **the magnetic attraction of two parallel currents is entirely consistent with Newton's Third Law.**

Q: What happens if the parallel wires bear currents flowing in opposite directions?

A: Reversal of one current flips the direction of the force without altering the magnitude. Reversing both returns us to the parallel case.

Parallel currents attract. $\Longleftrightarrow$ **Anti-parallel currents repel.**

Q: How large is the magnetic force acting between a pair of wires?

A: To illustrate, suppose that the current in each parallel wire is 1 A, and the separation distance is 1 m. The force per unit length acting between the wires has magnitude

$$f_{(12)} = \frac{\mu_0 \, I_1 \, I_2}{2 \, \pi \, a} = \frac{4 \, \pi \times 10^{-7} \times 1 \times 1}{2 \, \pi \times 1} \, \frac{\frac{\text{Wb}}{\text{A·m}} \, \text{A}^2}{\text{m}} = 2 \times 10^{-7} \, \frac{\text{Wb·A}}{\text{m}^2}.$$

As one tesla is equal to one weber per square metre, the units in the above expression reduce to T·A. We recall [Chapter 24] that one tesla is also equivalent to one newton per ampere per metre, and thus we infer that the units of $f_{(12)}$ are N/m. Finally, the strength of the force of interaction [per unit length] between two $1\,\text{A}$ currents separated by $1\,\text{m}$ is

$$f_{(12)} = 2 \times 10^{-7} \; \frac{\text{N}}{\text{m}}.$$

This rather small value indicates that, generally speaking, magnetic interactions are significantly weaker than electrical interactions.

> ASIDE: Were electricity and magnetism disjoint subjects, one might squint at the current–current formula, compare it with the expression for Coulomb's Law, and identify the permeability of free space with a magnetic coupling constant analogous to the Coulomb constant: $k \iff \mu_0$. In SI units, these couplings are:
>
> $$|k| \sim 9.0 \times 10^9 \; \frac{\text{N}\cdot\text{m}^2}{\text{C}^2} \qquad \text{and} \qquad \mu_0 \equiv 4\,\pi \times 10^{-7} \; \frac{\text{T}\cdot\text{m}}{\text{A}} = 4\,\pi \times 10^{-7} \; \frac{\frac{\text{N}}{\text{A}\cdot\text{m}}\cdot\text{m}}{\text{A}} = 4\,\pi \times 10^{-7} \; \frac{\text{N}}{\text{A}^2}.$$
>
> Glossing over the specifics of units (including the issue of inverse-linear *vs.* inverse-square dependence) and the "apples *vs.* oranges" comparison of point-like charges and infinitely extended currents, one can infer that the natural scale of magnetic forces is many orders of magnitude smaller than that of electric forces.

Q: Calculations involving magnetic fields are fun. Is there anything else we can compute?

A: Yes—the **circulation of the magnetic field**.

MAGNETIC CIRCULATION The circulation associated with the magnetic field is the line integral of the field evaluated along a closed path, γ_0. That is,

$$\text{Circ}\left(\vec{B}\right)_{\gamma_0} = \oint_{\gamma_0} \vec{B} \cdot d\vec{s}.$$

To make sense of this, let's effect this computation in a special case.

EXAMPLE [*Circulation of the Magnetic Field about a Long Straight Wire*]

An infinitely long straight wire carries an electric current, I. The magnetic field produced at field points located [perpendicular] distance a from the wire has magnitude

$$\left|\vec{B}(a)\right| = \frac{\mu_0\, I}{2\,\pi\, a},$$

and is directed so as to wrap around the wire. Two aspects of the axial symmetry exhibited by the magnetic field are here elaborated upon.

    o   The magnitude of the field depends only on the radial distance to the field point from [the centre of] the wire.

    o   The [local] direction of the field is such that the magnetic field lines form circles centred on the wire, in planes perpendicular to the wire.

Let the closed path γ_0, along which the circulation is to be computed, be a circle of radius r lying in a plane perpendicular to, and centred upon, the wire.

    [It is not accidental that this path follows precisely along a magnetic field line.]

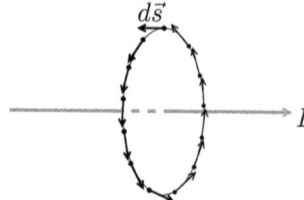

FIGURE 30.2   A Closed Path around an Infinitely Long Current-Carrying Wire

WLOG, we choose to traverse the path in the sense in which its local direction is everywhere aligned with the magnetic field.

Three steps are required to compute the magnetic circulation about this path.

1   The dot product of the local magnetic field and the infinitesimal arc length vector, appearing in the integrand, reduces to the product of their magnitudes,

$$\boldsymbol{B} \cdot d\vec{s} = |\boldsymbol{B}|\, ds = B(\mathcal{P})\, ds\,,$$

because the field everywhere along the circular path is parallel to the path's tangent vector.

2   On the locus of points comprising the circular path [the domain of integration], the magnetic field has constant magnitude, and thus

$$\mathrm{Circ}\big(\boldsymbol{B}\big)_{\gamma_0} = \oint_{\gamma_0} \boldsymbol{B} \cdot d\vec{s} = \oint_{\gamma_0} B(\mathcal{P})\big|_{a=r}\, ds = \frac{\mu_0\, I}{2\,\pi\, r} \oint_{\gamma_0} ds\,.$$

3   Integration of ds around the closed circular loop of radius r yields the arc length [circumference] of the loop, $2\,\pi\, r$.

Hence, the circulation of the magnetic field about this particular circular path is

$$\mathrm{Circ}\big(\boldsymbol{B}\big)_{\gamma_0} = \frac{\mu_0\, I}{2\,\pi\, r} \times \big[2\,\pi\, r\big] = \mu_0\, I\,.$$

Q: Wow! Is this astonishingly simplified result some sort of coincidence?

A: Nope.[2] It is a manifestation of [the TIME INDEPENDENT FORM OF] **Ampère's Law.**

AMPÈRE'S LAW   A statement of Ampère's Law, valid in steady state situations,[3] is that the line integral of $\boldsymbol{B} \cdot d\vec{s}$ around **any** closed loop is proportional to the net current piercing any surface having the closed loop as its boundary, *i.e.*,

$$\mathrm{Circ}\big(\boldsymbol{B}\big)_{\gamma_0} = \oint_{\gamma_0} \boldsymbol{B} \cdot d\vec{s} = \mu_0\, I_{\mathrm{encl.}}\,,$$

where the permeability of free space is $\mu_0 = 4\,\pi \times 10^{-7}$ T·m/A. The net current intersecting the surface is "enclosed" by the curve on which the circulation is computed. We emphasise this property by subscripting "encl." to the symbol for the current [appearing on the RHS].

AMPÈRE'S LAW is a fundamental LAW of NATURE.

---

[2]To appreciate this more fully requires some high-powered calculus.
[3]This restricted form is known as the "Time Independent (form of) Ampère's Law." In Chapter 32 we'll uncover its shortcoming and derive its proper generalisation.

Whew! Let's carefully determine enclosed currents for a variety of loops and wires.

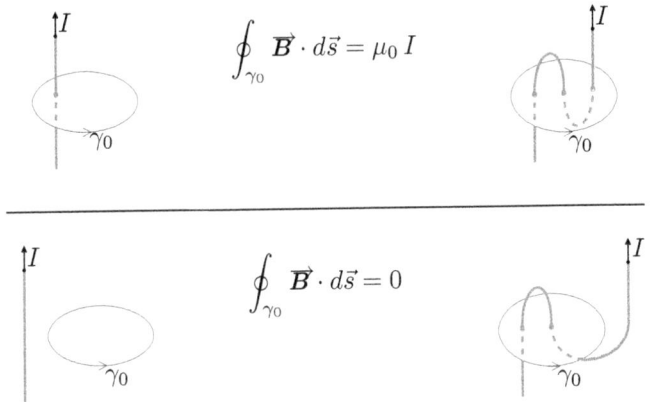

FIGURE 30.3 Determining the Enclosed Current: Various Scenarios

These pictures are meant to illustrate that the enclosed current is the net electric current passing through any surface bounded by the circulation path.

TOP LEFT   In this case, the circular loop is not centred on the current-carrying wire. Nonetheless, for the planar disk [the minimal surface bounded by the path], the enclosed current is exactly equal to the current in the wire, $I_{\text{encl.}} = I$, and hence the RHS of Ampère's Law reads $\mu_0 I$. Should one choose a non-minimal surface, it must be pierced by the wire an odd number of times.

TOP RIGHT   In this case, the wire bends in such a manner that the wire passes through the [minimal] surface thrice: first on the way up, then on the way down, and finally up again. Thus, the enclosed current is $I_{\text{encl.}} = +I[\text{up}] + I[\text{down}] + I[\text{up}] = +I$, and Ampère's RHS reads $\mu_0 I$ here also. Non-minimal surfaces capture the same enclosed current. Let's look at one in particular, obtained by stretching the minimal surface over the "hump" of wire. Here, the enclosed current is merely $I_{\text{encl.}} = I$, and the correct RHS is obtained more directly than in the case of the minimal surface.

LOWER LEFT   The wire fails to intersect the minimal surface at all, and hence $I_{\text{encl.}} = 0$. Should one take a non-minimal surface, bounded by the loop, which is sufficiently distorted to intersect the wire, it is inevitable that it be pierced an even number of times. Hence, the net enclosed current always vanishes in this scenario.

LOWER RIGHT   The wire pierces the minimal surface twice: up and then down, leading to $I_{\text{encl.}} = +I[\text{up}] - I[\text{down}] = 0$. Should one choose to consider a non-minimal surface stretched over the hump, then the enclosed current automatically vanishes, just as in the lower–left picture.

The LHS and the RHS of Ampère's Law are almost unambiguously defined. There are two complicating factors. One must await Maxwell's[4] elaboration of Ampère's Law, while the other shall be dispensed with immediately.

TIME   This formulation of Ampère's Law is only valid when all electric fields present are constant in time. The extension to time-dependent electric fields is developed in Chapter 32.

SIGNS   There are apparent sign ambiguities in our statement of Ampère's Law.

±   In the computation of the circulation on the LHS, we chose to traverse the path in such manner that the local field was everywhere parallel to $d\vec{s}$. Had the path been travelled in the other direction, the result would have been MINUS the one we obtained.

±   Concomitantly, there is also ambiguity in the assignment of positive and negative current contributions on the RHS.

The resolution of these ambiguities involves the invocation of another RHR.

RHR   The orientation in which the closed path is traversed determines the sense in which the flow of current piercing the surface bounded by the path is considered positive. [This RHR is much like the one which assigns direction to magnetic moments associated with current loops.]

  •   Align the digits in your right hand with the direction along which the path is traversed.
  •   Rotate your arm so that the palm of your hand faces "inward."
  •   Your thumb then indicates the sense in which current is assigned a POSITIVE value in the computation of net enclosed current.

Right Hand Rule [OPERATIVE]

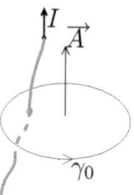

Left Hand Rule [BOGUS]

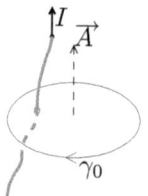

EXAMPLE [*Magnetic Field outside and inside a Thick Wire*]

A long straight uniform wire of finite radius a carries a total electric current, I_0, which is distributed uniformly over its cross-sectional area. We seek to determine the magnetic field produced by the current throughout all of space, that is, both outside and inside the wire, by application of Ampère's Law.

EXTERNAL   **Outside the wire, $r > a$.**

Consider γ_o, the circular Amperian path with radius $r > a$ lying in a plane perpendicular to the wire, as shown in Figure 30.4.

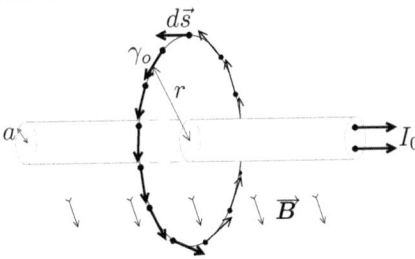

FIGURE 30.4   A Circular Amperian Path External to a Current-Carrying Wire

Formal evaluation of the LHS of Ampère's Law, the circulation of the magnetic field about the circular path, begins with recognition that [along the integration path] the field is parallel to each and every $d\vec{s}$ and the magnitude of the field is constant, due to the symmetry of the situation. Hence,

$$\text{LHS} = \oint_{\gamma_o} \vec{B} \cdot d\vec{s} = \oint_{\gamma_o} B(r)\,ds = B(r) \oint_{\gamma_o} ds = B(r)\left[2\,\pi\,r\right].$$

As all of the current flowing through the wire is enclosed by the path, $I_{\text{encl.}} = I_0$, and thus the RHS of Ampère's Law reads $\mu_0\,I_{\text{encl.}} = \mu_0\,I_0$. Setting LHS = RHS [invoking AMPÈRE'S LAW] fixes the form of the magnetic field:

$$B(r)\left[2\,\pi\,r\right] = \mu_0\,I_0 \qquad \Longrightarrow \qquad B(r) = \frac{\mu_0\,I_0}{2\,\pi\,r}.$$

Amazingly, the magnetic field external to the thick wire has exactly the same form as it would were the wire infinitesimally thin!

INTERNAL   **Inside the wire, $r < a$.**

Consider γ_i, the circular Amperian path with radius $r < a$, lying in a plane perpendicular to the wire, illustrated in Figure 30.5.

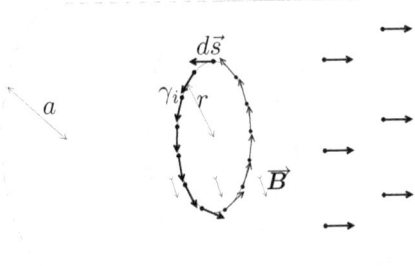

FIGURE 30.5   A Circular Amperian Path Internal to a Current-Carrying Wire

Construction of the LHS of Ampère's Law depends only on the axial symmetry of this system, irrespective of whether the path is external or internal to the wire. Hence,

$$\text{LHS} = \oint_{\gamma_i} \vec{B} \cdot d\vec{s} = B(r)\left[2\,\pi\,r\right].$$

---

[4] James Clerk Maxwell united the fields of electricity and magnetism in the 1860s.

Only a fraction of the total current flowing through the wire is enclosed by the path. Under the assumption of uniform current density, the enclosed current is

$$I_{\text{encl.}} = \frac{A_r}{A_a} I_0 = \frac{\pi r^2}{\pi a^2} I_0 = \frac{r^2}{a^2} I_0,$$

and thus the RHS of Ampère's Law reads $\mu_0\, I_{\text{encl.}} = \mu_0 \frac{r^2}{a^2} I_0$. Setting LHS = RHS [invoking AMPÈRE'S LAW] fixes the form of the magnetic field:

$$B(r)\,[2\pi r] = \mu_0 \frac{r^2}{a^2} I_0 \qquad \Longrightarrow \qquad B(r) = \frac{\mu_0\, I_0}{2\pi a^2}\, r.$$

Inside the wire, the strength of the magnetic field increases linearly with radial distance from the centre.

In summary, outside of a thick wire carrying a uniformly distributed current, the magnetic field is precisely the same as it would be were the entire current borne by an infinitesimally thin wire located at the centre of the thick wire. The field inside the thick wire grows linearly, from ZERO at the centre, to match the external field at the outermost edge of the wire.

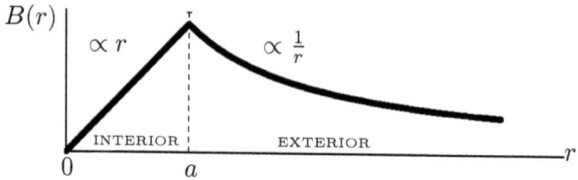

FIGURE 30.6   The Magnitude of the Magnetic Field inside and outside a Thick Wire

⊙   The magnetic field vanishes at the centre of the wire [as it must to respect the axial symmetry of the current].

○   The magnetic field is continuous [NOT differentiable] at the surface of the wire.

$$\lim_{r \to a^-} B(r) = \lim_{r \to a} \frac{\mu_0\, I_0\, r}{2\pi a^2} \qquad\qquad \lim_{r \to a^+} B(r) = \lim_{r \to a^+} \frac{\mu_0\, I_0}{2\pi r}$$
$$= \frac{\mu_0\, I_0}{2\pi a} \qquad\qquad\text{AND}\qquad\qquad = \frac{\mu_0\, I_0}{2\pi a}$$

[This Amperian analysis precisely paralleled the application of Gauss's Law in earlier chapters.]

Chapter 31

Ampère and Solenoids

A thin flat sheet of conductor occupies the entire yz-plane. A uniform **surface current** with density

$$\vec{J}_s(\vec{r}) = \begin{cases} J_s\,\hat{\jmath}\,, & x = 0 \\ \vec{0}\,, & x \neq 0 \end{cases}$$

flows in the conductor.

> ASIDE: This is distinguished from the generic current density of Chapter 18, by virtue of its being confined to the thin surface.

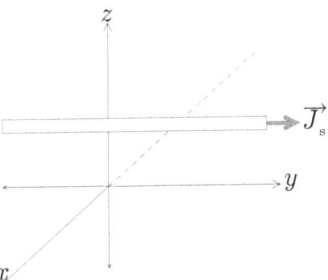

FIGURE 31.1 An Infinite Thin Plane Conductor Bearing Uniform Current Density

Each element of current flowing in a channel of width dz is the source of a differential magnetic field possessing axial symmetry [about the current element], as shown on the left in Figure 31.2. By superposition and symmetry, all of the contributions to the net magnetic field in the OFF-PLANE direction cancel identically. The net field on the $x > 0$ side of the plane of current points in the $+\hat{k}$ direction; on the $x < 0$ side, $-\hat{k}$.

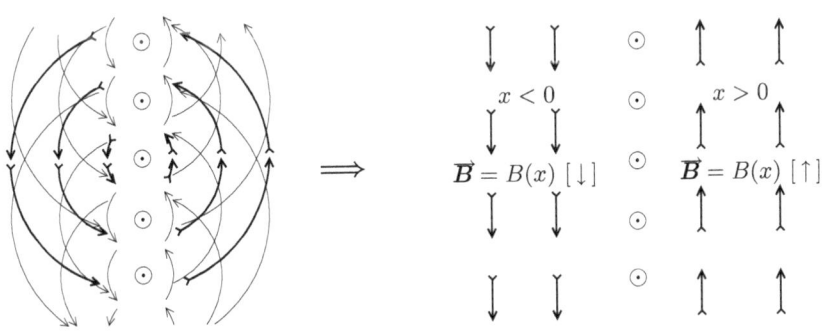

FIGURE 31.2 Superposition of Magnetic Fields from Uniform Planar Current

The symmetry and homogeneity of the infinite plane current compel the field to adopt the form

$$\boldsymbol{B}(\vec{r}) = \begin{cases} B(x)\,[\hat{k}]\,, & \forall\, x > 0 \\ B(x)\,[-\hat{k}]\,, & \forall\, x < 0 \end{cases},$$

and require $B(x)$ to be symmetric in its sole argument, *i.e.*, $B(-x) = B(x)$.

Such strong constraints on the field suggest that it might be possible to invert Ampère's Law in order to precisely determine $B(x)$. We effect this inversion below.

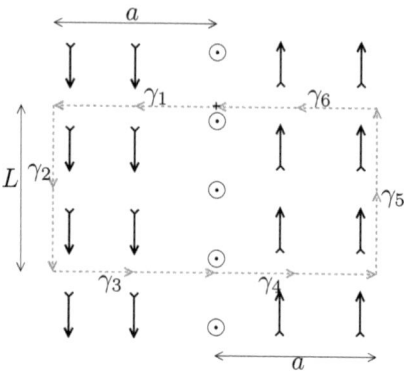

FIGURE 31.3 An Amperian Circuit Intersecting the Infinite Planar Current

Let's compute the circulation of the magnetic field generated by the planar current along the closed path comprised of six consecutive segments, $\gamma_0 = \bigcup_{i=1}^{6} \gamma_i$, illustrated in Figure 31.3. The first step is to realise that linearity permits each path segment to be considered separately. That is,

$$\mathrm{Circ}\left(\boldsymbol{B}\right)_{\gamma_0} = \oint_{\gamma_0} \boldsymbol{B} \cdot d\vec{s} = \sum_{i=1}^{6} \int_{\gamma_i} \boldsymbol{B} \cdot d\vec{s}.$$

Each contribution to the circulation is examined in turn.

PATH 1   The local magnetic field is perpendicular to each and every differential arc length element along γ_1. Thus the integrand, $\boldsymbol{B} \cdot d\vec{s}$, vanishes, and so too must the integral.

PATH 2   Everywhere along this section, the local magnetic field lies in perfect alignment with each step. Hence

$$\boldsymbol{B} \cdot d\vec{s}\big|_{\gamma_2} = B(x)\,ds\big|_{x=-a} = B(-a)\,ds = B(a)\,ds\,,$$

since the magnitude of the field is even in its argument, and therefore

$$\int_{\gamma_2} \boldsymbol{B} \cdot d\vec{s} = B(a) \int_{\gamma_2} ds = L\,B(a)\,.$$

PATH 3   The situation along γ_3 parallels[1] that of γ_1. Hence, this portion of the Amperian circuit provides no net contribution to the total magnetic circulation.

---

[1] "*Au contraire*," remarks the snark, "the third path is anti-parallel to the first!"

PATH 4   This path contributes exactly the same amount, ZERO, as γ_1 does.

PATH 5   Each step along this path lies fully in the direction of the local field, and hence

$$\int_{\gamma_5} \vec{B} \cdot d\vec{s} = \int_{\gamma_5} B(x)\, ds = B(a) \int_0^L ds = L\, B(a)\,.$$

The result simplified because the field was constant along the path.

PATH 6   No net circulation accumulates along γ_6.

Summing the contributions from all six path segments, the net circulation of the magnetic field about γ_0 is

$$\mathrm{Circ}\big(\vec{B}\big)_{\gamma_0} = 2\, L\, B(a)\,.$$

This result constitutes the LHS of Ampère's Law for this particular closed path, γ_0.

The enclosed current appearing on the RHS of Ampère's Law is the net current which flows through any surface bounded by the Amperian path. In the case at hand, the simplest surface to take is the rectangle whose edges are defined by the path segments. When this is done,

$$I_{\mathrm{encl.}} = J_s\, L\,,$$

i.e., the current per unit height in the plane multiplied by the height of the rectangle. Hence the RHS of Ampère's Law reads $\mu_0\, J_s\, L$.

Ampère's Law sets the LHS equal to the RHS:

$$\mathrm{Circ}\big(\vec{B}\big) = \mu_0\, I_{\mathrm{encl.}} \quad\Longrightarrow\quad 2\, L\, B(a) = \mu_0\, J_s\, L \quad\Longrightarrow\quad B(a) = \frac{\mu_0\, J_s}{2}\,.$$

The lack of a dependence is significant: **the magnitude of the magnetic field is constant throughout all of space!**

Déjà vu! This should be sounding familiar. Precisely analogous behaviour was observed for the electric field in the vicinity of an infinite sheet with uniform charge density.

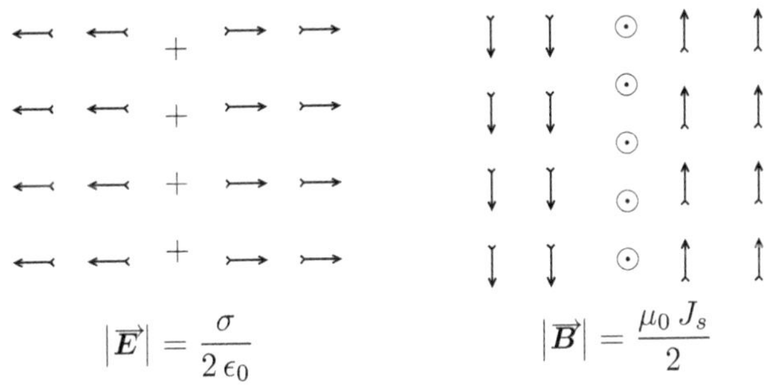

FIGURE 31.4   Infinite Sheet of Current *vs.* Infinite Sheet of Charge

ASIDE: It is also possible to argue for the constancy of the magnitude of the magnetic fields on each side of the infinite plane from the absence of an intrinsic length scale.

Having discussed the infinite plane of current, let's turn our attention to the idealised **solenoid**.

S   In all our investigations of the magnetic fields produced by current loops, it has been assumed that the current enters, follows the wire for one journey around the loop, and then exits. [However, in Chapter 29, it was noted that the wire might coil around more than once.]

S   Suppose that two loops, overlapping and in close proximity,[2] carry the SAME current, I. By LINEAR SUPERPOSITION, the net field produced by the two-loop system is double the field of either one alone.

While the two loops envisioned above are to be thought of as distinct loops of conductor, each bearing current, there is an economy to be gained by having a single current, I, be borne by a wire which is looped twice around.

FIGURE 31.5 Merging Distinct Current Loops into a Coil

S   IF the wire may be coiled around twice, THEN there ought to be no impediment to winding around $\mathcal{N}$ times, thus yielding a net field with magnitude equal to $\mathcal{N}$ times that of the field produced by the same current, I, traversing a single loop.

S   For a small number of turns, it is typically safe to assume that all of the loops lie in the same plane. However, as $\mathcal{N}$ grows large, it proves impossible to ignore the height of the stack of current loops.

Now we are ready to consider the magnetic field associated with an electric current in a solenoidal coil.

(IDEAL) SOLENOID   A solenoid is a tightly wound coil, comprised of $\mathcal{N}$ turns of wire arranged in the shape of a thin-walled helix. An **ideal solenoid** is infinitely long with tightly-spaced turns. The magnetic field associated with the current in an ideal solenoid vanishes everywhere external to the helical coil, and assumes a uniformly constant value inside. *I.e.*,

$$\vec{B}_{\text{Solenoid}}(r, \theta, z) = \begin{cases} \vec{0}, & r > R_{\text{Solenoid}} \\ \vec{B}_0, & r < R_{\text{Solenoid}} \end{cases}.$$

[Cylindrical polar coordinates centred on the solenoid are employed in this description.]

---

[2] "Overlapping" means that the orthogonal projection of one loop onto the other matches that of the other onto the one. "Close proximity" means that the separation of the loops is very small in terms of the scales set by the radius of the loop and the distance to the field point.

For further motivation [where none is needed] to investigate the solenoidal field configuration, consider the following arguments, arranged in order of sophistication.

1. [*A Rather Cheesy Argument*]

   One might consider the solenoid as though it were an infinite plane of uniform current that has been wrapped around and onto itself to form an infinite cylinder of winding current. This is tantamount to the imposition of **periodic boundary conditions** [in one direction[3]] on the infinite plane.

2. [*Somewhat Less Cheesy Argument*]

   A more precise version of the above argument considers a slice through the solenoid along its axis of symmetry, as exhibited in Figure 31.6.

   Along the left edge, a "wall" of uniform current emerges from the page. This wall is infinitely long in one direction (z), and differentially thin and wide in the other two. Even though the wall is not an infinite sheet, the arguments advanced earlier in this chapter apply. Hence, the wall produces a differential field, $d\vec{B}_{\text{left}}$, which is constant in magnitude throughout space, and whose direction is $\hat{k}$ to the right and $-\hat{k}$ to the left.

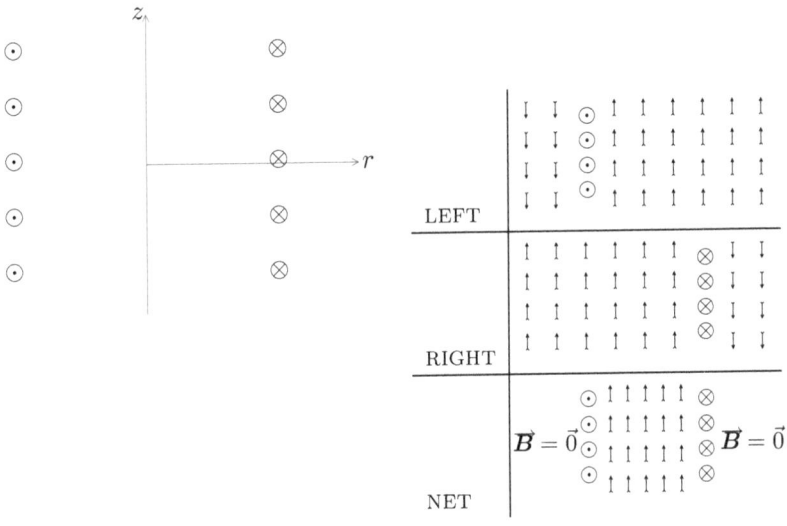

FIGURE 31.6 A Slice through a Solenoid

Meanwhile, the right wall has uniform surface current density directed into the page. It produces a differential magnetic field with the same magnitude as the field from the left wall, but with opposite orientation.

Everywhere outside the solenoid, the contributions from the two sources cancel precisely, whereas within the solenoid, they add to produce double the field of each alone. This is clearly seen in Figure 31.6. The overall axial symmetry of the solenoid ensures that this argument applies to all slices through the centre, and hence it must be the case that the magnetic field vanishes everywhere on the outside and is uniform inside the solenoid.

---

[3]If periodic boundary conditions are enforced in both possible directions, then the topology is that of a TORUS, rather than a cylinder.

3.   [*Time Independent Ampère's Law Argument*]

We can make definitive statements about the character of the magnetic field produced by the ideal solenoid by reverse application of Ampère's Law.

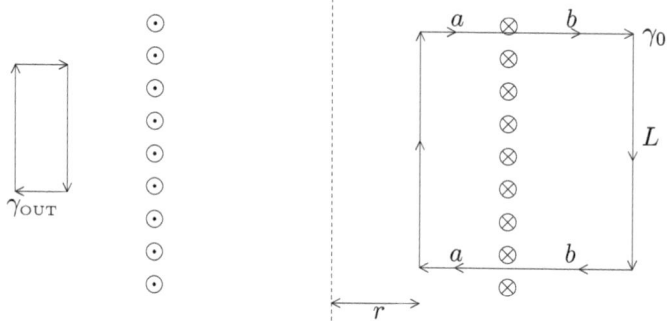

FIGURE 31.7   Amperian Analysis of the Ideal Solenoid

EXTERNAL   Let's first determine the magnetic field outside the solenoid.

In Figure 31.7, γ_{OUT} is an Amperian circuit lying entirely outside the solenoid.

[A feature of γ_{OUT} is that it respects the symmetries of the solenoid.]

Exactly zero net current passes through any surface which has this closed path as its boundary. Hence, the RHS of Ampère's Law vanishes, and thus

$$\oint_{\gamma_{\text{OUT}}} \boldsymbol{B} \cdot d\vec{s} = 0 \,.$$

Furthermore, since this must be true for all paths lying completely outside of the solenoid, we are forced to conclude that

$$\vec{\boldsymbol{B}}_{\text{Solenoid}}(\vec{r}) = \vec{0}\,, \ \forall \ \vec{r} \text{ outside the solenoid.}$$

INTERNAL   Next, consider an Amperian circuit, γ_0, which intersects the wall of the solenoid and respects its symmetries.

The leg completely outside contributes nothing to the total circulation, since the magnetic field vanishes outside (as shown just above). The parts which are inside the solenoid and perpendicular to the field also contribute zero. Thus, the entire circulation is provided by the segment of the path inside the solenoid and parallel to the axis. The cylindrical symmetry of the solenoid dictates that the field cannot possibly depend on z or θ. The field may depend on r, the radial distance from the central axis, and so the LHS of Ampère's Law reads

$$\text{Circ}\big(\vec{\boldsymbol{B}}_{\text{Solenoid}}\big) = \oint_{\gamma_0} \boldsymbol{B} \cdot d\vec{s} = \int_{z}^{z+L} ds\, B(r) = L\, B(r) \,.$$

The enclosed current [piercing the flat rectangular surface bounded by γ_0] depends on the number of times the current loops around, *i.e.*,

$$I_{\text{encl.}} = I\,\mathcal{N}.$$

The number of loops enclosed, $\mathcal{N}$, is equal to the product of η, the **winding density** of the coil [number of turns per metre], and L, the length of the path lying parallel to the axis. Hence, the enclosed current is $I_{\text{encl.}} = I\,\eta\,L$, and Ampère's Law implies that

$$\oint_{\gamma_0} \vec{B}_{\text{Solenoid}} \cdot d\vec{s} = \mu_0\, I_{\text{encl.}} \quad\Longrightarrow\quad L\,B(r) = \mu_0\, I\,\eta\, L \quad\Longrightarrow\quad B(r) = \mu_0\, I\,\eta,$$

a constant value, independent of position within the solenoid.

EXAMPLE [*A Solenoid*]

An extremely long piece of fine, insulated wire is twisted into a single-walled solenoidal coil with winding density $\eta = 2500$ turns per metre. A current, $I = 1/\pi$ A, is passed through the wire.

Q: What is the magnetic field strength along the central axis of the solenoid?

A: The magnetic field everywhere inside an ideal solenoid with these properties, bearing current I, is

$$B\big(\text{Inside}\big) = \mu_0\, I\,\eta = 4\,\pi \times 10^{-7} \times \frac{1}{\pi} \times 2500 = 10^{-3}\ \text{T}.$$

From this example, we see that it is not easy to produce strong magnetic fields.

For REAL solenoids, possessing finite length and comprised of coiled wire of finite thickness, the magnetic field outside, while weak, is NOT precisely zero, and the field inside varies slightly with position.

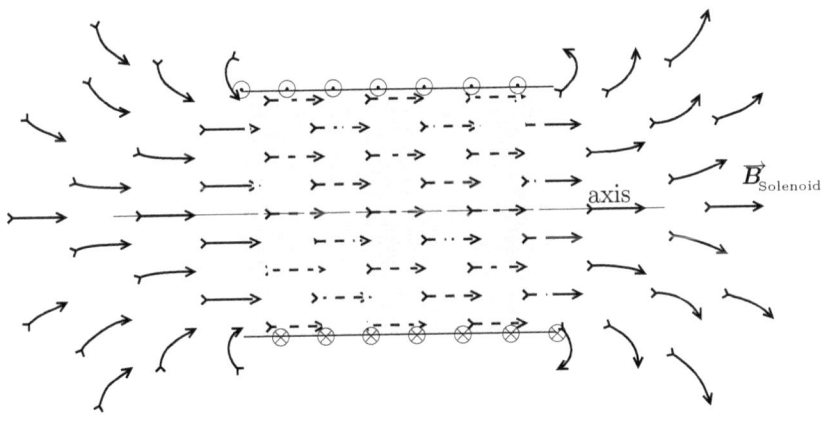

FIGURE 31.8 A Real Solenoid Suffers Magnetic Leakage and Non-Uniformity

Chapter 32

The General Form of Ampère's Law

Ampère's Law [in its TIME INDEPENDENT form],

$$\oint_{\gamma} \boldsymbol{B} \cdot d\vec{s} = \mu_0\, I_{\text{encl.}} \,,$$

relates the circulation of the magnetic field about a closed Amperian path to the current which pierces any surface bounded by that path. When the Amperian circuit lies entirely in a plane [as it has in all of the cases we have examined thus far], $I_{\text{encl.}}$ is commonly determined using the planar surface bordered by the path. However, any surface will do, and [as was noted in Chapter 30] employing a non-minimal surface sometimes proves advantageous.

Thus primed, let's reconsider the situation corresponding to the [infinitely] long straight wire carrying electric current I. In the analysis undertaken in Chapter 30, we chose a circular path of radius a in a plane perpendicular to the wire for the computation of the magnetic circulation. The electric current through the flat disk bounded by the path is I, as is the current through any sort of "windsock" path like the one shown in Figure 32.1. It is possible to prove[1] that Ampère's Law holds for any [continuous, open, non-self-intersecting] surface, as long as the Amperian path is its only boundary.

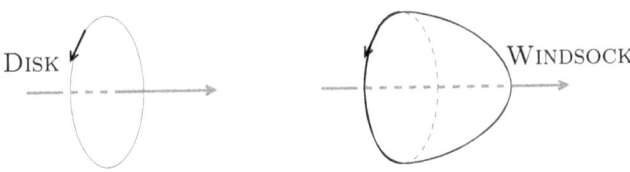

FIGURE 32.1 The Amperian Surface Is Bounded by the Amperian Circuit

The warm, fuzzy feeling of secure assurance that you might have acquired in our ability to apply Ampère's Law will prove illusory when we are confronted with the devastating counterexample below.

   ASIDE: We are dealing with a restricted [time-independent] form of Ampère's Law.

BREAKDOWN OF [THE TIME-INDEPENDENT FORM OF] AMPÈRE'S LAW

Consider two long straight [collinear] wires joined by a capacitor, as shown in the sketches collected in Figure 32.2. An electric current, I, is present in each of the wires, and yet **no current passes through the interstitial space in the capacitor!** Thus, the stage is set for a *reductio ad absurdum*.

---

[1] The proof is outside the scope of our analysis. We'll accept the claim and exploit the consequences.

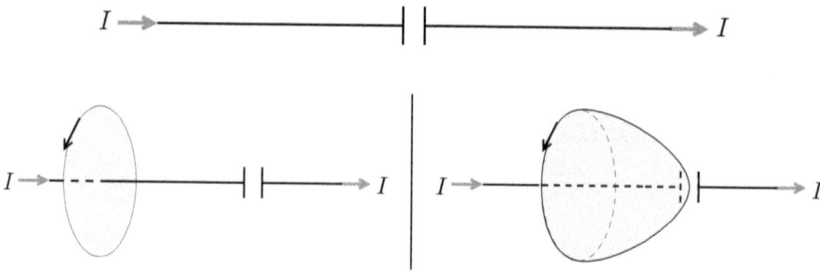

FIGURE 32.2   Counterexample to [Time Independent] Ampère's Law

CIRC   Consider the circulation of the magnetic field generated by the current along an Amperian circuit located far from the capacitor. For simplicity, the path is taken to lie in a plane perpendicular to the wire, to be circular, and to be centred on the wire. The LHS of Ampère's Law is

$$\text{LHS} = \text{Circ}(\vec{B})_\gamma .$$

RHS   The RHS of Ampère's Law is determined by the enclosed current [*i.e.*, the net current piercing surfaces which have γ_0, the Amperian circuit, as their boundary]. Two such surfaces, the disk and the windsock, are illustrated in Figure 32.2 and considered below.

DISK   The circular disk in the plane of the Amperian circuit is pierced by the wire and thus the current, I, flows through it. Hence, the RHS of Ampère's Law is

$$\text{RHS}_{(\text{Disk})} = \mu_0 \, I_{\text{encl.}(\text{Disk})} = \mu_0 \, I .$$

WINDSOCK   The particular windsock surface illustrated in the lower right panel of Figure 32.2 **has no electric current flowing through it**, whatsoever. Thus, the enclosed current is ZERO, and as a consequence

$$\text{RHS}_{(\text{Windsock})} = \mu_0 \, I_{\text{encl.}(\text{Windsock})} = 0 .$$

Ooops! The RHS of Ampère's Law ought not to vary when the LHS is fixed!

Q: How might this apparent conundrum be understood?

A: By careful consideration of the electric flux passing through the disk and windsock.

Let's suppose that the capacitor consists of two parallel plates. In this case, the magnitude of the interstitial field is constant everywhere in the region of space between the plates. The field vanishes everywhere else. Inside the capacitor,

$$\vec{E}_{\text{inside}} = \frac{\sigma}{\epsilon_0} = \frac{\frac{Q}{A}}{\epsilon_0} = \frac{Q}{\epsilon_0 \, A} ,$$

where Q is the charge on the plates and A is their area.

The rate at which the electric field is changing in time is

$$\frac{dE}{dt} = \frac{d}{dt} \left(\frac{Q}{\epsilon_0 \, A} \right) = \frac{1}{\epsilon_0 \, A} \frac{dQ}{dt} .$$

The current in the wire provides the electric charge which accumulates on the capacitor plates,

$$\frac{dQ}{dt} = I, \qquad \text{and therefore} \qquad \frac{dE}{dt} = \frac{I}{\epsilon_0 A}.$$

The changing electric field within the capacitor produces a concomitant change in the electric flux passing through the surface. Maxwell realised that this changing flux passing through the capacitor represents a continuation of the electric current flowing through each of the wires. Maxwell incorporated a **displacement current**, augmenting the conduction current, into a reformulation of Ampère's Law.

DISPLACEMENT CURRENT   The displacement current associated with the imaginary surface S, whose boundary is the Amperian circuit γ_0, is given by

$$I_{\text{disp.}} = \epsilon_0 \frac{\partial \Phi_{e,\text{S}}}{\partial t},$$

where $\Phi_{e,\text{S}}$ is the electric flux through the surface.

Three comments augment this definition.

o   The units work out consistently: $[\,I_{\text{disp.}}\,] = \dfrac{c^2}{\text{N}\cdot\text{m}^2}\,\dfrac{1}{\text{s}}\,\dfrac{\text{N}}{\text{C}}\cdot\text{m}^2 = \dfrac{\text{C}}{\text{s}} = \text{A}.$

o   The same surface is employed for the determination of both $I_{\text{encl.}}$ and $I_{\text{disp.}}$.

$I_{\text{encl.}} =$ the actual electric current piercing the surface

$I_{\text{disp.}} =$ the displaced electric current piercing the surface

o   The surface, S, is NOT closed. This is in contradistinction to the surfaces employed for the computation of electric fluxes in Gauss's Law. Rather, the Amperian circuit comprises the boundary of S.

To verify that Maxwell's modified version of Ampère's Law resolves the capacitor and windsock conundrum, let's reanalyse the example of two wires joined by a capacitor.

♮   Consider, as earlier, a circular Amperian path residing in a plane perpendicular to the wire and far from the capacitor. The path bounds a planar disk. The enclosed electric current through this surface is $I_{\text{encl.}} = I$, and the displacement current is $I_{\text{disp.}} = 0$.

> ASIDE: The vanishing of the displacement current may seem contestable on two grounds, both of which are specious. The first is the presence of the current in the wire, which may lead one to question whether some sort of electric field might leak out through the sides of the wire, contributing perhaps to electric flux through the surface bounded by the Amperian path. This may indeed be the case. However, for a steady state current, the field and flux will not be changing and thus will not contribute to a displacement current. The second is that the charge accruing on the capacitor plates may produce a field which leaks sufficiently far outside the capacitor to give rise to a changing flux through the planar disk-like surface. This possibility is excluded from consideration by our choice of an ideal parallel plate capacitor [which confines the electric field to the region of space between its plates].

Thus, the RHS of Ampère's Law is $\text{RHS}_{\text{Disk}} = \mu_0\left(I_{\text{encl.}} + I_{\text{disp.}}\right) = \mu_0\,I$, as was observed earlier.

♮   The windsock surface is bounded by the Amperian circuit and passes between the capacitor plates. The enclosed [conduction] electric current through the windsock surface is zero: $I_{\text{encl.}} = 0$. However, the displacement current is non-zero on account of the changing electric flux through the portion of the windsock residing within the capacitor. The assumptions which undergird our model for the [ideal] parallel plate capacitor ensure that there is no field [or flux] outside of the region lying between the plates. Inside the capacitor the field is uniform, and thus, irrespective of the shape of the surface, the electric flux at time t is $\Phi_{e,\text{S}} = E(t)\,A$, where $E(t)$ represents the uniform strength of the electric field between the plates at the specified instant in time. The displacement current is

$$I_{\text{disp.}} = \epsilon_0\,\frac{\partial \Phi_{e,\text{Windsock}}}{\partial t} = \epsilon_0\,A\,\frac{\partial E}{\partial t} = \epsilon_0\,A\left(\frac{I}{\epsilon_0\,A}\right) = I\,.$$

In this chain of equalities, we had recourse to the expression derived for the time rate of change of the electric field within the capacitor. The displacement current through the windsock turns out to be precisely equal to the conduction current bringing charge to the capacitor. As a consequence, the RHS of Ampère's Law for the windsock now reads

$$\text{RHS}_{\text{Windsock}} = \mu_0\,I\,.$$

♮   Maxwell's inclusion of the displacement current in the RHS of Ampère's Law has resolved the conundrum at the core of the counterexample.

AMPÈRE'S LAW (**Maxwell Improved**)   The magnetic circulation about a closed Amperian circuit, γ_0, is

$$\text{Circ}(\vec{B}) = \oint_{\gamma_0} \vec{B}\cdot d\vec{s} \equiv \mu_0\left(I_{\text{encl.}} + I_{\text{disp.}}\right) = \mu_0\,I_{\text{encl.}} + \mu_0\,\epsilon_0\,\frac{\partial \Phi_{e,\text{S}}}{\partial t}\,.$$

The Amperian path about which the circulation is computed, γ_0, is the boundary of the surface, S, used in determining both currents, enclosed and displacement.

AMPÈRE'S LAW [including the displacement current]
is a fundamental LAW of NATURE.

Chapter 33

Gauss's Law for Magnetism

In complete analogy with the computation of electric flux, one may compute the **magnetic flux** through a surface S, *i.e.*,

$$\Phi_{m,\mathrm{S}} = \int_{\mathrm{S}} \boldsymbol{B} \cdot d\overrightarrow{A}.$$

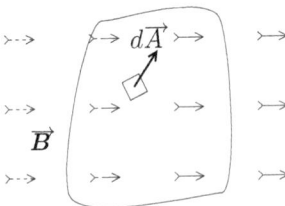

FIGURE 33.1   Magnetic Flux Passing through Surface S

The SI unit of magnetic flux is the weber [Wb]. Recall, from Chapter 23, that another name for the magnetic field, $\boldsymbol{B}$, is the **density of magnetic flux**, or **magnetic flux intensity**. Thus, one weber per square-metre equals one tesla:

$$1\,\mathrm{T} = 1\,\frac{\mathrm{Wb}}{\mathrm{m}^2}.$$

The magnetic flux may be computed for [almost[1]] any surface. Our experience with electric fluxes led us to consider the specialisation in which the surfaces were closed, and thence we obtained **Gauss's Law for Electric Fields**.

$$\Phi_{e,\mathrm{S}_0} = \oint_{\mathrm{S}_0} \boldsymbol{E} \cdot d\overrightarrow{A} \equiv \frac{q_{\mathrm{encl.}}}{\epsilon_0}.$$

Let's attempt this construction for magnetic fields. The LHS of this putative relation consists of the net magnetic flux through a closed, non-self-intersecting surface, S_0:

$$\mathrm{LHS} = \oint_{\mathrm{S}_0} \boldsymbol{B} \cdot d\overrightarrow{A}.$$

While the generalisation of the LHS was easily accomplished, the RHS is somewhat less so. The obvious candidate is:

> *a physical constant* [to get the units right] *times the net magnetic charge enclosed!*

---

[1]The same restriction to non-self-intersecting surfaces, mentioned in conjunction with electric flux computations in Chapter 5, applies here, too.

However, a [major] flaw in this proposal is that isolated magnetic monopolar charges have never[2] been observed! This obvious lack of symmetry [between electric and magnetic charges] does not daunt us. Along with Pierre Curie,[3] we accept on phenomenological grounds that the RHS equals ZERO! With this, all the ingredients for an expression of GAUSS'S LAW FOR MAGNETISM have been assembled.

GAUSS'S LAW for MAGNETIC FIELDS   The net magnetic flux through any closed [non-self-intersecting] surface, S_0, is precisely equal to zero. That is,

$$\Phi_{m,S_0} = \oint_{S_0} \boldsymbol{B} \cdot d\vec{A} \equiv 0 \,.$$

A number of important physical consequences follow from Gauss's Law for Magnetism. The two most significant are listed immediately below, while a few other related comments follow farther on.

- **Isolated magnetic monopoles do not exist in nature.**

  The only known sources of magnetic fields are magnetic dipoles [and higher-pole structures composed of dipoles] and current loops, which, as we have seen, act like dipoles.

- **Magnetic field lines never terminate; they form closed loops.**

GAUSS'S LAW IMPLICATIONS AND COMMENTARY

○   Samples of magnetite, a naturally occurring form of oxidised iron, often exhibit strong magnetic fields. This property, called **ferromagnetism**, will be briefly touched upon in Chapter 34.

   ASIDE: These "lodestones" have been highly valued for their exotic properties. Ancient navigators employed slender magnetised needles mounted on pivots (or floating in a liquid) to determine direction when out of sight of land. Ancient "scientists" (philosophers and charlatans, alike) used magnets to perform party tricks, as was mentioned in Chapter 1. Modern peoples use them to attach bits of paper to their refrigerators.

The simplest naturally occurring macroscopic magnetic objects are bar magnets. Bar magnets possess North, N, and South, S, poles. By convention, **sources** of magnetic field lines are N-poles, while **sinks** are S-poles.

Gauss's Law ensures that IF one cuts a bar magnet in half, THEN one does NOT get isolated N and S poles, but rather two smaller N-S bar magnets. True and false consequences of such a split are both illustrated in Figure 33.2.

---

[2]When the crew [chorus] of *H.M.S. Pinafore* asks: "Never?" our rejoinder is "Well, not yet to the level of scientific conviction," as we'll see in discussion later in this chapter.

[3]Marie Curie's spouse, Pierre (1859–1906), was a very talented physicist in his own right.

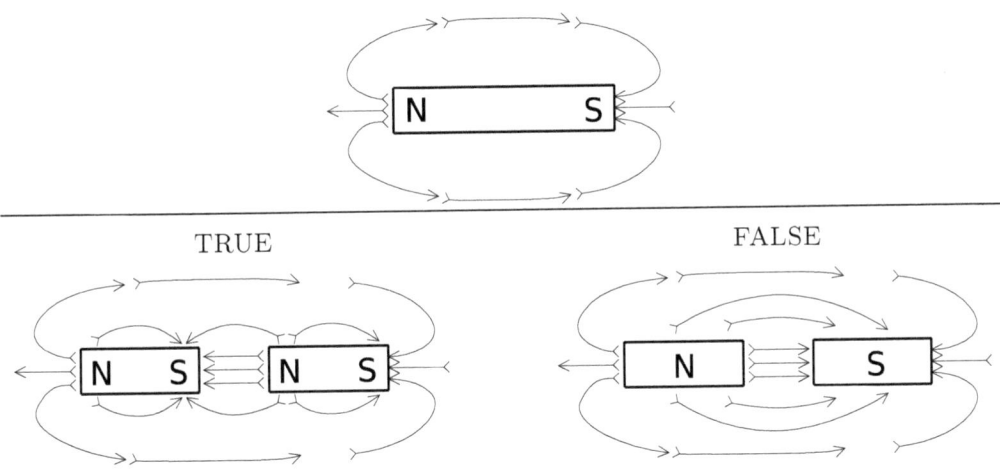

FIGURE 33.2 Fracturing a Bar Magnet Fails to Isolate Magnetic Poles

o   Each end of an ideal solenoid [separated by infinite distance] appears like a magnetic pole. These poles are not really ISOLATED, however, as the flux from one is conveyed to the other via the solenoid.

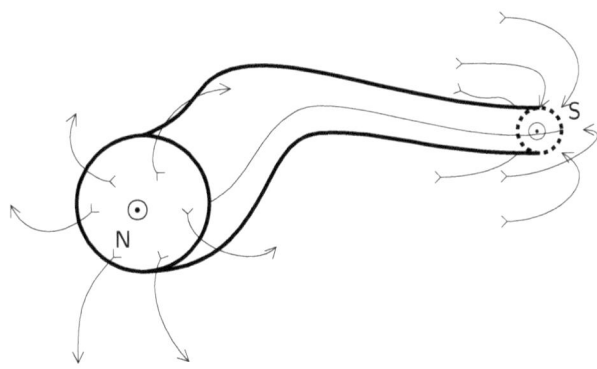

FIGURE 33.3 An Infinite Solenoid Has an Isolated Magnetic Pole at Each End

o   Paul A.M. Dirac[4] formulated (*circa* 1931) a rather general [relativistic quantum mechanical] argument that:

> IF there exists [at least] one magnetic monopole in the universe,
> THEN electric charge is quantised.

o   Spurred by Dirac's claim, and its later generalisations, people have been, and still are, searching for magnetic monopoles. Over the years there have been some number of claims of success. However, in each of these cases but one, subsequent investigation has not supported the initial claim. As yet unresolved is a putative detection claim made in 1982 by Blas Cabrera of Stanford University. Although

---

[4]Dirac (1902–1984) shared the 1933 Nobel Prize in Physics with Erwin Schrödinger.

Cabrera's monopole detector was in working order, and the detection event looked "picture perfect and suitable for framing," there were complicating factors.[5] Two of these were that:

—   the signal appeared in the evening on Saint Valentine's Day[6]
—   despite considerable efforts, no other experiments have managed to reproduce Cabrera's result

Hence, the scientific community has not [yet] accepted this experimental result as a genuine detection of a magnetic monopole.

> IF there is only ONE magnetic monopole in the entire universe [the minimum needed for electric charge quantisation, per Dirac], AND it was indeed detected in this single experiment, THEN Cabrera is simultaneously the *luckiest* and the *unluckiest* bloke in the universe!

o   Cosmologists, who study the large-scale structure and evolution of the universe, are avidly interested in magnetic monopoles for several reasons. The primary one is that current models of particle physics constrain such monopoles to be quite heavy, and weakly interacting with normal [*i.e.,* baryonic] matter. Recalling that inertial mass acts as gravitational charge,[7] monopoles situated in galactic halos could provide the [otherwise] "missing mass" needed to explain galactic rotation curves.

> ASIDE: Observers can infer the speeds of stars and clusters of stars in orbit about galactic centres. The measured speeds suggest centripetal accelerations grossly in excess of those which could be accounted for by [Newtonian] gravitational dynamics when the central-force-producing mass is inferred from measurements of the amount of starlight emitted from within the radius of the orbit. [The weakness of the non-gravitational interactions between normal matter and monopoles renders the latter difficult to illuminate. Thus, the light from stars is thought to reveal the presence of baryonic matter only.]

A second reason is that, in the very early universe, magnetic monopoles may have contributed to the initial density "seeds" [fluctuations] about which the galaxies formed.

> ASIDE: The issue of galaxy formation is still subject to considerable debate. The basic idea is not unlike formation of a raindrop, in which a dust particle acts as an initial nucleation site for accretion of water vapour. If the initial conditions are too homogeneous, and the nucleation sites too sparse, then galaxies do not form on time scales on the order of the 13.8-billion-year estimated age of our universe. If the galactic seeds are too large, or too densely distributed, then the galaxies evolve too quickly [often acquiring supermassive central black holes] and fail to last for [tens of] billions of years.

Magnetic monopoles may provide enough inertia to drive galaxy formation where they cluster, and be sufficiently scarce to prevent it from occurring too rapidly.

---

[5] *Isn't it the case that there are always complicating factors?*
[6] It was thought that this might have been part of an elaborate prank that spun out of control.
[7] This statement is the essence of the EQUIVALENCE PRINCIPLE introduced in VOLUME I, Chapter 47.

Chapter 34

Magnetism in Matter

At the atomic level, magnetic properties of materials arise via disparate **orbital** and **spin** mechanisms. The orbital effects are more amenable to classical interpretation and analysis, and so we shall consider these first. Spin is an inherently quantum mechanical effect, and we shall treat it, as best we can, by analogy with the orbital case.

Tiny loops of current may be associated with the motion of each of the electrons orbiting[1] the nucleus. For clarity and convenience, we shall adopt the simplistic model of a single electron, with charge $-e$, in circular orbit about an effective charge $+e$ which is fixed at the origin. This model is efficacious because the outermost electron moves through a region of space in which the effective potential receives contributions from the atomic nucleus and from all of the—more tightly bound—inner electrons.

A neutral atom with atomic number n has n protons confined within its [point-like] nucleus, and n rapidly moving electrons distributed in the nearby space. The total electric charge residing in the nucleus, $+ne$, produces an electric potential in the surrounding space. At radius r, corresponding to the [assumed circular] orbit of the outermost electron,

$$V_{\text{nucleus}} = \frac{k\,n\,e}{r}\,.$$

The $n-1$ electrons whose orbits lie closer to the nucleus than that of the outermost electron whiz around in such fashion as to produce an isotropic blur of electric charge, $-(n-1)\,e$. The inherent spherical symmetry of this distribution of charge suggests that its net [averaged] effect is the same as that of a single point-like charge coincident with the nucleus:

$$V_{\text{inner electrons}} = \frac{-k\,(n-1)\,e}{r}\,.$$

Thus, the net [effective] electric potential at radius r is

$$V_{\text{eff.}} = V_{\text{nucleus}} + V_{\text{inner electrons}} = \frac{k\,e}{r}\,.$$

The centripetal acceleration of the outermost electron arises from the electric force associated with the gradient of the effective potential:

$$m\,\vec{a}_c = (-e)\left[-\vec{\nabla}\,V_{\text{eff.}}\right] \qquad \Longrightarrow \qquad m\frac{v^2}{r} = \frac{k\,e^2}{r^2}\,.$$

And so, in effect, it is indeed as though the outermost electron, with charge $-e$, were alone in orbit about a fixed nucleus bearing charge $+e$.

At this juncture, one can continue the argument in either of two main directions. As is our wont, we shall do both.

---

[1]While this is a completely bogus [CLASSICAL PHYSICS] notion, it provides a convenient mental image for a very real physical effect.

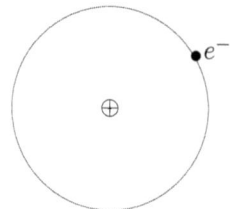

Object	Length Scale
Atom	Angstrom, Å
Nucleus	Fermi, fm
Electron	POINT-LIKE < 10 am

FIGURE 34.1   Crude Model of a Neutral Atom

METHOD ONE: [*Centripetal Force Analysis*]

The outermost electron moves uniformly in its orbit. To circle around precisely once each orbital period, it must move with speed $v = 2\pi r/T$, and hence

$$m\,\frac{\frac{4\pi^2 r^2}{T^2}}{r} = \frac{k\,e^2}{r^2} \qquad \Longrightarrow \qquad T^2 = \frac{4\pi^2 m}{k\,e^2}\,r^3\,.$$

Careful inspection shows that this relation is exactly like Kepler's Third Law [encountered in VOLUME I, Chapter 50], with the outer electron playing the rôle of the satellite, the nucleus [with effective charge $+e$] acting as the Sun, and the Coulombic electric force taking the place of Newtonian Universal Gravitation. Substituting approximate values $m = m_e = 9.11 \times 10^{-31} \simeq 10^{-30}$ kg, $e = 1.6 \times 10^{-19} \simeq 2 \times 10^{-19}$ C, $k = 8.99 \times 10^9 \simeq 10^{10}$ N·m^2/C^2, $\pi^2 \simeq 10$, and $r \simeq 10^{-10}$ m, one obtains

$$T^2 \sim \frac{4 \times 10 \times 10^{-30}}{10^{10} \times 4 \times 10^{-38}} \times 10^{-30} \sim 10^{-31}\,.$$

Thus, a rough estimate of the orbital period is 3×10^{-16} s.

> ASIDE: The steps taken just above to find the orbital period are similar to those taken in Chapter 27 to obtain the cyclotron frequency of a charged particle moving through a uniform magnetic field.

METHOD TWO: [*Virial Theorem*]

The VIRIAL THEOREM referred to in Chapter 49 of VOLUME I applies in this situation, too. In particular, the kinetic energy is minus one half of the potential energy, *i.e.*,

$$\frac{1}{2}\,m\,v^2 = \frac{k\,e^2}{2\,r}\,.$$

Since the virial approach reproduces the starting point of the centripetal force analysis, it is safe to conclude that the same approximate value for the period is obtained here.

Also, one can estimate the binding energy of the outermost electron,

$$E_{\text{binding}} = \frac{k\,e^2}{2\,r} \sim \frac{10^{10} \times 4 \times 10^{-38}}{2 \times 10^{-10}} \sim 2 \times 10^{-18}\ \text{J} \sim 10\ \text{eV}\,,$$

by employing the same approximate values of the physical parameters as were chosen in consideration of the period. The electron volt [eV] is a convenient atomic-scale unit of energy:

$$1\ \text{eV} = e\,(1\ \text{V}) = 1.6 \times 10^{-19}\ \text{J}\,.$$

Both approaches yield the same estimate for the orbital period of the outermost electron in a neutral atom. Getting back to the model for magnetism, the orbiting electron constitutes a moving charge and thus produces an electric current with approximate magnitude

$$ I_e \sim \frac{e}{T} \simeq \frac{1.6}{3} \times 10^{-3} \sim \frac{1}{2} \times 10^{-3} \text{ A} . $$

Associated with this atomic current loop is a magnetic dipole moment, $\vec{m} = I_e \vec{A}$. The area of the loop is that of the circular disk swept out by the orbiting electron. The radius of this disk is roughly the size of the atom, 10^{-10} m, and so $|\vec{A}| \sim \pi \times 10^{-20}$ m^2. Thus,

$$ |\vec{m}| = I_e \, |\vec{A}| \sim \frac{1}{2} \times 10^{-3} \times \pi \times 10^{-20} \sim 1.6 \times 10^{-23} \text{ A·m}^2 . $$

The magnetic potential energy of this putative current loop lying in a uniform applied magnetic field, $\vec{B}_0$, is

$$ U_m = -\vec{m} \cdot \vec{B}_0 . $$

IF the externally applied magnetic field has strength 1 T [a rather strong field, indeed], THEN the maximum magnetic potential energy of the orbiting electron is

$$ \text{Max}\big(-\vec{m} \cdot \vec{B}_0 \big) = |\vec{m}| \, |\vec{B}_0| \simeq 1.6 \times 10^{-23} \times 1 \text{ A·m}^2\text{·T} \simeq 1.6 \times 10^{-23} \text{ J} = 10^{-4} \text{ eV} . $$

Clearly the magnetic potential energy of the outermost electron is but a small fraction of its electric binding energy. For the sake of comparison, let's now consider the expected thermal energy of such an electron.

The EQUIPARTITION THEOREM[2] states that the expected thermal energy of a system at temperature T is equal to the energy per dynamical mode, $\frac{1}{2} k_B T$ [where $k_B = 1.38 \times 10^{-23}$ J/K is Boltzmann's constant], multiplied by its number of degrees of freedom. A point-like particle like the electron has three translational degrees of freedom. At room temperature, $T \sim 300$ K, the energy per mode is approximately

$$ \frac{1}{2} k_B T = \frac{1}{2} \times 1.38 \times 10^{-23} \, 300 \simeq 2.1 \times 10^{-23} \text{ J} = 1.3 \times 10^{-2} \text{ eV} . $$

Thus, there is a hierarchy of atomic energies:

$$ \text{Ionising} \, [= - \text{ Binding}] \gg \text{Thermal} \gg \text{Magnetic} , $$

under relatively unremarkable conditions.

RE-ANALYSIS IN TERMS OF QUANTISED ANGULAR MOMENTUM

A qualitatively and quantitatively stronger model is obtained by incorporating certain quantum mechanical features.

ASIDE: The crude semi-classical treatment discussed here is what led to the "filling rules" or "orbital structures" which are presented axiomatically in chemistry classes.

---

[2]Such thermodynamical matters are discussed in VOLUME II Chapters 49, 50, 45, and mentioned, not unlike here, in 33.

Seeking deeper meaning in the cheesy and still classical model, we write the magnitude of the outermost electron's magnetic moment in terms of its charge and its orbital parameters, *i.e.*,

$$|\vec{m}_l| = I\,A = \frac{e}{T}\,A = e\left(2\pi\frac{r}{v}\right)^{-1} \times \pi\,r^2 = \frac{1}{2}\,e\,v\,r,$$

and thereby discern an expression evocative of angular momentum! The subscripted "*l*" on the magnetic moment vector makes explicit this association. Two particular results from the investigation of classical angular momentum [in Chapters 42 and 49 of VOLUME I] are recalled here.

$\vec{L}$ The magnitude of the angular momentum of a point-like particle possessing mass m and moving with speed v in a circular orbit of radius r about a fixed axis is $L = m\,v\,r$.

> ASIDE: A little-noted deficiency in this model is the assumption that the electron moves in a circular orbit about the nucleus.
>
> [Ordinarily, two interacting objects rotate about their common CofM.]
>
> The model is perfectly valid in the limit in which the nucleus is infinitely more massive than the electron. Since the nucleus consists of protons [each roughly 1836 times more massive than an electron] and neutrons [roughly 1837 times more massive], the results obtained in this limit are pretty accurate. Nonetheless, a well-established means of relaxing this assumption is the introduction of the **reduced mass** of the electron,
>
> $$m_{e,\text{ reduced}} = \frac{m_{\text{nucleus}}\,m_e}{m_{\text{nucleus}} + m_e}.$$
>
> The justification for using reduced mass in central-force scenarios is a suitable subject for an Intermediate Mechanics class.

$\vec{L}$ In situations such as this, in which the force acts centripetally, angular momentum is CONSERVED:

$$\frac{d\vec{L}}{dt} = \vec{\tau}_{\text{net,ext'l}} = \vec{0},$$

and hence $L = m\,v\,r$ is **a constant of the motion.**

Therefore, the magnetic dipole moment of the orbiting charge may be expressed in terms of the [constant] angular momentum of the electron and its charge-to-mass ratio,

$$|\vec{m}_l| = \frac{1}{2}\frac{e}{m_e}\,L = \frac{e}{2\,m}\,L.$$

In the early 1910s, Niels Bohr[3] was able to explain [qualitatively and quantitatively] many features of the emission and absorption spectral lines observed in hydrogen [and other gases] by postulating that, on atomic scales, angular momentum is quantised in units of Planck's constant, $\hbar \simeq 1.055 \times 10^{-34}$ J·s:

$$L \equiv l\,\hbar \qquad \text{where } l \in \{0,1,2,\dots\} \text{ is a non-negative integer.}$$

> ASIDE: Bohr's ideas, refined in the 1920s, form the basis of what is now called *Old Quantum Mechanics*, which can explain a very great deal of chemistry!

---

[3]Bohr (1885–1962) had a profound influence in the development of quantum physics. He was awarded the Nobel Prize in Physics in 1922.

CLAIM: **Since the atomic magnetic moment is proportional to the angular momentum of the outermost electron, and angular momentum is quantised, atomic orbital magnetic moments must be quantised!**

BOHR MAGNETON   The Bohr magneton,

$$\mu_b = \frac{e\,\hbar}{2\,m_e} \simeq 9.27 \times 10^{-24} \ \text{J/T}\,,$$

is the elementary quantum of atomic magnetic moments and thus provides the natural unit for their measurement.

The electron mass appears in the expression for the Bohr magneton. *A priori*, there is nothing to prevent one from considering a nuclear Bohr magneton using the charge-to-mass ratio of the proton; however, it would be about 1836 times smaller than μ_b, and is thus neglected in our analysis of atomic magnetic moments.

ORBITAL GYROMAGNETIC RATIO   The orbital gyromagnetic ratio, g_l, is a factor introduced to describe the proportionality between the quantised orbital angular momentum of the outermost electron and the atomic magnetic moment. In the present circumstance, it assumes the value $g_l \equiv 1$. Colloquially, g_l is often dubbed "the orbital g-factor."

In terms of the Bohr magneton and the orbital gyromagnetic ratio, the magnitude of the atomic magnetic dipole moment is

$$|\vec{m}_l| = \frac{g_l\,\mu_b}{\hbar}\,L \qquad \Longrightarrow \qquad |\vec{m}_l| = \mu_b\,l\,.$$

ASIDE: One might recall from chemistry class, or books that one has read, that three quantum numbers, $\{n, l, m\}$, determine the extent [size] and shape of electron orbitals. These numbers are called

n the principle quantum number,

l the angular momentum quantum number, and

m the azimuthal quantum number.

Of these, l and m pertain to the CLASSICAL magnetic moment.

A quantum mechanical property called [intrinsic] **spin**[4] is manifest in certain experiments. Charged particles with non-zero spin possess an intrinsic magnetic moment.

---

[4]Spin has an essential association with the statistical properties of collections of particles. When the spin assumes a odd-half-integer value, $s \in \{\frac{1}{2}, \frac{3}{2}, \frac{5}{2}, \dots\}$, the particle is **fermionic** and the PAULI EXCLUSION PRINCIPLE is operative. Pauli exclusion prohibits double-occupancy of any single quantum state in a system. When the spin is integer-valued, $s \in \{0, 1, 2, \dots\}$, the particle is **bosonic**, and collective properties, such as condensation into a single macrostate, are permitted.

ASIDE: In a cheesy way, one might try to envision the spin magnetic moment as a consequence of the physical rotation of a distribution of charge. While this attempt at a model is completely bogus, it again provides a satisfactory mental picture to guide our intuition.

Q: How do we know that this "classical spinning top" model is bogus?

A: Were the charge on an electron to be distributed throughout a spherical region volume [fitting within the present-day experimental bounds on the spatial extent of the electron], AND were this sphere to rotate so as to generate the electron's spin magnetic moment, then points on the surface of the sphere would have to move at speeds in excess of the speed of light! OOoops!

In analogy with the orbital angular momentum, we define a spin gyromagnetic ratio and an intrinsic spin magnetic moment.

SPIN GYROMAGNETIC RATIO   The spin gyromagnetic ratio, *a.k.a.* the spin g-factor, g_s, is the proportionality factor relating [quantised] spin angular momentum to its associated magnetic moment. Curiously, $g_s \equiv 2$ in non-relativistic quantum mechanics.

In terms of the Bohr magneton, the spin gyromagnetic ratio, and the quantised spin angular momentum $S = s\,\hbar$, the magnitude of the atomic magnetic dipole moment is

$$ |\vec{m}_s| = \frac{g_s\,\mu_b}{\hbar}\,S \qquad \Longrightarrow \qquad |\vec{m}_s| = 2\,\mu_b\,s\,. $$

Three important questions arise.

Q: Why are the orbital and spin g-factors different?

A: The factor of 1 for the orbital case is tautological, arising from the definition of the Bohr magneton. In the case of spin, Dirac showed that in [non-relativistic] quantum mechanics [NRQM] g_s must be equal to 2.

> ASIDE: When very precise experimental measurements are made, g_s is found to be slightly greater than 2. This *anomaly* is accounted for—to about one part in a trillion—in the theory of quantum electrodynamics [QED].

Q: Why are spin and orbital angular momenta quantised in the same manner?

A: Classically, there is no *a priori* reason for anything other than an accidental relation. Take, for instance, a planet in the solar system. The [orbital] angular momentum associated with the planet's revolving about the Sun has little or no correlation with the [spin] angular momentum it possesses by virtue of its rotation about an axis. Nonetheless, the two species, orbital and spin, are both angular momenta, and hence it is not unlikely that they be related. This naturally leads to another question.

Q: Do the orbital and spin angular momenta simply add to yield the total, or net, angular momentum?

A: The surprising answer is that the algebraic structure of angular momenta is rather more complicated than that.

[Take a quantum mechanics class to find out the wonderful details!]

Magnetism in Materials

In most cases the macroscopic effect of these atomic-scale magnetic moments is very small [in the absence of an externally applied magnetic field]. This is because neighbouring magnetic moments are uncorrelated [*i.e.,* randomly directed]. Therefore, when a macroscopic volume of material is considered, the net magnetic moment tends toward ZERO.

The response of a particular material to the presence of an external magnetic field generally falls into one of three phenomenological classes: **diamagnetism, paramagnetism,** and **ferromagnetism.**

> ASIDE: In truth, there are FIVE classes. We shall pass over **antiferromagnetism** and **ferri-magnetism** since they are both weakly magnetic and rather complicated.

DIAMAGNETISM   All materials exhibit diamagnetism in weak and changing fields.

Recall [Chapter 15] that, when a lump of dielectric material is subjected to an applied electric field, its microscopic constituents polarise in such a way as to diminish the local field within the bulk of the material.

A similar effect occurs for diamagnetic materials, and thus we write

$$B_{\text{inside}} = B_0\left(1 + \chi\right),$$

where $\chi < 0$ is the **magnetic susceptibility** of the material. Typical susceptibilities are small, with magnitudes on the order of 10^{-5}.

On the other hand, superconducting substances behave as perfect diamagnets, $\chi = -1$, and as such exclude or expel magnetic field lines from their bulk, or quench those that penetrate into it.

PARAMAGNETISM   Materials whose microscopic constituents have non-zero net magnetic moment are typically paramagnetic.

Here is a *Gedanken* experiment to elucidate paramagnetic properties.

> **Part A:**   Consider a magnetic dipole situated in space and free to rotate. [In the case of paramagnetism, this is an atomic-scale dipole. For this *Gedanken* experiment, think of a current loop *à la* Chapters 25 and 26.]

> **Part B:**   Apply an external [uniform[5]] magnetic field to the dipole.

Let's recall, from Chapter 26, the manner in which a rectangular current-loop dipole responds to a uniform applied magnetic field $\vec{B_0}$. A torque,

$$\vec{\tau} = \vec{m} \times \vec{B},$$

attempts to rotate the magnetic moment of the current loop, $\vec{m}$, **into alignment with the applied field.** The details are sketched in Figure 34.2.

---

[5]This is only to ensure that the loop not be subjected to a net magnetic force. The field need only be approximately uniform over the spatial dimensions of the loop.

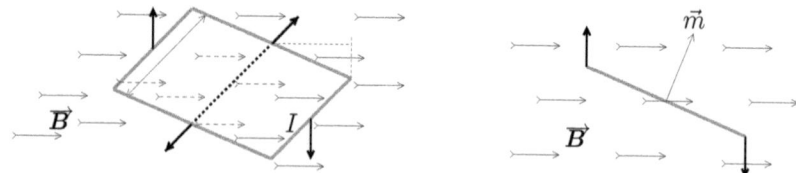

FIGURE 34.2   Response of a Current Loop to an Applied Field

Part C: Let us not forget that the electric current in the loop **produces** a magnetic field in the surrounding space. The field lines wrap around the wires in such manner that [in the plane containing the loop of wire] the field within the loop is oriented in the opposite direction to that outside of the loop. [Also, it is generally true that the field inside the loop is stronger than that outside. This is consistent with our knowledge that the total fluxes outside and inside of the loop, $\Phi_{m,\text{outside}}$ and $\Phi_{m,\text{inside}}$, must have the same magnitude, in accord with Gauss's Law for Magnetic Fields.] In effect, the current loop produces a magnetic field which is roughly aligned with $\vec{m}$ throughout the interior of the loop. The pattern of magnetic field lines passing through the plane of the loop is illustrated in Figure 34.3.

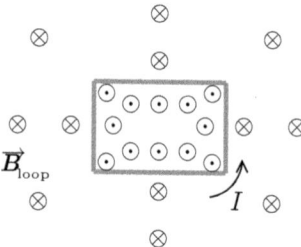

FIGURE 34.3   Production of a Magnetic Field by the Loop-Current

Part D: The torque rotates the dipole moment vector into alignment with the applied field, and thus causes there to be an increase[6] in the strength of the local magnetic field. We quantify this increase using the same magnetic susceptibility parameter as was introduced for diamagnetism:

$$B_{\text{inside}} = B_0 \left(1 + \chi\right),$$

for paramagnetic susceptibility $\chi > 0$. Typical values of χ are on the order of 10^{-4}, so paramagnetic effects, when they occur, usually trump those of diamagnetism.

The measure of coherence in the alignment of dipoles is provided by the **magnetisation,** which is, in effect, the local dipole moment density [*a.k.a.* the net dipole moment vector per unit volume]. When there is no applied field, the magnetisation of a paramagnetic substance is zero.

---

[6]This increase nearby is offset by a diminution of the field at points more distant.

FERROMAGNETISM   Even in the absence of an ambient magnetic field [provided that the temperature is not too high], micro-, meso-, and macroscopic regions of ferromagnetic materials exhibit significant SPONTANEOUS MAGNETISATION. Two comments are warranted.

$\langle \vec{m} \rangle$ Even though local regions are magnetised, the sample as a whole may be only weakly magnetic, owing to incoherent superposition of uncorrelated domains on the largest scales. On the other hand, when there is very-large-scale alignment, the system is strongly magnetic.

T_C Spontaneous magnetisation is a consequence of a symmetry-breaking quantum phase transition. Let's reflect on a few aspects of this statement.

+   That some materials possess a FERROMAGNETIC PHASE is no more mysterious than the notion that a material has solid, liquid, and gaseous phases. It only seems odd because we typically have less experience with the magnetisation set of thermodynamic state variables.[7]

+   SYMMETRY-BREAKING[8] refers to the idea that, *a priori*, local domains might magnetise in any direction without preference. That a single direction is chosen breaks [hides] the isotropy of space.

+   Associated with the phase transition is a context-dependent CRITICAL TEMPERATURE dubbed the "Curie Temperature," T_C, in honour of Pierre Curie.[9] Above its Curie temperature, a material is paramagnetic with a very large susceptibility, perhaps $\chi \sim 10^5$. Below the Curie Temperature, spontaneous magnetisation occurs, with asymptotically limited increase as the temperature approaches $0\,\mathrm{K}$.

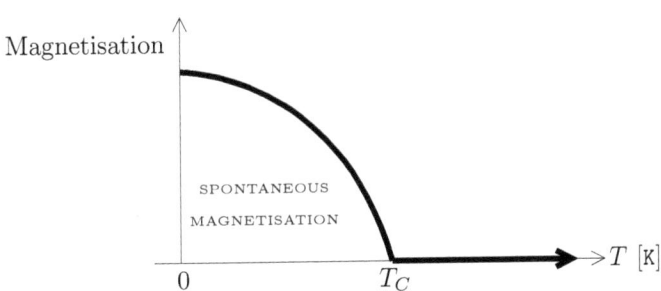

FIGURE 34.4   Spontaneous Magnetisation Occurring as a Phase Transition

It must be emphatically noted that the loss of long-range order [correlation] at the Curie temperature does not occur because of MELTING [crystalline order breaking down to an amorphous fluid structure]. The Curie and solidification temperatures [under typical conditions] for nickel and iron, two prominent ferromagnetic materials, are collected in the table which follows.

---

[7]Thermodynamic state variables are discussed, in the context of gases, in VOLUME II of this series.

[8]Some leading scientists, following the example set by Sidney Coleman, prefer to speak of HIDDEN, rather than BROKEN, symmetry.

[9]Pierre and Marie Curie shared the 1903 Nobel Prize in Physics with Henri Becquerel for the discovery of radioactivity.

Element	Ni	Fe
Curie Temperature	$T_{C_{Ni}} = 631\,\text{K}$	$T_{C_{Fe}} = 1043\,\text{K}$
Solidification Temperature	$T_{S_{Ni}} = 1728\,\text{K}$	$T_{S_{Fe}} = 1811\,\text{K}$

Three easy steps lead to the production of a **permanent magnet**.

1. Take a lump of ferromagnetic material, and heat it somewhat [but not above its Curie temperature] to increase the motility of the dipoles.

2. Apply a strong external magnetic field in a particular direction.

3. Cool the lump and reduce the applied field.

The result is a PERMANENT magnet, because the domains which are somewhat aligned with the applied field will grow, gobbling up their neighbours. Afterwards these domains do not relax to their previous states[10] and the lump retains a net magnetic dipole moment.

It takes but one easy step to destroy a permanent magnet.

1. Heat the sample of material beyond its Curie temperature, and then cool it in the absence of an applied field.

The spontaneously magnetised regions will be small and (it is expected) isotropically oriented. Hence, the total or net magnetisation may be expected to [nearly] vanish, and the lump of material will no longer exhibit strongly magnetic behaviours.

---

[10]This phenomenon is known as **hysteresis**.

Chapter 35

Faraday's Law

FARADAY'S LAW (non-technical)   Two common expressions of Faraday's Law are:

A changing magnetic field can induce a current in a circuit,

and

A change in magnetic flux induces an EMF.

Faraday's Law is often called THE LAW OF INDUCTION.

Before we can apply Faraday's Law, we need to know the sense in which the EMF is induced, or, equivalently, the sense in which the induced current flows. This important piece of information is provided to us by LENZ'S LAW.

LENZ'S LAW   Faraday's induced current is oriented so as to oppose the change that brought it about.

Let's retrace, as a *Gedanken* experiment, the steps that led Faraday to discover the law of nature named in his honour. First, we assemble a bar magnet, a wire hoop, and a galvanometer [G] to detect [and measure] electric current in the hoop.

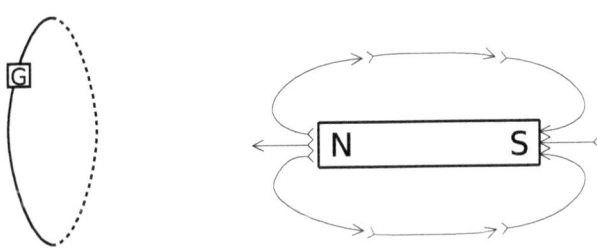

FIGURE 35.1   A Conducting Hoop, a Galvanometer, and a Bar Magnet

Part A: The hoop and magnet are at mutual rest.

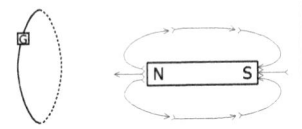   IF the magnet is at rest with respect to the hoop, THEN there is no induced current in the hoop.

Part B: The magnet is in motion toward the hoop.

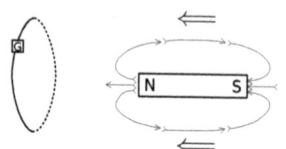

IF the magnet moves toward the hoop, **THEN** the amount of magnetic flux [equivalently, the number of magnetic field lines] passing through the hoop is [necessarily] increasing.

The hoop reacts to the increased flux [FARADAY'S LAW] by induction of a current which generates its own flux **in opposition to** [LENZ'S LAW] the increase coming from the bar magnet. In the present case, the increased flux is in the direction [←], *i.e.,* from right to left, in the sketch. Hence, the response flux through the face of the hoop passes from left to right. Thus, the induced magnetic field lines throughout the interior of the hoop are directed [→], *i.e.,* to the right. Consonant with this direction for the induced field, the induced current [registered by the galvanometer] flows out of the page at the top of the hoop, and into the page at its bottom.

Part C: The hoop is in motion toward the magnet.

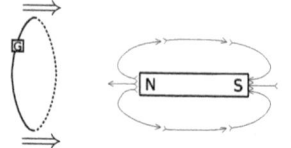

IF the hoop moves toward the magnet, **THEN** an increasing amount of magnetic flux passes through the hoop.

This situation is *de facto* identical to that in Part B. Only the relative motion of the hoop and magnet matters. Accordingly, the induced current, as registered by the galvanometer, flows from top to bottom on the part of the hoop emerging from the page.

Part D: The magnet is in motion away from the hoop.

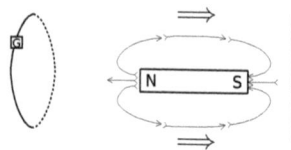

IF the magnet moves away from the hoop, **THEN** the amount of magnetic flux passing through the hoop is decreasing.

The hoop responds to the diminution of magnetic flux by the induction of a current which flows so as to generate compensatory flux. In this instance, the hoop strives to maintain the flux by induction of a current generating magnetic field lines passing from right to left through the interior of the hoop. The orientation of the induced [compensatory] magnetic field requires that the induced current in the hoop emerge from the page at the bottom and re-enter the page at the top.

Part E: The hoop is in motion away from the magnet.

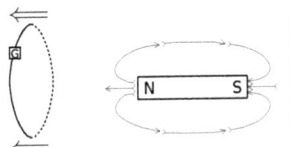   IF the hoop moves away from the magnet, **THEN** the amount of magnetic flux passing through the hoop decreases.

This case is the same as in Part D and must give rise to the same phenomenology. An induced current flows upwards through the galvanometer, generating an induced magnetic field whose flux through the loop compensates for the decline in flux emanating from the increasingly distant bar magnet.

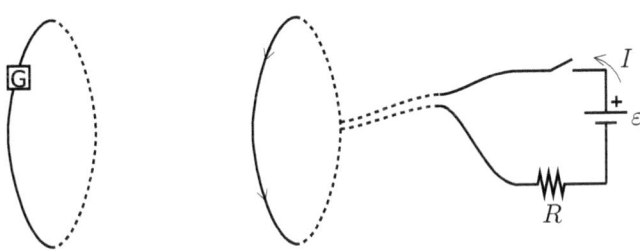

FIGURE 35.2   A Conducting Hoop, a Galvanometer, and a Driving Loop

Instead of a moving magnet, let's consider a **driving loop** [*a.k.a.* primary coil] connected to a source of EMF [and, inevitably, a resistor, too], as illustrated in Figure 35.2.

Part F: The switch on the primary circuit remains open.

As long as the switch remains open, there can be no current flowing in the primary coil, and hence ZERO magnetic field and flux associated with it. In this instance, the driving loop is NOT causing the amount of magnetic flux impinging upon the responder hoop to change, and consequently is NOT inducing a current in the hoop.

Part G: The switch on the primary circuit is suddenly closed.

As soon as the switch is closed, electric current begins to flow[1] in the primary loop. This current produces a magnetic field which is necessarily changing in time, since before the switch was closed there was no field. The orientation of the driving field is from left to right within the loop, and from right to left outside the loop, in accord with the RHR.

The secondary (responder) hoop feels the magnetic field, and thence flux, generated by the primary coil. Subsequent to the switch closing, an increasing amount of flux passes through the secondary hoop from left to right. This changing flux elicits a response current oriented so as to produce a counter-flux to cancel that coming from the primary loop. Inducing flux from right to left through the secondary hoop requires the induced current to flow upward through the galvanometer pictured in Figure 35.2.

A curious thing happens as time progresses.

---

[1]We will qualify this statement shortly. In our *current* working models [employed since Chapter 18] for circuits with only resistive and capacitive elements, the steady state current is instantly established. Stay tuned for the development of a more complete model.

P   The current in the primary loop attains its steady state value [as per Ohm's Law], at which time the field and flux produced by the driving loop approach constant values.

S   The current, initially induced in the secondary hoop to cancel the sudden increase in flux when the switch was closed, is dissipated by resistance in the hoop and is reduced to ZERO.

> ASIDE: This fate can be avoided only when the secondary hoop is superconducting. Induced currents in superconductors **persist**, as was noted in Chapter 19. Furthermore, this comports with a superconductor's being a perfect diamagnet, with $\chi = -1$, *cf.* Chapter 34.

While the induced current dies out, it is as though the secondary hoop becomes inured[2] to the [now] constant flux of the primary loop.

Part H: At some time after the current in the hoop has subsided, the switch on the primary circuit is suddenly opened. The current in the circuit is abruptly halted,[3] and hence the magnetic field and flux generated by the primary loop suddenly diminish.

The responding hoop strives to maintain the flux to which it had become accustomed. To this end, a current moving downward through the galvanometer is induced. Here, too, the current eventually dies out, on account of the resistivity of the hoop.

In practice, multiple loops and coils are employed to enhance the induction effects, and field lines and flux are guided through the coils by a ferromagnetic channel. Figure 35.3 shows PRIMARY and SECONDARY coils wound upon a common contiguous iron core.

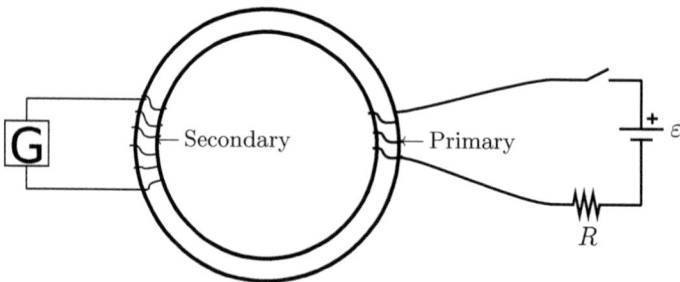

FIGURE 35.3   Primary and Secondary Coils Wound on an Iron Ring

---

We note that this is not yet a useful device—since the current is induced in the secondary coil only for brief times after the switch is opened or closed and not in a steady state situation—but we are headed toward the design of a **transformer**.

---

[2]One ought to beware anthropomorphising, but it does help one's intuition.

[3]This instantaneous response is an artifact of our circuit model. Although we shall extend the model to better describe the short time scale behaviour associated with opening and closing switches, the immediate full-stop approximation suffices for our present purposes.

FARADAY'S LAW   A quantitative expression of Faraday's Law is

$$\varepsilon = -\frac{d\Phi_{m,S}}{dt} = -\frac{d}{dt}\int_S \vec{B}\cdot d\vec{A}\,.$$

The quantity on the LHS is the induced EMF appearing in a circuit. On the RHS, $\Phi_{m,S}$ is the magnetic flux through an area bounded by the circuit, and the negative sign incorporates Lenz's Law. The EMF on the LHS is not localised. Rather, it is distributed along the boundary of the surface, S, appearing in the flux computation on the RHS.

When the circuit consists of an $\mathcal{N}$-coil [with fixed area and no magnetic flux leaking out of the sides], Faraday's Law becomes

$$\varepsilon = -\mathcal{N}\frac{d\Phi_{m,S}}{dt} = -\mathcal{N}\frac{d}{dt}\int_S \vec{B}\cdot d\vec{A}\,.$$

Recall that the computation of magnetic flux considers the surface once, which is practically equivalent to assuming that there is a single circuit boundary.

The total magnetic flux through a surface, S, spanning the interior of an $\mathcal{N}$-coil solenoid is

$$\mathcal{N}\,\Phi_{m,S} = \mathcal{N}\int_S \vec{B_0}\cdot d\vec{A} = \mathcal{N}\,\vec{B_0}\cdot\int_S d\vec{A}\,.$$

When S is a planar cross-section of the solenoid, it is evident that the differential area elements sum to a unique vector, $\vec{A}$, with magnitude equal to the interior cross-sectional area of the solenoid and direction parallel to the solenoidal axis.

ASIDE: Even for surfaces which are not flat, evaluating $\int_S d\vec{A}$ yields the result $\vec{A}$.

With $\vec{A}$ thus specified, expressions for the magnetic flux and its time rate of change, the induced EMF, are:

$$\mathcal{N}\,\Phi_{m,S} = \mathcal{N}\,B_0\,A_S\,\cos(\theta) \qquad \text{and} \qquad \varepsilon = -\frac{d}{dt}\left[\mathcal{N}\,B_0\,A_S\,\cos(\theta)\right]\,.$$

The induced EMF is non-zero if any [combination] of the following quantities are changing:

$\|\vec{B}\|$	the magnitude of the magnetic field
$\|\vec{A}\|$	the area through which the flux is computed
$\hat{B}\cdot\hat{A}$	the angle lying between the area vector and the magnetic field
$\mathcal{N}$	the number of coils [by addition or removal]

Reflecting on the non-localisable aspect of the induced EMF, and on the fundamental definition of potential difference in terms of the line integral of the local electric field, one is inspired to recast the LHS of Faraday's Law in terms of the circulation of the induced electric field, *i.e.*,

$$\varepsilon = \oint_{\gamma_0} \vec{E}_i \cdot d\vec{s} = \mathrm{Circ}\left(\vec{E}_i\right).$$

This is almost counterintuitive, in that γ_0 is a closed path, and one cannot help but expect the net potential difference around any closed loop to be zero [*à la* Kirchoff's Second Rule]. These concerns are ameliorated by realising two things:

  (**1**)   Kirchoff's Rules apply to steady state current flows [and stationary circuit components], and in such situations the magnetic fluxes are unchanging.

  (**2**)   In networks of resistors and batteries, the potential changes are localised, and single-valuedness of the potential function ensures the veracity of K2. In the present case, the induced potential is not localisable within a portion of the circuit; it is distributed. In other words, the electric field associated with this potential wraps around the circuit. Hence, the field is not irrotational, and it is perfectly reasonable that its integral about closed paths [which encircle the VORTEX] be non-zero.

With these final reflections in mind we quote the more general form of Faraday's Law.

Faraday's Law (general formulation)   The general expression of Faraday's Law is

$$\oint_{\gamma_0} \vec{E} \cdot d\vec{s} = -\frac{d}{dt} \int_{\mathrm{S}} \vec{B} \cdot d\vec{A},$$

where the closed loop path on the LHS, γ_0, forms the boundary of the surface, S, employed for the computation of the [changing] magnetic flux on the RHS.

FARADAY'S LAW is a fundamental LAW of NATURE.

Chapter 36

Motional EMF

Motional EMF exists whenever a conductor moves through a region of space in which there is an ambient magnetic field. In its simplest manifestation, consider a straight wire segment with length L moving with constant velocity $\vec{v}_0$, directed perpendicular to a uniform magnetic field, $\vec{B}_0$.

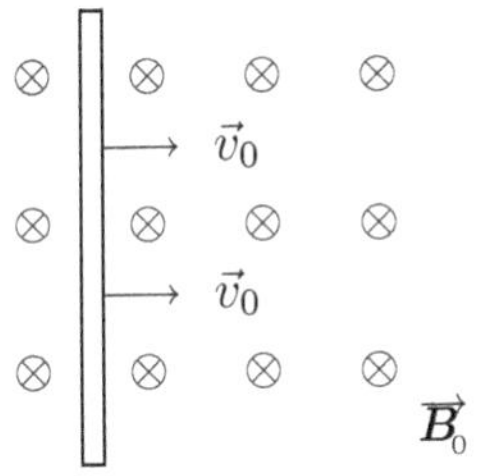

FIGURE 36.1 A Conducting Bar Moving through a Magnetic Field

All of the charged particles comprising the neutral wire experience a Lorentz force,

$$\vec{F}_{\text{M}} = q\,\vec{v}_0 \times \vec{B}_0\,,$$

by virtue of the bulk motion of the wire segment. The free conduction electrons, bearing $q = -e$, are somewhat displaced [downward in the instance shown in Figure 36.1] as they respond to the Lorentz force. In this manner, a net negative charge [an electron surplus over neutral matter] accumulates at the bottom of the wire segment, while an equal amount of positive charge [electron deficit] accrues at the top.

[This sounds rather like the Hall Effect, eh?]

The charge separation in the segment of wire produces an electric field which opposes further polarisation. A steady state situation ensues when the electric field within the bar and the external magnetic field combine in the manner of a velocity selector tuned to the speed of the wire segment. The condition for a particle to remain undeflected in the presence of **crossed** electric and magnetic fields is that the electric and magnetic forces exactly cancel, *i.e.*,

$$e\,E = e\,v_0\,B_0 \quad \implies \quad E = v_0\,B_0\,.$$

A [Hall-like] potential difference between the ends of the wire is associated with the electric field inside the wire:

$$\Delta V = E\,L = v_0\,B_0\,L\,.$$

Consider a conducting bar propelled along a pair of conducting rails, which are separated by a fixed distance, L, and joined by ideal wires and a resistor. The bar-rails-wire-resistor system forms a complete circuit. In Figure 36.2, the bar moves along the rails with velocity $\vec{v}$, under the influence of an applied force, $\vec{F_A}$, acting in the direction shown.[1] [We shall be primarily concerned with the constant velocity case.]

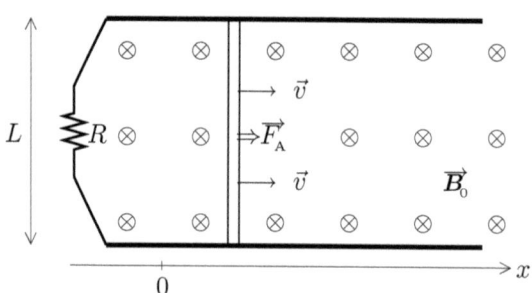

FIGURE 36.2   Motional EMF under an Applied Force

The free electrons in the bar experience a Lorentz force due to its motion. Unlike the case considered at the beginning of this chapter, the charge does not accumulate at the ends of the bar, but instead comprises a current flowing throughout the circuit. The magnitude of the induced current flowing through the conducting bar will be determined by consideration of Faraday's Law and will be confirmed by an energetic analysis.

The position of the bar may be characterised by the point at which it intersects with one of the side rails. Let us align the x-axis with the lower rail, and fix the origin at the point labelled 0 in Figure 36.2. The location of the bar is represented by x. The area enclosed by the circuit is given by

$$A = A_0 + L\,x\,,$$

where A_0 is a constant accounting for the area lying to the left of $x = 0$.

   [Whether this residual area is positive, zero, or negative is of no consequence.]

The magnetic flux passing through the area bounded by the circuit is

$$\Phi_{m,A} = \int_A \vec{B_0} \cdot d\vec{A} = \vec{B_0} \cdot \vec{A} = B_0\, A = B_0\left(A_0 + L\,x\right).$$

The magnetic flux changes as a consequence of the motion of the bar.
The induced EMF in the circuit, arising from the varying flux, is

$$\varepsilon = -\frac{d\Phi_{m,A}}{dt} = -\frac{d}{dt}\left[B_0\left(A_0 + L\,x\right)\right] = -B_0\,L\,\frac{dx}{dt} = -B_0\,L\,v\,.$$

The magnitude of the current induced in the circuit is obtained using Ohm's Law, where the potential difference is equal to the induced EMF and the resistance is R. That is,

$$I = \frac{|\varepsilon|}{R} = \frac{|B_0\,L\,v|}{R}\,.$$

---

[1] We'll argue, momentarily, that for the bar to continue in forward motion, the applied force must act in the forward direction.

We explicitly employ the absolute value of the EMF because, thus far, all we've been concerned with is that current will flow. In the present case [as illustrated] with the magnetic field directed into the page and the bar moving to the right, the induced current flows in an anti-clockwise sense around the loop.

  ASIDE: This follows from Lenz's Law. The movement of the bar produces an increase in the amount of magnetic flux directed into the page passing through the loop. The induced current produces an opposing flux by generating a magnetic field directed out of the page [within the loop].

Finally, let us suppose that the bar is moving at constant velocity $\vec{v}_0$, and, accordingly, that it is subject to ZERO net external force. The weight of the bar is precisely cancelled by the normal force of contact applied by the rail. Furthermore, frictional and drag forces are neglected in this analysis. Thus, there are but two forces, the applied force acting directly on the bar and the Lorentz force acting on the current conveyed by the bar:

$$\vec{F}_{\text{A}} = F_{\text{A}}\,[\rightarrow] \qquad \text{and} \qquad \vec{F}_{\text{M}} = I\,\vec{L} \times \vec{B}_0 = I\,L\,B_0\,[\leftarrow]\,.$$

For the bar to move with constant velocity, the applied force must cancel the magnetic force, and hence

$$\vec{F}_{\text{A}} = I\,L\,B_0\,[\rightarrow] = \frac{B_0^2\,L^2\,v_0}{R}\,[\rightarrow]\,.$$

We inserted the induced current, determined above, into the penultimate expression to arrive at the final result.

The MECHANICAL POWER input by the applied force is the rate at which work is being done on the system, *viz.*,

$$P_{\text{input}} = \vec{F}_{\text{A}} \cdot \vec{v} = \frac{B_0^2\,L^2\,v^2}{R}\,.$$

The input power is the rate at which energy is being added to the system. Meanwhile, electrical power is being dissipated in the resistor. [JOULE HEATING was first encountered in Chapter 19.] Recall that

$$P_{\text{output}} = I^2\,R = \left(\frac{B_0\,L\,v}{R}\right)^2 R = \frac{B_0^2\,L^2\,v^2}{R}\,.$$

All of the mechanical power being input to the system by the external applied force is being dissipated in the resistor. Of course, this does not surprise us, because the bar moving at constant velocity is neither gaining nor losing kinetic energy.[2]

  ASIDE: A classic problem inclines the rails at some fixed angle to the horizontal, thus yielding a *mash-up* of motional EMF and motion along an inclined plane.

---

[2] According to the WORK–ENERGY THEOREM, $W_{if}\left[\vec{F}_{\text{NET}}\right] = \Delta K$.

Suppose that the applied force drops to zero suddenly, at time $t = 0$, while the bar is in motion with speed v_0. The net force acting on the bar is the Lorentz force, and it acts in the direction opposite to the motion. Thus, the bar decelerates [observe the negative sign in the formula for acceleration]. Newton's Second Law, applied to the present situation, reads

$$m\,a = F_{\text{M}} = -I\,L\,B = -\frac{B_0^2\,L^2\,v}{R}.$$

The time rate of change of the velocity is directly proportional to the velocity. This state of affairs generically gives rise to exponential behaviour, and this case is no exception:

$$m\,a = -\frac{B_0^2\,L^2}{R}\,v \quad\Longrightarrow\quad \frac{dv}{dt} = -\frac{B_0^2\,L^2}{m\,R}\,v \quad\Longrightarrow\quad \frac{dv}{v} = -\frac{dt}{\tau}.$$

In this last form, the time constant for the motion of the bar,

$$\tau = \frac{m\,R}{B_0^2\,L^2}\,,$$

has been introduced.

> ASIDE: It is not so obvious that the units for τ indeed work out to be seconds. The SI unit for resistance, Ω, is equal to a V per A, which in turn may be expressed as $(\text{N·m/s})/\text{A}^2$. The combination, $B_0\,L$, has units T·m, which are equivalent to N/A. Combining these intermediate results into the expression for τ yields
>
> $$\left[\frac{m\,R}{B_0^2\,L^2}\right] = \frac{\text{kg}\cdot\text{N·m/s}}{\text{A}^2} \times \left(\frac{\text{A}}{\text{N}}\right)^2 = \frac{\text{kg·m/s}}{\text{N}} = \text{s}\,,$$
>
> as is required for consistency.

The solution to the differential equation for the speed,

$$\frac{dv}{v} = -\frac{1}{\tau}\,dt\,,$$

is exponential decay in time, with scale set by τ, *i.e.*,

$$v(t) = v_0\,e^{-t/\tau}\,.$$

Furthermore, the current in the circuit also undergoes exponential decay [with the same time constant]:

$$I = \frac{B_0\,L\,v}{R} = \frac{B_0\,L\,v_0}{R}\,e^{-t/\tau}\,.$$

Two comments finish off this chapter.

- The orientation of the current induced in the loop is dictated by Lenz's Law.

  IF the current went in the other direction, THEN the moving bar would experience a velocity-dependent force in the direction of its motion, AND consequently [in the absence of retarding forces] an exponential increase in its speed.

  [Such a scenario is utterly unphysical.]

- The EMF appearing in conjunction with changing magnetic flux via Faraday's Law is not localisable. Rather than appearing at a point [or a collection of points], it is generally distributed throughout the circuit or loop.

Chapter 37

Inductance

Recall the familiar example of the simple battery–resistor–switch circuit. An electric current flows once the switch is closed.

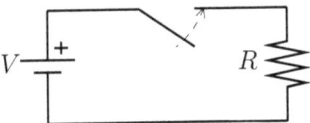

FIGURE 37.1   The Minimal Resistor–Battery–Switch Circuit

Long ago we invoked Ohm's Law to determine the steady state current: $I = V/R$. But this current had to grow from zero, and while it grew it produced an increasingly strong magnetic field, $\vec{B}_c$, in the space near the wires, and thus a changing magnetic flux through [almost[1]] any fixed area. The additional magnetic flux passing through any particular surface, S, whose boundary is defined by the circuit,

$$\Phi_{m,\mathrm{S}} = \int \vec{B}_c \cdot d\vec{A} \,,$$

was ZERO with the switch open, and later attained its steady state value. While this change occurred, an opposing **self-induced** EMF,

$$\varepsilon_i = -\frac{\partial \Phi_{m,\mathrm{S}}}{\partial t} \,,$$

was present in the circuit, preventing the current from instantaneously reaching its steady state value.

   ASIDE: The steady state current and flux are approached in the limit as t tends to infinity.

SELF-INDUCTANCE   The magnetic flux passing through the circuit is dependent on the magnetic field, which in turn is produced by the current [in the wires]. This militates for writing

$$\varepsilon_i = -L \frac{\partial I}{\partial t} \,,$$

where L is the circuit's self-inductance parameter. It turns out that L depends only on the geometry and physical properties of the circuit.

$$\left[\text{ If the current is confined to thin wires, then } \quad \frac{\partial I}{\partial t} \rightarrow \frac{dI}{dt}. \right]$$

---

[1] As usual, while noting that there can be exceptional cases, we focus on the generic.

The henry,[2] [H], is the SI unit of self-inductance. It is evident, from the formula in which L is first defined, that

$$1\,\mathrm{H} = 1\,\frac{\mathrm{V}}{\mathrm{A/s}} = 1\,\frac{\mathrm{V\cdot s}}{\mathrm{A}}\,.$$

In yet another instance of foreshadowing, we observe that one volt is one joule per coulomb, $1\,\mathrm{V} = 1\,\mathrm{J/C}$, and thus $1\,\mathrm{V\cdot s} = 1\,\mathrm{J/A}$, leading to the equivalence

$$1\,\mathrm{H} = 1\,\frac{\mathrm{J}}{\mathrm{A}^2}\,.$$

In the upcoming discussion of energetics [in Chapter 38], we shall come to a greater appreciation of the significance of this observation. Succumbing to the spirit of *"nothing exceeds like excess,"* we also point out that $1\,\mathrm{V/A}$ is, by definition, $1\,\Omega$, and thus

$$1\,\mathrm{H} = 1\,\Omega\cdot\mathrm{s}\,.$$

Implications of this seemingly innocuous observation first appear when we study RL circuits and shall *resonate* when we investigate AC voltage sources in Chapter 42 *et seq.*

The induced EMF from Faraday's Law,

$$\varepsilon_i = -\mathcal{N}\,\frac{\partial \Phi_{m,\mathrm{S}}}{\partial t}\,,$$

and the proposed self-induced EMF,

$$\varepsilon_i = -L\,\frac{\partial I}{\partial t}\,,$$

are one and the same. Thus,

$$\mathcal{N}\,\frac{\partial \Phi_{m,\mathrm{S}}}{\partial t} = L\,\frac{\partial I}{\partial t}\,.$$

Care must be taken, when integrating both sides of the above equation with respect to time, to allow for the presence of an additive position-dependent function of integration. Physical self-consistency demands that any [additional] flux vanish when the current does, setting this additional function to ZERO. Hence, for circuit circumstances in which the current is confined to thin wires,

$$\mathcal{N}\,\Phi_{m,\mathrm{S}} = L\,I\,.$$

This relation provides a means of determining L, as we'll see in the example below.

....

EXAMPLE [*Self-Inductance of a Solenoid*]

Q: What is the self-inductance of a solenoid of length l, with interior cross-sectional area $A_{\mathrm{x\text{-}c}}$, fashioned from $\mathcal{N}$ turns, or loops, of thin wire?

The **winding density** [*a.k.a.* the number of coils per unit length] is $\eta = \mathcal{N}/l$. Suppose that a steady current I flows through the wire. The field inside the solenoid is [to good

---

[2]This unit is named in honour of Joseph Henry (1797–1878), a scholar working contemporaneously with, though independent of, Michael Faraday.

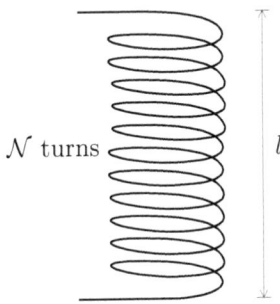

FIGURE 37.2 A Solenoidal Coil

approximation] uniform and directed parallel to the axis, with strength $B_0 = \mu_0\,\eta\,I$. The amount of magnetic flux through each loop in the coil is

$$\Phi_{m,\text{x-c}} = \left| \vec{B}_0 \cdot \vec{A}_{\text{x-c}} \right| = B_0\,A_{\text{x-c}} = \frac{\mu_0\,N\,I\,A_{\text{x-c}}}{l}\,.$$

Substituting these solenoidal results into the recently obtained expression for the self-inductance yields

$$L_{\text{solenoid}} = \frac{N\,\Phi_{m,\text{x-c}}}{I} = \frac{\mu_0\,N^2\,A_{\text{x-c}}}{l} = \mu_0\,\frac{N^2}{l^2}\,\left(A_{\text{x-c}}\,l\right) = \mu_0\,\eta^2\,(\text{Volume})\,.$$

The interior volume of the solenoid [cross-sectional area × length] appears as an overall multiplicative factor.

The self-inductance of a solenoidal coil depends only on its physical properties (winding density) and geometry (interior volume).

Let us go back to the original circuit and now explicitly account for its self-inductance by incorporating an inductive circuit element.

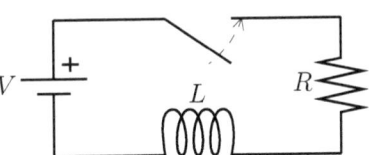

FIGURE 37.3 A Less Minimal Resistor–Battery–Switch–Inductor Circuit

This adds another device to the collection { battery, wire, resistor, switch, capacitor }.

INDUCTOR

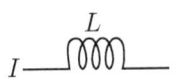

An **inductive device** generates a back-EMF whenever the current which passes through it varies:

$$\varepsilon = -L\,\frac{\partial I}{\partial t}\,.$$

The parameter L is independent of the [dynamical] properties of the circuit in which the inductive element resides.

In light of our experiences with combinations of capacitors and resistors, we are inspired to ask the following question.

Q: What effects are produced by combinations of two inductors?

A: Two inductors may be combined in **series** or in **parallel**. Alas, however, the inductors may not be presumed to act independently of one another. The remainder of this chapter is given over to an investigation of primitive models for the composition of inductive elements into a simple circuit.

EXAMPLE [*Two Inductors in Series*]

Q: What is the effective inductance of the series system comprised of two inductors, L_1 and L_2, illustrated in Figure 37.4?

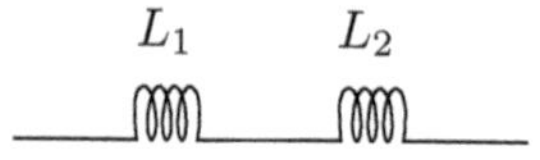

FIGURE 37.4 Two Inductors Joined in Series

A(ideal): Let's assume that the inductors are ideal solenoids in that the magnetic fields and fluxes are entirely confined to their respective coils and vanish everywhere else in space. IF the solenoids are non-overlapping and non-intersecting [*i.e.,* their magnetic fields and fluxes are not shared], THEN the inductors act independently.

Each inductor generates a back-EMF proportional to the time rate of change of the current passing through it:

$$\varepsilon_1 = -L_1 \frac{dI_1}{dt} \qquad \text{and} \qquad \varepsilon_2 = -L_2 \frac{dI_2}{dt}.$$

Recall that ordinary rather than partial derivatives may be employed whenever the currents are confined to [thin] wires joining the inductors to the rest of the circuit. Owing to the series structure, the currents I_1 and I_2 are each equal to $I_{(12)}$. Therefore,

$$\varepsilon_{(12)} = \varepsilon_1 + \varepsilon_2 = -L_1 \frac{dI_{(12)}}{dt} - L_2 \frac{dI_{(12)}}{dt} = -(L_1 + L_2)\frac{dI_{(12)}}{dt} = -L_{(12)}\frac{dI_{(12)}}{dt}.$$

ASIDE: In this instance, the SAME current passes through each inductor, and the SUM of the respective EMFs yields the total EMF.

The effective inductance of two ideal inductors arranged in series is the sum of their respective inductances:

$$L_{(12)} = L_1 + L_2.$$

Q: What about non-ideal cases?

A(non-ideal): Leakage of magnetic fields outside of the inductive elements almost inevitably entails some sharing of magnetic fluxes. This introduces a geometry-dependent non-linear coupling between the devices which is not captured by the simple additive series combinatorics exhibited by the ideal case.

Attempts to model parallel combinations of inductors, such as that illustrated in Figure 37.5, are fraught with difficulty. The first problem appears when we study the passage of an electric current through this network. The idealised solenoids have ZERO resistance, and thus the fractionation of the current at the junction is ambiguous. This difficulty is resolved by explicitly accounting for the actual resistances of the physical solenoids. In steady state, the current splits in inverse proportion to the ratio of these resistances, so as to ensure a consistent potential drop from one side of the network to the other.

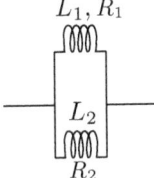

FIGURE 37.5 Two Inductors Arranged in Parallel

ASIDE: "Welcome to reality™," one might say. As all of the electric devices that we have encountered are composed of conducting material, one ought to incorporate inductive, resistive, and capacitive effects into their descriptive models. Presently, we shall show how one might do just this. However, we shall maintain our use of ideal components according to the notion of model-building as a sequence of approximations.

Whatever transitory relief may have been granted by this discovery of a method for assigning steady state currents through each of the network channels evaporates once one considers varying the overall current. While one might propose that the time-variation of each of the split-currents be in phase with, and in proportion to, its steady state fraction, the back-EMFs produced by the inductive devices will, in general, not be commensurate. Thus, it seems inevitable that *loop currents* will arise, further complicating the changes in network currents.

Whoah! Non-Linearity rears its head!

These last considerations militate for a reformulation of the treatment of combinations of inductors. Rather than thinking in terms of series and parallel current structures, one ought to consider the degree of flux linkage between inductors bearing *a priori* otherwise unconstrained currents.

$$\varepsilon_1 = L_1 \frac{\partial I_1}{\partial t} + M_{21} \frac{\partial I_2}{\partial t} + M_{31} \frac{\partial I_3}{\partial t}$$
$$\varepsilon_2 = M_{12} \frac{\partial I_1}{\partial t} + L_2 \frac{\partial I_2}{\partial t} + M_{32} \frac{\partial I_3}{\partial t}$$
$$\varepsilon_3 = M_{13} \frac{\partial I_1}{\partial t} + M_{23} \frac{\partial I_2}{\partial t} + L_3 \frac{\partial I_3}{\partial t}$$

The interactions among the elements in an $\mathcal{N}$-inductor system can be investigated in a pairwise fashion. For each pair, there is a range of physical behaviour lying between two extreme cases. At one extreme is the fully uncoupled case in which both of the off-diagonal [*a.k.a.* flux-linking] terms, M_{ij} and M_{ji}, are identically ZERO. The other extreme occurs when one member of the pair shares all of its flux with the other. The generic situation is known as **mutual inductance**,[3] and covers all possibilities (including the extremes). Mutual inductance will be the subject of Chapter 39, after we have disposed of the dynamics and energetics of the [single] inductor–resistor circuit.

---

[3]The symbol "M_{ij}" connotes the mutuality of the flux-linkage.

Chapter 38

RL Circuits

Consider the simplest RL circuit from Chapter 37, as recalled in Figure 38.1 [employing an ideal battery with EMF ε].

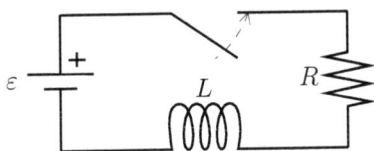

FIGURE 38.1   The Simplest RL Circuit

Analysis of the circuit's dynamics begins with application of Kirchoff's Rules.

K1   The junction rule applies trivially, as there is but one unique current.

K2   The loop rule gives rise to a dynamical equation for the current response,

$$0 = \varepsilon - IR - L\,\frac{\partial I}{\partial t}\,.$$

Several comments ought to be made. Since the current is confined to the wire, the partial derivative in the above expression may be replaced by the ordinary derivative. The loop-equation [expressing conservation of energy per unit charge flowing in the current] does indeed yield a dynamical equation for the current response, as will become evident when we explicitly construct its solutions. As is customary, the switch is thrown at time $t = 0$, and until then no current flows in the circuit, *i.e.*, $I(t) = 0$, $\forall\, t \le 0$. This sets the initial condition for $I(t)$, $t \ge 0$.

The current-dependent and time-dependent parts appearing in K2 may be separated, *i.e.*,

$$L\,\frac{dI}{dt} = \varepsilon - IR \qquad \Longrightarrow \qquad \frac{dI}{\varepsilon - IR} = \frac{dt}{L}\,.$$

One must self-consistently integrate both sides of this differential equation to determine the current, $I(t)$, for all times after $t = 0$. Care must be taken to avoid stumbling into the pathologies which ensue when the denominator on the LHS passes through zero. Also, while we seek $I(t)$, the I in the differential equation is a "dummy function," and hence it is proper to give it another symbol. Similar considerations also apply to t. So as to retain their connotations as current and time, we shall rewrite them as I' and t'. Also, we accomplish some additional simplification by multiplying both sides of the [dynamical] equation by $-R$. With these considerations in mind, the two sides of the rearranged Kirchoffian loop relation are transformed:

$$\text{LHS} = \int_0^{I(t)} \frac{-R \, dI'}{\varepsilon - I' R} = \int_0^{I(t)} \frac{-dI'}{\frac{\varepsilon}{R} - I'} = \ln\left(\frac{\varepsilon}{R} - I'\right)\Big|_0^{I(t)} = \ln\left[\frac{\frac{\varepsilon}{R} - I(t)}{\frac{\varepsilon}{R}}\right] = \ln\left[1 - \frac{I R}{\varepsilon}\right],$$

$$\text{RHS} = \int_0^t \frac{-R}{L} \, dt' = -\frac{R}{L} \, t' \Big|_0^t = -\frac{R}{L} \, t.$$

Setting these expressions for the LHS and RHS equal to one another provides an expression for $t(I)$,

$$\ln\left[1 - \frac{I R}{\varepsilon}\right] = -\frac{R}{L} \, t, \quad \text{which may be rearranged to read} \quad t = -\frac{L}{R} \ln\left[1 - \frac{I R}{\varepsilon}\right],$$

i.e., the time as a function of the current.

This relation, while correct and consistent, is not the sought-after expression for the current as a function of time. We desire to know the inverse of $t(I)$. Fortunately, $I(t)$ is readily obtained via exponentiation of its implicit expression before rearrangement, *viz.,*

$$\exp\left(\ln\left[1 - \frac{I R}{\varepsilon}\right]\right) = 1 - \frac{I R}{\varepsilon} \quad \text{and} \quad \exp\left(\frac{-R}{L} t\right) = e^{-\frac{R}{L} t}.$$

Solving for the current proceeds straightforwardly:

$$1 - \frac{I R}{\varepsilon} = e^{-Rt/L} \quad \Longrightarrow \quad I(t) = \frac{\varepsilon}{R}\left(1 - e^{-Rt/L}\right).$$

Hence, for $t \geq 0$, the current in the RL circuit is

$$I(t) = \frac{\varepsilon}{R}\left(1 - e^{-Rt/L}\right).$$

Several comments are to be made and an illustrative figure is to be examined.

Steady State   As $t \to \infty$, the current approaches an asymptotic steady state value, $I_\infty = \frac{\varepsilon}{R}$, which precisely matches that obtained using Ohm's Law [in Chapter 18].

Time Scale   The time scale on which the current grows toward I_∞ is determined by the resistive and inductive circuit properties, *i.e.,* $\tau = \frac{L}{R}$. In common parlance, τ is referred to as the inductive time constant. The SI units of τ are s, since $1\,\text{H} = 1\,\Omega \cdot \text{s}$, as was noted in Chapter 37, and therefore

$$[\tau] = \left[\frac{\text{H}}{\Omega}\right] = \text{s}.$$

Shape   A sketch of the current's dependence on time appears in Figure 38.2.

Slope   The time rate of change of the current [after the switch is thrown] is

$$\frac{dI}{dt} = \frac{\varepsilon - I R}{L} = \frac{\varepsilon}{L} e^{-\frac{R}{L} t}.$$

This rate is maximal initially, *i.e.,* at $t = 0$, when it assumes the value

$$\frac{dI}{dt}\Big|_{\text{max}} = \frac{dI}{dt}\Big|_{t=0} = \frac{\varepsilon}{L},$$

and thereafter decays exponentially.

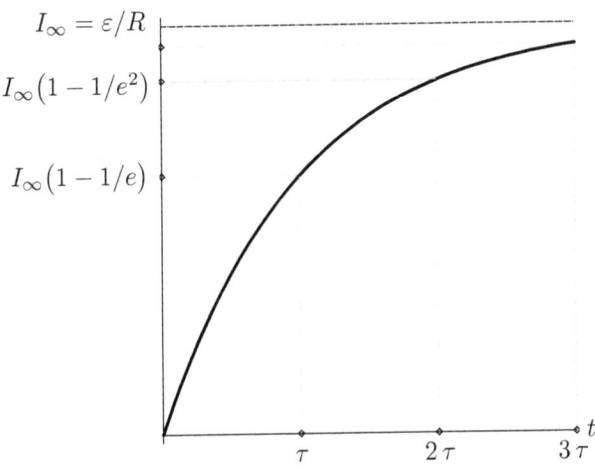

$I_\infty = \varepsilon/R$

$I_\infty(1 - 1/e^2)$

$I_\infty(1 - 1/e)$

$\tau \qquad 2\tau \qquad 3\tau$

t

FIGURE 38.2 Current in an RL Circuit

Suppose that sufficient time has elapsed for the current in the circuit shown in Figure 38.1 to have [for all intents and purposes] reached I_∞. Now suppose that the battery is suddenly removed from the circuit.

Q: How might the battery be removed without completely disrupting the circuit?

A: Quickly and carefully, as per Figure 38.3.

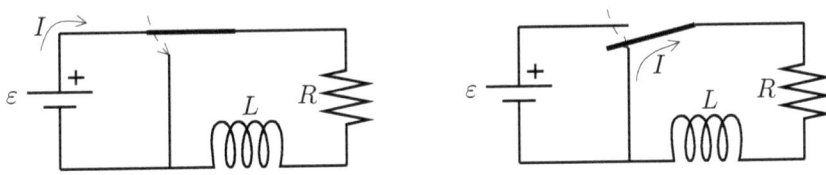

FIGURE 38.3 The Suddenly Interrupted RL Circuit

The idea is that the switcheroo takes place quickly enough that the current in the circuit keeps flowing even though it is suddenly cut off from its [driving] source of electric potential. Re-application of Kirchoff's Rules to this new situation is straightforward.

K1   The junction rule applies trivially again. There is one unique current after the reconfiguration of the wires.

K2   The loop rule provides the dynamical equation:

$$0 = -IR - L\frac{\partial I}{\partial t}.$$

As in the previous analysis, we convert the partial to an ordinary derivative, isolate the derivative, separate the I and t dependences, and introduce dummy variables:

$$\frac{dI}{dt} = -\frac{R}{L}I \quad \Longrightarrow \quad \frac{dI'}{I'} = -\frac{R}{L}dt'.$$

Self-consistently integrating both sides of this separated equation, from $I' = I_0$ at $t' = 0$ to $I' = I(t)$ at $t' = t$, we obtain

$$\text{LHS} = \int_{I_0}^{I(t)} \frac{dI'}{I'} = \ln\left[I'\right]\Big|_{I_0}^{I(t)} = \ln\left[\frac{I(t)}{I_0}\right],$$

$$\text{RHS} = \int_{0}^{t} -\frac{R}{L}\,dt' = -\frac{R}{L}\,t'\Big|_{0}^{t} = -\frac{R}{L}\,t.$$

Setting the integrated LHS and RHS equal to one another, and exponentiating in order to obtain an explicit expression for $I(t)$, results in

$$I(t) = I_0\,e^{-Rt/L} = I_0\,e^{-t/\tau}.$$

As usual, here come a few relevant comments.

Steady State   The asymptotic current vanishes, $\lim_{t\to\infty} I(t) = 0$.

> ASIDE: Don't be alarmed by the multiple uses of $t \to \infty$ in this analysis. Be assured that all is consistent.
>
> If a time interval with duration much larger than the inductive time scale elapsed while the current was building up, then owing to the rapid convergence of negative exponentials, the current may be said to have essentially attained its infinite limit. The clock may then be reset to read zero, again, at the instant the circuit is interrupted. Throughout the course of a time interval of comparable duration in the aftermath of interrupting the circuit, the magnitude of the current will decay to [effectively] ZERO.

Time Scale   The post-interruption time scale for current decay, $\tau = L/R$, is the same as that for its growth.

Growth and decay are symmetric processes.

Shape   A sketch of the current's dependence on time appears in Figure 38.4.

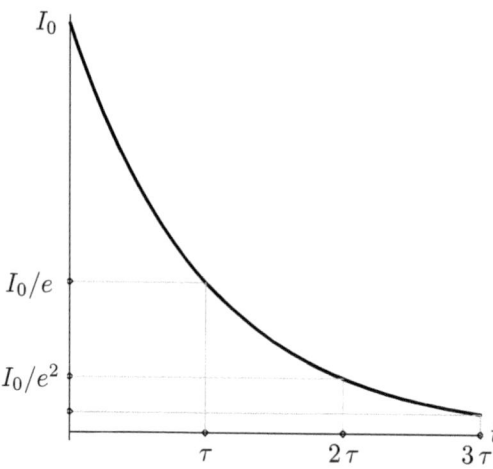

FIGURE 38.4   Current in an RL Circuit after Voltage Interruption

Slope   The time rate of change of the current [after the switcheroo] is

$$\frac{dI}{dt} = -\frac{R}{L}\,I\,.$$

The magnitude of this rate is maximised at $t = 0$, when the current is maximal:

$$\left|\frac{dI}{dt}\right|_{\mathrm{max}} = \frac{R}{L}\,I_0\,.$$

If you sense that something funny is going on here, you are certainly right to be suspicious! Consider the fact that the circuit has been disconnected from its voltage [energy] source, and yet current still flows! Furthermore, Joule heating continues while the current passes through the resistor, *i.e.*, energy is being dissipated!

Q: How might this be understood?

A: Prior to this [glaring] anomaly, there was another oddity: when the switch was first closed to complete the original circuit, there was a time lag while the steady state current was [asymptotically] established. This can be interpreted as an inductive effect. As the battery attempts to initiate a current throughout the circuit, the self-inductance property associated with the circuit creates a back-EMF acting in opposition to the battery. The net EMF is thus reduced, driving a smaller current and depositing less energy into the resistor. As time goes on the circuit becomes inured to its current and the back-EMF fades into irrelevance. Upon removal of the battery, with its driving potential, the circuit self-induces a forward-EMF in an attempt to maintain the flow of current.

This explanation in terms of electromotive forces is fine, as far as it goes, but an energetic analysis is ultimately richer and more informative. While the current is building-up in the circuit, **some of the work/energy provided by the battery is stored in the magnetic field** of the inductor. When the switcheroo eliminates the battery as the source of electrostatic potential energy, the current in the circuit can persist and continue to dissipate energy by drawing-down the magnetically stored energy. The current and the stored energy both tend to zero asymptotically.

ENERGETIC ANALYSIS

Let's re-analyse the original RL circuit in terms of its energetics. Rearranging Kirchoff's Loop Rule, as expressed above, yields

$$\varepsilon = I\,R + L\,\frac{\partial I}{\partial t}\,.$$

This relation holds at all times $t \geq 0$. Recall that ε, R, and L are all constant, while the current response, I, is time-dependent. Multiplying by $I(t)$ results in a valid relation among power terms,

$$\varepsilon\,I = I^2\,R + L\,I\,\frac{\partial I}{\partial t}\,.$$

Overall Conservation of Energy Constrains Instantaneous Power Flows.

The LHS of the above equation, $\varepsilon\,I$, is the [instantaneous] power input from the battery into the RL circuit. The first term on the RHS, $I^2\,R$, is precisely the rate at which energy is dissipated in the resistor. The remainder of the RHS,

$$LI\frac{\partial I}{\partial t}=\frac{1}{2}L\frac{\partial I^2}{\partial t}=P_M\,,$$

is interpreted as the power being stored in the magnetic field associated with the inductor. The amount of magnetic energy stored in the inductor, U_M, may be ascertained by integrating the magnetic power. As the expression for P_M is a total derivative, the integration is almost axiomatic:

$$U_M=\int P_M\,dt=\frac{1}{2}L\,I^2+\text{``irrelevant constant.''}$$

The constant of integration appearing in the formula for the magnetic energy is irrelevant because its conseqences are unobservable. We may elect to set this constant equal to zero and thereby enforce the [satisfyingly physical] condition that the stored magnetic energy vanishes when there is no current flowing in the circuit. We need not check to ensure that the units are consistent here as this has already been done [foreshadowed] in Chapter 37.

CLAIM:[1] **The energy stored in an inductor is held in the magnetic field.**

An ideal solenoid with self-inductance L bears a constant current I. Recall that

$$L_{\text{solenoid}}=\mu_0\,\eta^2\left(\text{Volume}\right),$$

while the strength of the uniform magnetic field directed along the solenoidal axis is

$$B_0=\mu_0\,\eta\,I\,.$$

Inverting the latter relation to express the current in terms of the field yields

$$I=\frac{B_0}{\mu_0\,\eta}\,.$$

Incorporating these results into the expression for the magnetic energy,

$$U_M=\frac{1}{2}L\,I^2=\frac{1}{2}\times\mu_0\,\eta^2\left(\text{Volume}\right)\times\frac{B_0^2}{\mu_0^2\,\eta^2}=\frac{B_0^2}{2\,\mu_0}\left(\text{Volume}\right).$$

Hence, the **magnetic energy per unit volume** within the inductor is

$$u_M=\frac{1}{2\,\mu_0}B_0^2\,.$$

MAGNETIC ENERGY DENSITY A magnetic energy density is associated with the presence of a non-zero magnetic field. Its value at a field point $\mathcal{P}$ is determined by the local magnetic field strength and the permeability of free space, according to

$$u_M(\mathcal{P})=\frac{1}{2\,\mu_0}B^2(\mathcal{P})\,.$$

The SI units are consistent: $\left[u_M\right]=\left[\frac{\text{A}}{\text{T·m}}\right][\text{T}]^2=\left[\frac{\text{T·A}}{\text{m}}\right]=\left[\frac{\text{N}}{\text{m}^2}\right]=\left[\frac{\text{N·m}}{\text{m}^3}\right]=\left[\frac{\text{J}}{\text{m}^3}\right].$

This result only seems less amazing than it is on account of our having discerned a precisely analogous relation for the electric field way back in Chapter 14!

---

[1]We shall not prove this claim in general; rather, its plausibility in a particular case is illustrated.

Chapter 39

Mutual Inductance

Thus far, we've considered how one circuit loop, or a single multi-loop coil,[1] reacts inductively to the current flowing through it. Now, let's consider two such coils in proximity. To facilitate our analysis we shall assume that Coil_1 has $\mathcal{N}_1$ turns and bears current I_1, while Coil_2 has $\mathcal{N}_2$ turns and carries I_2. Furthermore, these coils are assumed to share a common axis, as illustrated in Figure 39.1.

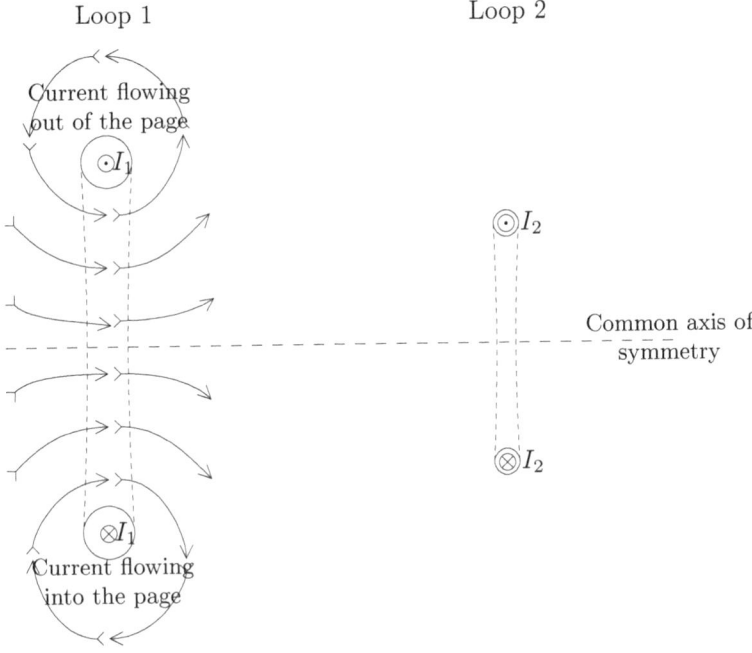

FIGURE 39.1 Two Current-Carrying Coils Sharing a Common Axis

The current flowing in Coil_1 produces a magnetic field throughout the surrounding space, including the region occupied by Coil_2. Thus, a magnetic flux, Φ_{12}, originating from Coil_1, impinges upon any surface spanning Coil_2. A moment's thought about field lines [and Gauss's Law for Magnetic Fields] makes it clear that Φ_{12} is proportional to the current I_1.

> ASIDE: The geometric factors appearing in the expression for the flux are daunting to compute, but the field, and hence the flux, is everywhere proportional to its current source.

There also arises a flux, Φ_{21}, spawned by Coil_2, experienced by Coil_1, that is expected to be linear in I_2.

---

[1] A coil amplifies the effect.

Consider a *Gedanken* experiment.

Part A: Effect a change in the current, I_1, flowing through Coil$_1$.

Part B: The contribution to the flux impinging on Coil$_2$ which is ascribable to Coil$_1$, Φ_{12}, will change.

Part C: Faraday's Law requires that there be a current response in Coil$_2$, in reaction to (B). This current response changes the value of I_2.

Part D: As a consequence of (C), Φ_{21} must vary.

Part E: Coil$_1$ responds to (D) by adjusting its current, I_1, to compensate.

This chain of argument, A → B → C → D → E (A), has become circular! Changing I_1 produces, via this circuitous mechanism, an additional change in I_1!

> ASIDE: It gives us scant comfort to appreciate that the circular argument persists under interchange of the labels 1 ↔ 2.

To resolve this perplexing situation, we introduce the **mutual inductance** of the coils.

MUTUAL INDUCTANCE   When two inductors are proximate, they can interact. The inductance of Coil$_2$ engendered by Coil$_1$ is

$$M_{12} = \frac{\Phi_{12}\,\mathcal{N}_2}{I_1} .$$

In this expression, Φ_{12} is the flux impinging on any surface bounded by Coil$_2$ produced by the field generated by the current passing through Coil$_1$. Dividing by I_1 eliminates the source-current dependence, while multiplying by $\mathcal{N}_2$ accounts for the fact that the flux through the imaginary surface bordered by the coils passes through each and every loop comprising Coil$_2$.

The SI unit for mutual inductance is the henry, $[\text{H}]$.

> ASIDE: This is most readily inferred by noting the structural similarity of the above definition, rewritten here as
>
> $$\Phi_{12}\,\mathcal{N}_2 = M_{12}\,I_1 ,$$
>
> and the formula relating flux to self-inductance,
>
> $$\mathcal{N}\,\Phi_{m,\text{S}} = L\,I ,$$
>
> developed in Chapter 37 [and employed to good effect in determining the self-inductance of a solenoid in terms of its geometry].

The mutual inductance M_{12} relates the back-EMF produced in Coil$_2$ to the time variation of I_1. That is,

$$\varepsilon_{2,1} = -\frac{\partial}{\partial t}\left(\Phi_{12}\,\mathcal{N}_2\right) = -M_{12}\,\frac{\partial I_1}{\partial t} .$$

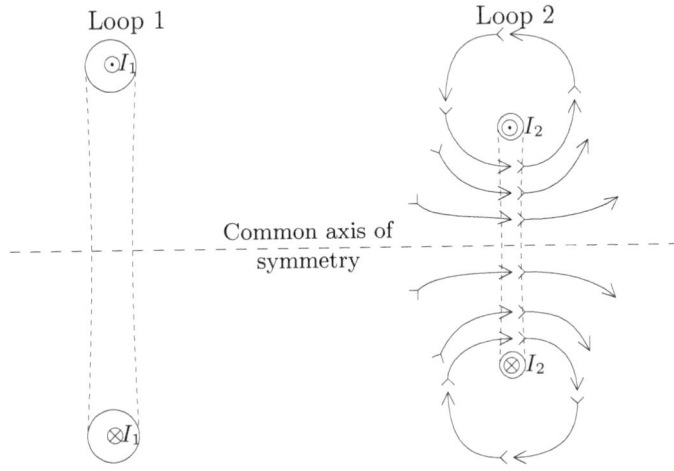

FIGURE 39.2 The Same Two Current-Carrying Coils Sharing a Common Axis

Let's not merely invoke the permutation symmetry of the coils' identifying labels, $1 \leftrightarrow 2$, but instead repeat [in an abbreviated form] the above analysis from the perspective of Coil$_2$. The current in Coil$_2$, I_2, produces a magnetic field throughout all of space, as is illustrated in Figure 39.2. An associated magnetic flux, Φ_{21}, penetrates any surface spanning Coil$_1$.

Consider a second *Gedanken* experiment.

Part a: Effect a change in I_2.

Part b: As a direct consequence of (a), Φ_{21} changes.

Part c: In response to (b), the value of I_1 must change.

Part d: As a consequence of (c), Φ_{12} must vary.

Part e: Coil$_2$ adjusts its current, I_2, to compensate for (d).

Once again, we have come full circle.

   ASIDE: This is not surprising, as the cycles

$$(C) \to (D) \to (A) \to (B) \to (C) \qquad \text{and} \qquad (a) \to (b) \to (c) \to (d) \to (a)$$

   are equivalent.

The mutual inductance M_{21} relates the back-EMF produced in Coil$_1$ to the time variation of I_2. That is,

$$\varepsilon_{1,2} = -\frac{\partial}{\partial t}\left(\Phi_{21}\,\mathcal{N}_1\right) = -M_{21}\,\frac{\partial I_2}{\partial t}\,.$$

The assertion that the mutual inductances M_{21} and M_{12} are each independent of their respective source currents [and presumably depend on geometrical factors] inspires one to conjecture that they might be equal.

CLAIM: Mutual inductances M_{ij}, $i \neq j$, are symmetric in their indices:

$$M_{ij} \equiv M_{ji}.$$

Note that this is simply stated (without proof). In what remains of this chapter we shall not enforce this symmetry, but will recognise when it occurs.

The effect of self-inductance is the generation of back-EMF in a coil, loop, or circuit in response to changes in the current within the same coil, loop, or circuit. Similarly, mutual inductance accounts for the generation of back-EMF in a coil, loop, or circuit occurring as a response to changes in currents in other coils, loops, or circuits. Mutual induction couples [nearby] inductive elements into interacting pairs.

In the simplest case of two proximate coils, labelled 1 and 2, the induced EMFs are

$$\varepsilon_{L1} = L_1 \frac{dI_1}{dt} + M_{21} \frac{dI_2}{dt}$$
$$\varepsilon_{L2} = M_{12} \frac{dI_1}{dt} + L_2 \frac{dI_2}{dt},$$

as was foreshadowed in Chapter 37. The off-diagonal flux-linking terms posited there are the [pairwise] mutual inductance factors.

EXAMPLE [*Coupled Circuits*]

Two nearby coils are each connected to voltage sources (batteries) and resistors, as illustrated in Figure 39.3. The circuit parameters associated with Coil$_1$ are $\{V_1, R_1, L_1, \mathcal{N}_1\}$, while those for Coil$_2$ are $\{V_2, R_2, L_2, \mathcal{N}_2\}$. The inductive interaction occurs via the mutual inductances M_{21} and M_{12}.

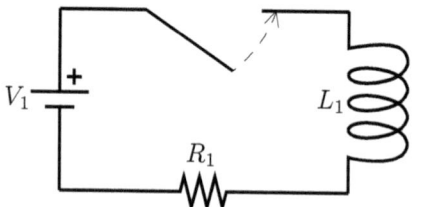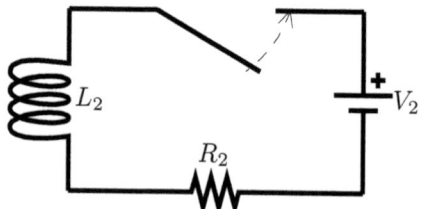

FIGURE 39.3 Two Inductively Interacting Circuits

Dynamical equations for both circuits arise from application of K2. These read:

$$0 = V_1 - I_1 R_1 - \varepsilon_{L1} \quad \Longrightarrow \quad V_1 - I_1 R_1 = L_1 \frac{dI_1}{dt} + M_{21} \frac{dI_2}{dt}$$
$$0 = V_2 - I_2 R_2 - \varepsilon_{L2} \quad \Longrightarrow \quad V_2 - I_2 R_2 = L_2 \frac{dI_2}{dt} + M_{12} \frac{dI_1}{dt}$$

in the present situation. **This is a set of coupled differential equations.**

ASIDE: A superficial glance at these,

"*Ah-ha, a system of linear equations,*"

might lead one to mistakenly think

"*This will be amenable to standard algebraic methods of solution.*"

Regrettably, such confidence is misplaced. Coupled systems of differential equations are generally non-linear, and nearly impossible to solve!

The detailed dynamics of this pair of simple circuits is extremely complicated and depends strongly on when, and in what order, the switches are thrown. Rather than get bogged down in the mathematical details, let's just consider the initial and asymptotic final states. Once we are secure in our knowledge of these, we can wave our hands and gloss over the specifics of how this system evolves from the former to the latter.

INITIAL   The switches are open to begin with, so the initial currents are both zero, *i.e.*,

$$I_{1,0} = 0 = I_{2,0} \,.$$

FINAL   After a very long time, the currents both tend toward [assume] their UNCOUPLED constant steady state values:

$$I_{1,\infty} = \frac{V_1}{R_1} \quad \text{and} \quad I_{2,\infty} = \frac{V_2}{R_2} \,.$$

We can reliably state this because (1) both sides of each dynamical equation for the currents vanish for this solution, and (2) the dissipative effects of the resistances will damp out transients. That the coupled circuits tend asymptotically toward independent (uncoupled) behaviour is remarkable.

EXAMPLE   [*Maximal Mutual Inductance*]

Two coaxial solenoids with the same length, l, different numbers of turns, $\mathcal{N}_1 \neq \mathcal{N}_2$, and differing radii, $a_1 \neq a_2$, overlap as completely as possible. [This is why we insist that they have the same overall length.] Without loss of generality, suppose that 2 lies within 1. Therefore, $a_2 < a_1$, and the respective cross-sectional areas are related by $A_2 < A_1$.

The magnetic field generated by Coil$_2$ is zero outside of its radius, and uniform with value

$$B_2 = \mu_0 \, \eta_2 \, I_2 = \mu_0 \, \frac{\mathcal{N}_2}{l} \, I_2$$

everywhere inside. The magnetic fluxes produced by this field are

COIL 2: $\Phi_{22} = A_2 \, B_2$, and

COIL 1: $\Phi_{21} = A_2 \, B_2 = \mu_0 \, \frac{\mathcal{N}_2}{l} \, I_2 \, A_2$.

The last result is not a misprint. The magnetic field B_2 is non-zero only over the portion of A_1 which intersects A_2.

The magnetic field generated by Coil$_1$ is zero outside of its radius, and uniform with value

$$B_1 = \mu_0 \, \eta_1 \, I_1 = \mu_0 \, \frac{\mathcal{N}_1}{l} \, I_1$$

everywhere inside. The magnetic fluxes associated with B_1 are

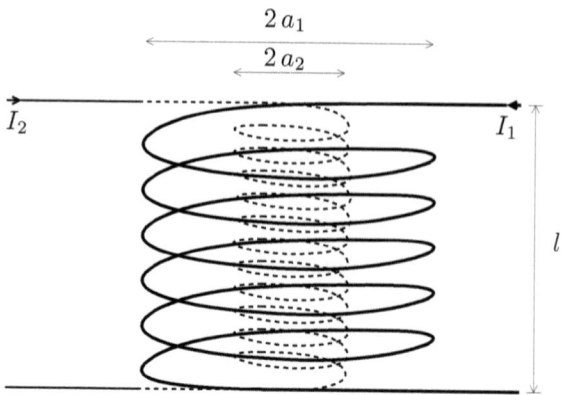

FIGURE 39.4   Maximal Flux Shared by Overlapping Solenoids

COIL 1: $\Phi_{11} = A_1 B_1$, and

COIL 2: $\Phi_{12} = A_2 B_1 = \mu_0 \frac{N_1}{l} I_1 A_2$.

The last result is also not a misprint.

Collecting the expressions determining the two linking-fluxes, Φ_{21} and Φ_{12}, and inserting these into the definitions of the mutual inductances, yields

$$M_{21} = \frac{\Phi_{21} N_1}{I_2} = \mu_0 N_1 N_2 \frac{A_2}{l}$$

and

$$M_{12} = \frac{\Phi_{12} N_2}{I_1} = \mu_0 N_1 N_2 \frac{A_2}{l}.$$

Not only have we successfully determined each of these mutual inductances, we have explicitly demonstrated their equality [in this case]!

For grins, we may also compute the self-inductances of the individual solenoids:

$$L_1 = \frac{N_1 \Phi_{11}}{I_1} = \mu_0 N_1^2 \frac{A_1}{l} \qquad \text{and} \qquad L_2 = \frac{N_2 \Phi_{22}}{I_2} = \mu_0 N_2^2 \frac{A_2}{l}.$$

IF the equal-length solenoids are interwoven [*i.e.,* possess the same radius and cross-sectional area, $A_1 = A_2 = A$], THEN

$$M_{21} = M_{12} = \mu_0 N_1 N_2 \frac{A}{l} = \sqrt{L_1 L_2}.$$

Chapter 40

LC Circuits

Consider a circuit comprised of a capacitor and an inductor, as illustrated in Figure 40.1. Prior to $t = 0$, the capacitor was charged to $\pm Q_0$. At $t = 0$ the switch is closed. Before we commence qualitative, energetic, and quantitative analyses of this simple circuit, let us acknowledge that the presumption of zero resistance is profoundly unphysical. Nevertheless, the results obtained here have bearing on situations in which the resistive effects are small *vis-à-vis* inductive and capacitive effects, and help pave the way for the general analysis of RCL circuits in Chapter 41.

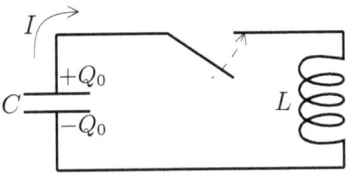

FIGURE 40.1   The Minimal Inductor–Capacitor Circuit

First we'll examine the behaviour of the system/circuit qualitatively.

LC   When the switch is first thrown, current begins to flow as the plates of the capacitor [attempt to] discharge. [Were this an RC circuit—neglecting inductive effects—the current would diminish monotonically as the capacitor discharges, because it is the residual charge that provides the potential difference responsible for driving the current. Instead, in this case the forward current starts from zero and grows.]

LC   The current generates a magnetic field in the region of the circuit, producing an additional magnetic flux [beyond that which might already have been present] through the circuit.

LC   A back-EMF is induced in the circuit to oppose this change in magnetic flux [in accord with Faraday's and Lenz's Laws]. The rate of increase of the current diminishes.

LC   The current cannot increase indefinitely, owing to the eventual exhaustion of the finite amount of charge initially on the capacitor.

LC   At some particular instant after the switch is thrown, all of the [excess] charge originally present on the capacitor will have drained from the plates. At this moment the capacitive potential which had been driving the current forward vanishes. Paradoxically, at this very same time, the current attains its MAXIMUM value, I_0.

LC   Despite the absence of capacitive potential to drive the current in the circuit, the inductive effect sustains it so as to maintain constancy of the magnetic flux.

LC   The forward current draws additional charge from the capacitor plates, above and beyond the initial charge, $\pm Q_0$, which has already been exhausted. As this occurs, an increasing negative potential difference is established across the capacitor. [By negative, we mean that it is oriented in opposition to the forward direction of the prevailing current.] The current, now being maintained by inductive effects and opposed by the increasing capacitive potential, diminishes at an increasing rate.

LC   In a resistanceless circuit [*i.e.*, one with no energy dissipation], as much additional charge is eventually pulled off the plates as was present at $t = 0$, *viz.*, $\mp Q_0$. At the instant this occurs, the current finally drops to zero, and the system/circuit is in the same state as it was at time $t = 0$, **except** that the polarity of the charge on the capacitor is reversed.

LC   The sequence of events described here will recur—**in reverse**—with the currents and EMFs in the opposite orientation.

The above analysis describes one **half-cycle** of an oscillatory [repeating in time] system. Sketches of $Q(t)$ and $I(t)$ in Figure 40.2 illustrate this behaviour through slightly more than one complete cycle.

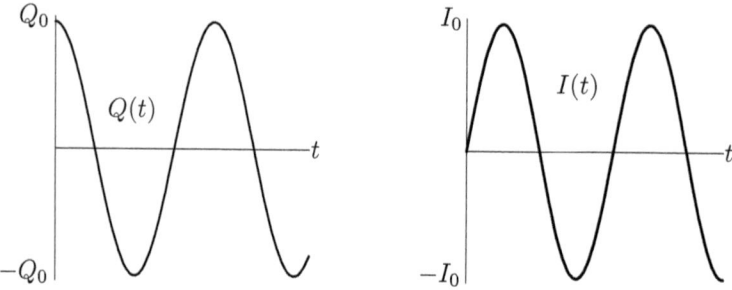

FIGURE 40.2   Sketches of Oscillatory Charge–Current Behaviour

---

Let's attempt to apprehend the workings of this circuit by analysis of energy flows.

Prior to the switch's being thrown, there is a finite quantity of energy stored in the [electric field inside the] capacitor because of the charge borne on its plates [$Q(0) = Q_0$], and no energy stored in the [magnetic field around the] inductor because of the absence of current [$I(0) = 0$]. These initial energies and their sum are:

$$
\left.
\begin{array}{ll}
\text{Initial Electric Energy} & E_E = \dfrac{Q_0^2}{2\,C} \\[2ex]
\text{Initial Magnetic Energy} & E_M = \dfrac{1}{2}\,L\,I^2(0) = 0
\end{array}
\right\}
\quad \Longrightarrow \quad
\text{Initial Total Energy} = \dfrac{Q_0^2}{2\,C} \,.
$$

Once the switch is thrown, energy flows from the capacitor. In the absence of resistance, there is no dissipation. Electrostatic energy stored in the electric field within the capacitor is converted into magnetic energy associated with the magnetic field generated by the current.

Once all of the charge has drained from the capacitor, the supply of stored electrostatic energy is completely depleted. It has been transformed into magnetic energy stored in the inductor:

$$
\text{Intermediate Value of the Total Energy} = \frac{1}{2}\,L\,I_0^2 \,.
$$

Since the total energy is conserved, the initial and intermediate total energies are equal,

$$\frac{Q_0^2}{2C} = \frac{1}{2} L I_0^2 ,$$

and therefore the maximal current is related to the circuit parameters and initial charge by $I_0 = Q_0/\sqrt{LC}$.

At this point, the system is one quarter of the way through its oscillatory cycle. The energy is passed back and forth twice in each complete cycle. [This is just like the transformation of kinetic and potential energies in a harmonic oscillator mechanical system, studied in Chapter 15 of VOLUME II.]

The total energy is constant [in time], since we have explicitly ruled out dissipation of electrical energy into heat. Thus, at all instants, the total energy is partitioned between the electric field in the capacitor and the magnetic field surrounding the wires, *i.e.*,

$$E_0 = U_E(t) + U_M(t) = \frac{Q^2(t)}{2C} + \frac{1}{2} L I^2(t) .$$

Quantitative analysis of the LC circuit shown in Figure 40.1 begins with the expression of Kirchoff's Rules. The circuit's single-loop renders K1 trivial, while K2 yields

$$0 = \frac{Q(t)}{C} - L\frac{dI(t)}{dt} \qquad \Longrightarrow \qquad \frac{dI}{dt} = \frac{Q}{LC} .$$

We explicitly indicated the time-dependence of the capacitor charge and circuit current in the expression of Kirchoff's Loop Rule, and rearranged the terms in preparation for the next steps in the analysis.

The source of the current in the circuit is the charge on the capacitor, *i.e.*,

$$I = -\frac{dQ}{dt} .$$

The minus sign associates a decrease in charge on the capacitor with POSITIVE current [flowing clockwise in the circuit diagram]. Increasing charge arises from NEGATIVE [anti-clockwise] current. Incorporating this relation into the Kirchoffian dynamical equation transforms it into

$$\frac{d^2 Q}{dt^2} = -\frac{1}{LC} Q .$$

This second-order linear ordinary differential equation is precisely of the form which governs [1-d] simple harmonic oscillation [SHO] with angular frequency

$$\omega = \sqrt{\frac{1}{LC}} = \frac{1}{\sqrt{LC}} .$$

The solutions of the SHO equation of motion are well-known and may be cast in a variety of forms:

$$Q(t) = A_c \cos\left(\omega\, t + \varphi_c\right)$$
$$= A_s \sin\left(\omega\, t + \varphi_s\right)$$
$$= A_c \cos(\omega\, t) + A_s \sin(\omega\, t)$$
$$= A_+\, e^{i\,\omega\, t} + A_-\, e^{-i\,\omega\, t}\,.$$

In each of these solutions, a pair of constants of integration

$$\{A_c\,,\varphi_c\} \quad \text{or} \quad \{A_s\,,\varphi_s\} \quad \text{or} \quad \{A_c\,,A_s\} \quad \text{or} \quad \{A_+\,,A_-\}\,,$$

appears to accommodate whatever particular initial conditions are in effect.

The various terms found in the [convenient] representation $Q(t) = Q_0 \cos\left(\omega\, t + \varphi_c\right)$ are:

- ○ Q_0 = maximum charge (amplitude)
- ○ ω = angular frequency[1]
- ○ φ = phase angle[2]

[WLOG, $\varphi = 0$ is assumed throughout the remainder of this chapter.]

Observe that

$$I = -\frac{dQ}{dt} = +\omega\, Q_0 \sin(\omega\, t + \varphi) = \left(\frac{Q_0}{\sqrt{L\,C}}\right) \sin\left(\frac{t}{\sqrt{L\,C}}\right) = I_0 \sin\left(\frac{t}{\sqrt{L\,C}}\right)\,,$$

specialised to the present case. This explicit determination of the current as a function of time conforms with all those aspects which were qualitatively and energetically inferred earlier.

Two final comments finish off this chapter.

⊥   The charge on the capacitor and the current in the circuit oscillate with the same frequency, while remaining exactly $\pi/2$, or $90°$, out of phase.

+   The quantitative solutions for the charge and current, quoted just above, exhibit the property of total energy conservation. This is verified by straightforward substitution:

$$E_0 = U_E(t) + U_M(t) = \frac{Q^2(t)}{2\,C} + \frac{1}{2}\, L\, I^2(t)$$
$$= \frac{Q_0^2}{2\,C} \cos^2(\omega\, t + \varphi) + \frac{1}{2}\, L\, I_0^2 \sin^2(\omega\, t + \varphi)$$
$$= \frac{Q_0^2}{2\,C} \cos^2(\omega\, t + \varphi) + \frac{1}{2}\, L\, \left(\frac{Q_0^2}{L\,C}\right) \sin^2(\omega\, t + \varphi)$$
$$= \frac{Q_0^2}{2\,C} \left(\cos^2(\omega\, t + \varphi) + \sin^2(\omega\, t + \varphi)\right)$$
$$= \frac{Q_0^2}{2\,C}\,,$$

which remains constant in time.

---

[1] The cycle frequency is $\nu = \frac{\omega}{2\,\pi}$.
[2] The value assigned to the phase angle is determined from the initial/boundary conditions.

Chapter 41

RCL Circuits

Consider the circuit pictured in Figure 41.1, which incorporates a switch, a fully charged capacitor, C, a resistor, R, and an inductor, L. The switch is closed at the instant $t = 0$. In this chapter, we shall perform the analysis of the circuit in terms of the charge on the capacitor, $q(t)$, which is equivalent [up to a factor] to the potential difference, V_C, across the capacitor. A current-based description of the circuit dynamics is presented in the Addendum.

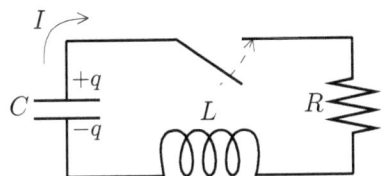

K1 $\Longrightarrow$ The current, $I(t)$, is uniquely defined at each instant.

K2 $\Longrightarrow 0 = \dfrac{q}{C} - I\,R - L\,\dfrac{dI}{dt}$.

FIGURE 41.1   A Simple RCL Circuit without a Voltage Source

The flow of current in the circuit depletes the charge held on the capacitor. Accordingly,

$$I = -\frac{dq}{dt}\,.$$

Augmented with this relation, the Kirchoffian loop equation becomes

$$0 = L\,\frac{d^2q}{dt^2} + R\,\frac{dq}{dt} + \frac{1}{C}\,q\,.$$

The dynamical equation governing the time-evolution of the charge on the capacitor is a **second-order homogeneous linear ordinary differential equation with constant coefficients.**

SECOND ORDER   A maximum of two derivatives appear in any single [additive] term.

HOMO-GENEOUS   Terms independent of q or its derivatives are NOT present.
[Homogeneity $\iff$ LHS $= 0$.]

LINEAR   IF $q_1(t)$ and $q_2(t)$ are distinct solutions, THEN any linear combination

$$q(t) = C_1\,q_1(t) + C_2\,q_2(t)\,,$$

with constant coefficients $\{C_1, C_2\}$, is also a solution.

ORDINARY   All of the derivatives appearing in the equation of motion are with respect to a common parameter, the time.

$\frac{\text{DIFF}}{\text{EQN}}$ The dynamical equation relates various rates of change of the charge held on the capacitor to one another.

$\frac{\text{CONST}}{\text{COEFFS}}$ No other time dependence appears, aside from that which is inherent in $q(t)$ and its derivatives. The factors R, C, and L are all constants.
[As an added bonus, these coefficients are all POSITIVE.]

The immediate and pressing task is to solve the equation of motion,

$$0 = L\frac{d^2q}{dt^2} + R\frac{dq}{dt} + \frac{1}{C}q \quad \Longleftrightarrow \quad 0 = \left[L\frac{d^2}{dt^2} + R\frac{d}{dt} + \frac{1}{C}\right]q(t).$$

Q: How does one begin to do so?
A: With an *Ansatz*[1] for the trajectory of the particle.
Suppose that the charge function, $q(t)$, is of exponential form,

$$q(t) = \exp(r\,t) \qquad [\text{ notation alert: } \exp(r\,t) \equiv e^{r\,t}\,],$$

where r is an as yet undetermined parameter.

ASIDE: Go ahead and ask: "Where did this *Ansatz* come from?" There are two answers. The first is that a fundamental property of the exponential function [its derivative is proportional to the function itself] makes it an ideal candidate for the solution of linear differential equations. The second is that, since *Ansatze* are "educated guesses," it is unlikely that we would report on an unsuccessful attempt.

Application of the differential operator to the *Ansatz* solution yields

$$0 = \left[L\frac{d^2}{dt^2} + R\frac{d}{dt} + \frac{1}{C}\right]\exp(r\,t) = \left[L\,r^2 + R\,r + \frac{1}{C}\right]\exp(r\,t).$$

Thus,

$$\text{EITHER} \quad 0 = L\,r^2 + R\,r + \frac{1}{C} \quad \text{OR} \quad r \to -\infty.$$

The latter choice leads to the extraneous solution $q(t) = 0$. Henceforth, we shall focus on the former option, with the intention of determining the particular value(s) of r for which the circuit's dynamical equation is satisfied.
Q: Isn't this a quadratic equation [*i.e.*, algebraic, rather than differential]?
A: Yes! This is the chief virtue of the *Ansatz* approach.
The quadratic formula yields those values of r for which the equation is satisfied [*a.k.a.* roots]:

$$r_\pm = \frac{-R \pm \sqrt{R^2 - 4\frac{L}{C}}}{2L} = -\frac{R}{2L} \pm \sqrt{\left(\frac{R}{2L}\right)^2 - \frac{1}{LC}} = -\gamma \pm \sqrt{\gamma^2 - \omega_0^2}.$$

For concision, the expression for the roots has been rewritten in terms of two particular combinations of $\{R, C, L\}$,

$$\gamma = \frac{R}{2L} \quad \text{and} \quad \omega_0^2 = \frac{1}{LC}.$$

---

[1] Recall that an *Ansatz* is a trial solution containing one or more free parameters. Particular values of these parameters are determined by substituting the *Ansatz* directly into the equation it is intended to satisfy and solving the constraints which then arise.

The γ term is the **damping coefficient**, while ω_0 is the **angular frequency** that the LC circuit would possess were the resistor not present.

Next, we discuss and resolve two sticky wickets.

DUO There are two solutions: $r = r_\pm$, labelled according to the sign preceding the square root term arising in the quadratic formula.

Resolution: Form an admixture of both solutions, *i.e.,*

$$q(t) = C_+ \exp(r_+ t) + C_- \exp(r_- t),$$

where $C_\pm$ are constants whose values are determined by the initial or boundary conditions of the system.

> ASIDE: In retrospect, one realises that a coefficient might well have been included in the original *Ansatz*. However, the linearity and homogeneity of the dynamical equation make it so that such an overall coefficient is entirely unconstrained, and thus it is best to elide it in the presentation of the *Ansatz*.

An unforeseen bonus is that [owing to existence and uniqueness theorems applicable to second-order ordinary differential equations] **the unconstrained superposition of two distinct solutions is a fully general solution** of the equation of motion.

COMPLEX Although all of the physical parameters $\{R, C, L\}$ and the agglomerations $\{\gamma, \omega\}$ are manifestly positive constants describing the physical attributes of the resistor–capacitor–inductor circuit, the roots may be real or complex, depending on the sign of the term appearing under the square root.

Resolution: Consider all of the mathematically consistent possibilities. These are distinguished by the sign of the DISCRIMINANT $\mathcal{D} = \gamma^2 - \omega_0^2$. Three cases occur. Two of these, **overdamped** $[\mathcal{D} > 0]$ and **underdamped** $[\mathcal{D} < 0]$, consist of ranges, while the third, **critically damped** $[\mathcal{D} = 0]$, exists as the boundary between the other two. Approached as a mathematical curiosity, these cases militate for consideration of the limits $\frac{\omega_0^2}{\gamma^2} \to 0$, $\frac{\omega_0^2}{\gamma^2} \to 1$, and $\frac{\gamma^2}{\omega_0^2} \to 0$, wherever they might be taken.

OVERDAMPED IF the value of the discriminant is POSITIVE,

$$\gamma^2 - \omega_0^2 > 0,$$

THEN the RCL circuit is overdamped. As a consequence, the square root term is real-valued, and thus $r_\pm$ are both real-valued and distinct. As γ and ω_0 are positive, it is incontrovertible that

$$r_- < r_+ < 0.$$

Thus, the general solution of the overdamped equation of motion is

$$q(t) = A_+ \exp\left((-\gamma + \delta)\,t\right) + A_- \exp\left(-(\gamma + \delta)\,t\right),$$

for constant coefficients $A_\pm$ and $\delta = +\sqrt{\gamma^2 - \omega_0^2} < \gamma$.

There are several important comments that must be made.

O-D   Both terms contributing additively to $q(t)$ are decaying exponentials, with differing rates of decay. *Ergo*, the trajectory tends [asymptotically[2]] toward ZERO charge on the capacitor, smoothly but non-trivially.

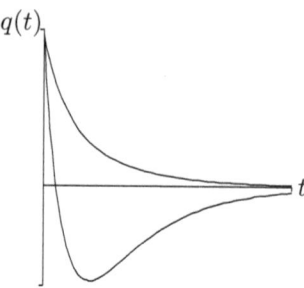

FIGURE 41.2   Trajectories of Two Overdamped Oscillators

O-D   After a somewhat long time, the term with the more negative exponential factor, $-\gamma - \delta$, will become insignificant *vis-à-vis* the less quickly attenuating term. When this occurs, the remaining charge will seem to decay exponentially at the rate $-\gamma + \delta$.

O-D   The general solution may be recast in several ways to highlight various features and to facilitate its application in specific cases. For example,

$$q(t) = e^{-\gamma t} \left[A_+ \, e^{\delta t} + A_- \, e^{-\delta t} \right]$$

shows quite clearly that γ acts as an overall damper, eventually driving the charge to zero with the passage of time. The additive factors within the brackets may be re-expressed in terms of hyperbolic trigonometric functions. Invoking the definitions

$$\cosh(x) = \frac{1}{2} \left(e^x + e^{-x} \right) \qquad \text{and} \qquad \sinh(x) = \frac{1}{2} \left(e^x - e^{-x} \right)$$

leads to the realisation that

$$A_+ \, e^{\delta t} + A_- \, e^{-\delta t} = A_c \, \cosh(\delta t) + A_s \, \sinh(\delta t) \, ,$$

where the coefficients are related by $A_c = A_+ + A_-$ and $A_s = A_+ - A_-$. Hence,

$$q(t) = e^{-\gamma t} \left[A_c \, \cosh(\delta t) + A_s \, \sinh(\delta t) \right] .$$

O-D   The relative sizes and signs of $A_\pm$ [or $A_{c,s}$] determine whether the function $q(t)$ crosses the $q = 0$ axis.

O-D   Consider the factorisation:

$$-\gamma + \delta = -\gamma \left[1 - \sqrt{1 - \frac{\omega_0^2}{\gamma^2}} \right] .$$

---

[2]Asymptotically, in this context, means "in the limit that the time parameter grows very large."

In the limit $\frac{\omega_0^2}{\gamma^2} \to 0$ [in which the effects of damping are certain to dominate over the system's attempt to oscillate], the argument of the more slowly decaying exponential becomes

$$-\gamma + \delta \simeq -\gamma \left[\frac{1}{2} \frac{\omega_0^2}{\gamma^2} \right] = -\frac{\omega_0^2}{2\gamma} = -\frac{\frac{1}{LC}}{\frac{R}{L}} = -\frac{1}{RC} \,.$$

Inductive effects are rather insignificant for this type of RCL circuit. After a very short time has elapsed, it behaves like an RC circuit.

The limiting case, $\frac{\omega_0^2}{\gamma^2} \to 1$, constitutes critical damping and is examined next.

CRITICALLY DAMPED   IF the value of the discriminant is ZERO,

$$\gamma^2 - \omega_0^2 = 0 \,,$$

THEN the RCL circuit is critically damped. Here $r_\pm = -\gamma$ is a real double root. The double root affects the claim made earlier about generality, and the solution in this particular instance[3] is

$$q(t) = \left(A + B\,t \right) \exp(-\gamma\,t) \,,$$

for constant coefficients A and B.

Peculiarities of the critically damped case are explored below.

C-D   The exponential suppresses both the constant and the linear terms, leading to a smooth and monotonic[4] descent toward $q = 0$.

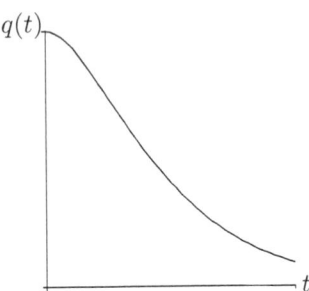

FIGURE 41.3   The Trajectory of a Critically Damped Oscillator

C-D   At criticality, the discriminant vanishes, and yet it remains factorisable:

$$0 = \gamma^2 - \omega_0^2 = (\gamma + \omega_0)(\gamma - \omega_0) \,.$$

Since both γ and ω_0 are positive, it must be that $\gamma = \omega_0$.

---

[3] Despite our not proving that this is the general solution, it is a valid solution, provided that $\gamma^2 = \omega_0^2$.
[4] It may be shown that, when the current is initially zero, the charge decreases to ZERO without crossing the $q = 0$ axis.

UNDERDAMPED IF the value of the discriminant is NEGATIVE,

$$\gamma^2 - \omega_0^2 < 0 \,,$$

THEN the RCL circuit is underdamped. Therefore, the square root term is imaginary-valued, and the $r_\pm$ are complex conjugates. Thus, the general solution of the underdamped equation of motion is

$$q(t) = A_+ \exp\left((-\gamma + i\,\widetilde{\omega})\,t\right) + A_- \exp\left((-\gamma - i\,\widetilde{\omega})\,t\right)$$
$$= \exp(-\gamma\,t)\left[A_+\,e^{+i\widetilde{\omega}t} + A_-\,e^{-i\widetilde{\omega}t}\right],$$

for constant coefficients $A_\pm$ and $\widetilde{\omega} = \sqrt{|\gamma^2 - \omega_0^2|} = +\sqrt{\omega_0^2 - \gamma^2} < \omega_0$.

There are several aspects of the underdamped case deserving of mention.

U-D   An overall exponential envelope, with decay constant γ, suppresses the two terms in the bracket. The exponentials with imaginary arguments are equivalent [via EULER'S THEOREM] to a linear combination of sinusoids.

> ASIDE: The proof of this statement is analogous to the re-presentation of the overdamped solutions in terms of hyperbolic trigonometric functions. In the underdamped case,
>
> $$\cos(x) = \frac{1}{2}\left(e^{ix} + e^{-ix}\right) \qquad \text{and} \qquad \sin(x) = \frac{1}{2i}\left(e^{ix} - e^{-ix}\right)$$
>
> enable re-expression of the charge, $q(t)$. The analysis is complicated in this instance by the presence of explicit factors of i [the imaginary unit]. A simple dodge is to declare, by *fiat*, that only the real-valued part of the full solution for $q(t)$ is meaningful. However, if one is not careful and naively discards an integration constant, one may be unable to accommodate the full range of possible initial conditions. Another approach is to allow the constant factors $A_\pm$ to be complex-valued. This presents the opposite problem: a surfeit of degrees of freedom (four real and imaginary parts in place of two real constants), and must be approached cautiously. One returns to two independent degrees of freedom by insisting that the components be related by complex-conjugation (supplying two algebraic constraints to the *a priori* four degrees of freedom). WLOG, we can gloss over the specific details of the analysis and simply claim that the solutions
>
> $$q(t) = \exp(-\gamma\,t)\left[A_+\,e^{+i\widetilde{\omega}t} + A_-\,e^{-i\widetilde{\omega}t}\right]$$
> $$= \exp(-\gamma\,t)\left[A_c\,\cos(\widetilde{\omega}\,t) + A_s\,\sin(\widetilde{\omega}\,t)\right]$$
> $$= \exp(-\gamma\,t)\,A_c\,\cos(\widetilde{\omega}\,t + \varphi_c)$$
> $$= \exp(-\gamma\,t)\,A_s\,\sin(\widetilde{\omega}\,t + \varphi_s)$$
>
> are equivalent. In each case above, γ and $\widetilde{\omega}$ are prescribed by the circuit parameters, and a pair of constants, $\{A_+, A_-\}$, $\{A_c, A_s\}$, $\{A_c, \varphi_c\}$, or $\{A_s, \varphi_s\}$, is present to accommodate the particular initial conditions. Should there be any doubt that these are solutions, one may simply plug them into the original [Kirchoffian] differential equation.

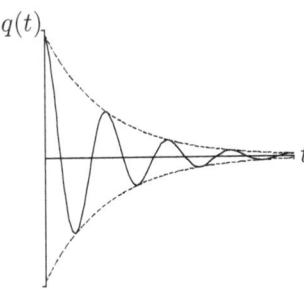

FIGURE 41.4 The Trajectory of an Underdamped Oscillator

Therefore, the trajectory oscillates within the confines of the [asymptotically] shrinking envelope, as illustrated in Figure 41.4.

U-D Consider the not-so-obvious expansion of the oscillatory frequency:

$$\widetilde{\omega} = \omega_0 \sqrt{1 - \frac{\gamma^2}{\omega_0^2}} \, .$$

The limit $\frac{\gamma^2}{\omega_0^2} \to 1$ was examined in the discussion of the critically damped case. In the limit $\frac{\gamma^2}{\omega_0^2} \to 0$ [in which the effects of damping are small in comparison to the underlying SHO dynamics], the effective frequency of the charge oscillation is only slightly shifted [down] from its natural frequency value. Equivalently, the oscillation period is slightly increased. To lowest non-trivial order, the shifted frequency is

$$\widetilde{\omega} \simeq \omega_0 \left[1 - \frac{1}{2} \frac{\gamma^2}{\omega_0^2} \right] = \omega_0 \left[1 - \frac{R^2 C}{8 L} \right] = \omega_0 \left[1 - \frac{1}{8} \frac{R C}{L/R} \right] = \omega_0 \left[1 - \frac{1}{8} \frac{\tau_{\mathrm{RC}}}{\tau_{\mathrm{RL}}} \right] .$$

Physically, this situation is realised when the inductive time scale is of much longer duration than that set by the RC dynamics.

ADDENDUM to Chapter 41

Here, we shall investigate the RCL circuit in terms of its current, $I(t)$. The Kirchoffian relation, considered earlier in this chapter, may be differentiated with respect to time. This produces another [valid] differential equation governing the system's dynamics:

$$\frac{d}{dt}\left[0 = \frac{q}{C} - RI - L\frac{dI}{dt}\right] \qquad \Longrightarrow \qquad 0 = \frac{1}{C}\frac{dq}{dt} - R\frac{dI}{dt} - L\frac{d^2I}{dt^2}.$$

The realisation that the current in the circuit bears charge away from the capacitor, $I = -\frac{dq}{dt}$, leads to a dynamical relation expressed solely in terms of the current and its derivatives:

$$0 = \frac{1}{C}I + R\frac{dI}{dt} + L\frac{d^2I}{dt^2}.$$

In form, this equation is identical to that obeyed by $q(t)$. The solutions fall into the same three classes: over-, under-, and critically damped, depending on the value of the discriminant [fixed by R, C, and L]. The differences in the physical solutions come from differences in the specified boundary conditions.[5]

OVERDAMPED   The overdamped solution is

$$I(t) = I_+ e^{(-\gamma+\delta)t} + I_- e^{-(\gamma+\delta)t} = e^{-\gamma t}\left[I_+ e^{\delta t} + I_- e^{-\delta t}\right].$$

Demanding that the initial current vanish, $I(0) = 0$, leads to $I_- = -I_+$, and hence

$$I(t) = C e^{-\gamma t} \sinh(\delta t),$$

for a constant, C, with the dimensions of current.

CRITICALLY DAMPED   The general solution is

$$I(t) = (I_0 + I_1 t) e^{-\gamma t}.$$

The "vanishing initial current" boundary condition forces $I_0 = 0$, and thus

$$I(t) = C t e^{-\gamma t},$$

where C is a constant.

UNDERDAMPED   The general solution for the current is

$$I(t) = e^{-\gamma t}\left[I_c \cos(\widetilde{\omega} t) + I_s \sin(\widetilde{\omega} t)\right].$$

Upon restricting to solutions with vanishing initial current, one obtains

$$I(t) = C e^{-\gamma t} \sin(\widetilde{\omega} t).$$

---

[5]For the charge, we chose $q(t)|_{t=0} = q_0 > 0$ and $I(t)|_{t=0} = 0$. For the current, we'll adopt the less restrictive condition $I(0) = 0$ in the mathematical analysis, all the while recognising that physically there must be complete concurrence between the flow of current and accumulation of charge.

Chapter 42

AC Circuits

All of the voltage sources, *i.e.*, batteries, that we've encountered until now have been assumed to provide an EMF that is constant in time.

> ASIDE: This is not quite true. In Chapter 21 we modelled the manner in which the output voltage of a real battery decreases with current draw by introducing an internal resistance acting in series with an otherwise constant source of EMF. So, the effective EMF of a real battery changes [somewhat] until the current passing through it attains its steady state value.

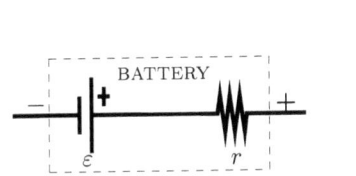
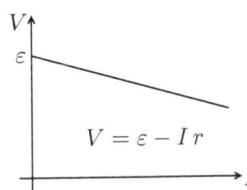

In this chapter we shall introduce so-called **Alternating Current** [AC] voltage sources.

[The terminology is historic and, regrettably, confusing.]

AC SOURCE   An AC voltage source is an electric device producing a harmonically oscillating EMF,

$$v(t) = V_0 \sin(\omega t + \phi),$$

between two terminals (points in space).

The [explicitly time-dependent] AC voltage and current are denoted by lowercase letters, in contrast with previous practice involving **Direct Current** [DC] circuits. Each of the factors on the RHS of the expression for the time-dependent potential retains its conventional interpretation.

V_0 The voltage amplitude is the maximum excursion from the average potential [here chosen to be ZERO, for simplicity], measured in volts, V, of course.

ω The source voltage oscillates harmonically, with a fixed angular frequency. The cycle frequency and period are

$$\nu = \frac{\omega}{2\pi} \quad \text{and} \quad T = \frac{1}{\nu} = \frac{2\pi}{\omega},$$

respectively.

ϕ The presence of the phase angle is concomitant with the freedom that one has to choose the instant deemed $t = 0$.

We add the AC voltage source to our toolbox and shall incorporate it, in conjunction with other idealised components, into a new class of models of simple electric circuits.

AC SOURCE

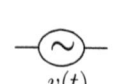

$v(t)$

An AC voltage source produces an EMF which oscillates in a simple harmonic manner:

$$v(t) = V_0 \sin(\omega t + \phi).$$

ASIDE: Here's how a sinusoidally varying EMF might be generated. Recall that the subject of Chapters 25 and 26 was the magnetic torque exerted on a current-carrying rectangular loop of wire bathed in a uniform magnetic field. From Faraday's and Lenz's Laws, one realises that if the loop is rotated [about an axis through its centre] with constant angular frequency ω, then an AC voltage is induced in the loop.

To begin our exploration of the phenomenology of AC voltage sources, we combine an AC source and a resistor, R, in series.

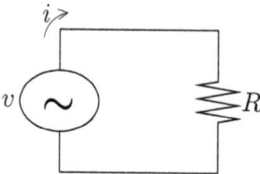

FIGURE 42.1   AC Source with a Purely Resistive Load

Kirchoff's Loop Rule implies that
$$0 = v - i\,R.$$

Provided that we choose to start our clocks at an instant when the AC voltage passes through zero [going from negative to positive], the voltage phase angle, ϕ, may be set to zero. In this case, Kirchoff's Loop Rule reads

$$0 = V_0 \sin(\omega t) - i\,R.$$

Solving for the current response yields

$$i(t) = \frac{v(t)}{R} = \frac{V_0}{R} \sin(\omega t) = I_0 \sin(\omega t).$$

In the final expression, the current amplitude, $I_0 = V_0/R$, has been introduced. The applied AC voltage, $v_R(t)$, and the current response, $i_R(t)$, are sketched [as functions of time] in Figure 42.2. The subscript R on the voltage and current serves as a reminder that this circuit has only a resistive aspect. Equivalently, one can think of v_R and i_R as the voltage across and the current through the resistor. It is apparent from the formulae and confirmed by the figures[1] that v_R and i_R are **in phase**, meaning that their variations are perfectly correlated and are alike.

---

[1]The plots are stacked one above the other, rather than presented side-by-side, to draw attention to their precise phase alignment.

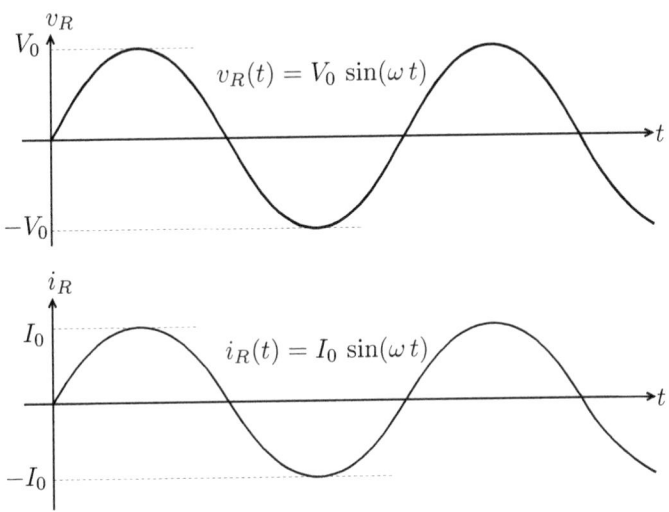

FIGURE 42.2 Correlated Voltage and Current Response in a Resistive AC Circuit.

Q: Though v_R and i_R are changing in time, is there a consistent way of characterising their values by means of a single number?

$\mathbf{A_{AMP}}$: The voltage and current amplitudes, V_0 and I_0, might serve in this capacity. Militating against this choice is that it amounts to a consistent overstatement, as the actual values of voltage and current are always less than or equal to these amplitudes.

$\mathbf{A_{AVE}}$: One might propose the **time-averaged** values of the voltage and current as representative of these changing quantities. Formally, the time average of any quantity $\mathcal{Q}(t)$ throughout an interval commencing at t_i and lasting until t_f is given by

$$\langle \mathcal{Q} \rangle_{\Delta t = t_f - t_i} = \frac{1}{t_f - t_i} \int_{t_i}^{t_f} \mathcal{Q}(t')\, dt' .$$

When the quantity is oscillating [*i.e.*, repeating in time], as is the case here, the long time average approaches the average value computed through a single oscillation period.

ASIDE: Suppose that the long interval, $\Delta t = t_f - t_i$, consists of $\mathcal{N}$ complete oscillation periods of duration $\mathcal{T}$, plus a remainder, δt, such that $\delta t < \mathcal{T}$. The multiplicative factor in the expression for the average is

$$\frac{1}{\Delta t} = \frac{1}{\mathcal{N}\mathcal{T} + \delta t} = \frac{1}{\mathcal{N}\mathcal{T}} \frac{1}{1 + \frac{\delta t}{\mathcal{N}\mathcal{T}}} \simeq \frac{1}{\mathcal{N}\mathcal{T}} \left(1 - \frac{\delta t}{\mathcal{N}\mathcal{T}} \right) .$$

For very long intervals, which perforce are comprised of a large number of complete periods, the approximation in the last step of the above analysis becomes more nearly exact. The integral can be partitioned into $\mathcal{N}$ identical copies of the integral over exactly one period, plus a residual term arising from the integrations[2] over the remainder time. Thus,

$$\int_{t_i}^{t_f} \mathcal{Q}(t')\, dt' = \mathcal{N} \int_0^{\mathcal{T}} \mathcal{Q}(t')\, dt' + \int_{\delta t} \mathcal{Q}(t')\, dt' = \mathcal{N}\mathcal{T}\, \langle \mathcal{Q}(t) \rangle_{\mathcal{T}} + \int_{\delta t} \mathcal{Q}(t')\, dt' .$$

---

[2]The plural is to allow the remainder [δt] to be split between the start and the end of the long interval, when this facilitates recognition of the periodic property of the quantity being averaged.

Recombining these terms into the expression for the average,[3]

$$\langle \mathcal{Q}\rangle_{\Delta t=t_f-t_i} = \frac{1}{\mathcal{N}\mathcal{T}}\left[1 - O\left(\frac{1}{\mathcal{N}}\right)\right] \times \left[\mathcal{N}\,\mathcal{T}\,\langle \mathcal{Q}(t)\rangle_{\mathcal{T}} + \int_{\delta t} \mathcal{Q}(t')\,dt'\right] = \langle \mathcal{Q}(t)\rangle_{\mathcal{T}} + O\left(\frac{1}{\mathcal{N}}\right),$$

and taking the long-duration limit $[\Delta t \to \infty \implies \mathcal{N} \to \infty]$ yields

$$\langle \mathcal{Q}\rangle_{\Delta t \to \infty} = \langle \mathcal{Q}(t)\rangle_{\mathcal{T}}\,.$$

Applying this averaging technique to the AC source potential leads to good and bad news.

GOOD   The integral over a single oscillation period exists and is calculable.

BAD   The average value of the AC source voltage vanishes:

$$\langle v_R\rangle_{\mathcal{T}} = \frac{1}{\mathcal{T}}\int_0^{\mathcal{T}} V_0 \sin(\omega\,t)\,dt = V_0\,\frac{1}{2\pi}\int_0^{2\pi/\omega} \sin(\omega\,t)\,d(\omega t) = V_0 \int_0^{2\pi} \sin(\theta)\,d\theta = 0\,.$$

The same fate, vanishing of its average value, befalls the current response. In hindsight, the nullity of these average values was entirely predictable, since both voltage and current oscillate symmetrically about zero.

———————————————————

In our search for a characterisation of the voltage and current response, we might be inspired to *follow the energy*.

The average power dissipated by the resistor is not zero, because the Joule heating that takes place throughout the cycle is insensitive to the direction of the current. At any given instant, the power being dissipated by the resistor is

$$P(t) = i_R^2\,R = I_0^2\,R\,\sin^2(\omega\,t)\,.$$

Averaged over one complete cycle [or, equivalently, over a long time], this becomes $\langle i_R^2\,R\rangle_{\mathcal{T}} = R\,\langle i_R^2\rangle_{\mathcal{T}}$, since the resistance, R, is constant. The average value of the squared-current may be computed:

$$\langle i_R^2\rangle_{\mathcal{T}} = \frac{1}{\mathcal{T}}\int_0^{\mathcal{T}} I_0^2\,\sin^2(\omega\,t)\,dt$$

$$= I_0^2\left[\frac{1}{2\pi}\int_0^{2\pi/\omega}\sin^2(\omega\,t)\,d(\omega t)\right]$$

$$= I_0^2\left[\frac{1}{2\pi}\int_0^{2\pi}\sin^2(\theta)\,d\theta\right].$$

The final integral, just above, has been encountered in Chapter 37 of VOLUME I. It evaluates to precisely $\frac{1}{2}$. Thus,

$$\langle i_R^2\rangle_{\mathcal{T}} = \frac{I_0^2}{2}\,.$$

———————————————————

[3]The symbol $O\left(\frac{1}{\mathcal{N}}\right)$ represents a collection of [unspecified] terms, all of which contain at least one overall [uncancelled] power of $\mathcal{N}^{-1}$.

Hence the average[4] power produced in the resistor is

$$\langle P \rangle = \frac{1}{2} I_0^2 R .$$

By analogous reasoning, one can readily conclude that

$$\langle v_R^2 \rangle = V_0^2 \langle \sin^2(\omega t) \rangle = \frac{1}{2} V_0^2 .$$

Pondering these results leads one to formalise the **RMS** prescription.

RMS   The Root-Mean-Square [RMS] method of characterising the size of a variable quantity consists of three distinct steps. [These steps are written in OPERATOR ORDER.[5]]

   SQUARE   First, square the quantity. This may involve squaring numerical values or working analytically with an algebraic expression.

   MEAN   Second, compute the uniformly weighted average of the square(s).

   ROOT   Third, take the square root(s) of the just-computed mean.

In symbolic mathematical form, the RMS prescription reads

$$Q_{\text{RMS}} \equiv \sqrt{\langle Q^2 \rangle} .$$

RMS AC VOLTAGE   Application of the RMS procedure to the simply varying sinusoidal source voltage leads [when we invoke the calculated results obtained just above] to:

$$V_{\text{RMS}} = \sqrt{\langle v_R^2(t) \rangle_T} = \sqrt{\frac{1}{2} V_0^2} = \frac{V_0}{\sqrt{2}} .$$

RMS AC CURRENT   When the current oscillates [about zero] in a sinusoidal manner,

$$I_{\text{RMS}} = \sqrt{\langle i_R^2(t) \rangle_T} = \sqrt{\frac{1}{2} I_0^2} = \frac{I_0}{\sqrt{2}} .$$

The common factor of $\frac{1}{\sqrt{2}} \simeq 0.7071 \ldots$ relating the RMS values to the voltage and current amplitudes is an artifact of the sinusoidal property of the time-varying quantities. Other forms of temporal dependence give rise to different numerical factors relating RMS and peak values.

---

[4]To make the expressions appear less cluttered, we shall often omit the subscript indicating that the average is taken throughout one complete period of oscillation. In situations such as the present case the oscillation period is the only relevant time scale and no ambiguity arises.

[5]Operators act on objects to their right in right-to-left sequential order.

EXAMPLE [*AC Resistive Circuit*]

An AC voltage source with peak strength $V_0 = 10\,$v, operating at 50 Hz, is connected in series to a 25 Ω resistor.

(a) Compute the

    (i)   peak current passing through the resistor,

    (ii)   maximum and   (iii) minimum   rates at which power is dissipated in the resistor,

    (iv)   average rate of power dissipation.

(b) Compute the RMS   (i) voltage across and   (ii) current through   the resistor.

(c) Verify that the DC power expressions, $P = I^2 R = V I = V^2/R$, retain their validity in an averaged sense when RMS quantities are employed.

(d) Determine how the circuit properties would change if the frequency of the AC source were to be   (i) doubled and   (ii) halved.

(a) (i) The peak current is $I_0 = \frac{V_0}{R} = \frac{10}{25} = \frac{2}{5}$ A.

    (ii) The maximum power occurs when the current is maximised: $P_{\text{max}} = I_0^2 R = 4\,$w.

    (iii) When the current vanishes, the power does too: $P_{\text{min}} = 0\,$w.

    (iv) The [TIME-]average rate of power dissipated in the resistor is $\langle P \rangle = \frac{1}{2} I_0^2 R = 2\,$w.

(b) (i) The RMS voltage across the resistor is $V_{\text{RMS}} = V_0/\sqrt{2} = 5\sqrt{2}\,$v.

    (ii) The RMS current through the resistor is $I_{\text{RMS}} = I_0/\sqrt{2} = \frac{\sqrt{2}}{5}$ A.

(c) Substituting RMS values into the three power expressions yields

$$I_{\text{RMS}}^2 R = \left(\frac{\sqrt{2}}{5}\right)^2 25 = 2\,, \qquad I_{\text{RMS}} V_{\text{RMS}} = \frac{\sqrt{2}}{5} 5\sqrt{2} = 2\,, \qquad \frac{V_{\text{RMS}}^2}{R} = \frac{\left(5\sqrt{2}\right)^2}{25} = 2\,.$$

All of these have units of watts, and each is equal to the average power per period deposited into the resistor.

(d) For a purely resistive element coupled to an AC source, the RMS values of the voltage and current in the circuit, and the average power deposited into the resistor, are not affected by changes in the frequency of the AC source.

PHASOR DIAGRAM FOR THE PURELY RESISTIVE AC CIRCUIT

A **phasor**[6] **diagram** offers an intuitive visualisation of the behaviour of an AC circuit and a convenient means for tracing its evolution. At its most primitive, the idea is that sinusoidal variation of any quantity can be envisioned as the 1-d projection of uniform circular motion in 2-d.

In the case at hand, let's consider the motion of a "rod" of fixed length V_0, rotating uniformly about an axis through one end [chosen to be the origin of coordinates]. Suppose that the angular speed of the rod is ω and that the reference direction is chosen to coincide with the alignment of the rod at $t = 0$. At any later time t, the projections of the rod onto the reference, x, and the perpendicular, y, directions are

$$v_x = V_0 \cos(\omega t) \qquad \text{AND} \qquad v_y = V_0 \sin(\omega t).$$

The y-axis projection of the rotating rod is instantaneously equal to the [time-dependent] voltage of the AC source.

The same construction involving a rod of length I_0 proceeds apace: its y-projection exhibits the same form as the AC current response function,

$$i_y = I_0 \sin(\omega t).$$

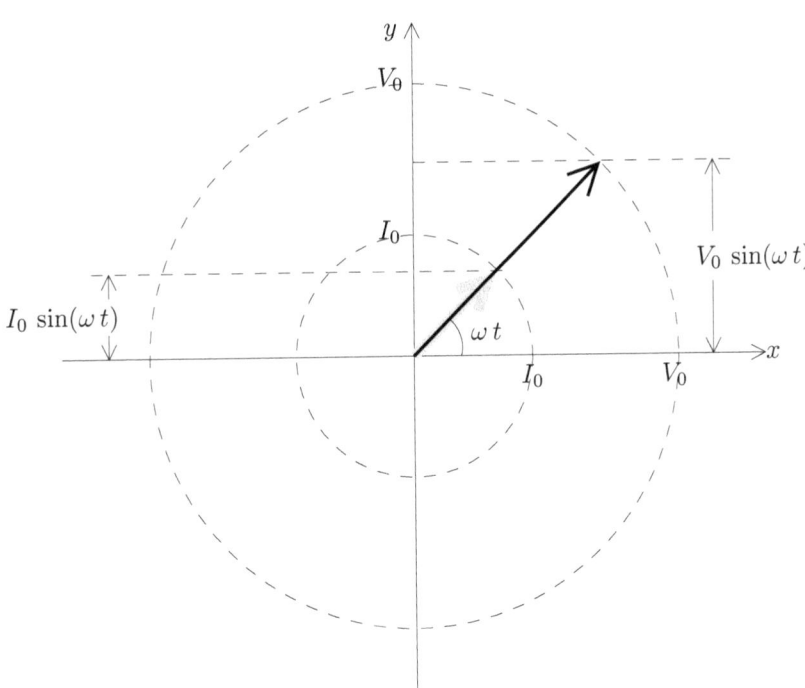

FIGURE 42.3 Phasor Diagram for a Simple Resistive AC Circuit

The benefits of the phasor approach appear when one considers the simultaneous uniform circular motion of both rods within a common plane [this despite their differing physical

---

[6] In the classic television series *Star Trek*, *phasors* are pistol-like weapons able to inflict calibrated degrees of harm. The order "Set phasors on STUN" is given in many episodes. Should you find yourself suddenly unconscious during our investigations of AC circuits, blame it on the phasors!

dimensions' causing them to inhabit different vector spaces]. The two rods rotate in lockstep, maintaining ZERO phase separation, as time advances. The values of the voltage and current in the circuit at any particular instant are given by the y-axis projections of the rotating voltage and current rods [vectors].

A final comment is in order. The absence of a phase difference, leading to overlap of the rods on the phasor plot, is peculiar to purely resistive circuits [and finely tuned inductive–capacitive–resistive circuits[7]] and is certainly not generic.

---

[7]This foreshadows developments in Chapters 45 and 46.

Chapter 43

Inductive and Capacitive AC Circuits

An AC voltage source is placed in series with an inductive element as shown in Figure 43.1.

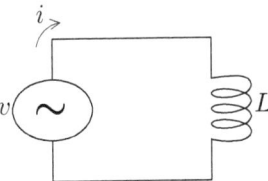

FIGURE 43.1 AC Source with a Purely Inductive Load

Kirchoff's Loop Rule applied to this circuit yields the expected differential equation:

$$0 = v - L\frac{\partial i}{\partial t} = V_0 \sin(\omega t) - L\frac{di}{dt} \qquad \Longrightarrow \qquad di = \frac{V_0}{L}\sin(\omega t)\,dt\,.$$

In passing from K2 to its mathematical expression, two specialisations were made. The sinusoidal form of the AC source voltage, with zero phase-offset, has been made explicit, and the partial derivative of the current has been replaced by $\frac{d}{dt}$ on account of the fact that the current is confined to the devices and the thin wires joining them. We can aspire to integrate both sides of the differential equation and thereby obtain an expression for the time-dependent current. Such an expression is not uniquely specified until a boundary/initial condition is imposed [to fix the value of the constant of integration]. The absence of a phase angle in the argument of the sinusoid describing the oscillation of the voltage source means that we have exercised our right to prescribe the zero of the time parameter [up to a shift by an integer number of whole periods]. Let us presume to know the value of the current at the time $t = 0$, and call it $i(0)$.

Integration of the LHS from $i(0)$ to $i(t)$ is trivial [FUNDAMENTAL THEOREM OF CALCULUS],

$$\text{LHS} = \int_{i(0)}^{i(t)} di' = i(t) - i(0)\,,$$

while integrating the RHS is straightforward:

$$\text{RHS} = \int_0^t \frac{V_0}{L}\sin(\omega t')\,dt' = \frac{V_0}{L}\left[\frac{-1}{\omega}\cos(\omega t')\right]\bigg|_0^t = -\frac{V_0}{\omega L}\Big(\cos(\omega t) - 1\Big)\,.$$

By requiring that the *a priori* equality of the left- and right-hand sides be preserved under self-consistent integration, we arrive at the following relation:

$$i(t) - i(0) = -\frac{V_0}{\omega L}\Big(\cos(\omega t) - 1\Big)\,.$$

The source of EMF which drives the current is assumed to be purely AC, with no DC component whatsoever, and thus the current in the circuit must oscillate about an average value of zero. Hence,

$$i(0) = -\frac{V_0}{\omega L} \quad \text{and} \quad i(t) = -\frac{V_0}{\omega L} \cos(\omega t) = -I_0 \cos(\omega t).$$

Let's pause to collect our thoughts and summarise.

✓   The sinusoidal driving voltage elicits a sinusoidal current response.

✓   The current oscillates at the same angular frequency as the driving voltage.

✓   The current and voltage oscillations, while correlated, are not in phase. This is best realised by inspection of the graphs of the trajectories, illustrated in Figure 43.2, and by examination of the functions $\{v_L(t), i_L(t)\}$.

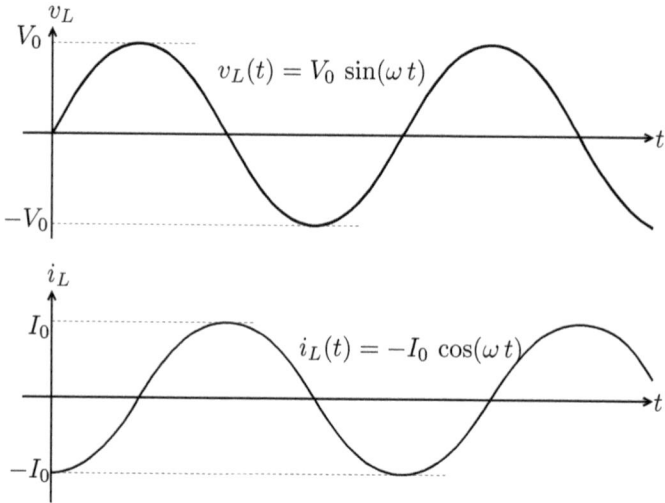

FIGURE 43.2   Correlated Voltage and Current Response in an Inductive AC Circuit.

To quantify the phase shift, one rewrites $i_L(t) = -I_0 \cos(\omega t)$ as

$$i_L(t) = I_0 \sin\left(\omega t - \tfrac{\pi}{2}\right),$$

from which it is evident that **the current through the inductor lags the voltage applied across it by one quarter cycle.**

✓   The amplitude of the current response is $I_0 = V_0/(\omega L)$. The current response is stronger, *i.e.*, has greater amplitude, at lower oscillation frequencies than at higher frequencies. It is not possible to take the limit $\omega \to 0$, because (1) this corresponds to DC, rather than AC, and (2) the small amount of electrical resistance, ignored in this analysis but necessarily present in the circuit, will limit the maximum steady state DC current attained.

As was shown in Chapter 42, the average values of the sinusoidal voltage and current vanish. The RMS prescription is adopted to obtain meaningful descriptive values of these variable quantities:

$$V_{\text{RMS}} = \sqrt{\langle v_L^2 \rangle_T} = \frac{V_0}{\sqrt{2}},$$

as was previously obtained in Chapter 42, and

$$I_{\text{RMS}} = \sqrt{\langle i_L^2 \rangle_T} = \frac{V_0}{\sqrt{2}\,\omega L} = \frac{V_{\text{RMS}}}{\omega L}.$$

The numerical factors of $\frac{1}{\sqrt{2}}$ are particular to the sinusoidal behaviour of the voltage source and the current response. The structure of the final expression for I_{RMS} inspires introduction of the INDUCTIVE REACTANCE.

INDUCTIVE REACTANCE   The inductive reactance, χ_L, is the constant factor relating the RMS and peak current responses to the driving voltages, *i.e.*,

$$I_{\text{RMS}} = \frac{V_{\text{RMS}}}{\chi_L} \quad \text{AND} \quad I_0 = \frac{V_0}{\chi_L} \qquad \Longleftrightarrow \qquad \chi_L = \omega L\,.$$

The inductive reactance has units $\left[\chi_L\right] = \left[\frac{\text{rad}}{\text{s}}\right]\left[\text{H}\right] = \left[\text{s}^{-1}\right] \times \left[\Omega \cdot \text{s}\right] = \Omega$, exactly as is necessary to relate volts to amperes. The identification of $1\,\text{H} = 1\,\Omega\cdot\text{s}$ comes from Chapter 37.

The inductive reactance is proportional to both the inductance of the circuit and the angular frequency of the driver. This latter dependence is consistent with Faraday's Law, since a more rapidly varying applied voltage will necessarily lead to a more rapidly varying current and thus a stronger back-EMF in the inductive element.

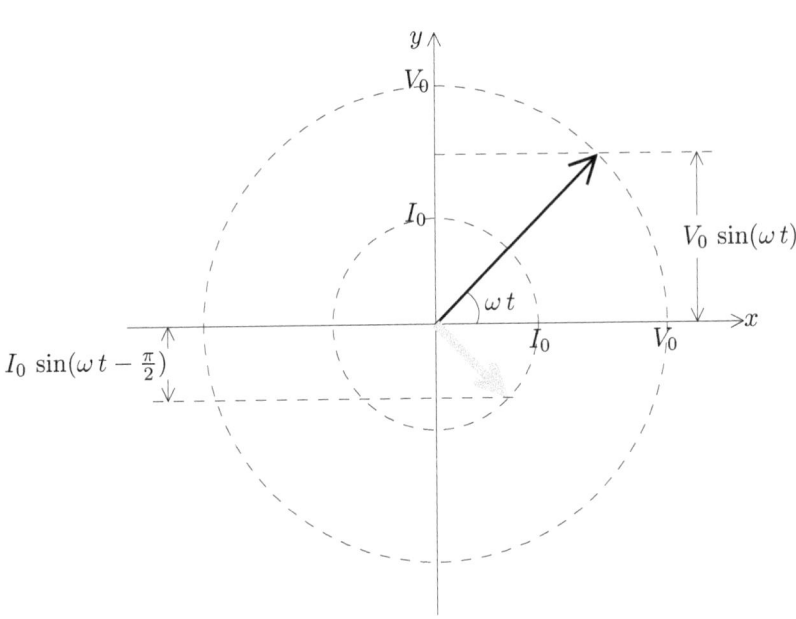

FIGURE 43.3   Phasor Diagram for a Simple Inductive AC Circuit

Figure 43.3 shows a phasor diagram appropriate to the present situation in which a purely inductive element lies in series with an AC voltage source. "Rods," of length V_0 and I_0

respectively, rotating rigidly about the origin at constant angular frequency ω and with fixed ($90°$) phase separation, are employed to indicate the instantaneous state of the circuit/system and to describe how it evolves in time. The source voltage [mirrored across the inductor] and the current in the circuit [passing through the inductor] are [with suitable/convenient choice of overall phase] identified with the y-axis projections of these voltage and current rods:

$$v_L(t) = V_0 \sin(\omega t) \quad \text{and} \quad i_L(t) = I_0 \sin(\omega t - \pi/2).$$

EXAMPLE [*AC Inductive Circuit*]

An AC voltage source with amplitude $V_0 = 10\,\text{V}$, operating at $50\,\text{Hz}$, is connected in series to a $200\,\text{mH}$ inductor.

(a) (i) Compute the inductive reactance of the system.

(ii) Determine the peak current passing through the inductor.

(b) Compute the RMS (i) voltage across and (ii) current through the inductor.

(c) Determine how the circuit properties change if the frequency of the AC source is (i) doubled and (ii) halved.

(a) (i) The inductive reactance is $\chi_L = \omega L = 50 \times 2\pi \times 0.2 = 2\pi\,\Omega$.

(ii) The peak current is $I_0 = \frac{V_0}{\chi_L} = \frac{10}{2\pi} = \frac{5}{\pi}\,\text{A}$.

(b) (i) The RMS voltage across the inductor is $V_{\text{RMS}} = V_0/\sqrt{2} = 5\sqrt{2}\,\text{V}$.

(ii) The RMS current passing through the inductor is $I_{\text{RMS}} = I_0/\sqrt{2} = \frac{5\sqrt{2}}{2\pi}\,\text{A}$.

(c) A change in the driving frequency of the source causes a proportional change in the inductive reactance. Hence, (i) doubling the frequency halves the peak and RMS currents, while (ii) halving the frequency results in a doubling of the current.

An AC voltage source is placed in series with a capacitive device as shown in Figure 43.4.

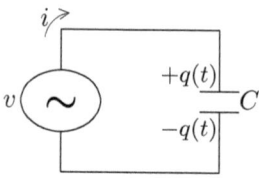

FIGURE 43.4 AC Source with a Purely Capacitive Load

Kirchoff's Loop Rule applied to this particular circuit gives rise to the following algebraic equation, solvable for the charge on the capacitor:

$$0 = v - \frac{q}{C} = V_0 \sin(\omega t) - \frac{q}{C} \quad \Longrightarrow \quad q(t) = C V_0 \sin(\omega t).$$

The current in the circuit is the sole source of the charge accumulated on the capacitor. Thus,

$$i_C(t) = \frac{dq}{dt} = \omega\, C\, V_0\, \cos(\omega\, t)\,.$$

The current response is sinusoidal, is at the source frequency although phase shifted (as is evident in Figure 43.5), and has amplitude $I_0 = \omega\, C\, V_0$.

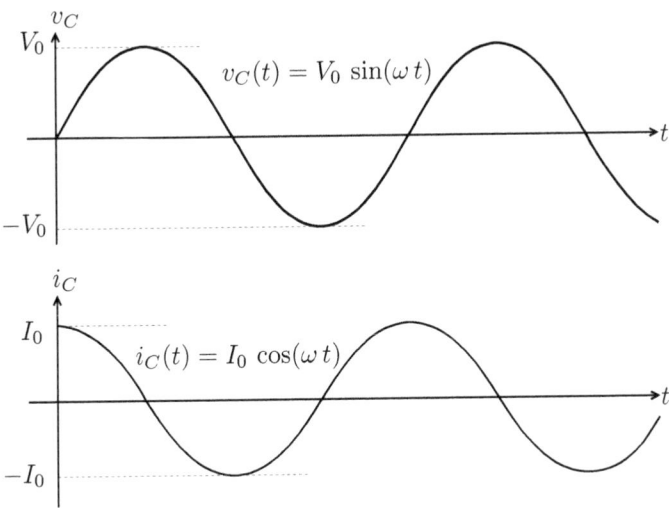

FIGURE 43.5 Correlated Voltage and Current Response in a Capacitive AC Circuit

To determine the phase shift, one rewrites $i_C(t) = I_0\, \cos(\omega\, t)$ as

$$i_C(t) = I_0\, \sin\left(\omega\, t + \tfrac{\pi}{2}\right),$$

thus making it evident that **the current through the capacitor appears to lead the voltage applied across it by one quarter cycle.**

> ASIDE: This provides a convenient way to describe the phase relation, but one must not ascribe this current change to conscious anticipation of the voltage change. Instead, its behaviour arises as a consequence of the circuit's ongoing response to the previous and present voltage conditions.

The amplitude of the current response is $I_0 = \omega\, C\, V_0$. It is smaller at lower oscillation frequencies than at higher frequencies. The DC-limit of zero frequency, in which the current vanishes, is attainable.

The RMS values of the voltage source and the current response are:

$$V_{\text{RMS}} = \sqrt{\langle v_L^2\rangle_T} = \frac{V_0}{\sqrt{2}} \quad\text{and}\quad I_{\text{RMS}} = \sqrt{\langle i_L^2\rangle_T} = \frac{\omega\, C\, V_0}{\sqrt{2}} = \frac{V_{\text{RMS}}}{\chi_C}\,.$$

The factor introduced into the final expression above is the CAPACITIVE REACTANCE.

CAPACITIVE REACTANCE   The capacitive reactance, χ_C, is the constant factor relating the RMS and peak current responses to the driving voltages, *i.e.*,

$$I_{\text{RMS}} = \frac{V_{\text{RMS}}}{\chi_C} \quad\text{AND}\quad I_0 = \frac{V_0}{\chi_C} \qquad\Longleftrightarrow\qquad \chi_C = \frac{1}{\omega\, C}\,.$$

The capacitive reactance has units

$$[\chi_C] = \left[\left(\frac{\text{rad}}{\text{s}}\right)^{-1}\right]\left[\text{F}^{-1}\right] = \left[\left(\frac{\text{C/s}}{\text{V}}\right)^{-1}\right] = \left[\left(\frac{1}{\Omega}\right)^{-1}\right] = \Omega\,,$$

exactly as is necessary to relate volts to amperes. [The identification of $1\,\text{F} = 1\,\text{C/V}$ occurs in Chapter 12, the definition of $1\,\text{C} = 1\,\text{A}\cdot\text{s}$ appears in Chapter 17, and the realisation that $1\,\Omega = 1$ V/A comes from Chapter 18.]

The capacitive reactance is inversely proportional to both the capacitance of the circuit element and the angular frequency of the driver. This stands to reason in both the low- and high-frequency domains.

LOW At low frequency, more time is available for the maximum charge to accumulate on the capacitor, and therefore this may be accomplished by a smaller average [and peak] current.

HIGH At high frequency, the charge on the capacitor must be loaded and offloaded with greater rapidity, and hence larger average [and peak] currents are necessary.

Figure 43.6 shows a phasor diagram in which a purely capacitive element lies in series with an AC voltage source.

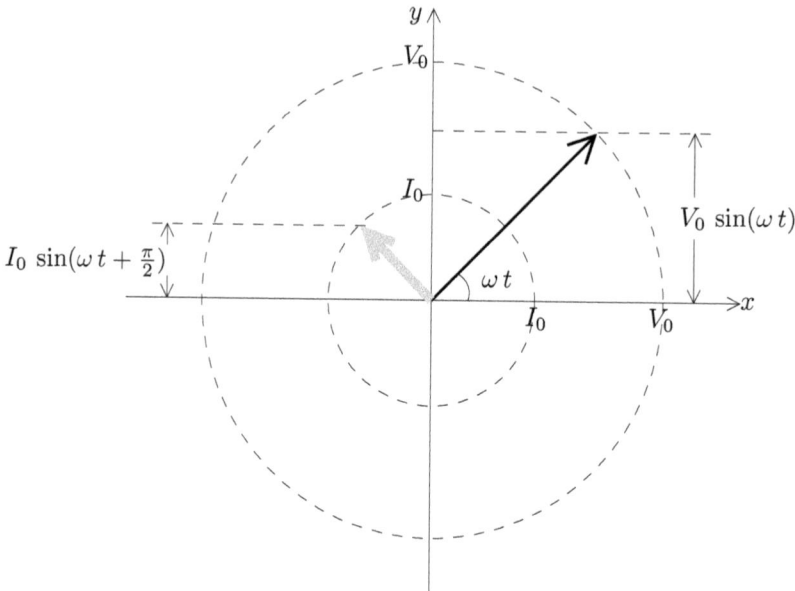

FIGURE 43.6 Phasor Diagram for a Simple Capacitive AC Circuit

Again employ "rods" of length V_0 and I_0, rotating rigidly at constant ω about the origin with fixed (90°) phase separation, to indicate the instantaneous state of the circuit/system and to describe how it evolves in time. The source voltage mirrored across the capacitor and the current in the circuit through the capacitor may be identified with the y-axis projections of the voltage and current rods:

$$v_C(t) = V_0 \sin(\omega t) \qquad \text{and} \qquad i_C(t) = I_0 \sin(\omega t + \pi/2)\,.$$

EXAMPLE [*AC Capacitive Circuit*]

An AC voltage source with peak voltage $V_0 = 10\,\text{V}$, operating at $50\,\text{Hz}$, is connected in series to a $2\,\text{mF}$ capacitor.

(a) (i) Compute the capacitive reactance of the system.

(ii) Determine the peak current passing through the capacitor.

(b) Compute the RMS   (i) voltage across and   (ii) current through   the capacitor.

(c) Determine how the circuit properties change if the frequency of the AC source is (i) doubled and   (ii) halved.

(a) (i) The capacitive reactance is $\chi_C = \frac{1}{\omega C} = \left(50 \times 2\,\pi \times 0.002\right)^{-1} = \frac{10}{2\,\pi} = \frac{5}{\pi} \simeq 1.59\;\Omega$.

(ii) The peak current is $I_0 = \frac{V_0}{\chi_C} = \frac{10}{\frac{5}{\pi}} = 2\,\pi\;\text{A}$.

(b) (i) The RMS voltage across the capacitor is $V_{\text{RMS}} = V_0/\sqrt{2} = 5\,\sqrt{2}\;\text{V}$.

(ii) The RMS current through the capacitor is $I_{\text{RMS}} = I_0/\sqrt{2} = \sqrt{2}\,\pi\;\text{A}$.

(c) A change in the frequency of the source causes an inversely proportional change in the capacitive reactance. Hence, (i) doubling the frequency doubles the peak and RMS currents, while (ii) halving the frequency halves the current.

ASIDE: The ever-observant reader will have noted that we did not consider energy flows in these purely inductive and capacitive AC circuits. In the absence of resistance it is difficult to conceive how net flow of energy might occur. Energetic issues are the subject of Chapter 45.

Chapter 44

RCL AC Circuits

Having considered purely resistive, inductive, and capacitive AC circuits in turn, let's go all out and place an AC voltage source in series with resistive, capacitive, and inductive elements characterised by R, C, and L.

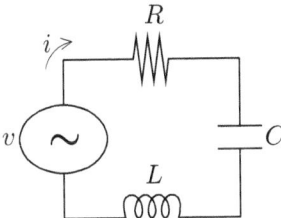

FIGURE 44.1   AC Source with Resistive, Capacitive, and Inductive Loads

Application of Kirchoff's Loop Rule is straightforward. Moreover, we recognise that the current is confined to the [thin] wires, and we exercise our freedom to choose $t = 0$ so that the phase angle associated with the AC voltage source vanishes. Hence,

$$0 = v - iR - \frac{q}{C} - L\frac{di}{dt} \quad \text{and} \quad v(t) = V_0 \sin(\omega t).$$

Taking the time derivative of this equation yields a valid expression. As R, C, and L are all [assumed to be] constant,

$$0 = \frac{dv}{dt} - R\frac{di}{dt} - \frac{1}{C}\frac{dq}{dt} - L\frac{d^2i}{dt^2}.$$

For this AC source, $\frac{dv}{dt} = \omega V_0 \cos(\omega t)$. The current in the circuit deposits charge on the capacitor, and thus

$$i = \frac{dq}{dt}.$$

Incorporating the explicit form of the voltage source and the capacitor's charge–current relation into the time-differentiated Kirchoff Loop relation yields a second-order differential equation for the current, $i(t)$:

$$\omega V_0 \cos(\omega t) = L\frac{d^2i}{dt^2} + R\frac{di}{dt} + \frac{i}{C}.$$

This looks rather like the differential equation[s] previously encountered when studying the DC RCL circuit in Chapter 41.[1] The substantive difference here is that this differential equation is NOT homogeneous, on account of the AC source of potential.

---

[1]Namely, the equation for the charge on the capacitor appearing in the body of Chapter 41, and the equation for the current in its Addendum.

As our interests lie in uncovering the physical phenomena [*i.e.*, solving the system, rather than investigating the mathematics of such equations], we propose an *Ansatz*,

$$i(t) = I_0 \sin(\omega t - \varphi).$$

The circuit dynamics following from the Kirchoff Loop Rule alone shall fix the values of the [constant] *Ansatz* parameters I_0 and φ. How one implements boundary conditions in order to express the complete and general solution of this equation is of secondary concern to us at this juncture.[2] The *Ansatz* already has a lot of physics built in:

- ω — The current response is assumed to **oscillate** about zero at the same frequency as the voltage source.
- I_0 — The amplitude of the current [an unknown parameter] is represented by I_0.
- φ — The parameter φ represents the relative phase of the current response with respect to the driving AC voltage.

  - IF $\varphi < 0$, THEN the current **leads** the voltage.
  - IF $\varphi = 0$, THEN the current and voltage are **in phase**.
  - IF $\varphi > 0$, THEN the current **lags** the voltage.

Given the *Ansatz*, the time derivatives of the current are easily evaluated:

$$i(t) = I_0 \sin(\omega t - \varphi) \implies \frac{di}{dt} = \omega I_0 \cos(\omega t - \varphi) \implies \frac{d^2 i}{dt^2} = -\omega^2 I_0 \sin(\omega t - \varphi).$$

Plugging these into the dynamical equation for the current response leads first to

$$\omega V_0 \cos(\omega t) = -\omega^2 L I_0 \sin(\omega t - \varphi) + \omega_0 R I_0 \cos(\omega t - \varphi) + \frac{1}{C} I_0 \sin(\omega t - \varphi),$$

and then to

$$0 = \omega V_0 \cos(\omega t) + \left(\omega^2 L - \frac{1}{C}\right) I_0 \sin(\omega t - \varphi) - \omega_0 R I_0 \cos(\omega t - \varphi).$$

This looks daunting until we remember the trigonometric identities which hold for sums and differences of angles:

$$\sin(\omega t - \varphi) = \sin(\omega t) \cos(\varphi) - \cos(\omega t) \sin(\varphi)$$
$$\cos(\omega t - \varphi) = \cos(\omega t) \cos(\varphi) + \sin(\omega t) \sin(\varphi).$$

First we expand, yielding a momentarily messy expression,

$$0 = \omega V_0 \cos(\omega t) + \left(\omega^2 L - \frac{1}{C}\right) I_0 \sin(\omega t) \cos(\varphi) - \left(\omega^2 L - \frac{1}{C}\right) I_0 \cos(\omega t) \sin(\varphi)$$
$$- \omega R I_0 \cos(\omega t) \cos(\varphi) - \omega R I_0 \sin(\omega t) \sin(\varphi),$$

---

[2] For those whose curiosity has been whetted, the solutions of the homogeneous [undriven] RCL-system, with their two *bona fide* constants of integration, may be added to the *Ansatz* solution in order to accommodate any initial conditions whatsoever. An astonishing feature of these systems is that the homogeneous solutions die off exponentially in time. Thus, there always exists a time scale on which the circuit will have [effectively] forgotten its initial conditions, and all that remains will be the *Ansatz* solution.

and then we regroup in terms of $\cos(\omega t)$ and $\sin(\omega t)$. This yields

$$
0 = \left[\omega V_0 - \left(\omega^2 L - \frac{1}{C} \right) I_0 \sin(\varphi) - \omega R I_0 \cos(\varphi) \right] \cos(\omega t)
$$
$$
+ \left[\left(\omega^2 L - \frac{1}{C} \right) I_0 \cos(\varphi) - \omega R I_0 \sin(\varphi) \right] \sin(\omega t)
$$
.

For the RHS of this expression to remain zero as time advances, it is necessary that the coefficients of the $\cos(\omega t)$ and $\sin(\omega t)$ terms separately vanish.

ASIDE: Formally, the coefficients vanish because **sine** and **cosine** are **orthogonal functions** [as discussed, within the context of Fourier analysis, in Chapter 22 of VOLUME II]. Here, we can assure ourselves of this by a heuristic argument.

Suppose that at some instant the expression vanishes. Achieving this perfect cancellation requires precise tuning of the [non-zero] coefficients of the **sine** and **cosine** terms. After the briefest of instants, both the **sine** and **cosine** terms will have changed slightly in manners which, although correlated, are not commensurate. Hence, there is no possible way the sum can remain zero at the later time. In fact, the only way to maintain the vanishing of the sum as time advances is to have each coefficient vanish separately.

The demand that both coefficients be identically equal to ZERO provides two algebraic equations for the *a priori* unknown quantities I_0 and φ:

$$
\#1 \qquad 0 = \omega V_0 - \left(\omega^2 L - \frac{1}{C} \right) I_0 \sin(\varphi) - \omega R I_0 \cos(\varphi),
$$

$$
\#2 \qquad 0 = \left(\omega^2 L - \frac{1}{C} \right) I_0 \cos(\varphi) - \omega R I_0 \sin(\varphi).
$$

#2 Let's first turn our attention to the second of these equations.

Cancelling the overall factor of I_0 and rearranging yields an equation for $\tan(\varphi)$ expressed in terms of ω, R, C, and L:

$$
\tan(\varphi) = \frac{\sin(\varphi)}{\cos(\varphi)} = \frac{\omega L - \frac{1}{\omega C}}{R} \ .
$$

The tangent of the phase angle is more elegantly formulated in terms of the inductive and capacitive reactances of the circuit, $\chi_L = \omega L$ and $\chi_C = (\omega C)^{-1}$, *i.e.*,

$$
\tan(\varphi) = \frac{\chi_L - \chi_C}{R} \ .
$$

A geometric mnemonic for the phase angle, φ, is provided by the IMPEDANCE TRIANGLE pertinent to the circuit.

IMPEDANCE TRIANGLE The impedance triangle provides a handy description of the form of the current response of an RCL circuit driven by an AC voltage source.

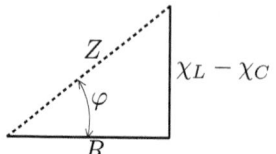

The impedance triangle has both RISE and RUN.

$$\text{rise} = \chi_L - \chi_C \,,$$
$$\text{run} = R \,.$$

The rise is the difference between the inductive and capacitive reactances, while the run is the resistance. The tangent of the slope angle is

$$\tan(\varphi) = \frac{\chi_L - \chi_C}{R} \,,$$

in exact accord with the expression for the phase shift. The hypotenuse, called the **impedance**, Z, is determined by Pythagoras:

$$Z = \sqrt{R^2 + \left(\chi_L - \chi_C\right)^2} \,.$$

The units of impedance are $[\Omega]$. It will become evident in a moment, when we analyse #1 to determine the amplitude of the current response, that the impedance provides the appropriate generalisation of simple DC resistance to AC circuits.

#1   Armed with the phase angle result from the solution of the second equation, we are now prepared to tackle the first equation [in order to determine the amplitude of the current]. Re-expressing the `cosine` and `sine` factors as

$$\cos(\varphi) = \frac{R}{Z} \qquad \text{and} \qquad \sin(\varphi) = \frac{\chi_L - \chi_C}{Z} \,,$$

substituting these into #1, and dividing by ω yields

$$0 = V_0 - I_0 \left[\left(\omega L - \frac{1}{\omega C} \right) \frac{\chi_L - \chi_C}{Z} + R \frac{R}{Z} \right] = V_0 - I_0 \left[\frac{R^2 + \left(\chi_L - \chi_C\right)^2}{Z} \right] \,.$$

The numerator in the bracketed term is precisely equal to Z^2, and thus

$$0 = V_0 - I_0 Z \qquad \Longrightarrow \qquad I_0 = \frac{V_0}{Z} \,.$$

Impedance is the generalisation of resistance.

There are two broad classes of behaviour, separated by a critical condition.

INDUCTIVE EFFECTS DOMINATE   IF $\chi_L > \chi_C$, THEN $\varphi > 0$, AND **current lags the voltage.** Consider the limit of a purely inductive circuit in which $C \to \infty$ [unbounded capacitance] and $R \to 0$ [no resistance], while the inductance has a finite positive value. In this instance, the impedance has magnitude $Z = \chi_L = \omega L$, and $\varphi = +\frac{\pi}{2}$.

CRITICALITY   IF $\chi_L = \chi_C$, THEN the impedance is minimised, $Z = R$, AND the voltage and the current in the circuit are **in phase.**

CAPACITIVE EFFECTS DOMINATE   IF $\chi_L < \chi_C$, THEN $\varphi < 0$, AND **current leads the voltage.** Consider the purely capacitive circuit in which both $L \to 0$ and $R \to 0$. Here, $Z = \chi_C = 1/(\omega C)$, and $\varphi = -\frac{\pi}{2}$.

Chapter 45

Power Dissipation in RCL AC Circuits

Prior to our conducting a rigorous analysis of the flows of power in RCL circuits driven by an AC source, let's think heuristically about the physics.

R   The current that flows in the circuit dissipates energy as it passes through the resistive element. [Recall that this conversion of electromagnetic energy into thermal energy is called Joule heating.]

C   The current flowing in the circuit deposits charge onto the capacitor until its stored energy reaches a maximum value of $U_E = \frac{1}{2} C V_0^2$. The potential drop across the capacitor is maximised[1] at the same instant[s] as the charge. When the voltage drop across the capacitor is reduced, charge flows off the plates and back into the circuit along with the stored electrostatic energy. Since the energetics is insensitive to the polarity of the capacitor, the energy *ebbs and flows* twice in each oscillation period, and hence at double the frequency.

L   The current in the circuit passes through the inductor and stores magnetic energy in the magnetic field in the space occupied by the magnetic field. The minimum amount of magnetic energy held in the inductor, *viz.*, zero, is present at those instants when the current vanishes. The maximum value, $U_M = \frac{1}{2} L I_0^2$, is attained when the current reaches an extremum, $\pm I_0$, since the amount of stored magnetic energy is insensitive to the direction of the current. Again, the *ebb and flow* of magnetic energy stored in the inductor occurs twice in each cycle of the driving voltage source.

Enough with the qualitative, let's get technical! The instantaneous power delivered to the circuit by the AC voltage source at time t is

$$
\begin{aligned}
P_{\text{in}}(t) &= i(t)\, v(t) \\
&= [I_0 \, \sin(\omega\, t - \varphi)]\, [V_0 \, \sin(\omega\, t)] \\
&= I_0\, V_0\, \sin(\omega t)\, \sin(\omega\, t - \varphi) \\
&= I_0\, V_0\, \sin(\omega\, t)\, \big(\sin(\omega\, t)\, \cos(\varphi) - \cos(\omega\, t)\, \sin(\varphi) \big) \\
&= I_0\, V_0\, \big(\sin^2(\omega\, t)\, \cos(\varphi) - \sin(\omega\, t)\, \cos(\omega\, t)\, \sin(\varphi) \big) .
\end{aligned}
$$

The AVERAGE power provided by the AC source throughout one complete cycle may be computed using the definition of the average value of any time-dependent quantity $f(t)$ over one oscillatory cycle:

$$
\langle f \rangle_{\mathcal{T}} = \frac{1}{\mathcal{T}} \int_0^{\mathcal{T}} f(t')\, dt' .
$$

---

[1] Actually, we mean that the magnitude of the potential difference is maximised, since both $\pm V_0$ are associated with maximal accrual of charge and energy.

In the computation of the average input power, SYMMETRIC INTEGRATION ensures that one contributing integral vanishes. The other evaluates to a constant.

$$\int_0^T \sin(\omega\, t')\, \cos(\omega\, t')\, dt' \equiv 0 \qquad \text{and} \qquad \frac{1}{T}\int_0^T \sin^2(\omega\, t')\, dt' \equiv \frac{1}{2}\,.$$

Thus, the average power input through one complete cycle [or, equivalently, a long time] is

$$P_{\text{av}} = \langle P \rangle_T = \frac{1}{2}\, I_0\, V_0\, \cos(\varphi) = I_{\text{RMS}}\, V_{\text{RMS}}\, \cos(\varphi)\,.$$

The quantity $\cos(\varphi)$ appearing in the above expression is called the **power factor** of the circuit. Let's muse on this latest result.

$\boldsymbol{P}_{\text{av}}$ Recall from the impedance triangle mnemonic, introduced in Chapter 44 and reproduced below, that

$$\tan(\varphi) = \frac{\chi_L - \chi_C}{R}\,.$$

In the limit that $R \to 0$ [within the generic situation where $\chi_L \neq \chi_C$], $\tan(\varphi) \to \pm\infty$. Thus, IF the resistance is zero [or vanishingly small], THEN the phase angle between the current and the driving voltage is $\varphi \to \pm\frac{\pi}{2}$. Also, $\cos\left(\pm\frac{\pi}{2}\right) = 0$, and hence **no net average electrical power is dissipated over a complete cycle when $R = 0$.**

$\boldsymbol{P}_{\text{av}}$ From the impedance triangle, it is tautological that

$$\cos(\varphi) = \frac{R}{Z}\,.$$

The additional fact that $V_{\text{RMS}} = I_{\text{RMS}}\, Z$ enables re-expression of the average power input to the circuit:

$$P_{\text{av}} = I_{\text{RMS}}\, V_{\text{RMS}}\, \cos(\varphi) = I_{\text{RMS}}\left(I_{\text{RMS}}\, Z\right)\left(\frac{R}{Z}\right) = I_{\text{RMS}}^2\, R\,.$$

This expression precisely mimics the form of the result for DC circuits obtained at the end of Chapter 19. This correspondence further bolsters the claim that the RMS characterisations of AC quantities admit interpretation as DC analogues. In this time-averaged sense, the intuitions that we have built up for DC circuits are applicable to AC situations.

> ASIDE: An added bonus[2] is that the expression for average power input to the circuit makes it apparent that the power vanishes as the load resistance approaches zero.

$\boldsymbol{P}_{\text{av}}$ Consider the special case arising when the load is purely resistive [or, more generally, when inductive and capacitive effects are precisely balanced].

> IF $\chi_L = 0 = \chi_C$ OR $\chi_L = \chi_C$,   THEN $\tan(\varphi) = 0$ AND   hence[3] $\varphi = 0$.

---

[2] A snark might suggest that we rephrase this as *addled* bonus.
[3] The other possible solution, $\varphi = \pi$, is extraneous because it corresponds to the unphysical situation of negative resistance, $R < 0$.

Under these conditions, the current response is precisely in phase with the driving AC voltage source, the POWER FACTOR is equal to 1, and the average power input to the circuit is $P_{\mathrm{av}} = I_{\mathrm{RMS}} V_{\mathrm{RMS}}$. The significance of this special case is that the input power [and hence its utilisation] is maximised [for the prescribed circuit elements and fixed amplitude AC voltage source]. Furthermore, the RMS and peak currents are also maximised.[4]

$$Z = \sqrt{R^2 + (\chi_L - \chi_C)^2} \quad \Longrightarrow \quad Z_{\min} = \lim_{\chi_L = \chi_C} Z = R$$

$$I_{\mathrm{RMS}} = \frac{V_{\mathrm{RMS}}}{Z} \quad \Longrightarrow \quad I_{\mathrm{RMS,max}} = \frac{V_{\mathrm{RMS}}}{Z_{\min}} = \frac{V_{\mathrm{RMS}}}{R}\,.$$

For the reactances to cancel,

$$\chi_L = \chi_C \quad \Longleftrightarrow \quad \omega\,L = \frac{1}{\omega\,C}\,.$$

This condition determines[5] the specific driving frequency,

$$\omega_0 = \sqrt{\frac{1}{LC}} = \frac{1}{\sqrt{LC}}\,,$$

for which the reactances cancel. Not surprisingly, this is the NATURAL FREQUENCY [a.k.a. the resonant frequency] of the LC part of the RCL circuit.

P_{av} While the RMS current is maximised for a given RCL circuit when the driving AC voltage is at the natural frequency of the circuit [determined by L and C only], the value of this maximum current depends on R, as is illustrated in Figure 45.1. Observe that the RMS current increases as the resistance decreases [at any fixed value of driving frequency], as expected.

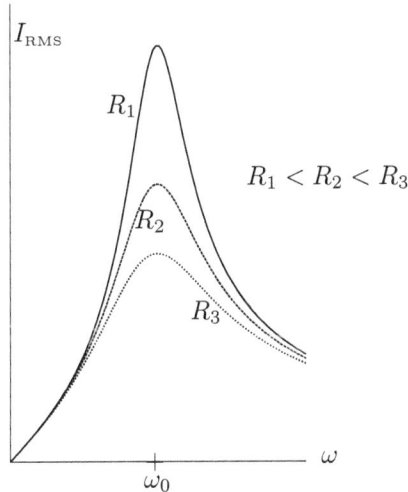

FIGURE 45.1 RMS Current Responses of AC RCL Circuits

---

[4]This occurs because the impedance is minimised when the reactances precisely cancel.
[5]The values of L and C are dictated by the circuit.

P_{av}   Expressions for the average power input to [and dissipated by] the circuit may be derived in terms of the circuit parameters R, L, and C, where $\omega_0^{-2} = LC$, along with the driving frequency and amplitude of the AC voltage source, ω and V_0. An example of such a derivation appears directly below:

$$P_{\text{av}} = I_{\text{RMS}}^2 \, R = \left(\frac{V_{\text{RMS}}}{Z}\right)^2 R = V_{\text{RMS}}^2 \, \frac{R}{Z^2} = V_{\text{RMS}}^2 \, \frac{R}{R^2 + \left(\chi_L - \chi_C\right)^2} \, .$$

To emphasise the rôle played by the frequency of the AC voltage source, we rewrite the power in terms of the circuit and voltage source parameters:

$$P_{\text{av}} = V_{\text{RMS}}^2 \, \frac{R}{R^2 + \left(\omega L - \frac{1}{\omega C}\right)^2} = V_{\text{RMS}}^2 \, \frac{R\,\omega^2}{R^2\,\omega^2 + \left(\omega^2 L - \frac{1}{C}\right)^2}$$

$$= V_{\text{RMS}}^2 \, \frac{R\,\omega^2}{R^2\,\omega^2 + L^2 \left(\omega^2 - \frac{1}{LC}\right)^2} = V_{\text{RMS}}^2 \, \frac{R\,\omega^2}{R^2\,\omega^2 + L^2\left(\omega^2 - \omega_0^2\right)^2} \, .$$

This result is best understood and appreciated by sketching it. Figure 45.2 shows the power input to the driven RCL circuit, as a function of the driving frequency ω, for three widely differing resistances.

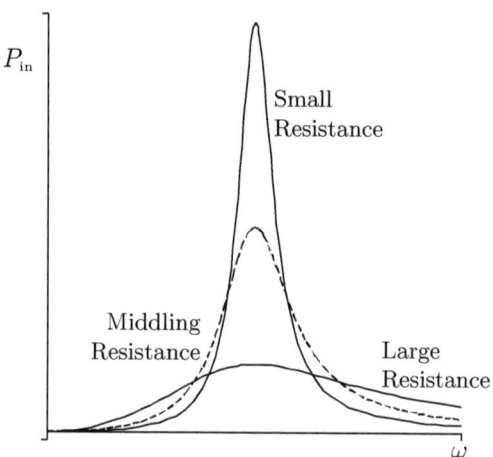

FIGURE 45.2   Input Power *vs.* Driving Frequency for Fixed $\{L, C, V_0\}$

The common features of these three curves are (looking first left, then right, and finally in the middle of the diagram):

0   As the driving frequency approaches zero the average power per cycle tends to zero. Suppose that a constant DC voltage[6] were to be applied to the RCL circuit. Current would flow[7] until the potential difference across the capacitor approached that applied

---

[6] The driving frequency is exactly zero for DC sources.

[7] Either onto or off the capacitor according to whether the applied voltage was greater than or less than the initial potential difference across the capacitor.

by the DC source. Recall our previous discussion in Chapters 22 and 23 of RC circuits in which the asymptotic current tends to zero irrespective of whether the system is charging or discharging.

∞   As the driving frequency increases without bound, the average power per cycle also diminishes toward zero. This is to be understood in terms of inductive effects. Increased driving frequency corresponds to more frequent switches in the direction of flow of the current in the circuit. The inductive component of the impedance increases along with the frequency. As the impedance grows, the peak and RMS currents diminish in inverse proportion. Since the average power depends on the square of the RMS current, and the [unchanging] resistance, R, it must diminish to zero in the limit of high driving frequency.

MID   Somewhere in the middle, the power must attain a maximal value [according to the MEAN VALUE THEOREM]. The figure suggests that the frequency at which the maximum occurs is insensitive to the value of the resistance in the circuit, even though the shape of the curve is strongly dependent upon R.

[Just ahead, we'll ascertain the location of the maximum of the power curve.]

Let's identify the frequency at which a particular power curve attains its maximum value by insisting that its derivative vanish at this point and solving for the corresponding value of the frequency. The mathematical details of this analysis are fraught with complication, but the procedure is completely straighforward.

ASIDE: To expedite matters, we shall have recourse to a trick which amounts to recognition that $df \equiv f \, d\ln(f)$, in cases where taking the logarithm is permitted, *i.e.*, $f > 0$.

$$
\begin{aligned}
\frac{dP_{\text{in}}}{d\omega} &= \frac{d}{d\omega}\left[\frac{V_{\text{RMS}}^2\, R\, \omega^2}{R^2\,\omega^2 + L^2\left(\omega^2 - \omega_0^2\right)^2}\right] \\
&= P_{\text{in}} \times \frac{1}{V_{\text{RMS}}^2\, R\, \omega^2} \times \frac{d}{d\omega}\left(V_{\text{RMS}}^2\, R\, \omega^2\right) \\
&\qquad - P_{\text{in}} \times \frac{1}{R^2\,\omega^2 + L^2\left(\omega^2 - \omega_0^2\right)^2} \times \frac{d}{d\omega}\left(R^2\,\omega^2 + L^2\left(\omega^2 - \omega_0^2\right)^2\right).
\end{aligned}
$$

The term deriving from the numerator simplifies dramatically,

$$
\frac{1}{V_{\text{RMS}}^2\, R\, \omega^2} \times \frac{d}{d\omega}\left(V_{\text{RMS}}^2\, R\, \omega^2\right) = \frac{V_{\text{RMS}}^2\, R\, 2\,\omega}{V_{\text{RMS}}^2\, R\, \omega^2} = \frac{2}{\omega},
$$

while the denominator-sourced term simplifies to a lesser extent,

$$
\begin{aligned}
\frac{\frac{d}{d\omega}\left(R^2\,\omega^2 + L^2\left(\omega^2 - \omega_0^2\right)^2\right)}{R^2\,\omega^2 + L^2\left(\omega^2 - \omega_0^2\right)^2} &= \frac{R^2\, 2\,\omega + L^2\, 2\left(\omega^2 - \omega_0^2\right) 2\,\omega}{R^2\,\omega^2 + L^2\left(\omega^2 - \omega_0^2\right)^2} \\
&= \frac{2\,R^2\,\omega + 4\,L^2\,\omega\left(\omega^2 - \omega_0^2\right)}{R^2\,\omega^2 + L^2\left(\omega^2 - \omega_0^2\right)^2}.
\end{aligned}
$$

Combining these to express the ω derivative of the input power spectral curve yields the following chain of equalities:

$$\frac{dP_{\text{in}}}{d\omega} = P_{\text{in}} \times \left[\frac{2}{\omega} - \frac{2R^2\omega + 4L^2\omega\left(\omega^2 - \omega_0^2\right)}{R^2\omega^2 + L^2\left(\omega^2 - \omega_0^2\right)^2} \right]$$

$$= P_{\text{in}} \times \frac{\frac{2}{\omega}}{R^2\omega^2 + L^2\left(\omega^2 - \omega_0^2\right)^2} \times$$

$$\left[R^2\omega^2 + L^2\left(\omega^2 - \omega_0^2\right)^2 - R^2\omega^2 - 2L^2\omega^2\left(\omega^2 - \omega_0^2\right) \right]$$

$$= P_{\text{in}} \times \frac{\frac{2L^2}{\omega}}{R^2\omega^2 + L^2\left(\omega^2 - \omega_0^2\right)^2} \times \left[-\left(\omega^2 - \omega_0^2\right)\left(\omega^2 + \omega_0^2\right) \right].$$

For an extremum of the power, we[8] insist that its derivative vanish. Hence,

$$\left. \frac{dP_{\text{in}}}{d\omega} \right|_{\omega = \omega_{\text{ex}}} = 0 \,.$$

Considering carefully the expression for the derivative, and recognising that the circuit parameters are all positive-valued (otherwise the system is unphysical or trivial), lead to the conclusion that the power curve has but one extremum [and thus a maximum] when the driving frequency matches the natural frequency of the LC part of the circuit, *viz.*,

$$\omega_{\text{ex}} = \omega_{\text{max}} = \omega_0 \,!$$

Thus, we have proven that the maximum power delivery to the circuit occurs when the driving frequency matches the natural frequency of the circuit. Furthermore, at this frequency [*a.k.a.* at resonance] this maximum power deposition attains the value

$$P_{\text{av,max}} = P_{\text{av}} \Big|_{\omega = \omega_0} = \frac{V_{\text{RMS}}^2\, R\,\omega_0^2}{R^2\,\omega_0^2 + \left(\text{ZERO}\right)^2} = \frac{V_{\text{RMS}}^2}{R} \,.$$

Although this is well and good, it is usually not quite sufficient to know merely the value of the resonant frequency, ω_0. It is also helpful to have a sense of the shape of the resonance power curve near its peak. A [conventional] measure of the sharpness of the peak is its FULL-WIDTH AT HALF-MAXIMUM, which is defined and illustrated just below.

FULL-WIDTH at HALF-MAXIMUM   The full-width at half-maximum is determined by ascertaining, via computation or inspection of the curve, the unique frequencies above and below the natural frequency at which the power is half of its peak value.

In Figure 45.3, the lower frequency at which the power attains half its maximal value is ω_l, while the upper frequency at which this occurs is ω_u. While these are situated on either side of ω_0, the resonant frequency, they are not symmetrically placed. The full-width at half-maximum is

$$\Delta\omega = \omega_u - \omega_l \,.$$

---

[8] Actually, it's not our idea. Fermat did it first.

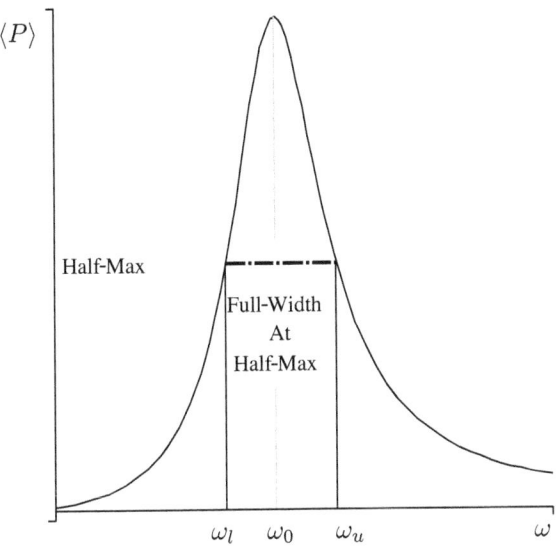

FIGURE 45.3 Full-Width at Half-Maximum for a Driven RCL Circuit

---

[CAVEAT: Some authors prefer to quote the half-width at half-maximum.]

IF the width is much smaller than the natural frequency [*i.e.,* $\Delta\omega \ll \omega_0$], THEN it turns out that $\Delta\omega \simeq \frac{R}{L}$. In this regime, the resonance can be made sharper and the maximum average power increased by decreasing the circuit's resistance. This trend continues as $R \to 0$, leading to a **delta-function**-like resonant spike, as illustrated in Figure 45.4.

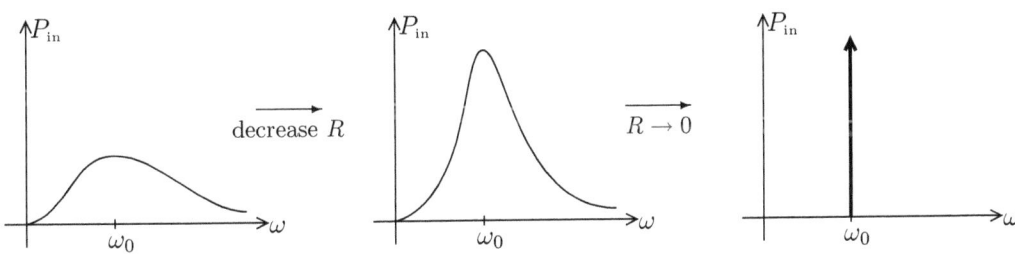

FIGURE 45.4 Decreasing R (with other parameters fixed) Sharpens the Resonance Peak

The full-width at half-maximum is a measure of the sharpness of the resonance. However, $\Delta\omega$ is dimensional [with SI units of radians per second]. Thus, it is acceptably useful for comparison of similar circuits, but less so for the comparison of rather different ones.

> ASIDE: A simple example illustrates the point. Suppose that two circuits each have $\Delta\omega = 10\,\mathrm{rad/s}$. These circuits are not at all alike if their resonant frequencies are $20\,\mathrm{rad/s}$ and 20 million $\mathrm{rad/s}$, respectively.

QUALITY FACTOR A relative and dimensionless characterisation of the shape of the resonance is afforded by the Quality Factor, which expresses the full-width at half-maximum in terms of the scale set by the natural frequency, *i.e.,*

$$Q = \frac{\omega_0}{\Delta\omega} \, .$$

High Q-factor circuits have resonant power curves which are strongly spiky, whereas low Q-factor circuits have broad and flat power curves.

> ASIDE: If $\Delta\omega \ll \omega_0$, then Q is very large and is well-approximated by
>
> $$Q \approx \frac{\omega_0}{(R/L)} = \frac{\omega\,L}{R} = \frac{\chi_L}{R} \, .$$

Many practical applications of RCL circuits exploit their resonant properties. For example, analogue radio and television receivers employ RCL tuning circuits. The wide variation in the cost of such tuners is often a reflection of the quality factors of the circuits.

> ASIDE: In the bad old days [before the advent of internet music streams] it was often essential for those with eclectic tastes in music to have a radio tuner with very narrow resonances. This was particularly true in large urban areas with a large number of stations "crowding" the spectrum. It was disconcerting [to say the least, and pardon the pun] to have one's source of good music overwhelmed by a high-powered station operating at a nearby frequency.

Continuing in this vein, **High-Q** behaviour is illuminated by the following poem.[9]

> *radio dial fill'd*
> *favour'd station crystal clear*
> *groovy tunes redux*

---

[9]Appropriately, by its 5–7–5 meter, this poem is a haiku.

Chapter 46

The Pinnacle: Maxwell's Equations

**Maxwell's Equations provide a four-equation summary
of all of classical electromagnetism!**

Note that we wrote ELECTROMAGNETISM rather than "electricity and magnetism." The reason for this is as simple as it is profound. In three spatial dimensions, the electric field, $\vec{E}$, and the magnetic field, $\vec{B}$, each have three *a priori* independent components. So, one might think that complete specification of both fields would require the solution of six [algebraic or differential] dynamical equations. That Maxwell achieves a complete description with four relations informs us that electricity and magnetism are but differing aspects of a single underlying natural phenomenon.

Maxwell's Equations [expressed in their integral representations]

Gauss's Law for Electric Fields reads

$$\oint_S \vec{E} \cdot d\vec{A} = \frac{q_{\text{encl.}}}{\epsilon_0} = \frac{1}{\epsilon_0} \int_V \rho_E \, dV \, , \quad \forall \, \{S, V\} \text{ such that } S = \partial V \, .$$

The LHS is the integral of the electric field normal to a closed[1] surface, S, enveloping a region of space V. The RHS is proportional to the integral of the local electric charge density over the region V. Both the LHS and RHS are equal to the net amount of electric charge enclosed by the surface, divided by the permittivity of free space. The *miracle* is that Gauss's Law holds for all pairs: $\{S, V\}$.

The implications of Gauss's Law for Electric Fields are that

    SOURCE   electric charges are sources for the electric field

    FIELDLINE   electric field lines may begin and end on charges, or form closed loops

Gauss's Law for Magnetic Fields reads

$$\oint_S \vec{B} \cdot d\vec{A} = 0 = \frac{1}{\mu_0} \int_V \rho_B \, dV \, , \quad \forall \, \{S, V\} \text{ such that } S = \partial V \, .$$

The LHS is the integral of the magnetic field normal to a closed[2] surface, S, bounding a region of space V. The RHS is the integral of the [putative] local magnetic charge density over the volume V. Both the LHS and RHS are identically equal to zero (for all pairs $\{S, V\}$)!

The implications of Gauss's Law for Magnetic Fields are that

---

[1] Not merely closed, but non-self-intersecting and orientable, too. We elide these additional conditions for concision and brevity.

[2] Ditto.

SOURCE   isolated magnetic charges [monopoles] are not found in nature

FIELDLINE   all magnetic field lines form closed loops, as there are no extant charges on which they might begin or end

Faraday's Law reads

$$\oint_\gamma \vec{E} \cdot d\vec{s} = -\frac{d\Phi_{m,S}}{dt} = -\frac{d}{dt} \int_S \vec{B} \cdot d\vec{A}, \quad \forall \ \{\gamma, S\} \text{ such that } \gamma = \partial S.$$

The LHS is the circulation of the electric field around a closed curve, γ. Recall that the line integral of an electric field along any path yields the potential difference between the endpoints of the path. The path is closed in this case, and yet the net circulation need not vanish. The integral appearing on the RHS yields the net magnetic flux passing through any[3] open surface S bounded by γ. According to Faraday [and Lenz], minus the time rate of change of the magnetic flux through S is equal to the circulation of the (induced) electric field about the boundary curve.

The primary implication of Faraday's Law is that changing magnetic flux through any surface [e.g., the disk bounded by a conducting loop] generates a back-EMF [which in turn may drive an electric current along the periphery in such a manner as to oppose the imposed change in magnetic flux].

Ampère's Law reads

$$\oint_\gamma \vec{B} \cdot d\vec{s} = \mu_0 \left(I_{\text{encl.}} + I_{\text{disp.}} \right) = \mu_0 I_{\text{encl.}} + \mu_0 \epsilon_0 \frac{d}{dt} \int_S \vec{E} \cdot d\vec{A}, \quad \forall \ \{\gamma, S\} \text{ with } \gamma = \partial S.$$

The LHS is the circulation of the magnetic field around a closed curve [Amperian circuit], γ. The RHS is comprised of two parts, both multiplied by μ_0 [the magnetic permeability of free space]. The first of these is $I_{\text{encl.}}$, the conduction current piercing through any surface S bounded by the Amperian circuit. The second part involves the DISPLACEMENT CURRENT, $I_{\text{disp.}}$, which is equal to the time rate of change of the electric flux through the same surface, S, multiplied by ϵ_0 [the electric permittivity of free space].

The implications of Ampère's Law are that a magnetic field is generated by the direct flow of electric charge [conduction current] or by a change in the electric flux passing through a surface [displacement current].

Maxwell's Equations are presented above in their integral form, and one might be concerned that these do not sufficiently adhere to the notion that all [CLASSICAL] physics ought to be local. Fortunately, any such worries are easily dispelled by quoting the differential form of Maxwell's Equations.[4]

Gauss's Law for Electric Fields reads

$$\vec{\nabla} \cdot \vec{E} = \frac{\rho_E}{\epsilon_0}.$$

---

[3]This is not quite accurate, as we again insist on non-self-intersection of both curve and surface and orientability of γ to avoid pathological cases.

[4]This is the manner in which they are most often expressed on T-shirts.

The dot product of the gradient operator acting upon a vector field is called the divergence [**Div**] of the field, and provides a measure of the degree to which the local field is *diverging* at a particular point in space. Hence, the LHS is **Div** $(\vec{E})$. The RHS consists of the electric charge density, ρ_E, at the same point in space where the divergence of the field is computed, divided by the permittivity of free space, ϵ_0. Electric field lines originate and terminate on electric charges.

Gauss's Law for Magnetic Fields reads

$$\vec{\nabla} \cdot \vec{B} = 0 \, .$$

The LHS is **Div** $(\vec{B})$, describing how the magnetic field lines are *diverging* at a point in space. That the RHS is zero everywhere informs us that there are no monopole charge sources for the magnetic field.

Faraday's Law reads

$$\vec{\nabla} \times \vec{E} = -\frac{d\vec{B}}{dt} \, .$$

The cross product of the gradient operator acting upon a vector field is called the **curl** of the field and measures the extent to which the field is wrapping around the point in space where the operator is applied. The LHS is **curl** $(\vec{E})$. The RHS is minus [owing to Lenz's Law] the time rate of change of the magnetic field at the same point in space.

A magnetic field which is changing in time at some point in space induces an electric circulation about that point.

Ampère's Law reads

$$\vec{\nabla} \times \vec{B} = \mu_0 \left(\vec{J} + \epsilon_0 \frac{d\vec{E}}{dt} \right) \, .$$

The LHS is **curl** $(\vec{B})$, and measures how the magnetic field lines are circulating around the particular point in space under consideration. The RHS is comprised of two types of sources.

DIRECT   The first is the electric current density[5] at the point in space, multiplied by μ_0. A term like this is absent from Faraday's Law as a consequence of Gauss's Law for Magnetic Fields. [The absence of magnetic monopoles nullifies the possibility of magnetic charge currents.]

DISPLACEMENT   The second term is the electric analogue of the Faradayan RHS, scaled by the constant factor $\mu_0 \, \epsilon_0$. The time rate of change of the electric field at a point in space gives rise to magnetic circulation about that point.

Both conduction current and changing electric fields induce magnetic circulation.

---

[5]Recall its introduction in Chapter 18.

Maxwell's unification of electricity and magnetism into ELECTROMAGNETISM has inspired many follow-on attempts to compose various disparate physical theories into—previously recondite—common structures. Einstein's searches for a unified field theory incorporating General Relativity and electromagnetism are emblematic of this urge. That Einstein's efforts were fruitless indicates the enormity and difficulty of the task of unification.

The understanding of electromagnetism achieved in the nineteenth century led to theoretical and technological advances in the early parts of the twentieth century which allowed glimpses of new scientific vistas. Chief among these were the discovery of various forms of radioactivity, and the accumulation of knowledge leading to the development of [Non-Relativistic] QUANTUM MECHANICS [NRQM] in the roaring '20s.

Subsequently, QUANTUM FIELD THEORY [QFT] emerged as the appropriate description of the subatomic constituents of matter. A significant, albeit partial, unification of electromagnetism with the WEAK NUCLEAR FORCES[6] was achieved by the so-called ELECTROWEAK THEORY.

In the 1970s the STANDARD MODEL OF PARTICLE PHYSICS took shape as a rather imperfect [and dare we say "inelegant"] amalgamation of the electroweak and STRONG NUCLEAR FORCE theories. To the present day, many physicists have grappled with the apparent immiscibility of General Relativity and QFTs. Many people hope that a theory of SUPERSTRINGS shall afford a complete and self-consistent unification of all of the fundamental forces appearing in nature.

Maxwell's Equations: *In Vacuo*

Let's take Maxwell's Equations, in their integral form, and apply them in a region of space devoid of electric charges and currents. In this instance, all of the source terms are entirely absent, and the equations specialise as quoted just below.

GAUSS ELECTRIC	$\oint_S \vec{E} \cdot d\vec{A} \;=\; 0$
GAUSS MAGNETIC	$\oint_S \vec{B} \cdot d\vec{A} \;=\; 0$
FARADAY	$\oint_\gamma \vec{E} \cdot d\vec{s} \;=\; -\dfrac{d}{dt} \oint_S \vec{B} \cdot d\vec{A}$
AMPÈRE	$\oint_\gamma \vec{B} \cdot d\vec{s} \;=\; \mu_0 \,\epsilon_0 \dfrac{d}{dt} \oint_S \vec{E} \cdot d\vec{A}$

The unimaginative might expect bland behaviour in this situation, reasoning that *"an absence of sources would suggest that nothing of note happens,"* but such sentiments are **wrong**. Instead, this set of equations admits propagating wavelike solutions! These are discussed in the remaining chapters and further *illuminated* in an Intermediate Electromagnetism class.

---

[6]The weak force is responsible for radioactivity.

Chapter 47

Analysis of Maxwell's Equations in Vacuum

At the end of the last chapter, we quoted the empty space [vacuum] versions of Maxwell's equations. It is not quite enough to have the equations which tell us HOW nature operates; we must also construct specific solutions of the dynamical equations in order to model particular phenomena. And so, we shall search for **wavelike solutions** to Maxwell's Equations. There is ample evidence, in mechanics and elsewhere, that wavelike solutions are important in their own right, and may be employed to construct[1] other time-dependent solutions. To begin, we'll make an *Ansatz* for the form of the electric and magnetic fields and the overall behaviour of the waves.

[This is not so much an *Ansatz* as it is a choice of favourable coordinates, as we shall see.]

Let us assume that the ELECTROMAGNETIC WAVE

TYPE   is **transverse**,[2] rather than **longitudinal**[3]

PLANE   is planar (invariant throughout parallel yz-planes)

DIRN   propagates in the x-direction (with wave number[4] $\vec{k} = k(t,\,x)\,\hat{\imath}$)

$\boldsymbol{E}$ has an electric field pointing exclusively in the y-direction, *viz.*, $\vec{\boldsymbol{E}} = E(t,\,x)\,\hat{\jmath}$

$\boldsymbol{B}$ and has a magnetic field directed along the z-axis, *viz.*, $\vec{\boldsymbol{B}} = B(t,\,x)\,\hat{k}$

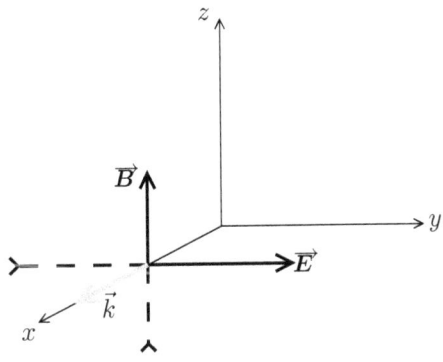

FIGURE 47.1   The Electric and Magnetic Components of a Travelling EM Wave

---

[1] Recall that the Joy of Fourier Analysis was experienced in Chapter 23 of VOLUME II.

[2] The disturbance characteristic of transverse waves lies orthogonal to the direction in which the wave propagates. Mechanical waves on a stretched string are transverse.

[3] Longitudinal wave disturbances are coincident in direction with the overall motion of these waves. Sound waves in air or another medium are longitudinal.

[4] Although it's called the "wave number," it really should be identified as a differential form.

In Figure 47.1 the arrows represent the magnitudes and directions of the electric and magnetic fields at the point in space at which they cross, *i.e.*, their midpoints. The concept of vector fields requires that we envision a crossed pairing of these fields resident at every point in space. The plane wave with fixed polarisation *Ansatz* assumptions enumerated above make it so that the vectors are uniform throughout parallel *yz*-planes. Consequently, there can be no explicit dependence on y and z appearing in the magnitude functions $E(t, x)$ and $B(t, x)$.

To apply Faraday's Law to this particular field configuration at the instant, t,

$$\oint_\gamma \vec{E} \cdot d\vec{s} = -\frac{d\Phi_{m,\mathrm{S}}}{dt},$$

consider a closed loop path in the form of a narrow rectangle of length l and width dx, as shown in Figure 47.2.

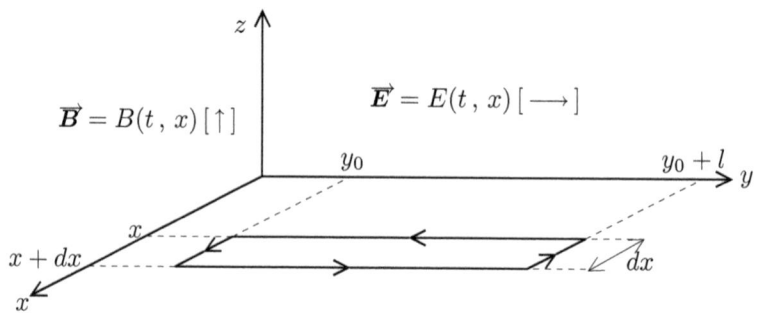

FIGURE 47.2 A Faradayan Circuit Lying in the *xy*-Plane

The LHS of Faraday's Law [the circulation of the electric field arising from a traversal of the loop] may be written as the sum of the contributions from each of the four straight segments:

$$\mathrm{LHS_{Far}} = \mathrm{Circ}\big(\vec{E}\,\big) = \left[\int_{\swarrow} + \int_{\rightarrow} + \int_{\nearrow} + \int_{\leftarrow}\right] \vec{E} \cdot d\vec{s}.$$

The electric field is expected to vary throughout the parts of the path running parallel to the *x*-axis, *viz.*, the [short] sides denoted by $\swarrow$ and $\nearrow$ in the path segmentation indicated in the computation of the circulation. However, the electric field is everywhere directed orthogonally to these short segments. Hence, the integrands for the *dx*-segment contributions to the LHS of Faraday's Law vanish, *i.e.*,

$$\vec{E} \cdot d\vec{s}\,\Big|_{\mathrm{short\ sides}} = \big[E(t, x)\,\hat{\jmath}\,\big] \cdot dx\,\hat{\imath} = E(t, x)\,dx\big[\hat{\jmath}\cdot\hat{\imath}\,\big] = 0,$$

and so too must the integrals:

$$\int_{\swarrow} \vec{E} \cdot d\vec{s} = 0 = \int_{\nearrow} \vec{E} \cdot d\vec{s}.$$

According to the plane-wave assumption, the electric field is constant everywhere along, and aligned with, the [long] segments of the path lying parallel to the *y*-axis. In this case, the line integrals along the two long sides do not vanish, but instead nearly cancel:

$$\int_{\rightarrow} \vec{E}(x + dx) \cdot d\vec{s} = \int_{\rightarrow} \big[E(t, x + dx)\,\hat{\jmath}\,\big] \cdot dy\,\hat{\jmath} = E(t, x + dx)\int_{\rightarrow} dy = E(t, x + dx)\,l,$$

$$\int_{\leftarrow} \vec{E}(x) \cdot d\vec{s} = \int_{\leftarrow} \big[E(t, x)\,\hat{\jmath}\,\big] \cdot dy\,\hat{\jmath} = E(t, x)\int_{\leftarrow} dy = -E(t, x)\,l.$$

Adding together these two [oppositely directed, slightly offset] contributions to the electric circulation yields

$$\text{Circ}\left(\vec{E}\right) = \left[E(t\,,\,x+dx) - E(t\,,\,x)\right]l$$

With a wee bit of squinting, or a second or two of pondering, one realises that the bracketed element on the RHS looks suspiciously ["auspiciously" is more apropos] like a differential,

$$E(t\,,\,x+dx) - E(t\,,\,x) \simeq \frac{\partial E}{\partial x}\,dx\,,$$

and thus, in the limit that the rectangle becomes increasingly narrow,

$$\text{LHS}_{\text{Far}} = \text{Circ}\left(\vec{E}\right) = \frac{\partial E}{\partial x}\,l\,dx\,.$$

Let us now consider the RHS of Faraday's Law. The time rate of change of the magnetic flux passing through surfaces bounded by the Faradayan circuit is

$$\text{RHS}_{\text{Far}} = -\frac{d\Phi_{m,\text{S}}}{dt} = -\frac{d}{dt}\int_{\text{S}} \vec{B} \cdot d\vec{A}\,.$$

The simplest [minimal] surface bounded by the rectangular loop is the slit-like rectangle, lying in the xy-plane, which appears shaded in Figure 47.2. The magnetic flux passing through this surface is

$$\Phi_{m,\text{S}} \simeq +B(t\,,\,x)\,l\,dx\,.$$

The details for this computation are expounded upon below.

Φ_m The magnetic field, $\vec{B}$, is assumed to point in the $+\hat{k}$ direction and to have magnitude dependent on t and x only.

Φ_m The area of the thin rectangular slit is $l \times dx$, and the **area vector** points perpendicular to the xy-plane, *i.e.*, in the direction of $\pm\hat{k}$. The orientation is determined by the sense in which the path is traversed, and assigned via a *handy* RHR.

> **RHR:** Align one's right hand along the Faradayan path, with fingers pointing in the forward direction and palm facing inward.
>
> [This recipe may be hard to follow for very complicated curves.]
>
> The general direction of one's right thumb then provides the sense in which the area vector has positive magnitude.

In this application of Faraday's Law, the area vector points in the $\hat{k}$ direction.

Φ_m Thus the field and area vectors are parallel, greatly simplifying the dot product involved in the calculation of the flux integrand. The **cosine** factor is $+1$ [and remains constant], contributing to the overall plus sign in the expression for flux.

Φ_m Over the very narrow width of the slit, dx, the magnetic field is effectively constant. This approximation becomes exact in the limit in which $dx \to 0$.

Taking all of these considerations into account, one arrives at the above-quoted expression for the magnetic flux, $\Phi_{m,S}$. It remains to take the partial derivative with respect to time. Here, unlike previous cases [*e.g.*, motional EMF], the loop dimensions and position are fixed and unchanging. The magnetic field alone is time-dependent. Hence,

$$\text{RHS}_{\text{Far}} = -\frac{d\Phi_{m,S}}{dt} \simeq -\frac{\partial}{\partial t}\Big[B(t,x)\,l\,dx\Big] = -\frac{\partial B}{\partial t}\,l\,dx\,.$$

The separate analyses of the LHS and RHS of Faraday's Law culminate when we set them equal to one another:

$$\text{LHS}_{\text{Far}} = \text{RHS}_{\text{Far}} \qquad \Longrightarrow \qquad \frac{\partial E}{\partial x}\,l\,dx = -\frac{\partial B}{\partial t}\,l\,dx\,.$$

As this relation holds irrespective of the specification of the dimensions of the loop, we may draw the succinct conclusion that

$$\text{VACUUM FARADAY} \qquad \Bigg| \qquad \frac{\partial E}{\partial x} = -\frac{\partial B}{\partial t}$$

relates the component functions of the electric and magnetic fields resident in the vacuum.

Now we apply Ampère's Law to this particular field configuration at time t:

$$\oint_{\gamma} \vec{\boldsymbol{B}} \cdot d\vec{s} = \mu_0\,\epsilon_0\,\frac{d\Phi_{e,S}}{dt}\,.$$

Let us choose a closed loop path in the form of the narrow rectangle of length l and width dx shown in Figure 47.3.

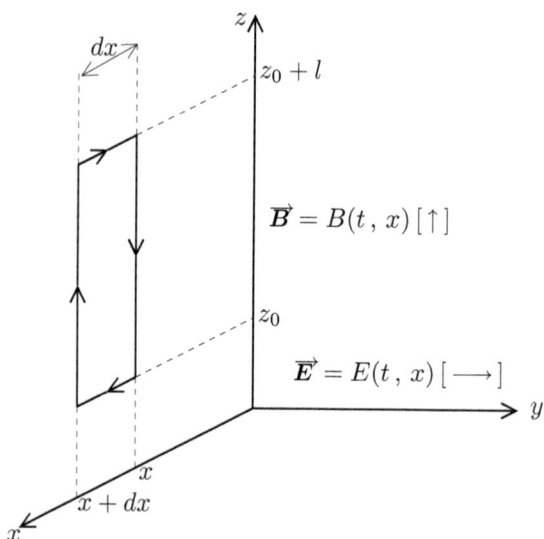

FIGURE 47.3   An Amperian Circuit Lying in the xz-Plane

The LHS of Ampère's Law [the circulation of the magnetic field when traversing the loop] is the sum of the contributions from the four straight segments, *i.e.*,

$$\text{LHS}_{\text{Amp}} = \text{Circ}\big(\vec{\boldsymbol{B}}\big) = \Bigg[\int_{\uparrow} + \int_{\nearrow} + \int_{\downarrow} + \int_{\swarrow}\Bigg]\vec{\boldsymbol{B}}\cdot d\vec{s}\,.$$

The magnitude of the magnetic field is expected to vary along the short path segments running parallel to the x-axis, $\nearrow$ and $\swarrow$. However, the direction of the magnetic field remains orthogonal to the path everywhere along these short segments. Hence these circulation integrands vanish, *i.e.*,

$$\vec{B} \cdot d\vec{s}\,\Big|_{\text{short sides}} = \left[B(t\,,x)\,\hat{k} \right] \cdot dx\,\hat{\imath} = B(t\,,x)\,dx\left[\hat{k} \cdot \hat{\imath} \right] = 0\,.$$

Thus, these contributions to the LHS of Ampère's Law are null:

$$\int_{\nearrow} \vec{B} \cdot d\vec{s} = 0 = \int_{\swarrow} \vec{B} \cdot d\vec{s}\,.$$

According to the plane-wave assumption, the magnetic field is constant everywhere along, and aligned with, the [long] segments of the path lying parallel to the z-axis. In this case, the line integrals along the two long sides do not vanish, but instead nearly cancel:

$$\int_{\uparrow} \vec{B}(x + dx) \cdot d\vec{s} = \int_{\uparrow} \left[B(t\,,x + dx)\,\hat{k} \right] \cdot dz\,\hat{k} = B(t\,,x + dx)\int_{\uparrow} dz = B(t\,,x + dx)\,l\,,$$

$$\int_{\downarrow} \vec{B}(x) \cdot d\vec{s} = \int_{\downarrow} \left[B(t\,,x)\,\hat{k} \right] \cdot dz\,\hat{k} = B(t\,,x)\int_{\downarrow} dy = B(t\,,x)\,(-l) = -B(t\,,x)\,l\,.$$

Added together, these two [oppositely directed, slightly offset] contributions to the magnetic circulation yield

$$\text{Circ}(\vec{B}) = \left[B(t\,,x + dx) - B(t\,,x) \right] l\,.$$

Again, the bracketed terms approximate a differential,

$$B(t\,,x + dx) - B(t\,,x) \simeq \frac{\partial B}{\partial x}\,\Delta x\,,$$

and, in the limit that the rectangle becomes increasingly narrow,

$$\text{LHS}_{\text{Amp}} = \text{Circ}(\vec{B}) = \frac{\partial B}{\partial x}\,l\,dx\,.$$

The RHS of Ampère's Law consists of the time rate of change of the electric flux passing through surfaces bounded by the Amperian circuit.

$$\text{RHS}_{\text{Amp}} = \mu_0\,\epsilon_0\,\frac{d\Phi_{e,\text{S}}}{dt} = \mu_0\,\epsilon_0\,\frac{d}{dt}\int_{\text{S}} \vec{E} \cdot d\vec{A}\,.$$

The simplest [minimal] surface is the shaded slit-like rectangle, lying in the xz-plane, which appears in Figure 47.3. The electric flux passing through this surface is

$$\Phi_{e,\text{S}} \simeq -E(t\,,x)\,l\,dx\,.$$

The details for this computation are enumerated as follows.

Φ_e The electric field, $\vec{E}$, is assumed to point in the $+\hat{\jmath}$ direction, and to have a magnitude which depends on t and x only.

Φ_e The area of the thin rectangular slit is $l \times dx$, and the **area vector** points perpendicular to the xy-plane. Application of the RHR mentioned earlier reveals that the area vector is aligned with $-\hat{\jmath}$.

Φ_e Thus, $\vec{E}$ and $\vec{A}$ are anti-parallel, producing a cosine factor of -1.

Φ_e Across the very narrow width of the slit, dx, the electric field is effectively constant. This approximation becomes exact in the limit in which $dx \to 0$.

Taking all of this into account in the generic formula for flux, we arrive at the above-quoted expression for $\Phi_{e,S}$. Inserting this into the RHS of Ampère's Law,

$$\mathrm{RHS}_{\mathrm{Amp}} = \mu_0\,\epsilon_0\,\frac{d\Phi_{e,S}}{dt} \simeq -\mu_0\,\epsilon_0\,\frac{\partial}{\partial t}\left[-E(t,\,x)\,l\,dx\right] = -\mu_0\,\epsilon_0\,\frac{\partial E}{\partial t}\,l\,dx\,.$$

The separate analyses of the LHS and RHS of Ampère's Law culminate when they are set equal:

$$\mathrm{LHS}_{\mathrm{Amp}} = \mathrm{RHS}_{\mathrm{Amp}} \qquad \Longrightarrow \qquad \frac{\partial B}{\partial x}\,l\,dx = -\mu_0\,\epsilon_0\,\frac{\partial E}{\partial t}\,l\,dx\,.$$

As this relation holds irrespective of the dimensions of the loop, we may conclude that

$$\textsc{Vacuum Ampère} \qquad \Bigg| \qquad \frac{\partial B}{\partial x} = -\mu_0\,\epsilon_0\,\frac{\partial E}{\partial t}$$

relates the component functions of the electric and magnetic fields resident in the vacuum.

Here is a quick recapitulation of this chapter's exciting developments.

♮   An *Ansatz* was chosen in which the electric and magnetic fields lay in fixed perpendicular directions, and only depended upon the mutually perpendicular coordinate and time.

♮   Plugging this *Ansatz* into the vacuum form of Maxwell's Equations led to a **coupled set of first-order partial differential equations.**[5]

♮   The **two** coupled equations are **linear** and **homogeneous**. Also, they involve **constant coefficients**.

In the next chapter we shall solve this pair of equations in general and construct a favoured class of explicit solutions.

---

[5]The keenly observant reader will note that the steps taken here amount to a derivation of the differential forms of Faraday's and Ampère's Laws [*cf.* Chapter 46] in the vacuum plane-wave context.

Chapter 48

Wavelike Solutions of Maxwell's Vacuum Equations

In the previous chapter, two of Maxwell's Equations, *viz.*, Faraday's and Ampère's Laws, were applied to the plane-wave *Ansatz*, and two coupled linear first-order partial differential equations with constant coefficients were obtained.

VACUUM FARADAY	$\dfrac{\partial E}{\partial x} = -\dfrac{\partial B}{\partial t}$
VACUUM AMPÈRE	$\dfrac{\partial B}{\partial x} = -\mu_0\, \epsilon_0\, \dfrac{\partial E}{\partial t}$

The standard trick, fruitfully employed to uncouple such sets of equations, is to

DIFFERENTIATE   Take the partial derivative [with respect to x, say] of one equation.

COMMUTE   Interchange the order of action of the distinct partial derivatives.

[For consistency, the functions must have continuous second-order derivatives.]

REPLACE   Identify a term in the commuted partially differentiated expression as being prescribed by the other equation and substitute.

voilà   A second-order differential equation for one function is obtained.

Let's put this algorithm into practice, acting upon the electric field. The first two steps produce the following sequence of results:

$$\frac{\partial}{\partial x}\Big[\text{VACUUM FARADAY}\Big] \implies \frac{\partial^2 E}{\partial x^2} = \frac{\partial}{\partial x}\left[-\frac{\partial B}{\partial t}\right] = -\frac{\partial^2 B}{\partial x\, \partial t} = -\frac{\partial}{\partial t}\left[\frac{\partial B}{\partial x}\right].$$

The rightmost term in the above contains a part which is identical to the LHS of the vacuum plane-wave Ampère equation. Substitution yields

$$\frac{\partial^2 E}{\partial x^2} = -\frac{\partial}{\partial t}\left[-\mu_0\, \epsilon_0\, \frac{\partial E}{\partial t}\right] = +\mu_0\, \epsilon_0\, \frac{\partial^2 E}{\partial t^2}.$$

Hence,

$$0 = \frac{\partial^2 E}{\partial x^2} - \mu_0\, \epsilon_0\, \frac{\partial^2 E}{\partial t^2}.$$

Two astonishing things have just occurred. The first is that the set of coupled first-order partial differential equations for $E(t, x)$ and $B(t, x)$ has been converted to a single, second-order partial differential equation for $E(t, x)$ alone. The second is that the particular equation obtained is **the wave equation**.

GENERAL PROPERTIES OF THE WAVE EQUATION

In the standard presentation of the 1-d wave equation,

$$0 = \frac{\partial^2 \Psi}{\partial x^2} - \frac{1}{c^2} \frac{\partial^2 \Psi}{\partial t^2},$$

$\Psi(t, x)$ denotes the **wavefunction**, while c is the [constant] **wave speed**. The wavefunction provides a complete description of the wave's salient features and its kinematics.

CLAIM: Any twice-differentiable function $f(z)$ of one argument, with either $z \equiv z_- = x - ct$ or $z \equiv z_+ = x + ct$, is a solution of the wave equation.

ASIDE: The PROOF of this claim relies on application of the chain rule. For the spatial part,

$$\frac{\partial f}{\partial x} = \frac{df}{dz} \frac{\partial z_\pm}{\partial x} = \frac{df}{dz} \qquad \Longrightarrow \qquad \frac{\partial^2 f}{\partial x^2} = \frac{\partial}{\partial x} \left[\frac{df}{dz} \right] = \frac{d^2 f}{dz^2} \frac{\partial z_\pm}{\partial x} = \frac{d^2 f}{dz^2}.$$

Meanwhile, for the temporal bit,

$$\frac{\partial f}{\partial t} = \frac{df}{dz} \frac{\partial z_\pm}{\partial t} = \frac{df}{dz} (\pm c) \qquad \Longrightarrow \qquad \frac{\partial^2 f}{\partial x^2} = (\pm c) \frac{d^2 f}{dz^2} \frac{\partial z_\pm}{\partial x} = (\pm c) \frac{d^2 f}{dz^2} (\pm c) = c^2 \frac{d^2 f}{dz^2}.$$

Assembling these results and substituting them into the wave equation shows that the RHS vanishes identically for all functions f, with the sole proviso that $f(z)$ possess continuous second-order partial derivatives.

This superabundance of solutions ought not to surprise us because one might imagine a very great many possible wave shapes and the solution set ought to accommodate them all.

Within this multitudinous splendour of shapes, we shall concentrate our attentions on the so-called **monochromatic harmonic waves**. The adjective **harmonic** connotes the property of being infinitely [and non-trivially] differentiable. A moment's thought leads to the realisation that only the `sine`, `cosine`, and `exponential` functions fit this bill. The exponentials are ruled out by their unbounded asymptotic behaviour.[1] This leaves only the `sine` and `cosine` options. Furthermore, as the natural inputs to these functions are measured in radians, we are inspired to rescale the z arguments to `rad` via

$$x - ct \quad \to \quad kx - \omega t, \qquad \text{and} \qquad x + ct \quad \to \quad kx + \omega t,$$

where $c = \omega/k$. **Monochromatic** specifies that only the pure `sine` or `cosine` functions with exact periodicity are to be considered. Rather than belabour this exposition,[2] we simply state that k, the **wave number**, and ω, the **angular frequency**, are each constant[3] and have dimensions of [`rad/m`] and [`rad/s`] respectively. The `sine` and `cosine` functions repeat with periodicity 2π in their arguments. Harmonic waves oscillate with a definite temporal **period**, $\mathcal{T}$, and spatial **wavelength**, λ. The angular frequency and the wave number convert specific time and space coordinates into [improper] phase fractions of the periodic wave:

$$\omega = \frac{2\pi}{\mathcal{T}} \qquad \text{and} \qquad k = \frac{2\pi}{\lambda}.$$

---

[1] We anticipate that phenomenologically interesting wave-like solutions might propagate to far distances. The exponential functions diverge in one x-direction or the other. As we have argued in Chapters 14 and 38, the electric and magnetic fields contribute to the local energy-density in proportion to their squared magnitudes. We would inevitably run afoul of energy-conservation principles should these fields diverge.

[2] A much more detailed presentation is found in VOLUME II.

[3] We shall not consider here more general situations in which k and/or ω are distribution-valued. The reader is invited to see for himself that linear superpositions of monochromatic harmonic waves are also valid solutions of the wave equation. This feature, essential to the Joy of Fourier Analysis, allows one to believe that restriction to just the monochromatic harmonic waves solutions is not overly limiting.

ASIDE: It's slightly more complicated than this. For instance, it is left unsaid that additional constant phase terms may be present, *i.e.*, $k\,x\pm\omega\,t+\varphi_\pm$. These are necessary for kinematical considerations, but irrelevant dynamically. We resolve to choose our coordinates and set our watches in such manner that $\varphi_\pm = 0$, so as to de-clutter the analysis.

It is oftentimes more practical to consider the **cycle frequency**,

$$\nu = \frac{1}{T} = \frac{\omega}{2\,\pi}\,.$$

Reconsideration of the rescaling of the generic function arguments, $z_\pm$, to make them amenable for input into harmonic functions reveals another expression for the wave speed,[4]

$$c = \frac{\omega}{k} = \frac{2\,\pi\,\nu}{\frac{2\,\pi}{\lambda}} = \nu\,\lambda\,.$$

Let us look, in some detail, at one of our proposed [monochromatic, harmonic, plane-wave] solutions:

$$\Psi(t\,,\,x) = \psi_L\,\cos(k\,x + \omega\,t) + \psi_R\,\cos(k\,x - \omega\,t)\,.$$

Coordinates were specifically chosen so as to cast the solution in terms of **cosine** functions alone, without constant additive phase terms. The amplitude factors, $\{\psi_L\,,\,\psi_R\}$, are *a priori* undetermined and correspond to the two constants of integration expected to appear in the solution of a second-order differential equation.

The two constituent parts of the monochromatic harmonic **travelling** plane wave are **leftmoving** and **rightmoving** respectively, and here they are distinguished by the labels *L* and *R*. Let's investigate the properties of the rightmover in some detail.

[All that we discuss will be effectively the same for the leftmovers, except for direction.]

Consider two "snapshots" of the wave, taken at instants t_s and $t_{s'} > t_s$, separated by a time interval which is much smaller than the period of the wave:

$$\begin{aligned}
\Psi_{Rs}(x) &= \Psi(t\,,\,x)\big|_{t=t_s} = \Psi(t_s\,,\,x) = \psi_R\,\cos(k\,x - \omega\,t_s) \\
\Psi_{Rs'}(x) &= \Psi(t\,,\,x)\big|_{t=t_{s'}} = \Psi(t_{s'}\,,\,x) = \psi_R\,\cos(k\,x - \omega\,t_{s'})
\end{aligned}\,.$$

These two functions are plotted on the same axes in Figure 48.1. The most noteworthy aspect of the pair of curves is that for every pair of corresponding phase points on the two waves, the later point lies to the right of its earlier counterpart.

[This is the origin of the adjective "rightmoving."]

---

[4]This form of the wave speed relation sets the stage for the classic joke found in VOLUME II.

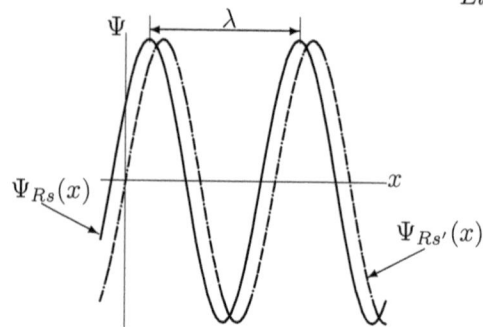

FIGURE 48.1   Two Snapshots of a Rightmoving Monochromatic Harmonic Wave

Considering the parametric dependence of the wave on both TIME and SPACE, we complement the snapshot [exhibiting the spatial dependence of the wave at fixed time] with a movie
close-up [illustrating the temporal dependence at a fixed point in space]. In fact, we'll take two
adjacent points, p and p', separated by a distance much smaller than one wavelength and
with $x_{p'} < x_p$ [*i.e.*, p' lies to the left of p]. The wave trajectories at these two points have
mathematical expressions given by:

$$\Psi_{Rp}(t) = \Psi(t,x)\big|_{x=x_p} = \Psi(t,x_p) = \psi_R \cos(k\,x_p - \omega\,t)$$
$$\Psi_{Rp'}(t) = \Psi(t,x)\big|_{x=x_{p'}} = \Psi(t,x_{p'}) = \psi_R \cos(k\,x_{p'} - \omega\,t)$$

These are plotted on the same axes in Figure 48.2. Here, the phase correspondence is such
that p' experiences each and every phase event slightly earlier than does the point p. This
is true for all pairs of points and, again, constitutes a consistent and coherent description
of what it means for a wave to be a rightmover.

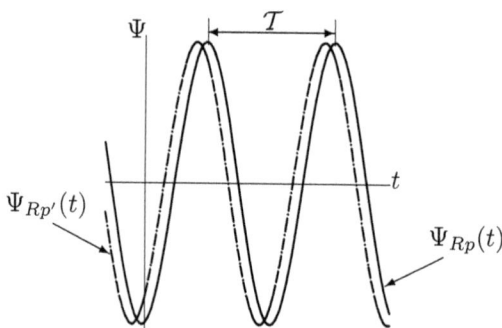

FIGURE 48.2   Two Point-Trajectories Associated with the Rightmoving Wave

---

Three comments are warranted before we return to the analysis of the electric field.

~   We avoid the need to explicitly consider leftmoving waves, too, because according to the *Ansatz* of Chapter 47 the wave solution propagates in the forward
direction along the positive x-axis.

~   "Snapshots" and "close-up movies" provide spatio-temporal projections of the actual wave motion. Even the simple case of monochromatic harmonic waves is rich with complexity.

~   More intricate wave patterns can be constructed from superpositions of waves with different frequencies/wave numbers and directions of travel.

MONOCHROMATIC ELECTRIC WAVE SOLUTIONS
OF MAXWELL'S VACUUM EQUATIONS

Matching the structure of the generic wave equation to the second-order differential equation for the magnitude of the electric field in the *Ansatz* subjected to Maxwell's equations *in vacuo* forces the identification:

$$\frac{1}{c^2} \equiv \epsilon_0 \, \mu_0 \qquad \Longrightarrow \qquad c = \frac{1}{\sqrt{\epsilon_0 \, \mu_0}} \, .$$

Taking this seriously, we input the SI values of the permittivity and permeability of free space into the expression on the RHS, obtaining

$$\frac{1}{\sqrt{\epsilon_0 \, \mu_0}} = \frac{1}{\sqrt{8.85 \times 10^{-12} \times 4\,\pi \times 10^{-7}}} \left[\frac{1}{\frac{C^2}{N \cdot m^2} \, \frac{T \cdot m}{A}} \right]^{1/2} \, .$$

The numerical factor is approximately $\left(\sqrt{0.1112124 \times 10^{-16}} \right)^{-1} \simeq 2.9986 \times 10^8$. Simplifying the units can be done in a variety of ways. For example, $1\,T = 1\,N/(A \cdot m)$, and therefore

$$\left[\frac{N \cdot m^2}{C^2} \, \frac{A}{T \cdot m} \right]^{1/2} = \left[\frac{m^2 \cdot A^2}{C^2} \right]^{1/2} = \left[\frac{m^2}{s^2} \right]^{1/2} = \left[\frac{m}{s} \right] \, .$$

In this chain of analysis, it was recognised that the unit of charge, the coulomb, is equal to one ampere flowing for one second, *i.e.*, $1\,C = 1\,A \cdot s$.

[Some small measure of relief is derived from the dimensional consistency of the units.]

To summarise, the speed of the electric field wave is

$$c \simeq 2.9986 \times 10^8 \ m/s \, .$$

Not only is this incredibly fast, but it conforms exactly[5] to the **speed of light!**

> ASIDE: This is a hugely significant result and constitutes circumstantial evidence for the claim that light is an electromagnetic wave.

---

[5] Any imprecision arises from the approximate value of the permittivity employed in the computation.

Although it might seem that we have had enough excitement already, the fact is that the magnetic field has been sorely neglected throughout all of this electric analysis. Remedying this oversight, we apply the algorithm employed earlier in this chapter to the magnetic field. Taking the spatial partial derivative of the VACUUM AMPÈRE relation and exchanging the order of derivatives result in

$$\frac{\partial}{\partial x}\Big[\text{VACUUM AMPÈRE}\Big] \implies$$

$$\frac{\partial^2 B}{\partial x^2} = \frac{\partial}{\partial x}\left[-\mu_0\,\epsilon_0\,\frac{\partial E}{\partial t}\right] = -\mu_0\,\epsilon_0\,\frac{\partial^2 E}{\partial x\,\partial t} = -\mu_0\,\epsilon_0\,\frac{\partial}{\partial t}\left[\frac{\partial E}{\partial x}\right].$$

The rightmost term in the above is identical to the LHS of the vacuum plane-wave Faraday equation. Substitution yields

$$\frac{\partial^2 B}{\partial x^2} = -\mu_0\,\epsilon_0\,\frac{\partial}{\partial t}\left[-\frac{\partial B}{\partial t}\right] = +\mu_0\,\epsilon_0\,\frac{\partial^2 B}{\partial t^2}.$$

Hence,

$$0 = \frac{\partial^2 B}{\partial x^2} - \mu_0\,\epsilon_0\,\frac{\partial^2 B}{\partial t^2}.$$

Two more astonishing things have occurred.

The magnetic and electric fields obey exactly the same wave equation! Both electric and magnetic waves travel at the speed of light!

Let's think a wee bit more about the significance of these astonishing results.

The speed of the waves is a physical constant, since both the permittivity and the permeability of free space are physical constants.

"No problem," one might think, *"there are plenty of examples of waves, like those on a string, which move at a constant speed."*

"Hold on," replies Einstein, *"the issue isn't that the speed is constant, but rather that it is a* universal constant*."*

Furthermore, the speed of a wave on a string is uniquely determined [by the string's mass-per-unit-length and tension, as shown in Chapter 26 of VOLUME II] in the IRF in which the string is at rest.

ASIDE: One must be a bit more careful. The preferred IRF is that in which the string's endpoints are fixed, since the propagation of the wave produces motion of the string. In addition, the amplitude of the wave ought not to be so large as to appreciably change the tension or the mass per unit length, because this would affect the wave speed.

Q: What is the medium in which the electric and magnetic Maxwell waves propagate?

A: People made tremendous efforts in the latter half of the nineteenth century trying to verify the existence of a substance, dubbed *luminiferous ether*, within which the electric and magnetic waves propagated. However ingenious their methods, the experimental results were always ambiguous, inconclusive, or contradictory!

The apparent speed of a wave on a string is IRF-dependent. Waves can be made to appear to move forward, backward, or stand still.

[Recognition of this was essential for deriving the wave speed in VOLUME II, Chapters 26 and 27.]

Einstein was cognizant of these issues and pondered the following question.

Q_E: How would one of Maxwell's waves appear to an observer running alongside it at the same speed as the wave [or at any other speed]?

A_E: Einstein's answer to this seemingly elementary question was revolutionary.[6]

EINSTEIN'S CLAIMS: **The speed of the waves has exactly the same value in every vacuum IRF. One can never run alongside a light wave.**

That is, the speed of the electric and magnetic waves is truly a physical constant and is independent of the observer and the inertial reference frame in which it is measured!

NOT MERELY ELECTRIC AND MAGNETIC

Thus far, we have shown that the electric and magnetic fields in the vacuum both satisfy exactly the same wave equation, and thereby propagate with the same speed. Even when we restrict attention to monochromatic harmonic magnetic field solutions, there is no *a priori* reason to assume that the $\{k, \omega\}$ pair for the electric wave has anything to do with the corresponding pair for the magnetic wave, *except* that we insist that the speeds of the waves, $c = \omega/k$, be equal.

There is also the issue of relative phase. By employing [and exhausting] the coordinate degrees of freedom we were able to write explicit electric field monochromatic harmonic wave solutions with zero additional constant phase shift. When analysing the magnetic solutions, the coordinate origins have been fixed (in the electric analysis), and one must anticipate that constant phase shifts will be needed to describe the left- and rightmoving constituents of the magnetic field. Otherwise unconstrained rightmoving solutions to the [separate, but identical] wave equations for the electric and magnetic fields have the forms:

$$E(t, x) = E_R \cos(k_E x - \omega_E t) \qquad \text{and} \qquad B(t, x) = B_R \cos(k_B x - \omega_B t + \phi_{BE}),$$

for constants $\{k_E, \omega_E; k_B, \omega_B; \phi_{BE}\}$.

CLAIM: **The monochromatic harmonic magnetic field must have the same angular frequency and wave number as, and must be precisely in phase with, the corresponding electric field.**

PROOF: Recall that our showing that the electric and magnetic fields satisfy the same wave equation began with separating the linkages present in Faraday's and Ampère's Laws. Let us revisit Faraday [or Ampère] and insist that it hold for all values of t and x [within the applicable spatio-temporal domain]. Explicit construction of the LHS and RHS of Faraday's Law yields the following:

$$\text{LHS}_{\text{Far}} = \frac{\partial E}{\partial x} = -k_E E_R \sin(k_E x - \omega_E t),$$

$$\text{RHS}_{\text{Far}} = -\frac{\partial B}{\partial t} = -\omega_B B_R \sin(k_B x - \omega_B t + \phi_{BE}).$$

---

[6]Depending on whether one choses to attribute the origins of the scientifically formalised concepts of space, time, and motion to Newton, Galileo, the Medievals, or the Ancient Greeks, Einstein's resolution of the SPEED OF LIGHT PROBLEM overthrew either centuries or millenia of received scientific wisdom.

For the moment, let us neglect the amplitude factors and focus our attention on the arguments of the **sine** functions. The only possible way that the equality of the LHS and RHS can be preserved for differing values of t and x is for

$$k_E = k_B = k, \quad \text{and} \quad \omega_E = \omega_B = \omega, \quad \text{and} \quad \phi_{BE} = 0.$$

These wave-like electric and magnetic fields do not propagate independently; rather, they are coupled.

The final task in this chapter is to expose the electromagnetic constraint which comes from equating the amplitude coefficients in the Faradayan relation, *i.e.*,

$$k E_R = \omega B_R \quad \Longrightarrow \quad E_R = \frac{\omega}{k} B_R = c B_R.$$

The amplitudes of the correlated electric and magnetic waves occur in fixed proportion.

The irrefutable implication of this analysis of the vacuum solutions of Maxwell's Equations is that electric and magnetic fields ought not to be considered in distinct isolation, but as separate manifestations of an overall electromagnetic field.[7]

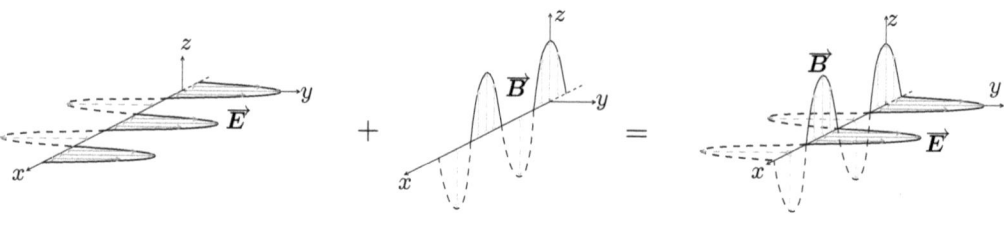

---

[7] The six components of $\boldsymbol{E}$ and $\boldsymbol{B}$ find their proper homes in the electromagnetic field 4×4-tensor, which, as it happens to be anti-symmetric, is just capacious enough to accommodate them. The fundamental electromagnetic object is a four-potential (vector), from which $\boldsymbol{E}$ and $\boldsymbol{B}$ may be consistently derived.

Chapter 49

The Poynting Vector

In the previous chapter we derived the explicit form of the monochromatic harmonic plane-wave solutions to Maxwell's Equations in vacuum. This achievement tells us two additional things: that Maxwell's Equations reduce to wave equations in this limit [by virtue of the fact that they admit propagating wave solutions], and that our *Ansatz* was reasonable [by virtue of the phenomenological relevance of these solutions]. Several conditions specific to our *Ansatz*, *viz.*, that the electric and magnetic fields are mutually orthogonal and that each lies perpendicular to the direction in which the wave propagates, turn out to be more generally valid.

The right-moving harmonic electromagnetic plane-wave solutions to Maxwell's Equations [discovered in Chapter 48] have the form

$$\left\{ \begin{array}{c} \vec{E}(t, x) \\ \vec{B}(t, x) \end{array} \right\} = \left\{ \begin{array}{c} E_R \cos(k\,x - \omega\,t)\,\hat{\jmath} \\ B_R \cos(k\,x - \omega\,t)\,\hat{k} \end{array} \right\}.$$

Numerous relations hold among the parameters and factors appearing in these solutions. A few of the most significant are listed here:

$$E_R = c\,B_R, \qquad c = \frac{\omega}{k} = \frac{1}{\sqrt{\mu_0\,\epsilon_0}} = \lambda\,\nu, \qquad \lambda = \frac{2\,\pi}{k}, \qquad \nu = \frac{\omega}{2\,\pi}, \qquad etc.$$

An amazing feature of these wave solutions is that they carry energy and momentum. The transport of electromagnetic energy is often characterised by the **Poynting vector**.

POYNTING VECTOR The Poynting vector [field], $\vec{S}(\mathcal{P})$, "points" in the direction in which electromagnetic energy is flowing at a point in space, $\mathcal{P}$. Its magnitude is the instantaneous power per unit cross-sectional area (orthogonal to the direction of flow) at $\mathcal{P}$. It so happens that $\vec{S}(\mathcal{P})$ is simply expressed in terms of the local electric and magnetic fields:

$$\vec{S}(\mathcal{P}) = \frac{1}{\mu_0}\,\vec{E}(\mathcal{P}) \times \vec{B}(\mathcal{P}).$$

The Poynting vector describes the intensity of the flux[1] of electromagnetic energy.

The SI units associated with $\vec{S}$ are

$$\left[\vec{S}\right] = \left[\frac{\text{T·m}}{\text{A}}\right]^{-1} \left[\frac{\text{N}}{\text{C}}\right] \left[\text{T}\right] = \left[\frac{\text{A·N}}{\text{m·C}}\right] = \left[\frac{\text{N·m}}{\text{m}^2\text{·s}}\right] = \left[\frac{\text{J/s}}{\text{m}^2}\right] = \left[\frac{\text{W}}{\text{m}^2}\right].$$

[Recall that the units of $\vec{J}$ are A/m^2.]

---

[1] In search of an analogy, we recollect that the electric current density, $\vec{J}$, introduced in Chapter 18, is the charge flux per unit cross-sectional area at points in space.

Constructing the Poynting vector associated with the monochromatic harmonic electromagnetic plane-wave solutions developed in the previous chapter is straightforward because the electric and magnetic fields are known precisely. Substituting yields

$$\vec{S}_{\text{P-W}} = \frac{1}{\mu_0} \left[E_R \cos(k\,x - \omega\,t)\,\hat{\jmath} \right] \times \left[B_R \cos(k\,x - \omega\,t)\,\hat{k} \right]$$

$$= \frac{E_R\,B_R}{\mu_0} \cos^2(k\,x - \omega\,t) \left[\hat{\jmath} \times \hat{k} \right]$$

$$= \frac{E_R\,B_R}{\mu_0} \cos^2(k\,x - \omega\,t)\,\hat{\imath}.$$

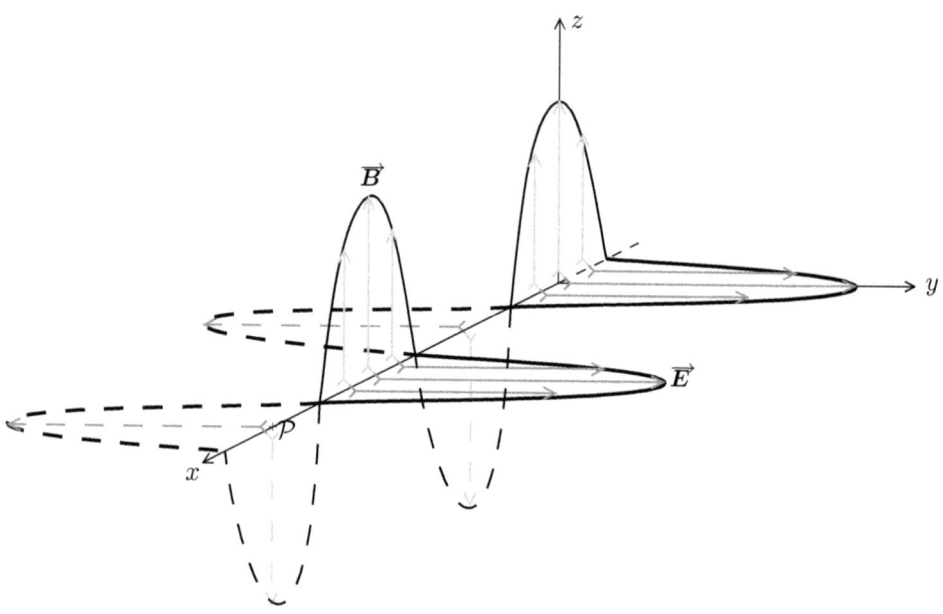

FIGURE 49.1   The Electromagnetic Plane Wave Fields

There are several comments to make about this determination of the Poynting vector.

DIRECTION   According to the computation, the flow of energy is everywhere directed along the positive x-axis. Explicit evaluation of the cross product of the paired electric and magnetic fields lying at the origin and at the point labelled $\mathcal{P}$ along the x-axis in Figure 49.1 provides quick corroboration of this general result.

[A flux of electromagnetic energy is directed forward along the path of the plane wave.]

ASIDE: Another aspect of the phase correlation between the electric and magnetic fields is revealed by this analysis. If either were phase-shifted by π the flow of energy would occur in the opposite direction!

The relative orientation of the fields presented in the *Ansatz* [proposed in Chapter 47] and our exclusive treatment of rightmoving waves were not coincidental.

DEPENDENCE The energy intensity varies with both position and time. Where and when the electric and magnetic fields are stronger, so too is the energy flow. This is not altogether surprising, since we have already observed that in static fields the electric and magnetic energy densities are proportional to the [squared] magnitudes of the respective fields. Here, the novel aspect is that the fields bear this energy with them as they propagate.

[This provides us with all the more reason to take the field concept seriously!]

Q: Wait! The static energy densities were proportional to the squares of the field amplitudes. Here, the Poynting vector depends on the product of the electric and magnetic amplitudes. Is this problematic?

A: Good question! One might invoke the fixed proportionality of the electric and magnetic components of the electromagnetic wave to re-write the product of their amplitude factors:

$$E_R \, B_R = c \, B_R^2 = \frac{E_R^2}{c} \, .$$

OSCILLATION The magnitude of the Poynting vector is positive indefinite and oscillates at twice the frequency [and wave number] of the electric and magnetic fields. This doubling arises because the electric and magnetic field energy densities are insensitive to the directions [orientations] of the fields.

These aspects of the plane-wave Poynting vector can be made explicit by invoking the identity

$$\cos^2(\theta) = \frac{1}{2}\big(1 + \cos(2\,\theta)\big)\,,$$

which applies even when the argument is $\theta = k\,x - \omega\,t$. The cosine-squared term oscillates sinusoidally, with half-amplitude about the central value of $\frac{1}{2}$, at twice the original frequency and wave number.

Faced with a [sinusoidally!] varying Poynting vector, we cannot help but wish to determine its AVERAGE value. *I.e.*,

$$\langle \vec{S} \rangle = \langle S_0 \, \cos^2(k\,x - \omega\,t)\,\hat{\imath} \rangle = S_0 \, \langle \cos^2(k\,x - \omega\,t) \rangle \, \hat{\imath}\,,$$

where

$$S_0 = \frac{E_R \, B_R}{\mu_0} = \frac{c\,B_R^2}{\mu_0} = \frac{E_R^2}{c\,\mu_0}\,.$$

The average may be taken over the spatial extent of one wavelength at fixed time, or throughout an interval of duration one period at a fixed point in space.

[These are the options which most clearly provide the meaningful average value.]

Either way, one inevitably obtains

$$\langle \cos^2(k\,x - \omega\,t) \rangle = \frac{1}{2}\,.$$

ASIDE: Should one be disinclined to perform the calculations, and steadfastly refuse to admit that this result follows by analogy with the averaging undertaken in the analysis of AC circuits in Chapters 42 through 45, then a third line of argument stems from the observation, made just above, that the variation in the squared-cosine term is sinusoidal [averaging to zero] about the central [residual] value of one-half.

Combining these results allows expression of the averaged Poynting vector in simplified form:

$$\langle \vec{S} \rangle = \frac{S_0}{2}\, \hat{i} \,.$$

The magnitude of the averaged Poynting vector [the magnitude of the average energy intensity] admits a variety of expressions:

$$\left| \langle \vec{S} \rangle \right| = \langle |\vec{S}| \rangle = \frac{E_R\, B_R}{2\,\mu_0} = \frac{c\, B_R^2}{2\,\mu_0} = \frac{E_R^2}{2\,c\,\mu_0} \,.$$

[The first equality owes to the plane wave's propagating purely in one direction.] The major result of this chapter, that the Maxwellian electromagnetic waves transport the energy content of the fields, is made incontrovertible by the argument which follows.

In the midst of averaging the Poynting vector one is inclined to reminisce about things RMS. Since the electric and magnetic fields are both pure sinusoidal,

$$E_{R,\text{RMS}} = \frac{E_R}{\sqrt{2}} \quad \text{and} \quad B_{R,\text{RMS}} = \frac{B_R}{\sqrt{2}} \,.$$

The generic local electric and magnetic energy densities [recalled from Chapters 14 and 38] are

$$u_E = \frac{1}{2}\, \epsilon_0\, E^2 \quad \text{and} \quad u_M = \frac{1}{2\,\mu_0}\, B^2 \,.$$

When averaged over the variable extent of the electric and magnetic wave fields, these forms are preserved, with replacement of the local fields by their RMS characteristic values,

$$\langle u_E \rangle = \frac{1}{2}\, \epsilon_0\, E_{R,\text{RMS}}^2 \quad \text{and} \quad \langle u_B \rangle = \frac{1}{2\,\mu_0}\, B_{R,\text{RMS}}^2 \,.$$

A curious and deeply significant property of nature is made evident in this last pair of relations when the amplitude constraint is invoked. Fixing $E_R = c\, B_R$ also sets $E_{R,\text{RMS}} = c\, B_{R,\text{RMS}}$, and hence

$$\langle u_E \rangle = \frac{1}{2}\, \epsilon_0\, E_{R,\text{RMS}}^2 = \frac{1}{2}\, \epsilon_0\, c^2\, B_{R,\text{RMS}}^2 = \frac{1}{2\,\mu_0}\, B_{R,\text{RMS}}^2 = \langle u_B \rangle \,.$$

[That the wave speed squared is equal to the reciprocal of the product of the permittivity and permeability was employed in the third step just above.]

The averaged electric and magnetic energy densities associated with the monochromatic electromagnetic plane wave are identically equal.[2]

---

[2]This feature may be viewed as another manifestation of the thermodynamic property of EQUIPARTITION of the energy of a system among its accessible [equivalent] degrees of freedom. Equipartition is discussed in VOLUME II.

Returning to the expression for the averaged Poynting vector, it is possible to write

$$\left| \langle \vec{S} \rangle \right| = \frac{c\, B_R^2}{2\,\mu_0} = \frac{c\, B_{R,\mathrm{RMS}}^2}{\mu_0} = 2\, c \left\langle u_B \right\rangle .$$

With a wee bit of *legerdemain* $[\, c^2\,\mu_0 = 1/\epsilon_0\,]$, it is also evident that

$$\left| \langle \vec{S} \rangle \right| = \frac{E_R^2}{2\,c\,\mu_0} = \frac{E_{R,\mathrm{RMS}}^2}{c\,\mu_0} = c\, \frac{E_{R,\mathrm{RMS}}^2}{c^2\,\mu_0} = c\,\epsilon_0\, E_{R,\mathrm{RMS}}^2 = 2\, c \left\langle u_E \right\rangle .$$

Tautologically, the total electromagnetic energy is formed by summing the electric and magnetic energies, *i.e.*,

$$u_{\mathrm{Total}} = u_E + u_B ,$$

and, since the operation of averaging is linear,

$$\left\langle u_{\mathrm{Total}} \right\rangle = \left\langle u_E + u_B \right\rangle = \left\langle u_E \right\rangle + \left\langle u_B \right\rangle = \frac{1}{2\,c} \left| \langle \vec{S} \rangle \right| \times 2 = \frac{1}{c} \left| \langle \vec{S} \rangle \right| .$$

Therefore,

$$\left| \langle \vec{S} \rangle \right| = c \left\langle u_{\mathrm{Total}} \right\rangle .$$

In other words, the magnitude of the average energy intensity of the propagating electromagnetic plane wave is equal to the average total electromagnetic energy density multiplied by the forward speed of the wave.

[This is perfectly generic wave behaviour, *cf.* Chapters 26 and 28 in VOLUME II.]

ASIDE: It is a relief to note that the Poynting vector's units are those of energy intensity:

$$\left[\, \left| \langle \vec{S} \rangle \right| \,\right] = [\, c\,]\, [\, u_{\mathrm{Total}}\,] = \left[\, \frac{\mathrm{m}}{\mathrm{s}}\, \frac{\mathrm{J}}{\mathrm{m}^3}\,\right] = \left[\, \frac{\mathrm{J/s}}{\mathrm{m}^2}\,\right] = \left[\, \frac{\mathrm{W}}{\mathrm{m}^2}\,\right] .$$

In this chapter, we have convincingly demonstrated that

plane-wave solutions of Maxwell's Equations (*in vacuo*) transport energy.

In the next chapter, we'll see that they bear momentum, too.

EXAMPLE [*Fun and Games with Electromagnetic Plane Waves*]

Three beams of electromagnetic [harmonic] plane waves are sent out from the origin in the positive x-direction. One is a soft x-ray, with frequency $\nu_X = 3.0 \times 10^{17}$ Hz. Another is a medium-high frequency radio wave, with frequency $\nu_R = 3.0$ MHz. The third is an ultra-low frequency (U) wave at 300 Hz. The electric field component of each wave has amplitude 600 V/m, and is aligned with the z-axis.

Q: What are some of the properties of these electromagnetic waves?

A: These properties fall into one (or more) of three classes: wave, electromagnetic, and energetic.

WAVE: Each wave has a distinct angular frequency, wavelength, and wavenumber.

$\omega \quad \omega = 2\pi\nu$

$$\omega_X = 6\pi \times 10^{17} \simeq 1.885 \times 10^{18} \text{ rad/s}$$
$$\omega_R = 6\pi \times 10^6 \simeq 1.885 \times 10^7 \text{ rad/s}$$
$$\omega_U = 600\pi \simeq 1885 \text{ rad/s}$$

$\lambda \quad \lambda = c/\nu$

$$\lambda_X = 10^{-9} = 1 \text{ nm} \qquad \lambda_R = 10^2 = 100 \text{ m} \qquad \lambda_U = 10^6 \text{ m}$$

$k \quad k = 2\pi/\lambda$

$$k_X = 2\pi \times 10^9 \simeq 6.283 \times 10^9 \text{ rad/m}$$
$$k_R = 0.02\pi \simeq 0.06283 \text{ rad/m}$$
$$k_U = 2\pi \times 10^{-6} \simeq 6.283 \times 10^{-6} \text{ rad/m}$$

ELECTROMAGNETIC: The magnetic component of the wave is highly correlated with the electric. For example, the frequency, wavelength, *etc*, are precisely the same; the amplitude is proportional; and the direction is prescribed by the direction of the electric component and the direction in which the wave propagates.

$B_0 \quad B_0 = E_0/c$. Therefore, $B_0 = 2 \times 10^{-6}$ T is the same for all three waves.

$\hat{B}_0 \quad \vec{B} \perp \vec{E}$ and $\vec{B} \perp \hat{\imath}$

ENERGETIC: The flow of energy associated with the wave is described by the Poynting vector. This may be averaged to obtain the *constant* rate of flow which bears an equivalent amount of energy per complete oscillation of the wave. The electric and magnetic field components of the wave possess an average [spatial] energy density. The average energy intensity of the wave is consistent with the forward passage of the average total energy in the fields at the wave speed c.

$\langle \vec{S} \rangle \quad \vec{S} = \frac{1}{\mu_0}\vec{E} \times \vec{B}$. The energy intensity is exactly the same for all three beams, *i.e.*, $|\langle \vec{S} \rangle| = \frac{1}{\mu_0} E_{\text{RMS}} B_{\text{RMS}} = \left(4\pi \times 10^{-7}\right)^{-1}\left(300\sqrt{2}\right)\left(\sqrt{2} \times 10^{-6}\right) = \frac{1500}{\pi} \simeq 477.5 \text{ W/m}^2$.

$\langle u_E \rangle \quad$ The average amount of energy stored in the electric component of the wave is $\langle u_E \rangle = \frac{1}{2}\epsilon_0 E_0^2 \simeq 7.965 \times 10^{-7} \text{ J/m}^3$.

$\langle u_B \rangle \quad$ The average energy stored in the magnetic component of the wave is $\langle u_B \rangle = \frac{1}{2\mu_0} B_0^2 \simeq 7.958 \times 10^{-7} \text{ J/m}^3$. That the numerical values of $\langle u_E \rangle$ and $\langle u_B \rangle$ differ slightly stems from the imprecision of the values that we use for c and ϵ_0.

SYNTHETIC   The stored energy [density] moves forward at c. Thus, the energy intensity is $[\langle u_E \rangle + \langle u_B \rangle]\, c \simeq (7.965 + 7.958) \times 30 \simeq 477.7 \text{ W/m}^2$. Again, the slight discrepancy observed between the energy intensity values arises from our using approximations to physical constants.

Upon studying this example, one might be tempted to reason *falsely* thus:

All electromagnetic waves (with a particular amplitude) possess the same energy density and provide the same energy intensity irrespective of their wave properties.

This [erroneous] conclusion has disastrous thermodynamical implications when one tries to reconcile the EQUIPARTITION PRINCIPLE with the realisation that objects [at $T > 0$ K] radiate electromagnetic energy. [*Cf.* the "ultraviolet catastrophe" mentioned in VOLUME II, Chapter 41.] The granularity of allowed energies inherent in quantum mechanics saves the day.

Chapter 50

Electromagnetic Waves Carry Momentum, Too

At this juncture we encounter a limitation of the edifice of CLASSICAL MECHANICS which has been so painstakingly mapped[1] throughout these three volumes.

First, we shall recount a bit of history. Then, we'll state the special relativistic expression for the momentum associated with light [a type of massless particle]. With the *correct* answer in hand [thanks to Einstein], we will paraphrase an analysis attributed to Maxwell which effectively yields the same result.

ASIDE: The greatest scientists are somehow able to deduce TRUTHS about nature, even while working in a context of incorrect or deficient models!

To end the chapter and this VOLUME, we'll take a brief look at a scientific gizmo.

Immediately after Maxwell unified electricity and magnetism into electromagnetism and discovered the vacuum plane-wave solutions [studied in the past few chapters], it was realised that Maxwell's Equations and Newtonian Mechanics were incompatible. The universality claimed for the speed of electromagnetic waves is in glaring conflict with the GALILEAN RELATIVITY PRINCIPLE assuring the equivalence of inertial observers.

The scientific consensus was that Maxwell's Equations were in need of modification to bring them into accord with the rest of [Classical] Physics. The physics community was acting in an appropriate, properly conservative manner by requiring the newer, less well explicated theory to conform to the structures [and strictures] of that day's STANDARD MODEL.

ASIDE: By Maxwell's time (*circa* 1860), Newtonian Mechanics had been in place for about 200 years, during which its domain of application had steadily increased. Physical models to describe particular phenomena were in good agreement with increasingly precise experimental observations. Synergies with Mathematics, especially in the study of differential equations, were particularly fruitful. THERMODYNAMICS, long thought to be a separate branch of science, was subsumed into mechanics through application of statistical methods to systems comprised of large numbers of [classical] particles.

[Maxwell was a leading luminary in the development of statistical mechanics.]

In this context, it seemed unthinkable that Newtonian Classical Mechanics should be supplanted, especially given the absence of any credible alternative. However, to be fair, scientific investigations carried out in the latter years of the nineteenth century [spurred in no small part by the incorrigibility of Maxwell's Equations] suggested the presence of "clouds on the horizon," in Lord Kelvin's memorable turn of phrase.

---

[1] MAPSED, SPAMMED, and AMPSED, one might quip!

Einstein realised that LIGHT—Maxwell's electromagnetic waves—was the key to reconciling Newton with Maxwell. The result of his labours was the SPECIAL THEORY OF RELATIVITY [SRT], published in 1905.

> ASIDE: In 1905,[2] Einstein produced FOUR spectacularly important papers, addressing different aspects of physics. The title of the RELATIVITY paper is [usually translated as] *On the Electrodynamics of Moving Bodies.*

In SRT, Einstein postulated that the speed of light assumes the same value in all inertial reference frames and deduced the main consequences of this radical proposal. Although controversial at the time, SRT is self-consistent and its predictions have been borne out in countless experiments, including those in which the pre-Einsteinian Galileo-Newton model of nature is falsified.

One consequence of Einstein's discoveries is the realisation that space and time are more properly viewed as observer-dependent aspects of **spacetime**. In analogous ["dual"] fashion, energy and momentum are components of a generalised **four-momentum**. The geometric structure at the heart of spacetime enacts a constraint,

$$E^2 = \left|\vec{p}\right|^2 c^2 + m_0^2 c^4\,,$$

among the various components of the four-momentum.[3] The LHS is the square of the total energy of the moving particle as measured by an observer in a particular IRF. Two summed terms appear on the RHS. The first contains the squared magnitude of the momentum of the particle, as measured in the same IRF as the energy. The second term is the square of the rest-mass energy. It is present for all observers, in all frames, with the same numerical value!

> ASIDE: This constraint is consonant with the famous mass–energy[4] relation $E = mc^2$. Here, the mass is $m = \gamma m_0$, in terms of the Lorentz scale-factor $\gamma = 1/\sqrt{1 - \frac{v^2}{c^2}}$ and the rest-mass m_0. The [linear 3-] momentum is $\vec{p} = m\vec{v} = \gamma m_0 \vec{v}$. A quick derivation starts with the square of Einstein's mass–energy relation:
>
> $$E^2 = \gamma^2 m_0^2 c^4 \quad \Longrightarrow \quad E^2\left(1 - \frac{v^2}{c^2}\right) = m_0^2 c^4 \quad \Longrightarrow \quad E^2 = \frac{E^2 v^2}{c^2} + m_0^2 c^4\,.$$
>
> Remembering that $E = \gamma m_0 c^2$ allows us to transform the first term on the RHS to
>
> $$\frac{E^2 v^2}{c^2} = \gamma^2 m_0^2 v^2 c^2 = p^2 c^2\,.$$

The relativistic four-momentum constraint is applicable to light, even though light particles [in a very loose sense of the term] possess a rest-mass which is identically zero and travel with speed c in all inertial frames. For light moving solely in the x-direction, it is thus determined that

$$E^2 = p_x^2 c^2 \quad \Longrightarrow \quad E = p_x c \quad \Longrightarrow \quad p_x = \frac{E}{c}\,.$$

That is, the magnitude of the forward momentum associated with light is equal to the energy that the light bears, divided by its forward speed.

---

[2]This year is considered Einstein's *Annus Mirabilis*, or miracle year, as 1666 was Newton's.

[3]A snarky fellow might remark that in this particular expression, E is not the magnitude of an electric field vector, $\vec{p}$ is not an electric dipole moment vector, nor is m a magnetic moment. We shan't.

[4]This member of 1905's "big four" bore the title: *Does the Inertia of a Body Depend on Its Energy Content?*

Maxwell "derived" the same [sort-of] result decades before Einstein! Despite his argument's being "fake" there is an element of *truthiness* in it which deserves our attention.

A microscopic particle of mass m, bearing charge q, is released from rest at the point $(x_0, 0, 0)$ at the instant $t = 0$, in the presence of a rightmoving electromagnetic plane wave of the type studied in Chapters 47–49. Both the electric and magnetic fields associated with the wave are time- and space-dependent:

$$\vec{E} = E_R(t, x)\, \hat{j} \qquad \text{and} \qquad \vec{B} = B_R(t, x)\, \hat{k}.$$

Just after the particle is released, it is subject to electric and magnetic fields acting with strengths $E_0 = E_R(0, x_0)$ and $B_0 = B_R(0, x_0)$. We presume that no other fields are present and that no other forces act on the particle.

ASIDE: This analysis is applicable for short timescales only: $t \in [0, \epsilon]$, where ϵ is a very small number. Also, as the particle is originally at rest, it remains within a small distance of its original position.

Owing to the electric field, the particle experiences a net force (at $t = 0$),

$$\vec{F}_{\mathrm{E}} = q\,\vec{E} = q\,E_0\,\hat{j}.$$

In response to this force, the particle begins to move, with initial acceleration

$$\vec{a}_0 = \frac{q\,E_0}{m}\,\hat{j}.$$

This [transverse] electric acceleration preserves the value of the x-coordinate of the particle's position, x_0. The local electric field at any instant is constant throughout all planes parallel to the yz-plane. For very short timescales, as are assumed in this analysis, the magnitude of the field does not change appreciably, and hence the initial force and acceleration may both be deemed constant. Two related effects ensue. One stems from the kinematical relation between constant acceleration and linearly increasing component(s) of velocity:

$$v_y = v_{0y} + a_y\,t \qquad \Longrightarrow \qquad v_y(t) \simeq \frac{q\,E_0\,t}{m}$$

in the present case. The other comes from the observation that the net change in a component of the momentum, the [projected] impulse, is produced by the average force acting in that direction throughout the time during which the change occurred, multiplied by the time interval. Hence,

$$\Delta p_y = F_{\mathrm{av}, y}\,\Delta t \qquad \Longrightarrow \qquad p_y(t) \simeq q\,E_0\,t.$$

[It is a relief to see that these two related results are consistent: $p_y(t) = m\,v_y(t)$.]

Once the particle begins to move in response to the electric field, it begins to experience a Lorentz force owing to the presence of the magnetic field.[5] The Lorentz force, $\vec{F}_{\mathrm{M}} = q\,\vec{v} \times \vec{B}$, increases in proportion with the velocity of the particle. In the situation considered here, the magnetic force can be approximated as

$$\vec{F}_{\mathrm{M}} \simeq q\,\frac{q\,E_0\,t}{m}\,B_0\,\left[\hat{j} \times \hat{k}\right] = \frac{q^2\,E_0\,B_0\,t}{m}\,\hat{i}.$$

Two aspects of this force prove to be noteworthy.

---

[5] "*Don't be cross*," might smirk the snark.

$+\hat{\imath}$ The magnetic force [effectively] acts in the forward x-direction only throughout the time interval being considered. [Given more time, the particle would gain forward speed, causing there to be a magnetic force acting in the $-\hat{\jmath}$ direction.]

~ 0 The magnitude of the magnetic force is very small, irrespective of the strength of the field, B_0, owing to the condition that t be very small.

> ASIDE: The electromagnetic plane-wave field amplitudes are fixed in proportion to the speed of light. This is a large factor, and helps ensure that the magnetic effect on the particle is small compared to the electric effect.

Another way to see this is that the particle, starting from rest, does not have sufficient time to acquire a velocity significant enough to produce an appreciable magnetic force.

The magnetic force produces an impulse in the x-direction, and, since $p_x(0) = 0$,

$$p_x(t) = \Delta p_x(t) = \int_0^t F_{\mathrm{M}}(t')\, dt' = \frac{q^2 \, E \, B}{m}\left[\int_0^t t'\, dt'\right] = \frac{q^2 \, E \, B}{m}\left[\frac{t^2}{2}\right] = \frac{q^2 \, E \, B \, t^2}{2\, m}\,.$$

To verify that the forward motion is subordinate to the transverse, let's compare the components of the momentum of the particle. The momenta occur in proportion, *i.e.*,

$$\frac{p_x(t)}{p_y(t)} = \frac{\frac{1}{2\,m}\, q^2 \, E_0 \, B_0 \, t^2}{q\, E_0 \, t} = \frac{1}{2}\,\frac{q}{m}\, B_0\, t\,.$$

The numerical, charge-to-mass ratio, and magnetic field strength factors are innocuous, as, alone or in conjunction, they do not significantly constrain the relative magnitudes of the momentum components. In contrast, the factor of t ensures that the forward momentum scale is much smaller than the transverse momentum scale. [This is quite consistent with all the claims made thus far.] In summary, the momentum acquired by the particle, in the time interval from 0 to t, is

$$\vec{p}(t) = \left(p_x\,,\, p_y\,,\, p_z\right) = \left(\frac{q^2 \, E_0 \, B_0 \, t^2}{2\, m}\,,\, q\, E_0\, t\,,\, 0\right) = \left(O(t^2)\,,\, q\, E_0\, t\,,\, 0\right)\,.$$

Since the particle is in motion, it possesses a certain amount of kinetic energy,

$$K(t) = \frac{\left|\vec{p}(t)\right|^2}{2\, m} \approx \frac{q^2 \, E_0^2 \, t^2}{2\, m} + O(t^4)\,.$$

With a wee bit of squinting, it proves possible to rewrite the forward momentum gained by the particle in terms of its kinetic energy, *i.e.*,

$$p_x(t) = \frac{1}{2}\,\frac{q^2}{m}\, E_0 \, B_0 \, t^2 = K(t)\,\frac{B_0}{E_0} = \frac{K}{c}\,.$$

The [Maxwellian] constraint relating the electric and magnetic field amplitudes in the plane-wave solutions, $E_0 = c\, B_0$, was used to arrive at the final equality.

Thus, we have determined that, in the short time interval in which the Maxwell wave has interacted with the [originally stationary] charged particle, the particle has gained kinetic energy and momentum.

[Now comes the *Gestalt* (reversing figure and ground) trick!]

**The energy and momentum gained by the particle
via interaction with the wave must be offset by
reductions in the energy and momentum borne by the wave.**

Adapting our perspective to consider the momentum and energy lost by the electromagnetic field, we see that

$$\Delta p_{\text{lost}} = \frac{\Delta E_{\text{lost}}}{c},$$

or, as this might readily be re-imagined, $E = pc$!

> ASIDE: While Maxwell's "derivation" is rather bogus, this does not detract one whit from his achievement. On the contrary, it takes a special kind of genius to obtain the correct conclusions from an incorrect theory or model!

Q: What physical effects exist owing to the momentum carried by electromagnetic waves?

A: Radiation pressure is a manifestation of the momentum transfer accompanying absorption and reflection of electromagnetic waves. To see how this works, consider the following pair of *Gedanken* experiments.

Part A: Suppose that a *packet of light* strikes a surface whereupon it is completely absorbed.

[A surface that is painted flat black acts as a pretty good absorber.]

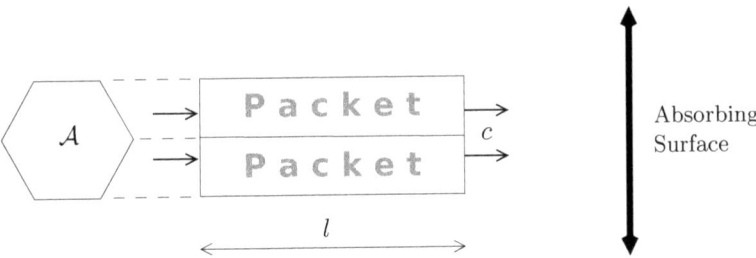

FIGURE 50.1 A Packet of Light about to Strike an Absorbing Surface

The following assumptions are made in a physical model of the absorption of the light packet.

— The light being absorbed consists of plane waves bearing energy and momentum striking the surface at normal [perpendicular] incidence. These can be partitioned into little packets.

— The light packet has cross-sectional area $\mathcal{A}$.

  The packets are hexagonal so that they can pack closely (in a space-filling manner, leaving no gaps) when assembled into the plane waves. [Just kidding!]

— The packet is deemed to have spatial extent [length] l.

  The process of absorption occurs during the brief time interval $\Delta t = l/c$.

ASIDE: The notion of a light packet employed here is merely an artifice of partitioning to localise the interaction of the light wave with the surface in both space and time. This analysis does not propose a particulate or granular aspect to light. We leave that to Einstein, again! The title of his 1905 PHOTO-ELECTRIC EFFECT paper is *On a Heuristic Point of View about the Creation and Conversion of Light.*

The average force per packet, acting perpendicular to the absorbing surface, is equal to the change in momentum of the packet divided by the time through which the change occurred. I.e.,

$$F_{\mathrm{av}} = \frac{\Delta p}{\Delta t}$$

[*cf.* Chapter 29 of VOLUME I]. All of the packet's momentum, $p = E/c$, is transferred to the surface, and $\Delta t \approx l/c$. Thus, an estimate of the average perpendicular force exerted by the light packet upon the absorber is

$$F_{\mathrm{av}} \approx \frac{E/c}{l/c} = \frac{E}{l} \, .$$

The average pressure is obtained by dividing the cross-sectional area of the packet into the average force. I.e.,

$$P_{\mathrm{av}} = \frac{F_{\mathrm{av}}}{\mathcal{A}} \approx \frac{E}{l \, \mathcal{A}}$$

[*cf.* Chapter 4 in VOLUME II]. The quantities E and $l \, \mathcal{A}$ are the total energy of and the volume associated with the light wave packet. Hence their ratio is the average electromagnetic energy density, $\langle u_{\mathrm{tot}} \rangle$, associated with the Maxwellian wave. Therefore,

$$P_{\mathrm{av}} = \langle u_{\mathrm{tot}} \rangle \, .$$

By means of this analysis, the pressure exerted by each [local] packet is expressed entirely in terms of the large-scale properties of the electromagnetic wave!

In Chapter 49, the Poynting vector $\vec{S}$ [a field-based measure of electromagnetic energy intensity] was expressed in terms of the average energy density of the wave. There it was determined that

$$\langle |\vec{S}| \rangle = c \langle u_{\mathrm{tot}} \rangle \, ,$$

and hence, when the wave [at normal incidence] is completely absorbed, the average pressure exerted on the absorber is

$$P_{\mathrm{av}} = \frac{\langle |\vec{S}| \rangle}{c} \, .$$

Part R: Suppose that a light packet is perfectly reflected from a surface, as in Figure 50.2.

[A shiny surface (*e.g.*, a mirror) makes a good reflector.]

Employing the same physical model as before, it is evident that the momentum transferred to the wall, and hence the pressure, is twice that which was determined in the completely absorbing case.

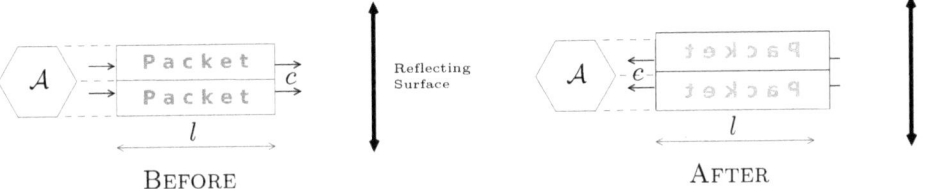

FIGURE 50.2   Reflection of a Light Packet from a Shiny Surface

A **radiometer**, a device for measuring the intensity of light, can be constructed using a paddlewheel-type arrangement of surfaces with one side dark [absorbing] and the other shiny [reflecting]. Place such a device in a spot where light is strongly shining, and the wheel will turn as a consequence of the differing amounts of radiation pressure acting on the fronts and backs of the paddles.

[One might place the paddlewheel in a "vacuum" to reduce the effects of drag on the system.]

One can purchase devices that purport to do this at *Science Shoppes*. More often than not, this is false advertising as, when in operation, the little wheel spins the WRONG way! The likely reason for this contrarian behaviour is that the dark side absorbs light energy and heats up, while the reflecting side stays cool. When gas molecules hit the dark side they tend to gain kinetic energy as they bounce off the HOTTER surface. Along with the additional kinetic energy, there comes a corresponding gain in momentum. In order that the total momentum be conserved, the hotter surfaces must endure stronger recoil kicks from the gas collisions. Gas molecules bounce off the shiny [COOLER] side without commensurate gains in energy and momentum. Unless the paddlewheel's vacuum is very good [which is typically expensive to produce and difficult to maintain], the collision rate with gas molecules can be large enough that momentum transfers owing to the thermal effect can exceed those coming from radiation pressure.

Caveat emptor, eh?

It is some small consolation that, even when the device fails to perform as expected, it still exhibits interesting physics [conservation of momentum and energy within a gas system being slowly and unevenly heated].

> ASIDE: Although lacking direct relevance to the motion of [macroscopic] paddlewheels, the remaining member of Einstein's 1905 quartet, *On the Motion of Small Particles Suspended in a Stationary Liquid, as Required by the Molecular Kinetic Theory of Heat*, informs our discussion of its anomalous behaviour. This paper explicated BROWNIAN MOTION, a phenomenon in which small grains, viewed under a microscope, appear to jostle randomly about without any evident means of propulsion. The social effect of this paper was to prod the scientific community into unreservedly accepting: (1) the atomic hypothesis as the appropriate sub-microscopic description of matter[6] and (2) the kinetic theory of heat. As an added bonus, Einstein's model explaining Brownian Motion provides a way of estimating Avogadro's Number in terms of quantities that can be measured in the laboratory.

---

[6] Pre-1905, the chemical evidence for atoms was circumstantial and combinatorial. Thus, it was logically possible to view the atomic hypothesis as merely a convenient fiction for ordering one's knowledge of matter. Einstein argued that Brownian Motion arose through the agency of, and hence implied the necessity for, the atomic/molecular structure of matter.

Epilogue

In this volume we've undertaken a thorough survey of classical electromagnetism, culminating in Maxwell's Equations.

Having begun with the physical existence of electric charge and the empirical properties of the Coulombic interaction between static charges, we were led to introduce the electric field and its associated potential. **Gauss's Law for Electric Fields** emerged from the analysis of special cases, while the fieldline perspective, along with heuristic arguments, assured us of its generality. With the introduction of capacitance, it became necessary to regard the electric field as a physical object, rather than a mere formal construction, owing to its ability to store and release energy. "Planar" combinations of capacitors were shown to be amenable to reductionist analysis employing well-defined combinatorial rules.

Study of a charge-current confined to a thin wire moved us away from strict electrostatics to steady-state and time-dependent circuit electrodynamics. Kirchoff's Rules, along with combinatorial rules for planar networks of resistors, facilitated the analyses.

While we were concentrating solely on electric aspects, Nature had a double surprise in store for us:

$$\text{Moving Charges} \quad \left\{ \begin{array}{c} \text{Respond to} \\ \text{Produce} \end{array} \right\} \quad \text{Magnetic Fields!}$$

We tackled this complication methodically. The responses of charged particles and electric currents to an existing background static magnetic field were examined. The repertoire of physical cases was expanded to include crossed electric and magnetic fields, and currents flowing through loops and coils of wire. Tremendous simplification occurred when the background magnetic field was uniform. With the response part having been investigated, we then studied the production of magnetic fields. The Law of Biot–Savard was introduced and employed in a number of simple, highly symmetric situations. More general analysis led to the explication of the remaining three Maxwell's Equations: **Ampére's Law**, **Gauss's Law for Magnetic Fields**, and [in a return to electric phenomena] **Faraday's Law**.

Energetic considerations militate for ascribing physicality to the magnetic field. Magnetic effects in electric circuits are modelled by incorporation of inductive elements which adhere to the Kirchoffian principles. The combinatoric rules for inductors are more complicated because of (long-range) interaction effects.

The dynamical responses of circuit systems [possessing resistive, capacitive, and inductive elements] to applied electromotive forces are entirely determined by the circuit and source descriptive parameters. The class of sinusoidally varying EMFs is of particular interest because it forms a complete basis, via linear superposition, for generic drivers of circuits. From a practical perspective, the net average flux of energy, the power, passing through the system has special significance.

The final topic of these notes is the propagating plane-wave vacuum solutions of Maxwell's Equations. These illustrate that Maxwell's Equations, although only four in number, can completely describe electrodynamic phenomena. These also illuminate the essential rôle of LIGHT in Einstein's reformulation of the basis of mechanics in SPECIAL RELATIVITY.

Now, one might be tempted to object that our claim that Maxwell's Equations are fundamental is disjoint from our analysis of circuits. To overturn this [false] impression, we appeal to a final *Gedanken* experiment in four Parts: E, B, S, and P.

Part E: Consider a length of straight wire with uniform geometry: circular cross-section (radius P) and length L, composed of a homogeneous conducting material. Such a segment of conductor was investigated in the development of the microscopic model of electric current in Chapters 18 and 19.

Supposing that the resistance ascribable to the segment is R, and the current passing through it is I, then the potential difference and [magnitude of the uniform] electric field are:

$$V = I R \qquad \Longrightarrow \qquad E_0 = \frac{V}{L} = \frac{I R}{L}.$$

The uniform electric field persists to the edge of the wire and it points in the same direction as that in which the [positive] current flows.

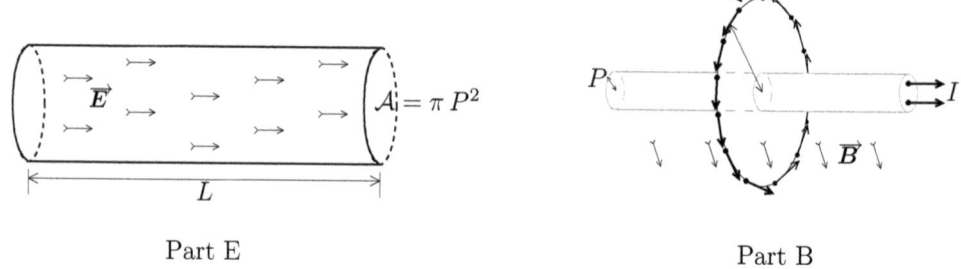

Part E   Part B

Part B: The current produces a magnetic field everywhere in space [both outside and inside of the wire] circulating in planes perpendicular to the length of the segment. At points approaching the surface of the wire the magnitude of the magnetic field converges to

$$B_0 = \frac{\mu_0 I}{2 \pi P},$$

provided that the length of the segment is much greater than its radius [Chapters 28 and 30].

Part S: The electric and magnetic fields on the side surface of the cylindrical segment are everywhere well-defined. [They also happen to be perpendicular.] Therefore the Poynting vector is specified everywhere on the surface:

$$\vec{S} = \frac{1}{\mu_0} \vec{E} \times \vec{B} = \frac{1}{\mu_0} \left[\frac{I R}{L}\right] \left(\frac{\mu_0 I}{2 \pi P}\right) [-\hat{r}] = \frac{I^2 R}{2 \pi P L} [-\hat{r}].$$

Note that the Poynting vector has constant magnitude and points inward throughout the side surface of the wire segment. Recall that $\vec{S}$ is the electromagnetic energy intensity.

Part P: The electromagnetic energy flux [*i.e.*, power] through the side surface of the wire is

$$\int \vec{S} \cdot d\vec{A} = -\frac{I^2 R}{2 \pi P L} A_{\text{cyl.}} = -\frac{I^2 R}{2 \pi P L} 2 \pi P L = -I^2 R.$$

The minus sign arose because we chose the vector area elements to be directed radially outwards. The significance of the overall sign is that **this amount of energy is flowing into the wire from the fields in the adjacent space.** Recall that, in Chapter 19, it has been established that the current I flowing through a resistance R dissipates energy [via Joule heating] at the rate $P = I^2 R$.

**The energy being deposited (as heat) in the resistive wire
is flowing into the wire from the fields on its boundary!**

As in previous VOLUMES, there is no escape without enduring a physics joke.

Late one evening after long hours spent studying, Bertal went for a walk. Striding along a quiet and rather deserted street, Bertal saw a man approach a lamp post, drop to his hands and knees, and begin clambering about on the pavement. Not one to abandon a neighbour in distress, Bertal said, "Hi, what're you looking for?"

"Car key," replied the man, without looking up.

Bertal quickly surveyed the empty street and sidewalk nearby and saw no evidence of keys anywhere within view. Nonetheless, he joined in the search. After several minutes of fruitless effort, Bertal asked, "Is it with other keys on a fob, or just one loose key?"

"On a ring with five or six other keys," the man said.

After a few more minutes spent going over the area again, Bertal asked, "Where is your car? And where were you before you noticed that your keys were missing?"

"I was at the pub watching the hockey game. My car is parked behind, in the parking lot," sighed the man, with resignation tingeing his voice.

Trying not to sound exasperated, Bertal asked, "If your car is way over and beyond there, and you were over there, ..." pointing. "Why are we looking here?"

"Well," replied the fellow, *"This is where the light is."*

Part II

Electricity and Magnetism Problems

E

Electric Field and Potential Problems

Useful Data

Unless otherwise specified, here and throughout the other collections of problems we employ the following conventions.

SI   Times are measured in seconds, s, distances in metres, m, masses in kilograms, kg, forces in newtons, N, and angles in radians, rad.

e   The quantum of elementary charge is $e \simeq 1.6 \times 10^{-19}$ C.

k   Coulomb's constant has the approximate value $k \simeq 8.99 \times 10^9$ N·m^2/C^2. Here we shall use $k = 9 \times 10^9$ for simplicity.

ϵ_0   The permittivity of free space is $\epsilon_0 = (4\pi k)^{-1} \simeq 8.85 \times 10^{-12}$ C^2/(N·m^2).

m   The electron mass is $m_e \simeq 9.1 \times 10^{-31}$ kg. The proton mass is $m_p \simeq 1.67 \times 10^{-27}$ kg. The atomic mass unit is $1\,\text{AMU} \simeq 1.66 \times 10^{-27}$ kg.

E.1   Two equal point-like charges, $q = +1$ C, are separated by a distance of 1 kilometre. Determine the Coulombic force acting between the two charges. Comment.

E.2   Two point electric charges, $q_L = 3\,\mu$C and $q_R = 12\,\mu$C, lie along an axis in the xy-plane, at $\vec{r}_L = (-4, 0)$ and $\vec{r}_R = (8, 0)$. A third charge, $q_0 = 2\,\mu$C, is placed at the origin.

  Compute the electric force acting upon q_0, produced by   (a) q_L and   (b) q_R.

  (c) Determine the net force acting on q_0.

  (d) Discuss how the result for (c) would change if q_0 were displaced to $(0, 0.01)$.

E.3   Two point-like electrons belonging to different atoms might be approximately one Ångstrom [$1\,\text{Å} = 10^{-10}$ m] apart.

  (a) Estimate the magnitude of the Coulombic force that each electron experiences owing to the presence of the other.

  (b) Were it the case that this was the only force acting on the electrons, compute the magnitude of their respective accelerations.

E.4   Three charged particles reside at fixed positions in a common plane. Charge $q_0 = 2\,\mu$C is at the origin, $q_1 = 5\,\mu$C has coordinates $(3, -1)$, and $q_2 = 10\,\mu$C is at $(8, 6)$.

  Compute the Coulombic force exerted by   (a) q_1,   (b) q_2, and   (c) both   acting on q_0.

  (d) Suppose that the particle bearing q_0 has mass $m_0 = 2.5$ kg. Determine the acceleration that q_0 would experience were it to be suddenly released from rest.

E.5 [See also E.44, E.45.] Three point-like charges are arrayed in the xy-plane. Their magnitudes and coordinates are: $Q_1 = 1\,\mu\text{C}$, $\vec{r}_1 = (0,1)$, $Q_2 = 2\,\mu\text{C}$, $\vec{r}_2 = (0,2)$, and $Q_3 = 3\,\mu\text{C}$, $\vec{r}_3 = (0,3)$. Compute the magnitude and direction of the net electric field produced by these three charges at   (a) the origin, $(0,0)$,   (b) $(0,4)$, and   (c) $(1,0)$.

E.6

A uniform electric field, $\vec{E}_o = -50\,\text{N}/\text{C}\,[\hat{\jmath}]$, fills a region of space. A particle with mass $m = 0.025\,\text{kg}$, bearing charge $q = 1\,\text{mC}$, is fired into the region with initial velocity $\vec{v}_0 = 3\,\text{m/s}\,[\hat{\imath}]$, at an initial height of $4\,\text{m}$ above the x-axis. Determine where and when the particle crosses the x-axis.

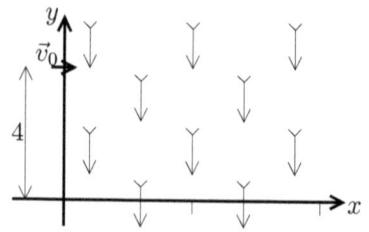

E.7 [See also E.68.]

A wire filament, heated by the passage of an electric current through it, emits electrons. These emerge from the metal with very little (effectively zero) kinetic energy. Once outside the filament, these electrons experience a uniform electric field of strength $E_0 = 4000\,\text{N}/\text{C}$, directed toward the left. The region of constant field extends $5\,\text{mm}$ from the wire filament.

filament

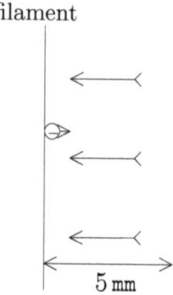

Determine the   (a) electric force acting on and   (b) acceleration of   the electron.   (c) Compute the speed of the electron when it reaches the edge of the field region.

E.8

An electric dipole, comprised of two charges, $\{+q, -q\}$, separated by distance $2\,a$, is shown in the neighbouring figure. This dipole is subjected to a uniform electric field, $\vec{E}_o = E_0\,[\hat{\imath}]$.

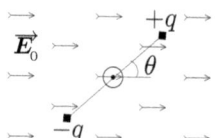

(a) Compute the electric force (magnitude and direction) acting on the   (i) positive and   (ii) negative   charge.

(b) Ascertain the net force acting on the dipole.

(c) Compute the torque (about the midpoint of the dipole) produced by the electric force acting on the   (i) positive and   (ii) negative   charge.

(d) Ascertain the net torque acting on the dipole.

(e) Compare your result for (d) with $\vec{p} \times \vec{E}_o$. Comment.

(f) Compute the net force and net torque when θ equals   (i) 0,   (ii) $\pi/4$,   (iii) $\pi/2$, and   (iv) π.

E.9 [See also E.46.] Two equal-magnitude electric charges, $|q_1| = |q_2| = 2\,\mu\text{C}$, are fixed at positions $10\,\text{cm}$ apart along the x-axis.

(a) Compute the magnitude of the Coulombic force acting on each charge when the charges have   (i) the same and   (ii) opposite   sign.

Compute the magnitude of the net electric field produced by the charges at the point which lies   (b) midway and   (c) one-quarter of the distance   between the charges, when they have   (i) the same and   (ii) opposite   sign.

E.10 [See also E.47.] Two equal-magnitude electric charges, $|q_1| = |q_2| = 2\,\mu\text{C}$, are fixed at positions 10 cm apart along the x-axis. Compute the magnitude and direction of the net electric field produced by the charges at the point along the x-axis located   (a) 5 cm from q_1 and 15 cm from q_2 and   (b) 10 cm from q_1 and 20 cm from q_2,   when the charges have (i) the same and   (ii) opposite   sign.

E.11 [See also E.48.] Two equal-magnitude electric charges, $|q_1| = |q_2| = 2\,\mu\text{C}$, are fixed at positions 10 cm apart along the x-axis. Compute the magnitude and direction of the net electric field produced by the charges at the point which lies   (a) 5 cm and   (b) 10 cm   in the y-direction from q_1, when the charges have   (i) the same and   (ii) opposite   sign.

E.12 [See also E.49.]

A dipole is comprised of charges, $\{+q, -q\}$, separated by a distance of $2\,a$, as described in Chapter 2 and illustrated in the adjacent figure. WLOG, the x-axis is chosen to align with the dipole arrangement of charges, and hence $\vec{p} = p\,\hat{\imath} = 2\,a\,q\,\hat{\imath}$.

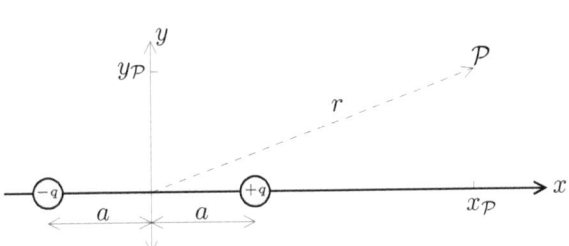

A field point, $\mathcal{P}$, is at $(x_\mathcal{P}, y_\mathcal{P})$ with respect to the midpoint of the dipole. The distance between the dipole and the field point is $r = \sqrt{x_\mathcal{P}^2 + y_\mathcal{P}^2}$. Under the assumption that the field point is distant [on the scale set by the size of the dipole, *i.e.*, $r \gg a$], compute the components of the electric field at $\mathcal{P}$ directed   (a) parallel and   (b) perpendicular   to the axis of the dipole.

E.13 One type of electric quadrupole is comprised of two equal-magnitude antiparallel dipoles sharing a common point. This configuration consists of three charges, $\{+q, -2\,q, +q\}$, arrayed along a straight line with length $4\,a$, as illustrated in the figure below.

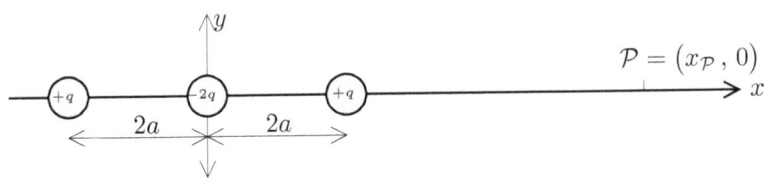

Defining the magnitude of the quadrupole moment of this configuration to be $Q = 8\,a^2\,q$, show that the magnitude of the electric field at point $\mathcal{P} = (x_\mathcal{P}, 0)$, with $x_\mathcal{P} \gg a$, is approximately equal to $\frac{3\,k\,Q}{x_\mathcal{P}^4}$.

E.14
Two identical electric dipoles, with dipole moment $p = 2\,a\,q$, are placed side-by-side and anti-aligned, as shown in the adjacent figure. The distance between the dipoles is $2\,a$. [This is the other type of quadrupole.]

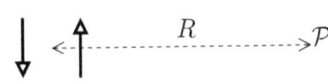

Determine the leading-order non-trivial behaviour of the electric field at the field point $\mathcal{P}$, a distance R from the centre of this double-dipole system. Assume that $R \gg a$.

E.15 [See also E.52, E.60.]

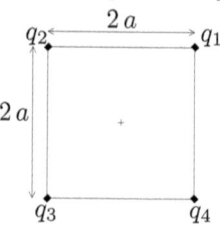

Four electric charges, $\{q_1, q_2, q_3, q_4\}$, are affixed to the corners of a square with edge length $2\,a$, as shown in the adjacent figure. Here, the signs and relative magnitudes of the charges are: $q_1 = 2\,q$, $q_2 = -q$, $q_3 = q$, and $q_4 = -2\,q$.

(a) Determine the net electric field produced by the charges at the centre of the square.

(b) Set $q = 1\,\mu\text{C}$ and $a = 1\,\text{m}$. Compute the net electric force that would act on a point(-like) particle, bearing charge $q_0 = 1\,\text{nC}$, placed at the centre of the square.

E.16 Suppose that the signs of both q and q_0 appearing in E.15 were reversed. Without recalculating, predict the effect that this has on the electric force acting on the test charge.

E.17 [See also E.15.] The configuration of charges appearing in E.15 may be modelled as two adjacent antiparallel dipoles, $\vec{p}_{41} = 4\,a\,q\ [\ \uparrow\ ]$ and $\vec{p}_{23} = 2\,a\,q\ [\ \downarrow\ ]$, separated by a distance of $2\,a$. Employ the dipole approximation to estimate the electric field produced at the centre of the square by the (a) 41 and (b) 23 dipoles. (c) Add these contributions to estimate the net electric field at the centre of the square. (d) Comment.

E.18 [See also E.53, E.61.] Repeat E.15 when $q_1 = q$, $q_2 = -q$, $q_3 = q$, and $q_4 = -q$.

E.19 [See also E.54, E.62.] Repeat E.15 when $q_1 = q$, $q_2 = 2\,q$, $q_3 = 3\,q$, and $q_4 = 4\,q$.

E.20 [See also E.55, E.63.] Repeat E.15 when $q_1 = q$, $q_2 = -2\,q$, $q_3 = 3\,q$, and $q_4 = -4\,q$.

E.21 [See also E.56, E.64.]

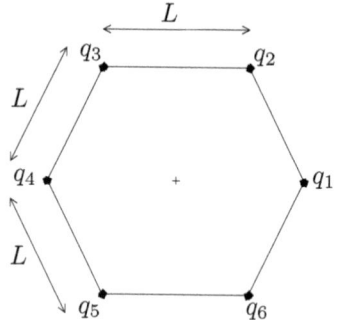

Six electric charges, $\{q_1, q_2, q_3, q_4, q_5, q_6\}$, are affixed to the vertices of a regular hexagon with edge length L, as shown in the adjacent figure. Here, the signs and relative magnitudes of the charges are: $q_1 = 2\,q$, $q_2 = -q$, $q_3 = q$, $q_4 = -2\,q$, $q_5 = q$, and $q_6 = -q$.
[HINT: First convince yourself that the distance from each point(-like) charge to the centre of the hexagon is L.]

(a) Determine the net electric field produced by the charges at the centre of the hexagon.

(b) Set $q = 1\,\mu\text{C}$ and $L = 1\,\text{m}$. Compute the net electric force that would act on a point(-like) particle, bearing charge $q_0 = 1\,\text{nC}$, placed at the centre of the hexagon.

E.22 Suppose that the signs of both q and q_0 appearing in E.21 were reversed. Without recalculating, predict the effect that this has on the electric force acting on the test charge.

E.23 [See also E.57, E.65.] Repeat E.21 when $q_1 = q$, $q_2 = -q$, $q_3 = q$, $q_4 = -q$, $q_5 = q$, and $q_6 = -q$.

E.24 [See also E.58, E.66.] Repeat E.21 when $q_1 = q$, $q_2 = 2\,q$, $q_3 = 3\,q$, $q_4 = 4\,q$, $q_5 = 5\,q$, and $q_6 = 6\,q$.

E.25 [See also E.59, E.67.] Repeat E.21 when $q_1 = q$, $q_2 = -2\,q$, $q_3 = 3\,q$, $q_4 = -4\,q$, $q_5 = 5\,q$, and $q_6 = -6\,q$.

E.26 [See also E.69.] A thin uniformly charged rod has length $L = 2\,\text{m}$ and bears charge per unit length $\lambda_0 = 0.5\,\mu\text{C/m}$. A field point is located along the axis formed by the rod, a distance of $5\,\text{m}$ from its centre.

(a) Compute the electric field produced by the rod at the field point, employing the [appropriate] formula derived in Chapter 3.

(b) Crudely model the rod as though it were a single point charge located at its centre. Compute the electric field at the field point produced by this single point charge.

(c) Model the rod as two [equal-magnitude] point charges, each situated at the centre of its respective half of the rod, and compute the electric field produced by these two point charges at the field point.

(d) Model the rod as three [equal-magnitude] point charges, each situated at the centre of its respective third of the rod, and compute the net electric field at the field point.

(e) Model the rod as four point charges, each centred in its respective quarter of the rod, and compute the net field produced by this collection of charges at the field point.

(f) Comment.

E.27   [See also E.28, E.70.]

Two uniformly charged rods are assembled to form a sideways T. Rod 1 has length $L_1 = 1.2\,\mathrm{m}$ and total charge $Q_1 = 0.6\,\mathrm{C}$. Rod 2 has length $L_2 = 1.5\,\mathrm{m}$ and $Q_2 = 0.4\,\mathrm{C}$. The field point is $1.5\,\mathrm{m}$ from the junction of the T. Compute the net electric field at $\mathcal{P}$.

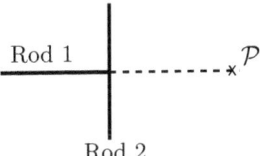

E.28   [See also E.71.] Two uniformly charged rods are arranged in a sideways T, as shown in E.27. The first rod is twice the length of the second, *i.e.*, $L_1 = 2L$ and $L_2 = L$, and they bear equal-magnitude charges $Q_1 = -Q$ and $Q_2 = +Q$. Determine the net electric field produced by this system of rods at the field point, $\mathcal{P}$, located distance L from the rod junction.

E.29   [See also E.30, E.72.]

Two uniformly charged rods are assembled to form an L. Rod 1 has length $L_1 = 1.2\,\mathrm{m}$ and total charge $Q_1 = 0.6\,\mathrm{C}$. Rod 2 has $L_2 = 1.5\,\mathrm{m}$ and $Q_2 = 0.4\,\mathrm{C}$. The field point is at the intersection of the perpendicular bisectors of the rods. Compute the net electric field at $\mathcal{P}$.

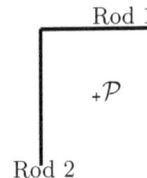

E.30   [See also E.73.] Two uniformly charged rods are arranged in an ell shape, as shown in E.29. The first rod is twice the length of the second, *i.e.*, $L_1 = 2L$ and $L_2 = L$, and they bear equal-magnitude charges $Q_1 = -Q$ and $Q_2 = +Q$. Determine the net electric field produced by this system at the intersection point of the perpendicular bisectors of the rods.

E.31   [See also E.74.]

Four uniformly charged thin rods comprise the edges of the rectangle shown in the adjacent figure. The rectangle has length $2\,a$ and width $2\,b$, with $0 < b < a$. The charges on each rod are (in anticlockwise order starting at the top): $Q, 2\,Q, -Q, 4\,Q$.

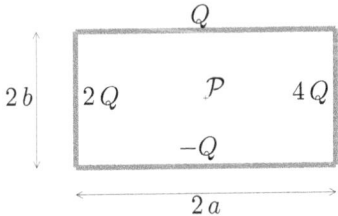

Determine the electric fields produced at the geometric centre of the rectangle, *i.e.*, at $\mathcal{P}$, by each straight segment, and the net field.

E.32   [See also E.75.] A thin rod of length L bears lineal charge density. $\lambda = \lambda_0 \frac{x^2}{L^2}$, where λ_0 is constant. We choose the origin of coordinates to lie at the midpoint of the rod.

(a) Compute the total charge on the rod.

(b) Compute the electric field produced by the charged rod at points lying along the axis of the rod, *viz.*, $\mathcal{P} = (x_\mathcal{P}, 0)$, where $x_\mathcal{P} > L/2$.

(c) Take the far-field limit, $x_\mathcal{P} \gg L/2$, of your result in (b).

(d) Compute the electric field produced by the charged rod at points lying along the perpendicular bisector of the rod, $\mathcal{P} = (0, y_\mathcal{P})$.

(e) Take the far-field limit, $y_\mathcal{P} \gg L/2$, of your result in (d).

E.33 [See also E.76.] Repeat E.32 when $\lambda = \lambda_0 \frac{|x|}{L}$.

E.34 [See also E.77.] Repeat E.32 when $\lambda = \lambda_0 \frac{x^3}{L^3}$.

E.35 [See also E.78.] Repeat E.32 when $\lambda = \lambda_0 \cos\left(\frac{\pi x}{2L}\right)$.

E.36 [See also E.79.] Repeat E.32 when $\lambda = \lambda_0 \cos^2\left(\frac{\pi x}{2L}\right)$.

E.37 [See also E.80.] The on-axis electric field associated with a uniformly charged ring (centred in the yz-plane) was determined in Chapter 3. Determine the point(s) along the x-axis at which the field assumes its maximum value.

E.38
A uniform electric field, $\vec{E}_0 = E_0\, \hat{\jmath}$, fills the region of space near the origin of coordinates. Compute the field's flux through the planar triangular surface with vertices at $\{(0,0,0), (1,1,0), (0,1,1)\}$. [The coordinate axes are scaled in metres.] The surface is illustrated by the two views shown alongside.

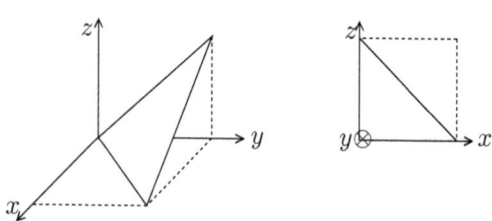

E.39
An electric field varies according to $\vec{E} = c\,z^2\, \hat{\jmath}$, where c is a constant. Ascertain the electric flux through a rectangular surface, lying in the zx-plane, with vertices at $\{(0,0,0), (a,0,0), (a,0,b), (0,0,b)\}$.

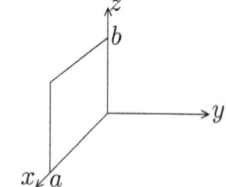

E.40 Everywhere on a closed Gaussian surface, the local electric field points outward. What might one conclude?

E.41 A cube's six sides are coloured red, orange, yellow, green, blue, and indigo. The edge lengths of the cube are all 1.5 m. The electric field on each face is locally constant, and expressed below in units of N/C. The [local] outward and inward directions on each surface are denoted by $\hat{o}$ and $\hat{\imath}$, respectively.

$E_{\text{red}} = 2\,[\hat{o}]$ $E_{\text{orange}} = 3\,[\hat{o}]$ $E_{\text{yellow}} = 4\,[\hat{o}]$ $E_{\text{green}} = 2\,[\hat{\imath}]$ $E_{\text{blue}} = 2\,[\hat{\imath}]$ $E_{\text{indigo}} = 2\,[\hat{\imath}]$

Compute the net electric flux through   (a) each of the sides and   (b) the cube.

(c) Ascertain the net electric charge located within the cube.

(d) Suppose that the region of space surrounding the cube out to a radius of ten metres contains no regions of non-zero net electric charge. What amount of electric flux flows through a sphere of radius 5 m centred on the cube?

(e) What can be said about the electric flux through a Gaussian sub-cube of edge length 0.1 m lying entirely within the original cube?

E.42 [See also E.81, C.16.] Two concentric charged spherical conducting shells are in electrostatic equilibrium. The outer shell, with radius R_o, carries charge $+Q$, while the inner shell, with radius R_i, bears $-q$ net charge. Furthermore, $Q > q$.

Apply Gauss's Law to determine the electric field   (a) outside of the larger concentric shell [*i.e.*, for $r > R_o$],   (b) between the concentric shells [*i.e.*, for $R_o > r > R_i$],   and (c) within the innermost shell [*i.e.*, for $r < R_i$].

(d) Sketch a graph of the radial component of the electric field as a function of the distance from the centre of the concentric shells when   (i) $Q = 2\,q$ and $R_o = 3\,R_i$ and (ii) $Q = 3\,q$ and $R_o = 2\,R_i$.

E.43 [See also E.82, C.12.] Three co-axial cylindrical conducting shells with common length, L, have radii $\frac{R}{2}$, R, and $2\,R$, respectively. Positive charge, $\frac{Q}{2}$, is deposited on the innermost and outermost shells, and an amount of negative charge, $-Q$, is placed on the middle cylinder. The charges are assumed to be evenly distributed. [Edge effects are taken to be small.]

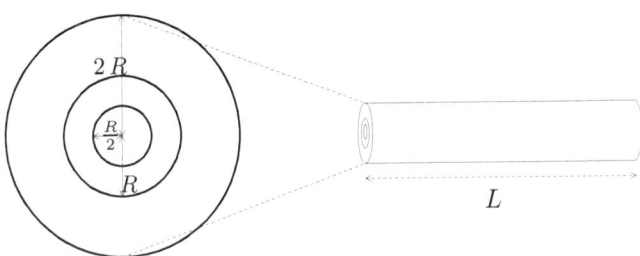

Apply Gauss's Law to determine the electric field everywhere   (a) outside the larger shell [*i.e.*, $r > 2\,R$],   between the (b) outer [*i.e.*, $R < r < 2\,R$]   and   (c) inner [*i.e.*, $\frac{R}{2} < r < R$]   pairs of coaxial shells,   and   (d) within the innermost shell [*i.e.*, $r < \frac{R}{2}$].

(e) Sketch a graph of the radial component of the electric field as a function of the radial distance from the common axis of the cylinders.

E.44 [See also E.5, E.45.] Compute the electrostatic potential energy held in the configuration of point charges considered in E.5.

E.45 [See also E.5, E.44.] Compute the net electric potential produced by the three point charges considered in E.5 at   (a) the origin, $(0\,,0)$,   (b) $(0\,,4)$,   and   (c) $(1\,,0)$.

E.46 [See also E.9.] Two equal-magnitude electric charges, $q = 2\,\mu$C, are fixed at positions 10 cm apart along the x-axis.

(a) Compute the electrostatic potential energy of this two-charge system when the charges have   (i) the same and   (ii) opposite   sign.

Compute the electric potential at the point which lies   (b) midway   and   (c) one-quarter of the distance   between the charges when they have   (i) the same and   (ii) opposite   sign.

E.47 [See also E.10.] Two equal-magnitude electric charges, $q = 2\,\mu$C, are fixed at positions 10 cm apart along the x-axis. Compute the electric potential produced by these charges at the point along the x-axis located   (a) 5 cm from q_1 and 15 cm from q_2 and   (b) 10 cm from q_1 and 20 cm from q_2 when the charges have   (i) the same and   (ii) opposite   sign.

E.48 [See also E.11.] Two equal-magnitude electric charges, $|q_1| = |q_2| = 2\,\mu$C, are fixed at positions 10 cm apart along the x-axis. Compute the magnitude and direction of the net electric field produced by the charges at the point which lies   (a) 5 cm and   (b) 10 cm   in the y-direction from q_1, when the charges have   (i) the same and   (ii) opposite   sign.

E.49 [See also E.12.] The dipole considered in E.12 [comprised of charges, $\{+q, -q\}$, separated by distance $2\,a$] has $\vec{p} = p\,\hat{\imath} = 2\,a\,q\,\hat{\imath}$. A field point, $\mathcal{P}$, is located at $(x_\mathcal{P}, y_\mathcal{P})$ with respect to the midpoint of the dipole. Let r denote the distance between the dipole and the field point. Under the assumption that the field point is distant [on the scale set by the size of the dipole, *i.e.*, $r \gg a$], compute the approximate value of the electric potential at $\mathcal{P}$.

E.50
Two identical point(-like) charges, q, are fixed in space and separated by distance $2\,a$. Field points, lying along the circular arc of radius a centred on the charge distribution, may be uniquely specified by the angle, θ, shown in the figure. [We insist that the field point not lie directly upon either charge.]

For the (a) left and (b) right charge, determine the (i) distance to and (ii) the electric field at the field point, $\mathcal{P}$.

(c) Superpose these fields to obtain the net field at $\mathcal{P}$.
[The circular geometry constrains the angle between the constituent fields to be $\frac{\pi}{2}$.]

E.51 Repeat E.50 when the charges are of equal magnitude and opposite sign. [Without loss of generality, take the charge on the left to be $-q$; that on the right $+q$.]

E.52 [See also E.15, E.60.] Compute the electrostatic potential energy of the four-charge system considered in E.15.

E.53 [See also E.18, E.61.] Compute the electrostatic potential energy of the four-charge system considered in E.18.

E.54 [See also E.19, E.62.] Compute the electrostatic potential energy of the four-charge system considered in E.19.

E.55 [See also E.20, E.63.] Compute the electrostatic potential energy of the four-charge system considered in E.20.

E.56 [See also E.21, E.64.] Compute the electrostatic potential energy of the six-charge system considered in E.21.

E.57 [See also E.23, E.65.] Compute the electrostatic potential energy of the six-charge system considered in E.23.

E.58 [See also E.24, E.66.] Compute the electrostatic potential energy of the six-charge system considered in E.24.

E.59 [See also E.25, E.67.] Compute the electrostatic potential energy of the six-charge system considered in E.25.

E.60 [See also E.15, E.52.] Recall the system of four electric charges affixed to the corners of a square studied in E.15.

(a) Determine the net electric potential at the centre of the square.

(b) Set $q = 1\,\mu\text{C}$ and $a = 1\,\text{m}$. Compute the electric potential energy of a point(-like) particle, bearing charge $q_0 = 1\,\text{nC}$, placed at the centre of the square.

(c) Without recalculating, determine the electric potential energy of the system if the signs of q and q_0 were both reversed.

E.61 [See also E.18, E.53.] Repeat E.60 for the four charges of E.18.
E.62 [See also E.19, E.54.] Repeat E.60 for the four charges of E.19.
E.63 [See also E.20, E.55.] Repeat E.60 for the four charges of E.20.
E.64 [See also E.21, E.56.] Recall the six-charge system considered in E.21.

(a) Determine the electric potential at the centre of the hexagon.

(b) Set $q = 1\,\mu\text{C}$ and $L = 1\,\text{m}$. Compute the electric potential energy of a point(-like) particle, bearing charge $q_0 = 1\,\text{nC}$, placed at the centre of the hexagon.

(c) Without recalculating, determine the electric potential energy that the system would have if the signs of q and q_0 were both reversed.

E.65 [See also E.23, E.57.] Repeat E.64 for the six charges of E.23.

E.66 [See also E.24, E.58.] Repeat E.64 for the six charges of E.24.

E.67 [See also E.25, E.59.] Repeat E.64 for the six charges of E.25.

E.68 [See also E.7.] A uniform electric field, $\vec{E}_0 = 4000\,\text{N/C}\,[\hat{\imath}]$, exists throughout a region of space with width $\Delta x = 5\,\text{mm}$.

(a) Determine the electric potential difference across the region.

(b) Compute the electrostatic potential energy lost by an electron as it crosses the region.

(c) Suppose that the electron started from rest. Compute its kinetic energy at the moment it has crossed the region.

E.69 [See also E.26.] A uniformly charged rod has length $L = 2\,\text{m}$ and bears charge per unit length $\lambda_0 = 0.5\,\mu\text{C/m}$. A field point is located along the axis formed by the rod, a distance of $5\,\text{m}$ from its centre.

(a) Compute the electric potential produced by the rod at the field point.

(b) Crudely model the rod as though it were a single point charge located at its centre. Compute the electric potential at the field point produced by this single point charge.

(c) Model the rod as two [equal-magnitude] point charges, each situated at the centre of its respective half of the rod, and compute the electric potential at the field point produced by these two point charges.

(d) Model the rod as four point charges—each centred in its respective quarter of the length of the rod—and compute the potential produced by this collection of charges at the field point.

(e) Comment.

E.70 [See also E.27.] Two uniformly charged rods are assembled into a sideways T as shown in the figure alongside E.27. Rod 1 has length $L_1 = 1.2\,\text{m}$ and total charge $Q_1 = 0.6\,\text{C}$. Rod 2 has length $L_2 = 1.5\,\text{m}$ and $Q_2 = 0.4\,\text{C}$. The field point is $1.5\,\text{m}$ from the junction of the T. Compute the net electric potential at $\mathcal{P}$.

E.71 [See also E.28.] Consider the two uniformly charged rods arranged in a sideways T from E.28. Determine the net electric potential at the field point located a distance L from the rod junction (shown in the figure alongside E.27).

E.72 [See also E.29.] Determine the electric potential produced at $\mathcal{P}$ by the uniformly charged rods considered in E.29.

E.73 [See also E.30.] Determine the electric potential produced at $\mathcal{P}$ by the uniformly charged rods considered in E.30.

E.74 [See also E.31.] Determine the electric potential at the centre of the rectangle considered in E.31, produced by   (a–d) each of the four rods individually and   (e) all four together.

E.75 [See also E.32.] A thin rod of length L bears lineal charge density $\lambda = \lambda_0 \frac{x^2}{L^2}$, where λ_0 is constant and the origin of coordinates lies at the midpoint of the rod.

(a) Compute the total charge on the rod.

(b) Compute the electric potential produced by the charged rod at points lying along the axis of the rod, *viz.*, $\mathcal{P} = (x_P, 0)$, where $x_P > L/2$.

(c) Take the far-field limit, $x_P \gg L/2$, of your result in (b).

(d) Compute the electric potential produced by the charged rod at points lying along the perpendicular bisector of the rod, $\mathcal{P} = (0, y_P)$.

(e) Take the far-field limit, $y_P \gg L/2$, of your result in (d).

E.76 [See also E.33.] Repeat E.75 when $\lambda = \lambda_0 \frac{|x|}{L}$.

E.77 [See also E.34.] Repeat E.75 when $\lambda = \lambda_0 \frac{x^3}{L^3}$.

E.78 [See also E.35.] Repeat E.75 when $\lambda = \lambda_0 \cos\left(\frac{\pi x}{2L}\right)$.

E.79 [See also E.36.] Repeat E.75 when $\lambda = \lambda_0 \cos^2\left(\frac{\pi x}{2L}\right)$.

E.80 [See also E.37.] The on-axis electric potential associated with a uniformly charged ring (centred in the yz-plane) was determined in Chapter 10. Determine the point(s) along the x-axis at which the potential assumes its maximum value.

E.81 [See also E.42, C.16.] Two concentric charged spherical conducting shells are in electrostatic equilibrium. The outer shell, with radius R_o, carries charge $+Q$, while the inner shell, with radius R_i, bears $-q$ net charge. Furthermore, $Q > q$. Our goal is to determine the electric potential established throughout space by this configuration of charges.

Determine the electric potential   (a)   outside of the larger concentric shell [*i.e.*, for $r > R_o$],   (b)   between the concentric shells [*i.e.*, for $R_o > r > R_i$],   (c)   within the innermost shell [*i.e.*, for $r < R_i$].

(d) Sketch a graph of the electric potential as a function of radial distance.

E.82 [See also E.43, C.12.] Three co-axial cylindrical conducting shells with common length, L, have radii $\frac{R}{2}$, R, and $2R$, respectively, and are shown in the sketch alongside E.43. An amount of positive charge, $\frac{Q}{2}$, is deposited on the innermost and outermost shells, and negative charge, $-Q$, is placed on the middle cylinder. The charges are assumed to be evenly distributed.

(a) Apply Gauss's Law to determine the electric field in the spaces between the cylinders.

Determine the electric potential   (b)   outside of the larger concentric shell [*i.e.*, $r > 2R$],   (c)   between the outer pair of concentric shells [*i.e.*, $R < r < 2R$],   (d)   between the inner pair of concentric shells [*i.e.*, $\frac{R}{2} < r < R$],   (e)   within the innermost shell [*i.e.*, $r < \frac{R}{2}$].

(f) Sketch a graph of the electric potential *vs.* radial distance for this system of coaxial cylinders.

E.83 Three charges are arrayed in the plane shown in the adjacent figure. Let $q = 3 \times 10^{-4}$ C, $q_0 = 2\,\mu$C, and $a = 1$ m.

(a) Determine the magnitude and direction of the electric field at the point $\mathcal{P}$ produced by the (i) q charge lying along the x-axis,   (ii) q charge on the y-axis, and   (iii) $2q$ charge situated at the origin.

(b) Determine the electric   (i) field at $\mathcal{P}$ and (ii) force acting on q_0 were it placed at $\mathcal{P}$.

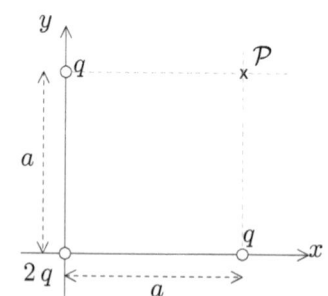

(c) Determine the electric potential at $\mathcal{P}$ produced by the   (i) q charge lying along the x-axis,   (ii) q charge on the y-axis, and   (iii) $2\,q$ charge situated at the origin.

(d) Determine the   (i) electric potential at $\mathcal{P}$ and   (ii) electrostatic potential energy of q_0 were it placed at $\mathcal{P}$.

(e) Compute the electrostatic potential energies of the   (i) q-q and   (ii) $(2\,q)$-q pairs of charges.   (iii) Determine the total electrostatic potential energy of this three-charge system.   (iv) How much energy was required to assemble the system?

E.84 Repeat the analysis in E.83 when the values of q and a remain unspecified.

E.85 Four charges are arrayed in the plane shown in the adjacent figure. Let $q = 3\times10^{-4}$ C, $q_0 = 2\,\mu$C, and $a = 1$ m.

(a) Determine the magnitude and direction of the electric field at the origin produced by the (i) q charge lying along the x-axis,   (ii) q charge on the y-axis,   (iii) q charge on the $-x$-axis, and   (iv) $-q$ charge on the $-y$-axis.

(b) Determine the electric   (i) field at $\mathcal{P}$ and (ii) force acting on q_0 were it placed at $\mathcal{P}$.

(c) Determine the electric potential at the origin produced by the   (i) q charge lying along the x-axis,   (ii) q charge on the y-axis,   (iii) q charge on the $-x$-axis, and (iv) $-q$ charge on the $-y$-axis.

(d) Determine the   (i) electric potential at $\mathcal{P}$ and   (ii) electrostatic potential energy of q_0 were it placed at $\mathcal{P}$.

(e) Compute the electrostatic potential energies of the   (i) q-q and   (ii) $(-q)$-q pairs of charges. [HINT: There are six different pairwise combinations and *a priori* each might be expected to have a different value of energy.]   (iii) Determine the total electrostatic potential energy of this four-charge system.   (iv) How much energy was required to assemble the system?

E.86 Repeat the analysis in E.85 when the values of q and a remain unspecified.

E.87

Four equal pointlike electric charges, q, are distributed symmetrically in the xy-plane as shown to the left in the nearby figure.

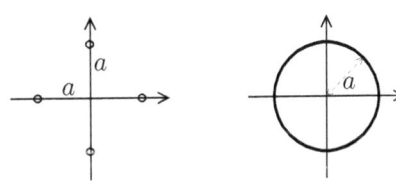

(a) (i) Determine the electric field produced by this arrangement of charges at a point along the z-axis, $\mathcal{P} = (0\,,0\,,z)$.   Take the limits   (ii) $z \to 0$ and   (iii) $z \gg a$ of your result in (i).

(b) (i) Determine the electric potential produced by the four charges at $\mathcal{P}$.   Take the limits   (ii) $z \to 0$ and   (iii) $z \gg a$ of your result in (i).

(c) Compute the electrostatic potential energy held in the four-charge system.

(d) A thin ring of radius a bearing a uniformly distributed charge of $Q = 4\,q$ is centred in the xy-plane, as shown above. In Chapter 3, the electric field produced along the axis of such a charged ring was determined to be $\frac{k\,4\,q\,z}{(z^2+a^2)^{3/2}}$ $[\hat{k}]$. Comment.

E.88
A thin insulating rod is bent in the shape of a half-circle, as shown nearby. The top and bottom portions of the rod bear equal and opposite amounts of uniformly distributed electric charge. The total amounts of positive and negative charge are $\pm Q$, the radius of the semi-circle is a, and thus the charge densities are $\pm \lambda_0 = \pm \frac{Q}{\pi a}$.
Compute the electric   (a) field and   (b) potential at the field point located at the centre of curvature of the rod.

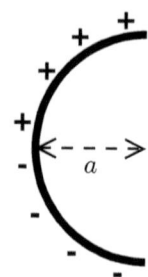

E.89   Consider a thin insulating rod in the shape of a semi-circle as in E.88, with radius of curvature equal to 5 m, and bearing 100 nC of electric charge uniformly distributed along its length. Compute the magnitude and direction of the electric field produced at the centre of curvature of the rod.

E.90   A thin insulating rod is bent in the shape of a quarter circle [*i.e.*, the upper POSITIVE half of the semi-circular rod shown in E.88]. Supposing that 100 nC of electric charge is uniformly distributed on the rod, and that its radius of curvature is 5 m, compute the magnitude of the electric field at the centre of curvature.

E.91   A thin rod segment is bent to form one-eighth of a circle of radius 0.5 m [subtending $45°$]. A total charge of $Q = 10$ nC is uniformly distributed along the length of the rod. Compute the electric   (a) field and   (b) potential   at the rod's centre of curvature.

E.92
A thin cylindrical shell of radius a and length b, shown in the adjacent figure, possesses uniform areal charge density σ_0. The z-axis is aligned with the symmetry axis of, and centred on, the cylinder. Determine the electric potential difference between the origin and the point on the axis directly in line with the end of the cylinder, *i.e.*, $z = b/2$.

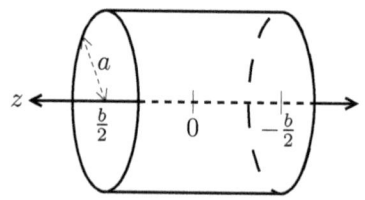

E.93   The electric potential in a region of two-dimensional space $\{x > 1, y > 1\}$ is given by: $V(x,y) = \frac{x^2 y}{(x^2+y^2)^2}$.

   (a) Compute the electric field everywhere in the region.

   (b) Are there points in the region at which the electric field vanishes? Comment.

E.94   Two thin concentric spherical shells with radii $R_i = 0.1$ m and $R_o = 0.5$ m share a common polar axis. Net charges of $Q_i = 2\,\mu C$ and $Q_o = 3\,\mu C$ are uniformly distributed on the shells. As usual, the electric potential vanishes infinitely far away from the shells. Determine the   (a) magnitude and direction of the net electric field   and   (b) value of the electric potential   at the following points along the polar axis: (i) $r = 1.0$ m on the north side,   (ii) $r = 0.3$ m on the south side, and   (iii) $r = 0.05$ m on the north side.

E.95   Reanalyse E.94 with spherical shells of radii $R_i = 0.05$ m and $R_o = 0.15$ m, bearing charges $Q_i = -1.6\,\mu C$ and $Q_o = 5.1\,\mu C$. Determine the   (a) magnitude and direction of the net electric field and   (b) electric potential   at the following points along the polar axis: (i) $r = 0.20$ m on the north side,   (ii) $r = 0.10$ m on the south side, and   (iii) $r = 0.01$ m on the north side.

E.96   Reanalyse E.94 with spherical shells of radii $R_i = a$ and $R_o = 3\,a$, bearing charges $Q_i = -1.6\,\mu C$ and $Q_o = 5.1\,\mu C$. Determine the   (a) magnitude and direction of the net electric field   and   (b) electric potential   at the following points along the polar axis: (i) $r = a/5$ on the north side,   (ii) $r = 2\,a$ on the south side, and   (iii) $r = 5\,a$ on the north side.

E.97   A spherical insulator of radius R has charge distributed throughout its bulk according to $\rho(r) = \rho_0 \exp(-\lambda r/R)$ for radial distances $0 \le r \le R$, and $\rho(r) = 0$ for all $r > R$. The constant factors have appropriate units (to ensure consistency): $[\rho_0] = $ C/m^3, while λ is dimensionless.

(a) Compute the total amount of charge borne by the insulator.

(b) Determine the electric   (i) field and   (ii) potential   for $r > R$.

(c) Determine the electric   (i) field and   (ii) potential   for $r < R$.

E.98   [See also E.99.] A point charge, Q, lies at the centre of a thick spherical conducting shell. A charge, q, is distributed upon the shell. Both Q and q are positive. The inner and outer edges of the shell are at radii R_i and R_o.

(a) Assume that the system is in electrostatic equilibrium, and ascertain the manner in which the electric charges are distributed throughout space.

(b) Infer the electric field throughout the following regions: (i) exterior to the shell [*i.e.*, for $r > R_o$],   (ii) within the thick shell [*i.e.*, for $R_o > r > R_i$], and   (iii) inside the cavity enclosed by the shell [*i.e.*, for $r < R_i$].

(c) From knowledge of the electric field acquired in (a), determine the electric potential throughout the regions: (i) $r > R_o$,   (ii) $R_o > r > R_i$, and   (iii) $r < R_i$.

E.99   [See also E.98.]   Repeat E.98 when $-q$ is held on the thick shell. Suppose that $Q > |-q|$.

E.100   A long thin straight wire of length L bears a uniformly distributed charge of Q.

(a) Apply Gauss's Law to estimate the electric field in the region of space adjacent to the wire and far from its endpoints.

(b) Equipotential surfaces very near to the wire are concentric cylindrical shells. Choosing the shell at radial distance $r_0 = 0.25$ m from the centre of the wire to be at 0 V, determine the potential function.

E.101   A small spherical solid lump of conductor with radius $R = 0.1$ m bears a charge of 10 mC.

(a) Describe the manner in which the electric charge on the sphere is distributed.

(b) Determine the electric potential associated with the charge on the sphere at points at radial distance $r > R$ from the centre of the sphere.

(c) Suppose that a point charge of 1 μC is placed 3 m away from the centre of the sphere. Compute the electrostatic potential energy of this system.

(d) How does the electric potential associated with the charge on the sphere differ between any pair of points located within the sphere?

E.102   Two small spherical lumps of conductor, both with radius $R = 0.1$ m, are separated by distance 0.7 m, *i.e.*, the closest distance between the two surfaces is 0.5 m. Equal-magnitude, opposite-sign charges of 10 mC are placed on the two spheres. This system is an electric dipole.

(a) Ascertain the magnitude of the system's dipole moment.

(b) Suppose instead that 5 mC charges were placed on the spheres. How ought one adjust the spacing of the spheres to yield the dipole moment determined in (a)?

(c) Sketch the pattern of electric field lines in the region of space surrounding the spheres.

(d) Sketch the pattern of electric field lines just outside the surface of the positively charged sphere.

E.103   Three point-like charges, $\{q_l, q_c, q_r\}$, lie along the x-axis, along with a distant field point, $\mathcal{P}$, as shown in the figure below. Here, we shall assume that $q_c > q_r > q_l$.

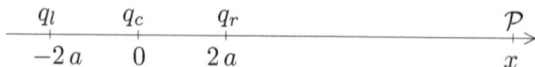

(a) The total charge is $Q_0 = q_l + q_c + q_r$. The monopole contribution to the field at $\mathcal{P}$ is obtained by assuming that the total charge is found at the origin. Compute the monopole field.

(b) An improved model augments the monopole in (a) with two dipoles: $\vec{p}_l = 2\,q_l\,a\;[\leftarrow]$ centred at $x_l = -a$, on the left; and $\vec{p}_r = 2\,q_r\,a\;[\rightarrow]$.   (i) Verify that the superposition of the monopole [in (a)] plus the two dipoles reproduces the original charge distribution. Making the approximation that both dipoles reside at the origin, compute the dipole field at $\mathcal{P}$ produced by   (ii) $\vec{p}_l$ and   (iii) $\vec{p}_r$.   (iv) Sum the monopole and two dipole contributions to obtain a better approximation to the field at $\mathcal{P}$.

(c) A still better model treats the system as a superposition of monopole, dipole, and quadrupole terms. Let $Q_1 = \frac{1}{2}(-q_l + q_r)$ and $Q_2 = \frac{1}{2}(q_l + q_r)$ be dipole and quadrupole charges, respectively. The dipole consists of charges $\pm Q_1$ separated by distance $4\,a$. Thus the associated dipole moment vector is $\vec{p} = 2\,a\,(-q_l + q_r)\;[\rightarrow]$. The quadrupole moment [*cf.* E.13] is $\mathcal{Q} = 4\,a^2\,(q_l + q_r)\;[\rightarrow]$.   (i) Verify that the superposition of the monopole [in (a)] dipole and quadrupole reproduces the original charge distribution. Compute the   (ii) dipole and   (iii) quadrupole   contributions to the field at $\mathcal{P}$.

E.104

An infinitely long thin wire bent in the shape of a parabola bears electric charge with uniform density λ_0. The mathematical expression of the parabola which places its focus at the origin of coordinates is $y = \alpha\,x^2 - \frac{1}{4\alpha}$, where α is the parameter governing its width.

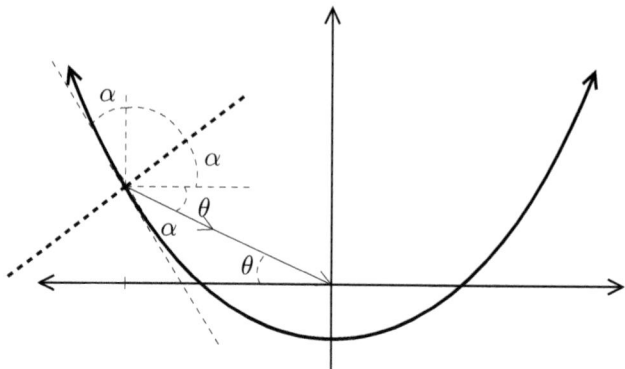

(a) Predict whether the magnitude of the electric field at the origin is greater than, equal to, or less than that which would be produced by an infinitely long straight wire lying parallel to the x-axis and passing through the parabola's vertex, $(0, -\frac{1}{4\alpha})$.

(b) Compute the electric field at the origin [*viz.*, the parabola's focal point] produced by the charge on the parabola.

[HINTS: By symmetry, only one component of the field is non-zero. The differential arclength is $ds = \sqrt{1 + \left(\frac{dy}{dx}\right)^2}\;dx = \sqrt{1 + 4\alpha^2 x^2}\;dx$. It may prove helpful to break the integral along the parabola into the part(s) for which $y(x)$ is positive and those for which $y(x)$ is negative. It is a property of all parabolas that at any point the angle lying between the normal and the vertical is the same as that between the normal and the ray to the focus. With this, one can argue that θ, in the figure, is given by $\frac{\pi}{2} - 2\alpha$. The trig identities $\cos(a+b) = \cos(a)\cos(b) - \sin(a)\sin(b)$ and $\sin(a+b) = \cos(a)\sin(b) + \sin(a)\cos(b)$, may prove useful, along with the definition of the dot product. For all parabolas the distance from any point to the focus is equal to the distance from the point to the directrix. Here the directrix is the line $y_{\mathrm{d}} = -\frac{1}{2\alpha}$.]

E.105 It is expected that the parabolic line of charge examined in E.104 should produce a potential at the origin [its focal point].

(a) Without doing any computations infer the value of the potential at the origin.

(b) Justify your result in (a) with a calculation.

E.106 Suppose that the parabolic wire in E.104 instead has position-dependent charge density: λ_0 for $x > 0$, 0 for $x = 0$, and $-\lambda_0$ for $x < 0$. [*I.e.*, constant positive charge density on the right; constant negative to the left.] Determine the electric field at the origin.

E.107 Repeat E.105 for the parabolic wire considered in E.106.

E.108 Three identical conducting spheres are labelled A, B, and C. At $t = t_0$, the charge on each sphere is: $Q_{A0} = 10\,\mu\text{C}$, $Q_{B0} = 0$, and $Q_{C0} = -6\,\mu\text{C}$. At $t_1 > t_0$, spheres A and B are brought into contact and at $t_2 > t_1$ they are separated. Subsequently, at $t_3 > t_2$, A touches C until t_4, after which they are moved apart. Finally, at t_4, C is put in contact with B until t_5 and afterward separated. Determine   (a) the total charge on all three spheres together   (i) at time t_0 and   (ii) at later times,   (b) the charge on each of the spheres after t_5.

E.109 Repeat E.108 when $Q_{A0} = 20\,q$, $Q_{B0} = -16\,q$, $Q_{C0} = 0$, and the spheres are brought into pairwise contact and separated in the following order: first B–C, then A–B, and finally B–C again.

E.110
A small sphere made of insulating material is suspended by [conducting] ideal rope from a ceiling anchor.   The mass of the insulating sphere is 60 g. A small conducting rod bearing net charge $1\,\mu\text{C}$ is brought near the sphere (at the same height), as shown in the adjacent figure. When the rod is 0.2 m from the centre of the sphere, the ideal rope is angled at $\theta = \pi/12 = 15°$ with respect to vertical. [Take the acceleration due to gravity to be $10\,\text{m/s}^2$, acting directly downward.]

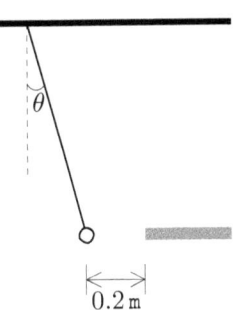

(a) Explain, briefly, why the sphere is attracted to the rod.

(b) Compute the tension force in the ideal rope.

(c) Estimate the electric charge induced on the sphere.

E.111 A particle bearing charge $Q_1 > 0$ is held fixed in place at the origin. A second particle, with charge $Q_2 > 0$ and mass M_2, is moving with speed v_0 directly toward the origin as it passes the point located a distance r_0 from the origin. Ascertain the distance of closest approach for the pair of particles.

E.112 A particle bearing charge $-5\,\mu\text{C}$ is held fixed in place. A second particle, with charge $-9\,\mu\text{C}$ and mass 0.9 g, is moving with speed 50 m/s directly toward the first as it passes a point 45 mm away. Ascertain the distance of closest approach for the pair of particles.

E.113 A particle bearing charge $+5\,\mu\text{C}$ is held fixed in place. A second particle, with charge $-9\,\mu\text{C}$ and mass 0.9 g, is moving with speed 50 m/s directly away from the first as it passes a point located 15 mm away. Ascertain the distance from the first particle at which the moving particle turns around.

E.114 A particle bearing charge $Q_1 < 0$ is held fixed in place at a point in space. A second particle, with charge $Q_2 < 0$ and mass M_2, is moving with speed v_0 directly toward the first as it passes a point located a distance r_0 away. The incoming particle stops at distance R_0 from the origin. Express the initial speed, v_0, in terms of the masses, charges, and distances.

E.115   A particle bearing charge $-5\,\mu$C is held fixed. A second particle, with charge $-9\,\mu$C and mass $0.9\,$g, is moving with speed v_0 directly toward the first as it passes the point $50\,$mm away. Later, it is noted that the second particle was instantaneously stopped $25\,$mm from the first. Infer the value of the speed v_0.

E.116   The dielectric strength of dry air is approximately 3×10^6 V/m. By rubbing your arms whilst wearing a wool sweater, it is not overly difficult to produce a spark across a gap of about $5\,$mm from your hand to a door knob, appliance, or unsuspecting victim. Estimate the potential difference between your hand and the recipient of the spark.

C

Capacitance Problems

C.1 A pair of conductors carry equal and opposite charges of magnitude 1.5 mC. The potential difference between the conductors is measured to be 1500 V. Compute the capacitance.

C.2 Two lumps of conductor are placed near each other. A charge of $+1.5\,\mu C$ is present on the first lump and $-1.5\,\mu C$ is on the second. Careful measurement reveals that the potential difference between the two lumps is 0.3 V.

(a) Determine the capacitance of this system of conductors.

(b) Compute the amount of electrostatic potential energy stored in the system.

C.3 Two small lumps of conductor are separated in space. When charges ± 20 mC are deposited on the lumps, a potential difference of 4 V is measured between them.

(a) Compute the capacitance of this system of conductors.

(b) Suppose that the charge on each lump is reduced to ± 15 mC. What then would be the electric potential difference between the lumps?

C.4 Show that   (a) $[\,\epsilon_0\,] = F/m,$ (b) $\left[\,\frac{1}{2}\,\epsilon_0\,E^2\,\right] = J/m^3.$

C.5 Parallel conducting plates have area 1 cm^2 and separation 0.25 mm. Compute the capacitance.

C.6 A parallel plate capacitor has area 2.5 cm^2. What must be the separation of the plates for its capacitance to be 15 pF?

C.7 Two parallel conducting plates are 5.9 mm apart. What must be their area for the capacitance of the system to be 1.5 pF?

C.8 A capacitor consists of two parallel plates with area $A = 0.6$ cm^2, separated by 0.59 mm. Equal and opposite charges of $1\,\mu C$ are distributed (uniformly) on the plates.

(a) Compute   (i) the capacitance of the system,   (ii) the electric potential difference between the plates, and   (iii) the amount of stored electrostatic potential energy.

(b) How do (a.i–iii) change if the amount of charge is doubled?

C.9 Repeat C.8 with area $A = 1.2$ cm^2, separation 0.59 mm, and a deposited charge of $1\,\mu C$.

C.10 A parallel plate capacitor has face area 5 mm^2 and separation $d = 0.1$ mm.

(a) Compute the capacitance of this system of conductors.

(b) Ascertain the factor by which the capacitance changes if   (i) the area is doubled and the separation halved and   (ii) the area is halved and the separation doubled.

C.11
A capacitor consists of two hollow thin-walled concentric conducting tubes of length $L = 0.01\,\text{m}$. The inner tube has radius $a_i = 0.0005\,\text{m}$ [half of a millimetre], while the outer tube has radius $a_o = 0.001\,\text{m}$. Equal and opposite charges of $\pm Q = \pm 0.01\,\mu\text{C}$ are placed on the inner and outer tubes. [Assume that the electric charge is uniformly distributed on the (effectively two-d) surfaces, and neglect fringing of the electric field near the cylinder ends.]

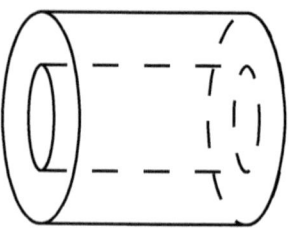

(a) Determine the electric field in the regions of space   (i) outside of the outer tube, (ii) between the tubes, and   (iii) within the inner tube.

(b) Comment on your results for (a).

(c) Compute the electric potential difference between the two tubes. [HINT: Integrate the field along a radial path from the outer to the inner tube.]

(d) Compute the capacitance of this system of conductors.

C.12   [See also E.43, E.82.] Compute the capacitance of the system of coaxial cylindrical conducting shells studied in E.43 and E.82.

C.13   Two very long thin-walled concentric cylindrical conductors have radii a and $2\,a$ respectively. Compute the capacitance per unit length for this system of conductors.

C.14   A capacitor is comprised of three identical conducting plates, each with face area A, arranged as shown in the nearby figure. The outer plates may be considered thin, while the central plate is imagined to have enough thickness to make meaningful the notion that it has distinct faces (l and r). The outer two plates are joined by an insignificant amount of conducting wire. The distance between the left and centre plates is d_l, while the gap on the right is d_r.

Self-consistently determine the charge distribution, the electric fields in the two gaps, the potential difference, and then the capacitance of this configuration of conductors. [Assume that fringing effects are negligible. If you find it convenient, let $d_{lr} = d_l + d_r$.]

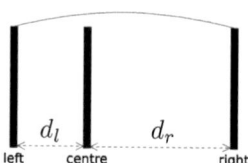

C.15   Two concentric thin-walled spherical shells composed of conducting material have radii a and b, with $b > a$.

(a) Determine the capacitance of this arrangement of conductors.

(b) Take the limit: $b \to \infty$ of the result in (a). Compare this with the capacitance of an isolated spherical shell [as determined in Chapter 12].

C.16   [See also E.42, E.81.] Compute the capacitance of the system of concentric conducting spherical shells investigated in E.42 and E.81. Specialise to the cases: (i) $R_o = 3\,R_i$ and (ii) $R_o = 2\,R_i$. [HINT: Set $q = Q$.]

C.17   Two capacitors have $C_1 = 10\,\mu\text{F}$ and $C_2 = 20\,\mu\text{F}$. Determine the effective capacitance when C_1 and C_2 are connected in   (a) parallel and   (b) series.

C.18   [See also R.17, R.45.] Two sets of conducting plates have capacitances $C_1 = 5\,\text{pF}$ and $C_2 = 20\,\text{pF}$. For C_1 and C_2 combined in   (a) series and   (b) parallel,   and then subjected to an applied potential difference of $2000\,\text{V}$, determine   (i) the effective capacitance of and (ii) the amount of energy stored in   the system.

C.19 [See also R.18, R.46.] Repeat C.18 when $C_1 = 10\,\text{pF}$ and $C_2 = 40\,\text{pF}$.

C.20
A $0.01\,\mu\text{F}$ capacitor is initially charged to 11 V.
A $0.1\,\mu\text{F}$ capacitor has zero initial charge.
These capacitors are connected via a pair of
switches [as discussed in Chapter 13]. At $t = 0$
the switches are closed and shortly thereafter
electrostatic equilibrium is established.
Determine the equilibrium potential difference
across the pair of connected capacitors.

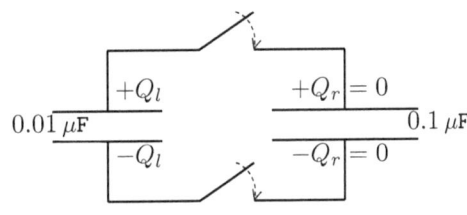

C.21 Repeat C.20 with $C_l = 20\,\text{nF}$, $V_l = 3\,\text{V}$, $C_r = 60\,\text{nF}$, and $V_r = 2\,\text{V}$.
C.22 Repeat C.20 with $C_l = 40\,\text{nF}$, $V_l = 3\,\text{V}$, $C_r = 120\,\text{nF}$, and $V_r = 1\,\text{V}$.
C.23 A parallel plate capacitor with face area $1.5\,\text{cm}^2$ and plate separation $0.1\,\text{mm}$ is sub-
jected to an applied potential of 10 V.

(a) Determine the capacitance.

(b) Determine the   (i) energy stored in,   (ii) interior volume of, and   (iii) [average]
energy density within   the capacitor.

(c) Determine the   (i) charge and   (ii) [constant] charge density   on each plate.
(iii) Ascertain the strength of the electric field in the interstitial region and verify your
result.   (iv) From the field strength, ascertain the local electric energy density within
the capacitor.

(d) Comment.

C.24 Two identical $4\,\text{mF}$ capacitors are joined in series. This combination is placed in
parallel with a $1\,\text{mF}$ capacitor. Compute the effective capacitance of this network.
C.25 [See also R.19, R.47.]

Three $1\,\text{mF}$ capacitors (small) and three $4\,\text{mF}$ capacitors (large) are linked to form the network shown
nearby. A 3 V battery is connected to this system
and sufficient time has elapsed for electrostatic equilibrium to be established.

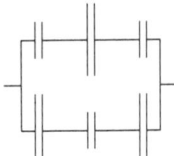

Determine   (a) the effective capacitance of, and   (b) the charge on, and   (c) the
electrostatic potential energy stored in   (i) the upper and   (ii) the lower   branch,
and   (iii) the entire network.

(d) From (b.i), determine the potential difference across   (i) each of the small capacitors and   (ii) the large capacitor   in the upper branch.   (iii) Infer the potential
difference across the entire upper branch.

(e) From (b.ii), determine the potential difference across   (i) the small capacitor and
(ii) each of the large capacitors   in the lower branch.   (iii) Infer the potential difference across the entire lower branch.

(f) Consider the upper branch. Determine the energy stored in   (i) each of the small
capacitors and   (ii) the large capacitor.   (iii) Infer the amount of energy stored in
the upper branch.

(g) Consider the lower branch. Determine the energy stored in   (i) the small capacitor
and   (ii) each of the large capacitors.   (iii) Infer the amount of energy stored in the
lower branch.

(h) From (f–g), determine the amount of energy stored in the entire network. Comment.

C.26 [See also R.20.]

Compute the effective ca-
pacitances of the adja-
cent capacitor networks,
(a) and (b). Each of the
large [small] capacitors has
capacitance $2\,C$ $[C]$.

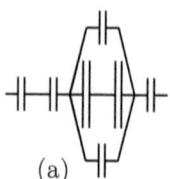
(a)

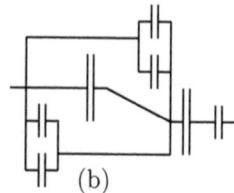
(b)

C.27 [See also R.21.] Repeat C.26 with each of the large capacitors having four times the
capacitance of the smaller ones.

C.28 [See also R.22, R.48, I.65, S.44.]

A network of identical $1\,\mu$F capacitors is illustrated in the adjacent figure. Determine the
effective capacitance of this network.

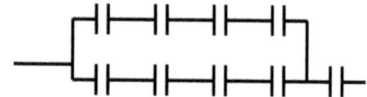

C.29

A network of identical $24\,\mu$F capacitors is illustrated in the adjacent figure. A battery with
EMF $12\,$V is connected across this network.

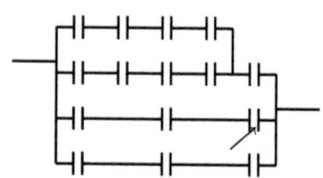

(a) Determine the effective capacitance.

(b) Determine the amounts of   (i) charge and   (ii) energy   that are held on the
network [assuming that the battery has been connected for a long time].

(c) Determine the amounts of   (i) charge and   (ii) energy   that are held on the
particular capacitor indicated by the arrow.

C.30 [See also R.24.]

Eight identical $0.5\,\mu$F capacitors are connected as
shown. A total charge of $Q = 1\,\mu$C is held on the
network.

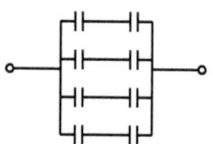

(a) Compute the effective capacitance.

(b) Determine the   (i) potential difference across and   (ii) amount of electrostatic
energy stored in   the network.

(c) How much of the total charge resides on each of the capacitors?

(d) Determine the   (i) potential difference across and   (ii) energy held within   each
of the capacitors.

(e) Comment.

C.31 [See also R.25.] A number of identical $10\,\mu$F capacitors are assembled into the network
shown nearby. The terminals of an ideal battery sustaining a potential difference of $9\,$V are
connected to the two ends of this network. A sufficient time for the system to attain
electrostatic equilibrium is assumed to have elapsed.

(a) Compute the effective capacitance of this network.

(b) How much electrical energy is stored in the system?

(c) How much charge is held on the capacitor indicated by the arrow?

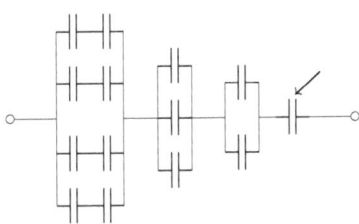

C.32   [See also R.26.] A network of eight capacitors, each with $C = 0.5\,\mu\text{F}$, appears in the nearby figure. A total charge of $1\,\mu\text{C}$ is stored on this network.

(a) Compute the effective capacitance of this network.

(b) Determine the   (i) potential difference across and (ii) amount of electrostatic potential energy stored in the network.

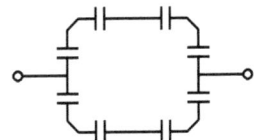

(c) Determine the   (i) charge on,   (ii) potential difference across, and   (iii) energy stored in   each of the capacitors.

C.33   [See also R.27, I.47, S.40.] A network of capacitors, each with $C = 10\,\mu\text{F}$, appears in the figure just below. An ideal battery supporting a potential difference of 5 V was connected to this network long ago.

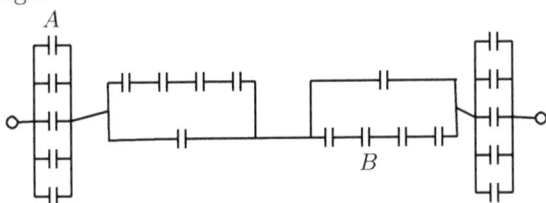

(a) Compute the effective capacitance of this network.

(b) Ascertain the amount of electrostatic energy held in   (i) the entire network, (ii) Capacitor A, and   (iii) Capacitor B.

C.34   [See also R.28.] Ten identical capacitors, C, are assembled into the networks described below.

(a) Five are connected in parallel. Determine the capacitance of this combination, C_{5P}.

(b) The remaining five are combined in series. Determine C_{5S}.

(c) Ten-capacitor networks are constructed by linking C_{5P} and C_{5S} in   (i) parallel and   (ii) series.   Determine the effective capacitances of these networks.

(d) Determine the amount of energy stored on the network: (i) C_{5P},   (ii) C_{5S}, (iii) (c-i), and   (iv) (c-ii)   when it is subjected to a 20 V potential difference. Take $C = 1\,\mu\text{F}$.

C.35   A certain type of $60\,\mu\text{F}$ capacitor has a breakdown voltage of 3 V.

(a) Design a network with an effective capacitance of $120\,\mu\text{F}$, capable of operating at up to 12 V.

(b) Determine the amount of   (i) charge and   (ii) energy   stored on your network when it is fully charged to 12 V.

C.36 A dielectric substance with $k = 5$ fully occupies the space between the two metallic plates in a parallel plate capacitor. The plates have area $A = 8\,\text{mm}^2$ and are separated by $d = \frac{1}{2}$ mm. Compute the capacitance of this device.

C.37 [See also C.38.] Two slabs of dielectric material with face area, thickness, and dielectric constant $\{A_1, d, k_1\}$ and $\{A_2, d, k_2\}$ respectively are placed side-by-side between the metallic plates of a parallel-plate capacitor. The face area and thickness of the capacitor are $\{A = A_1 + A_2, d\}$. Compute the effective capacitance of this arrangement.

C.38 [See also C.37.] Two slabs of dielectric material with face area, thickness, and dielectric constant $\{A, d_1, k_1\}$ and $\{A, d_2, k_2\}$ are placed one-atop-the-other between the metallic plates of a parallel-plate capacitor. The face area and thickness of the capacitor are $\{A, d = d_1 + d_2\}$. Compute the effective capacitance of this arrangement.

C.39
Three slabs of dielectric material with face area, thickness, and dielectric constant $\{A_{12}, d_1, k_1\}$, $\{A_{12}, d_2, k_2\}$ and $\{A_3, d, k_3\}$ respectively are placed between the metallic plates of a parallel-plate capacitor as shown in the figure. The face area and thickness of the capacitor are: $\{A = A_{12} + A_3, d = d_1 + d_2\}$. Compute the effective capacitance of this system.

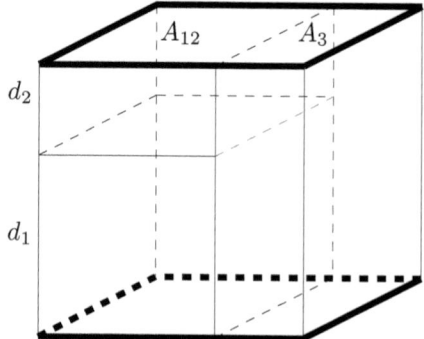

C.40 Two parallel conducting plates with area A and separated by distance D bear charges $\pm Q$. A block of conductor having the same face area as the plates and thickness $d < D$ is fully interposed between the plates.

(a) Suppose that the conductor lies symmetrically, in the middle of the gap [as suggested in Figure 15.5]. *I.e.*, the gaps on either side are $(D - d)/2$. (i) Argue for a distribution of charge, and ascertain the magnitude and direction of the electric field everywhere in the region between the plates. (ii) Compute the potential difference between the two parallel plates. (iii) Compare this potential difference with that obtained when the space between the plates is filled with vacuum or air. (iv) Compute the capacitance of this system with the conductor present. (v) Compare this capacitance to that obtained when the space between the plates is filled with vacuum or air. (vi) Propose a combinatorial model for re-interpreting this result.

(b) Repeat (a) supposing that the lump of conductor is off-centre. In this case the left- and right-hand gaps are d_l and $d_r = D - d - d_l$, respectively. [Both d_l and d_r are greater than zero.]

(c) Formally repeat (b) when the conductor is jammed against the left plate, *i.e.*, $d_l = 0$.

(d) Comment on your results for (a–c).

C.41 A parallel plate capacitor with plate area A and separation d is filled to one third of its thickness, *i.e.*, $d/3$, with a substance with dielectric constant k. The other two thirds is filled with air [treat this as though it were vacuum]. Determine the capacitance of this arrangement of conductors and dielectric.

C.42 Repeat C.41 when the dielectric has thickness $d/2$ and thus occupies one half of the volume between the plates.

C.43 Two concentric metallic spheres have radii R_i and R_o. The inner shell is coated with a dielectric material (coefficient k) extending to radius R_d, where $R_o > R_d > R_i$. Determine the capacitance of this pair of conducting spheres.

R

DC Circuits Problems

Useful Data

Resistivities of Selected Elements		$[\Omega \cdot m]$
GRAPHITE (Carbon)	ρ_C =	8.0×10^{-6}
DIAMOND (Carbon)	ρ_C =	1.0×10^{12}
ALUMINIUM	ρ_{Al} =	2.7×10^{-8}
COPPER	ρ_{Cu} =	1.7×10^{-8}
GOLD	ρ_{Au} =	2.4×10^{-8}

Charge Carrier Densities for Selected Elements		$[\,\text{number}/\text{m}^3\,]$
ALUMINIUM	η_{Al} =	18.1×10^{28}
COPPER	η_{Cu} =	8.5×10^{28}
GOLD	η_{Au} =	5.9×10^{28}

R.1 A toaster draws 10 A of current through a copper wire with cross-sectional area $0.25\,\text{mm}^2$. Estimate the drift speed of the electrons in the wire.

R.2 Repeat R.1 for an aluminium wire with cross-sectional area $0.5\,\text{mm}^2$.

R.3 Repeat R.1 for a gold wire with cross-sectional area $0.1\,\text{mm}^2$.

R.4 The resistivity of copper quoted in the table at the start of this section was measured at 20 C. Near 100 C, copper's resistivity increases to $2.2 \times 10^{-8}\,\Omega \cdot \text{m}$. Estimate the temperature coefficient of resistivity for copper.

R.5 A long coil of wire had a measured resistance of $22.00\,\Omega$ at 20 C. When remeasured at 50 C, the resistance was $25.30\,\Omega$. Determine the wire's temperature coefficient of resistivity.

R.6 A uniform rod of graphite (such as that found in a mechanical pencil) has cross-sectional area $A = 0.20\,\text{mm}^2$ and length $L = 5\,\text{cm}$.

(a) Compute the electrical resistance between the ends of the graphite rod.

Determine the resistance were the   (b) area doubled and the length halved and (c) area halved and the length doubled.

R.7 Suppose that a copper wire with uniform cross-sectional area $\mathcal{A} = 1\,\text{cm}^2$ runs from the North Pole to the Equator, $i.e.$, $L = 10^4\,\text{km}$. The density of metallic copper is $8940\,\text{kg}/\text{m}^3$.

(a) Compute the volume of copper in the wire.

(b) The world spot market price of copper hovers roughly around $8000 per tonne [1000 kg]. Estimate the dollar value of this copper wire.

(c) Compute the electrical resistance of this wire.

R.8 Repeat R.7 for an aluminium wire with the given dimensions. The density of aluminium is 2700 kg/m^3 and its price is approximately $2000 per tonne.

R.9 For each of the copper wires whose geometries are specified below, compute its (i) resistance, (ii) volume, and (iii) mass. [Take the density of copper to be 8.94 g/cm^3.]

CASE	a	b	c	d
LENGTH	5 m	30 m	5 km	500 km
CROSS-SECTIONAL AREA [mm^2]	1	5	1000	100,000

R.10 Compute the resistance of a uniform cylindrical copper wire with length 10 m and radius 0.2 mm.

R.11 Repeat R.10 when the wire is made of (a) aluminium and (b) gold.

R.12 [See also R.13. Gold is often used in switches and connectors, because it does not easily corrode.] Compute the electrical resistance of gold wires with the following dimensions: (a) length = 1.25 cm and cross-sectional area = 0.5 mm^2 and (b) length = 0.5 mm and cross-sectional area = 1.25 cm^2.

R.13 [See also R.12. Diamond is an exceedingly poor conductor.] Compute the electrical resistance of diamond "wires" with the same dimensions as in R.12.

R.14 Two wires have the same length and the same resistance. One is composed of aluminium and the other of copper. Determine the ratio of the cross-sectional areas of these two wires and state which one is larger.

R.15 An aluminium wire has cross-sectional area 0.2 mm^2 and length 15 cm.

(a) Determine the resistance of this wire.

(b) Suppose that this wire is accidentally connected across the terminals of a 1.5 V battery. [This has the effect of "shorting out" the battery and is, generally speaking, not a good idea.] (i) Determine the power consumed in the wire. (ii) What is the source of this energy? (iii) To what form is this energy converted?

R.16 Repeat R.15 for a copper wire with the same dimensions.

R.17 [See also C.18, R.45.] Two devices have resistances $R_1 = 5\,\Omega$ and $R_2 = 20\,\Omega$. For R_1 and R_2 combined in (a) series and (b) parallel, and then subjected to an applied potential difference of 20 V, determine (i) the effective resistance of and (ii) the power consumed in the system.

R.18 [See also C.19, R.46.] Repeat R.17 when $R_1 = 10\,\Omega$ and $R_2 = 40\,\Omega$.

R.19 [See also C.25, R.47, I.42.]

Three 10 Ω resistors (small) and three 40 Ω resistors (large) are linked to form the network shown nearby. A 3 V battery is connected to this system and sufficient time has elapsed for steady state to be established.

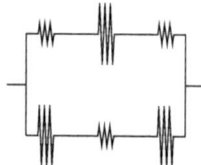

(a) Determine (a) the effective resistance of, (b) the current flowing through, and (c) the power dissipated in (i) the upper and (ii) the lower branch and (iii) the entire network.

(d) From (b.i), determine the potential difference across   (i) each of the small resistors and   (ii) the large resistor   in the upper branch.   (iii) Infer the potential difference across the entire upper branch.

(e) From (b.ii), determine the potential difference across   (i) the small resistor   and (ii) each of the large resistors   in the lower branch.   (iii) Infer the potential difference across the entire lower branch.

(f) Consider the upper branch. Determine the power dissipated in   (i) each of the small resistors   and   (ii) the large resistor.   (iii) Infer the power dissipated in the upper branch.

(g) Consider the lower branch. Determine the power dissipated in   (i) the small resistor and   (ii) each of the large resistors.   (iii) Infer the power dissipated in the lower branch.

(h) From (f–g), determine the power dissipated in the entire network. Comment.

R.20   [See also C.26.]

Compute the effective re-
sistances of the adjacent
resistor networks, (a) and
(b).   Each of the large
[small] resistors has resistance $2R$ $[R]$.

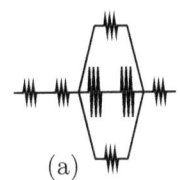
(a)

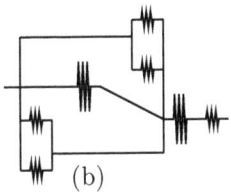
(b)

R.21   [See also C.27.] Repeat R.20 with each of the large resistors having four times the resistance of the smaller ones.
R.22   [See also C.28, R.48, I.43, S.44.]

Determine the effective resistance of the
network of resistors in the nearby figure.
Each resistor is $12\,\Omega$.

R.23
Determine the effective resistance of the
network of resistors shown in the nearby
figure.

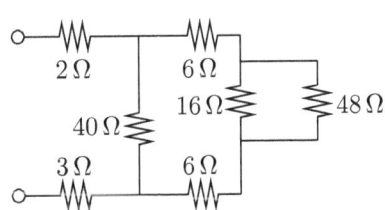

R.24   [See also C.30.]

Eight identical $20\,\Omega$ resistors are assembled into the network shown nearby. A
total current of $I = 1\,\text{mA}$ flows through
the network.

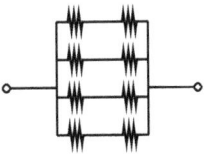

(a) Compute the effective resistance.

(b) Determine the   (i) voltage drop across and   (ii) power dissipation throughout the network.

(c) How much current passes through each of the resistors?

(d) Determine the   (i) potential difference across and   (ii) power dissipated within each of the resistors.

(e) Comment.

R.25   [See also C.31.] A number of identical $10\,\Omega$ resistors were assembled into the network shown nearby. The terminals of an ideal battery with EMF $9\,\text{V}$ were connected to the two ends of this network, and sufficient time has elapsed for the system to be in steady state.

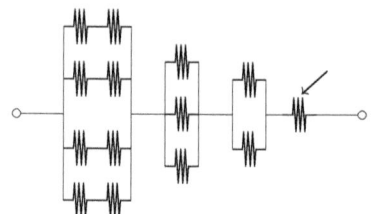

(a) Compute the effective resistance.

(b) How much electrical power is dissipated by the system?

(c) How much current passes through the resistor indicated by the arrow?

R.26   [See also C.32.] A network of eight resistors, each with $R = 20\,\Omega$, appears in the nearby figure. A total current of $1\,\text{mA}$ flows through this network.

(a) Compute the effective resistance of this network.

(b) Determine the   (i) potential difference across and (ii) amount of electric power dissipated by   the network.

(c) Determine the   (i) current through,   (ii) potential difference across, and   (iii) power dissipated in   each resistor.

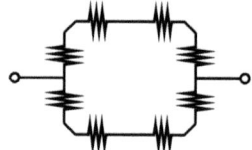

R.27   [See also C.33, I.44, S.42.] A network of resistors, each with $R = 10\,\Omega$, appears in the figure below. An ideal battery with EMF $\varepsilon = 5\,\text{V}$ was connected to this network long ago.

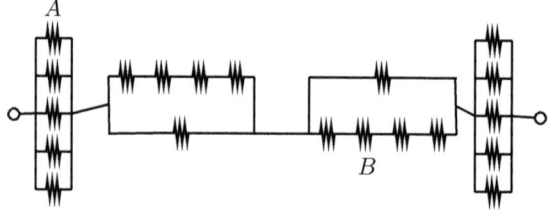

(a) Compute the effective resistance of this network.

(b) Compute the electric power dissipated by   (i) the entire network,   (ii) Resistor A, and   (iii) Resistor B.

R.28   [See also C.34.] Ten identical resistors, R, are assembled into the networks described below.

(a) Five are connected in parallel. Determine the resistance of this combination, R_{5P}.

(b) The remaining five are combined in series. Determine R_{5S}.

(c) Ten-resistor networks are constructed by linking R_{5P} and R_{5S} in (i) parallel and (ii) series. Determine the effective resistances of these networks.

(d) Compute the rate at which electrical energy is converted to heat when 2 A current flows through the resistor network: (i) R_{5P}, (ii) R_{5S}, (iii) (c-i), and (iv) (c-ii). Take $R = 50\,\Omega$.

R.29 An ideal battery is connected in series with a resistor, R_o, and a 10 A steady state current flows in the circuit. When an 8 Ω resistor is inserted into the circuit, in series with the battery and original resistor, the current drops to 8 A. Determine the original resistance, R_o, present in the circuit.

R.30 A real battery drives a current of 150 mA through a 100 Ω resistor. When the resistance is increased to 135 Ω, the current is reduced to 120 mA. Determine the ideal EMF and internal resistance of the battery.

R.31 A battery causes a current of 5.00 mA to flow through a 1000 Ω resistor and 7.50 mA to flow through a 650 Ω resistor. Determine the ideal EMF and internal resistance of the battery.

R.32 How much current passes through a 170 Ω resistor connected to the battery in R.30?

R.33 How much current passes through a 1200 Ω resistor connected to the battery in R.31?

R.34 Various batteries and identical $R = 5\,\Omega$ resistors are assembled as shown below.

(a) Write down the relation among currents I_1, I_2, and I_3 which follows from Kirchoff's Junction rule.

(b) Write two independent relations among the currents using Kirchoff's Loop rule.

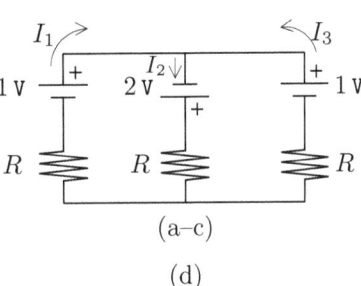

(a–c)

(c) Solve for the three currents.

(d)

(d) Another network, consisting of the same batteries and resistors, is obtained by switching the connections to the central battery, as illustrated in the second figure. Write down Kirchoffian Loop relations for this [revised] circuit configuration.

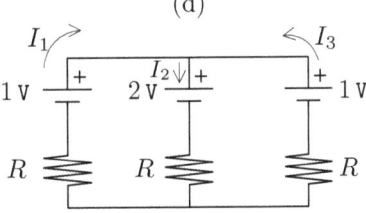

R.35

Apply Kirchoff's Rules to determine the three independent currents found in the circuit pictured nearby. The battery voltages are as indicated in the figure. The resistances are $R_1 = 30\,\Omega$ and $R_2 = 10\,\Omega$, respectively. [Do not be alarmed should one or more of these currents happen to equal zero.]

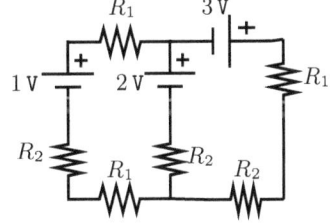

R.36
Apply Kirchoff's Rules to determine the three independent currents found in the circuit pictured nearby. The batteries have potential difference 5 V, and all of the resistors are 10 Ω.

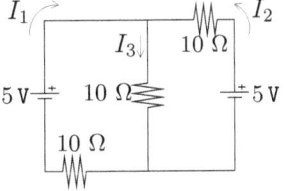

R.37 [See also R.38.] The nearby figure below shows a multi-resistor two-battery network. Assume that each resistor is $2\,\Omega$ and that each battery provides $3\,\text{V}$ electromotive force. The currents in the three segments are labelled $\{I_1, I_2, I_3\}$, as shown.

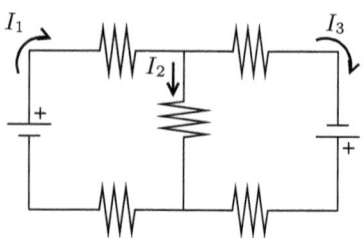

(a) Apply Kirchoff's Junction Rule to express a relation among the currents.

(b) Apply Kirchoff's Loop Rule to obtain two more such relations.

(c) Solve for the three currents in the circuit.

R.38 [See also R.37.] Repeat R.37 where the resistors are all $R = 3\,\Omega$ and the batteries have EMF $2\,\text{V}$.

R.39

Apply Kirchoff's Rules to determine the three independent currents found in the circuit pictured nearby. The batteries and resistors are noted on the circuit diagram.

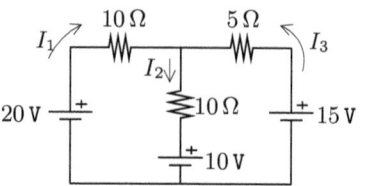

R.40

A circuit with resistors and batteries is shown in the nearby schematic diagram. Apply Kirchoff's Rules to determine the current, I_3, through the $4\,\text{k}\Omega$ resistor.

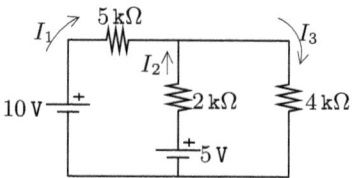

R.41 *Crossing the Wheatstone Bridge*

The adjacent circuit consisting of four resistors and a pair of batteries has a non-trivial circuit topology because of the "bridge" across its middle. Apply the constraint that there be no current flowing across the bridge to find a relation among the resistors, $\{R_1, R_2, R_3, R_4\}$.

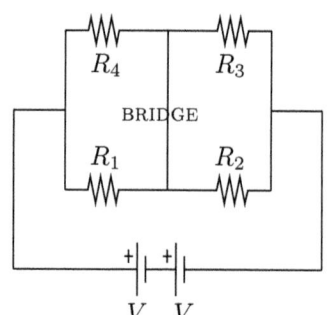

R.42 *Crossing the Wheatstone Bridge, Encore*

Three known resistors, one unknown resistor, a battery, and a voltmeter, are configured in a wheatstone bridge arrangement as shown in the adjacent figure. In this particular case, $R_1 = 100\,\Omega$, $R_2 = 50\,\Omega$, $R_3 = 10\,\Omega$, $R_4 = ?$, and the voltmeter measures exactly zero potential difference across the bridge linking one side of the circuit to the other. Infer the value of the unknown resistance.

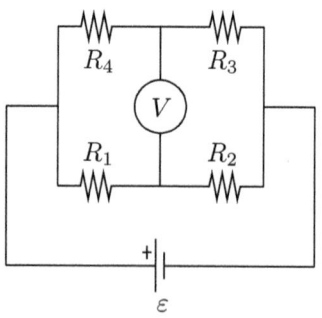

R.43 A network of resistors has effective resistance $R_{\text{eff}} = 5\,\Omega$ while a separate network of capacitors has effective capacitance $C_{\text{eff}} = 3\,\text{F}$. These two networks are to be connected in series, forming a single combined network. Compute the time constant associated with this combined network.

R.44 Repeat R.43 for $R_{\text{eff}} = 50\,\text{k}\Omega$ and $C_{\text{eff}} = 3\,\text{nF}$.

R.45 [See also C.18, R.17.] Two capacitors, $C_1 = 5\,\text{F}$ and $C_2 = 20\,\text{F}$, may be combined in series, C_s, or parallel, C_p. Similarly, two resistors, $R_1 = 5\,\Omega$ and $R_2 = 20\,\Omega$, may be arranged in series, R_s, or parallel, R_p. Determine the time constant for the RC circuit comprised of   (a) R_s–C_s,   (b) R_s–C_p,   (c) R_p–C_s, and   (d) R_p–C_p.

R.46 [See also C.19, R.18.] Repeat R.45 for capacitors $C_1 = 10\,\text{mF}$ and $C_2 = 40\,\text{mF}$ and resistors $R_1 = 10\,\Omega$ and $R_2 = 40\,\Omega$.

R.47 [See also C.25, R.19, I.62, S.43.] The networks of capacitors and resistors considered in C.25 and R.19 are combined in series to form an RC circuit.

(a) Compute the time constant for this circuit.

(b) A 3 V battery was connected to this combined network a relatively long time ago. (i) Which time intervals, chosen from the set { 1 ns, 1 μs, 1 ms, 1 s, 1 h }, constitute a relatively long time for this particular RC circuit?   Determine the [approximate] magnitude of   (ii) the current in the circuit and   (iii) the charge on the capacitor after this long time has elapsed.

(c) Long after the conditions described in (b) have been established, the battery is suddenly excised and replaced with a short length of ideal wire. Another long interval then passes.   (i) Which time intervals, chosen from the set { 1 ns, 1 μs, 1 ms, 1 s, 1 h }, constitute a relatively long time for this particular RC circuit?   Determine the [approximate] magnitude of   (ii) the current in the circuit and   (iii) the charge on the capacitor   after this (second) long time interval.

R.48 [See also C.28, R.22, I.65, S.44.] Repeat the analysis described in R.47 for the capacitors considered in C.28 and the resistors from R.22.

R.49 An RC circuit consists of a resistive element, $R = 200{,}000\,\Omega$, and a capacitive element, $C = 10\,\mu\text{F}$, connected together in series with a switch. For times before $t = 0\,\text{s}$, the switch is open, no current flows, and careful measurements reveal that there is a 5 V potential difference across the capacitor. At $t = 0\,\text{s}$ the switch is closed and the capacitor discharges through the resistor.

Determine the initial   (a) charge on the capacitor and   (b) current in the circuit.

(c) Determine the RC time constant for this circuit.

At the instants   (d) $t = 1\,\text{s}$ and   (e) $t = 3\,\text{s}$ after the switch is closed, determine the   (i) charge remaining on and   (ii) voltage drop across   the capacitor, and (iii) ascertain the current in the circuit.

R.50 Repeat R.49 with $R = 8\,\text{M}\Omega$ and $C = 50\,\text{nF}$.

R.51 A network containing both resistors and capacitors is shown in the figure just below. Each resistor has $R = 30{,}000\,\Omega$, and each capacitor has $C = 30\,\text{nF}$.

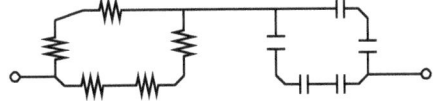

(a) Simplify the network and obtain effective values for the resistance and capacitance.

(b) Ascertain the value of the RC time constant for this network.

(c) Suppose that the capacitors are all uncharged prior to time $t = 0$, at which instant a battery with EMF 3 V is connected across the entire network. Determine the asymptotic [*i.e.*, $t \to \infty$] values of the (i) current through and (ii) charge held on the network. Sketch graphs showing the (iii) total current through and (iv) total charge on the network *vs.* time.

R.52

An RC circuit is made by combining the two networks shown nearby in series with a nine volt battery. Each of the capacitors is 100 mF; the resistors 20 Ω.

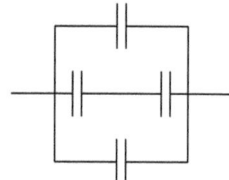
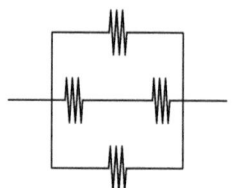

Compute the effective (a) capacitance, (b) resistance, and (c) time constant of the combined network.

(d) It takes roughly three time constants for an initially uncharged capacitor to reach 95% of its full charge. How long must one wait in this case?

R.53 Repeat R.52 with $R = 100\,\mathrm{M\Omega}$ and $C = 250\,\mathrm{nF}$.
R.54 An RC circuit consists of an initially charged capacitor, $C = 0.1\,\mu\mathrm{F}$, in series with a 20 MΩ resistor. Supposing that at $t = 3\,\mathrm{s}$ the voltage across the capacitor is $V_C = 10\,\mathrm{V}$, predict the time at which $V_C = 5\,\mathrm{V}$.
R.55 Repeat R.54 with the capacitance changed to $C = 15\,\mu\mathrm{F}$.
R.56 Repeat R.54 with the clock reading 50 s at the instant at which the capacitor voltage is equal to 10 V.
R.57 Repeat R.55 with the clock reading 50 s at the instant at which the capacitor voltage is equal to 10 V.
R.58 An RC circuit consists of an initially charged capacitor, a resistor, and a switch. At $t = 0\,\mathrm{s}$ the switch is closed. Measured values of the potential difference across the capacitor at equal time intervals are collected in the table below.

t [s]	0.0	0.5	1.0	1.5	2.0	2.5	3.0	3.5	4.0	4.5
V_C [V]	3.00	2.56	2.18	1.86	1.58	1.35	1.15	0.98	0.83	0.71
t [s]	5.0	5.5	6.0	6.5	7.0	7.5	8.0	8.5	9.0	9.5
V_C [V]	0.61	0.52	0.44	0.37	0.32	0.27	0.23	0.20	0.17	0.14

(a) Plot $\ln(V_C)$ *vs.* t. [Or plot the data directly on 2-cycle semi-log paper.]

(b) (i) What does it mean for the points to fall on a[n approximate] straight line?
(ii) Determine the slope of the best-fit straight line on your graph. (iii) Estimate the time constant for the RC circuit.

(c) Determine the capacitance when (i) $R = 625\,\mathrm{k\Omega}$ and (ii) $R = 125\,\mathrm{M\Omega}$.

R.59
The RC circuit shown nearby has $R = 20\,\mathrm{M\Omega}$, $C = 0.1\,\mu\mathrm{F}$, and $\varepsilon = 9\,\mathrm{V}$. The capacitor is initially uncharged. The switch is closed at $t_0 = 0\,\mathrm{s}$.

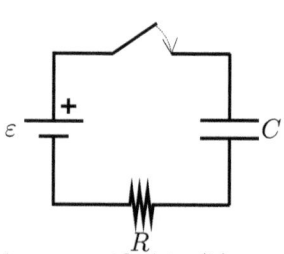

(a) Ascertain the time constant for this circuit.

(b) Predict the time at which the voltage across the capacitor is (i) 2 V, (ii) 5 V, and (iii) 8 V.

(c.i-iii) Determine the current in the circuit at each of the times specified in (b).

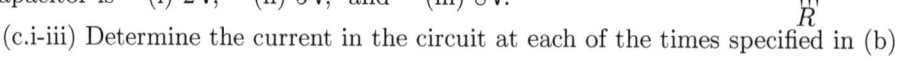

R.60   Repeat R.59 under the condition that, at t_0 when the switch is thrown, the capacitor bears an initial charge of $0.1 \, \mu\text{C}$. [Assume that the positive [negative] plate faces the positive [negative] terminal of the battery.]

R.61   An RC circuit consists of an initially uncharged capacitor, a resistor $(R = 20 \, \Omega)$, a battery $(\varepsilon = 9 \, \text{V})$, and a switch. At $t = 0 \, \text{s}$ the switch is closed. Measured values of the current in the circuit at equal time intervals are collected in the table below.

t [s]	0.0	1.0	2.0	3.0	4.0	5.0	6.0	7.0	8.0
I [mA]	450	381	322	273	231	196	166	140	119

(a) Plot $\ln (I)$ vs. t. [Or plot the data directly on one-cycle semi-log paper.]

(b) (i) What does it mean for the points to fall on a[n approximate] straight line?
(ii) Determine the slope of the best-fit straight line on your graph.   (iii) Estimate the time constant for the RC circuit.

(c) Determine the capacitance.

R.62   An RC circuit consists of an initially uncharged capacitor, a resistor, a battery $(\varepsilon = 12 \, \text{V})$, and a switch. At $t = 0 \, \text{s}$ the switch is closed. Measured values of the potential across the capacitor at equal time intervals are collected in the table below.

t [s]	0.00	0.05	0.10	0.15	0.20	0.25	0.30	0.35	0.40	0.45	0.50
V_C [V]	0.00	2.18	3.96	5.41	6.61	7.59	8.39	9.04	9.58	10.02	10.38

At times beyond $t = 2.00 \, \text{s}$, the potential across the capacitor stabilises at $12 \, \text{V}$.

(a) Plot $\ln (1 - V_C/12)$ vs. t. [Or use semi-log paper.]

(b) (i) What does it mean for the points to fall on a[n approximate] straight line?
(ii) Determine the slope of the best-fit straight line on your graph.   (iii) Estimate the time constant for the RC circuit.

(c) Determine the capacitance when   (i) $R = 10 \, \text{k}\Omega$ and   (ii) $R = 50 \, \text{M}\Omega$.

R.63   The timing for a particular type of heart pacemaker is controlled by an RC circuit in which an effective capacitor is repeatedly charged and discharged through an effective resistance of $1.5 \, \text{M}\Omega$. The time interval between pulses is equal to the time that it takes the capacitor to drain from being fully charged to $\frac{1}{e}$ of its original charge. Given that the pacemaker is intended to direct a pulse to the heart muscle 80 times per minute, determine the required effective capacitance.

R.64   A resistor R, a capacitor C, and a battery with EMF ε are arranged with switches so that the battery may be [instantaneously] inserted into and removed from an RC circuit. Prior to $t = 0$, the battery is switched out and there is zero charge on the capacitor. At $t = 0$, the battery is suddenly switched in. After exactly one time constant has elapsed, the battery is switched out and remains out. Determine the charge on the capacitor as a function of time for $t \geq 0$.

R.65   Repeat R.64, with the battery switched in for an interval of one time constant, then out for one time constant, then in again for another time constant, then out and remaining out. Determine the charge on the capacitor as a function of time for $t \geq 0$.

R.66   Determine the long-time behaviour of the circuit in R.64 if the battery were to be continuously switched in and out (repeating every two time constants). [This RC circuit is being DRIVEN in an oscillatory manner.] We will show that there is an iterative relation between successive charge maxima and identify a fixed point (toward which the system's dynamical trajectory converges).

(a) Suppose that at the end of the mth interval with the battery inserted, *i.e.*, $t = (2m-1)\tau$, the charge on the capacitor is $Q_{2m-1} = \varepsilon C \xi$, where $0 < \xi < 1$. Show that   (i) at $t = 2m\tau$, (the end of the mth interval with the battery switched out), the charge of the capacitor is $Q_{2m} = \varepsilon C \xi/e$, and   (ii) subsequently, at $t = (2m+1)\tau$, the charge on the capacitor is $Q_{2m+1} = \varepsilon C \left[1 - \frac{1}{e}\left(1 - \frac{\xi}{e}\right)\right]$.

(b) The result in (a) iteratively relates the subsequent maximum charges appearing on the capacitor. Find the fixed point of this relation by solving $\bar{\xi} = 1 - \frac{1}{e}\left(1 - \frac{\bar{\xi}}{e}\right)$.

(c) Comment.

R.67   A set of $\mathcal{N}$ light bulbs is wired purely in parallel across a battery. Each bulb has resistance $R = 15\,\Omega$. The battery has (ideal) EMF ε and internal resistance $r = \frac{3}{4}\,\Omega$. It happens that when all $\mathcal{N}$ bulbs are connected the voltage output at the terminals of the battery is precisely equal to $\varepsilon/2$. Determine the number of bulbs present in this circuit.

R.68   The mechanism underlying a voltmeter is a galvanometer. When a potential difference is measured using voltmeter, a small amount of current is diverted through it. This current leakage has an effect on the measured value of the potential difference. Let's examine how this works in a specific context.

(a) Two $1000\,\Omega$ resistors are in series across a $40\,\mathrm{V}$ battery. [All circuit connections are effected with ideal wires.] Ascertain the potential difference across one of the resistors.

Suppose that a voltmeter with an internal resistance of   (b) $10\,\mathrm{M}\Omega$ and   (c) $250\,\Omega$ is connected across one of the resistors. Determine   (i) the effective resistance facing the battery,   (ii) the current in the circuit, and   (iii) the voltage drop across the voltmeter.

R.69   [See also S.25.] While waiting for his toast to pop and his coffee to brew at breakfast, PK fell to thinking about the wonders of modern electric devices. Ignoring, for the moment, that the power supply in his kitchen is AC [coming soon in Chapter 42 *et seq.*], he sought to estimate the electric current flowing silently and invisibly through the kitchen outlet and into his appliances. The nominal voltage at the outlet is $120\,\mathrm{V}$. According to their product manuals, his toaster and coffee pot have power consumption ratings of $1300\,\mathrm{W}$ and $1100\,\mathrm{W}$ respectively. [HINT: The appliances are in parallel, mirroring the same voltage source [assumed to be constant].]

R.70   [See also S.26.] The circuit breaker for one of PK's kitchen outlets, with nominal $120\,\mathrm{V}$ supply, is set to trip at $20\,\mathrm{A}$. Supposing that PK plugs his $1200\,\mathrm{W}$ waffle iron, $1440\,\mathrm{W}$ microwave oven, and $240\,\mathrm{W}$ laptop into the same outlet and tries to run them all simultaneously. [Neglect, for the time being, that these are all AC appliances.]

(a) Determine the resistance of the   (i) waffle iron,   (ii) microwave, and   (iii) laptop.

(b) Determine the effective resistance of the three appliances arranged in parallel.

(c) Compute the total current drawn through the outlet. Does the breaker trip?

R.71   An old-fashioned incandescent light bulb consumes $60\,\mathrm{W}$ of power input when connected to a $120\,\mathrm{V}$ source. Pretending for the moment that the source is a battery maintaining a constant potential, determine   (a) the resistance in the bulb and   (b) the electric current draw when two of these bulbs are switched on in parallel.

R.72   Convert the following base-10 numbers to hexadecimal.
          (a) 96,   (b) 108,   (c) 2304, and   (d) 4321.

R.73   Convert the following hexadecimal numbers to base-10.
          (a) $D9$,   (b) 100,   (c) $B3D$, and   (d) $ABCD$.

L

Lorentz (Magnetic) Force Problems

Useful Data

eV   The electron volt is a unit of energy. One eV $\simeq 1.6 \times 10^{-19}$ J. One MeV is 10^6 eV.

Properties of Various Particles		
PARTICLE	CHARGE	MASS
electron	$-e = -1.6 \times 10^{-19}$ C	$m_e = 9.11 \times 10^{-31}$ kg
proton	$e = 1.6 \times 10^{-19}$ C	$m_p = 1.67 \times 10^{-27}$ kg
alpha	$2\,e = 3.2 \times 10^{-19}$ C	$m_\alpha = 6.64 \times 10^{-27}$ kg

Properties of Elements		
ELEMENT	ATOMIC NUMBER	ATOMIC MASS
Helium	$2\,e$	6.64×10^{-27} kg
Gold	$79\,e$	3.27×10^{-25} kg

L.1 Consider the Cartesian unit vectors, $\{\hat{\imath}, \hat{\jmath}, \hat{k}\}$, along with the five listed below.

$\vec{r} = 5\,\text{m}$ [in the xy-plane, $45°$ from the x-axis (toward the $+y$ axis)]

$\vec{v} = 5\,\text{m/s}$ [in the xy-plane, $53.13°$ from the x-axis (toward the $+y$ axis)]

$\vec{F_b} = 10\,\text{N}$ [directed along the y-axis]

$\vec{F_c} = 10\,\text{N}$ [directed along the z-axis]

$\vec{B} = 10\,\text{T}$ [directed along the y-axis]

Employ the geometric expression of the cross product to evaluate each of the following:

(a) $\hat{\imath} \times \hat{k}$,   (b) $\vec{r} \times \vec{F_b}$,   (c) $\vec{r} \times \vec{F_c}$,   (d) $\vec{v} \times \vec{B}$.

L.2 Evaluate the cross products in L.1 using component expressions.

L.3 Compute   (a) cross and   (b) dot   products for the following pairs of [3-d] vectors:

(i) 13 [in the xy-plane, at $22.62°$ anti-clockwise w.r.t. $x-$ axis],

        5 [in the xy-plane, at $53.13°$ anti-clockwise w.r.t. $y-$ axis],

(ii) $(12\,,\,5\,,\,0)$, $(-4\,,\,3\,,\,0)$,   and   (iii) $6\,[\,\downarrow\,]$, $5\,[\,\rightarrow\,]$.

L.4 A constant magnetic field, $\vec{B_0} = B_0\,[\,\odot\,]$, is found to exist in a region of space. A particle with mass M and charge q is moving with velocity $\vec{v} = v_\parallel\,[\,\odot\,] + v_\perp\,[\,\rightarrow\,]$ within this region at some time t. Compute the magnitude of the component of the magnetic force acting on the particle in the direction of its velocity at this instant.

L.5   The electric dipole considered in E.8 is comprised of equal and opposite point charges, $\pm q$, separated by distance $2\,a$. Suppose that such a dipole, with dipole moment $\vec{p} = 2\,a\,q\,\hat{\jmath}$, translates with initial velocity $\vec{v}_0 = v_0\,\hat{\imath}$ within a region of space filled with a uniform magnetic field, $\vec{B}_0 = B_0\,\hat{k}$. Compute the net magnetic   (a) force   and   (b) torque acting on the dipole at the initial instant.

L.6   The Earth's magnetic field at points on its surface may be decomposed into a vertical (radial) component and a horizontal component (directed toward magnetic north). The values of these components vary with latitude. Suppose that the local horizontal component has magnitude 2.5×10^{-5} T, and that a proton, moving due East, is observed to not fall on account of the magnetic force cancelling its weight. Argue that this scenario is possible and determine the speed of the proton.

L.7   Considering the same scenario as in L.6, what must the velocity of an electron be for its weight to be cancelled?

L.8   Three beams comprised of neutral, singly ionised $[+e]$, and doubly ionised $[+2\,e]$ gold atoms, respectively, are prepared with common beam velocity $\vec{v}_0 = 1500\,\text{m/s}\,[\hat{\imath}]$, and directed into a region of space filled with a uniform magnetic field, $\vec{B}_0 = 0.25\,\text{T}\,[\hat{k}]$.

For the   (a) neutral,   (b) singly ionised,   and   (c) doubly ionised   beam, compute the   (i) magnetic force acting on   and   (ii) acceleration experienced by   the atoms.

L.9   A collection of small spherical ceramic (insulating) beads are carefully manufactured with mass $m = 0.01\,\text{g}$, and a net positive charge of $q = 4\,\text{mC}$ is uniformly distributed throughout the interior of each bead.

(a) The beads are fed one-by-one into a two-metre-long vacuum tube in which there is a uniform electric field of $1000\,\text{V/m}$. Assuming that the beads are effectively at rest when they enter the tube, determine the speed with which they exit.

(b) The (now) fast-moving beads enter a large region of space filled with a uniform magnetic field with strength $B_0 = 0.5\,\text{T}$. The particles enter the region of space at $90°$ to the magnetic field. Determine the cyclotron   (i) radius   and   (ii) frequency   of the beads moving through the magnetic field.

L.10   Repeat L.9 with $m = 0.1\text{g}$.

L.11   Repeat L.9 with $q = 40\text{mC}$.

L.12   Various particles moving with velocity $(3\,,0\,,0) \times 10^5$ m/s $[0.1\%$ of the speed of light$]$ enter a uniform magnetic field, $\vec{B} = (0\,,0\,,1.5)$ T.

(a) If the particle is a proton, compute the   (i) magnetic force   and   (ii) acceleration that it experiences just as it enters the field. Also, compute the cyclotron   (iii) radius and   (iv) frequency   associated with the subsequent motion of the proton.

(b) Repeat (a) for a gold nucleus, with charge $79\,e$.

(c) Repeat (a) for an electron.

L.13   A beam of protons moves in a circle with radius $0.30\,\text{m}$ lying perpendicular to a uniform magnetic field with strength $0.25\,\text{T}$. Determine   (a) the speed of the protons in the beam   and   (b) the centripetal force acting on each proton.

L.14   Repeat L.13 with the circle radius and magnetic field strength equal to $0.50\,\text{m}$ and $0.15\,\text{T}$, respectively.

L.15   Singly ionised helium atoms (equivalent to alpha particles bearing $+e$ charges) are injected into a region of space containing a uniform magnetic field with strength $\frac{1}{8}$ T, and subsequently observed to move in a circle of radius $1.25\,\text{m}$.

(a) Compute the speed of the helium ions.

(b) Ascertain the magnitude of the magnetic force acting on each ion.

L.16 Uniform electric and magnetic fields, $\vec{E}_0 = 5$ V/m $\hat{\jmath}$, $\vec{B}_0 = 3$ T $\hat{k}$, fill a region of space. A particle with mass 2 kg and charge $+4$ C is injected into the region with an initial velocity of $\vec{v}_0 = v_0\,\hat{\imath}$ at the instant $t = 0$ s.

Determine the (i) magnitude and (ii) direction of the net electromagnetic force acting on the particle at the initial instant when it is injected with speed
(a) $v_0 = 0$ m/s, (b) $v_0 = 1$ m/s, and (c) $v_0 = 2$ m/s.

(d) How are your results for (a–c) affected if the particle bears charge -4 C?

(e) How are your results for (a–c) affected if the field is reversed, *i.e.*, $\vec{B}_0 = -3$ T $[\hat{k}]$?

L.17 Crossed electric and magnetic fields have magnitude $E_0 = 1500$ V/m and $B_0 = 0.6$ T, respectively. With what speed might a particle of mass 1.25 g bearing charge $0.5\,\mu$C move through these fields and remain undeflected?

L.18 Repeat L.17 when the particle has mass 2.50 g and bears charge $0.25\,\mu$C.

L.19 Repeat L.17 when the particle has mass 0.75 g and bears zero net electric charge.

L.20 A particle of mass $M = 0.003$ kg, bearing charge $Q = 5\,\mu$C, is moving with initial velocity $\vec{v}_0 = (0,\,100,\,0)$ m/s in a region of space filled with a uniform magnetic field, $\vec{B} = (0.7,\,0,\,0)$ T.

Determine the initial (a) kinetic energy of and (b) magnetic force [magnitude and direction] acting on the particle.

(c) Ascertain the magnitude and direction of the electric field which must be applied in order to select a particle with this initial velocity.

L.21
A piece of wire 7 m long is bent to form two sides of a $5:4:3$ right-triangle and immersed in a uniform magnetic field with strength $B_0 = 3$ T, directed perpendicular to the plane of the wire, as shown in the nearby figure.

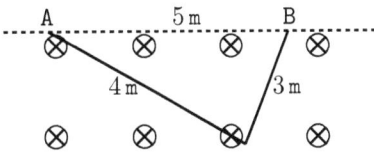

(a) Suppose that the wire bears no electric current, *i.e.*, $I = 0$ A. Compute the magnetic force acting on the (i) 4 m and (ii) 3 m segment of the wire. (iii) Sum these to obtain the net magnetic force acting on the wire.

(b) Repeat (a) for a current of 5 A entering at A and exiting at B.

(c) Repeat (a) for a current of 5 A entering at B and exiting at A.

(d) Reconsider (a–c) for a single straight segment of wire joining A and B. Comment.

L.22

A wire consisting of two straight segments carries current $I = 1.2$ A from point $\mathcal{A}$ to $\mathcal{B}$, as shown in the figure. The wire is immersed in a uniform magnetic field, $\vec{B} = 0.8$ T $[\odot]$.

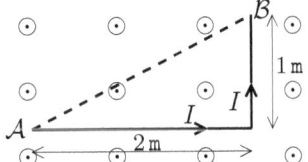

(a) Compute the magnetic force acting on the (i) 1 m and (ii) 2 m wire segment.
(iii) Determine the total magnetic force acting on the wire.

(b) Suppose instead that the bent wire is replaced with a straight wire, of length $\sqrt{5}$ m, passing directly from $\mathcal{A}$ to $\mathcal{B}$. Determine the magnetic force acting on this wire.

L.23 Repeat L.22 with the direction of the current reversed.

L.24 A region of space contains a uniform magnetic field, $\vec{B}_0 = B_0\,[\hat{k}]$, where $B_0 = 5$ T. Three straight thin wires lie in the xy-plane [perpendicular to the magnetic field] and meet at a junction, as shown in the upper figure. The lengths of these wires are: $L_1 = 1$ m, $L_2 = 2$ m, $L_3 = 3$ m. The currents that they bear are: $I_1 = 1$ A, $I_2 = 2$ A, $I_3 = 3$ A.

(a) Suppose that I_3 flows toward the junction, *i.e.*, $\hat{L}_3 = (-1/\sqrt{2}, 1/\sqrt{2}, 0)$, and compute the net magnetic force acting on the wires.

(b) Same as (a), except in this case wire segment 2 is bent so that it lies at $45°$ in the zx-plane, *i.e.*, $\hat{L}_2 = (-1/\sqrt{2}, 0, 1/\sqrt{2})$. [The three wires are no longer co-planar.] Compute the magnetic force acting on the wires.

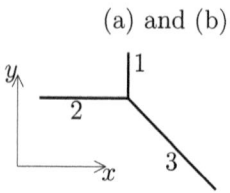

(a) and (b)

(c) In this case, wire segment 3 is lengthened and follows a curved arc within the xy-plane. Its endpoints are the same as in (a). Compute the magnetic force acting on this collection of wires.

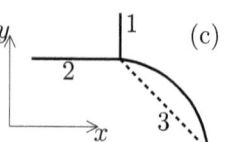

(c)

L.25
A thin conducting bar, 93.75 cm long, lies on a horizontal frictionless plane as illustrated nearby. Two identical thin wire springs, each with spring constant $k = 75\,\text{N/m}$, carry electric current to the bar. A uniform magnetic field with strength $B_0 = 0.16\,\text{T}$ is directed perpendicular to the plane. When an electric current flows through the wires and bar, the springs are observed to stretch by 1 cm. [The springs are otherwise held in place.] Determine the magnitude and direction of the current flowing through the bar.

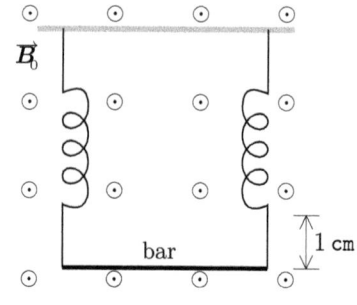

L.26
A planar rectangular region is bounded by the lines $y = 0$, $y = L$, $x = a$, and $x = b$, with $0 < a < b$ AND $0 < L$. Within the bounds of the rectangle, there is a magnetic field directed perpendicular to the plane, *i.e.*, along $[\hat{k}]$. The magnitude of the field varies according to $B(x, y) = \frac{B_0 L}{x}$, where B_0 is a constant with units of tesla.

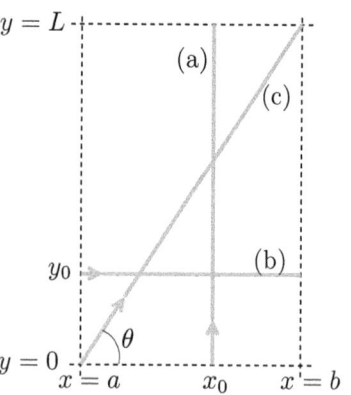

By integrating $d\vec{F} = I\, d\vec{L} \times \vec{B}$, determine the magnetic force acting on a straight wire bearing current I extending from:
(a) $(x_0, 0)$ to (x_0, L) [with length L],
(b) (a, y_0) to (b, y_0) [with length $b - a$], and
(c) $(a, 0)$ to (b, L) [with length $\sqrt{(b-a)^2 + L^2}$].

[HINTS: For (a) use $d\vec{L}_{(a)} = dy\,\hat{j}$. For (b) use $d\vec{L}_{(b)} = dx\,\hat{i}$. For (c) employ $d\vec{L}_{(c)} = ds\cos(\theta)\,\hat{i} + ds\sin(\theta)\,\hat{j}$, where s denotes the arc length along the diagonal wire, so $s \in [0, \sqrt{(b-a)^2 + L^2}]$. Also, $\cos(\theta) = (b-a)/\sqrt{(b-a)^2 + L^2}$, $\sin(\theta) = L/\sqrt{(b-a)^2 + L^2}$. Use $x = a + s\cos(\theta)$ while $y = s\sin(\theta)$ to express the force-component integrands entirely in terms of arclength and constant parameters.]

L.27 Repeat the analysis in L.26 when the magnitude of the magnetic field is described by $B(x, y) = B_0 L y/x^2$. [HINTS: As above, and for (c) use $\int \frac{u\,du}{(p+q\,u)^2} = \frac{1}{q^2}\left[\ln(p + q\,u) + \frac{p}{p+q\,u}\right]$. This integral is easily verified using the technique of partial fractions.]

L.28 Three triangular current loops { (i), (ii), (iii) }, are pictured below. Each loop has base 4 m and height 6 m, and bears current $I = 2\,\text{A}$. The arrows indicate the sense in which the current flows. An applied magnetic field with uniform strength $B_0 = \frac{1}{3}$ T fills the region containing the loops.

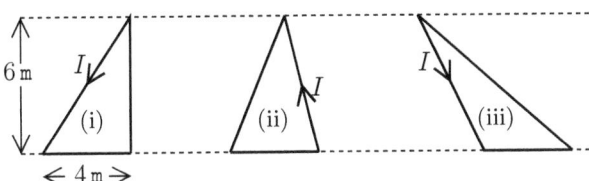

(a) (i–iii) Compute the magnetic dipole moment of each of the three current loops.

(b) (i–iii) Compute the magentic force exerted on each loop when $\vec{B}$ is directed [→].

Compute the magnetic torque exerted on each loop when the applied field acts in the direction   (c) (i–iii) [⊙]   and   (d) (i–iii) [←].

Compute the magnetic potential energy of each loop when the applied field acts in the direction   (e) (i–iii) [↑]   and   (f) (i–iii) [⊗].

L.29 Consider the three wire loops pictured below. An electric current, I, circulates in each loop as indicated.

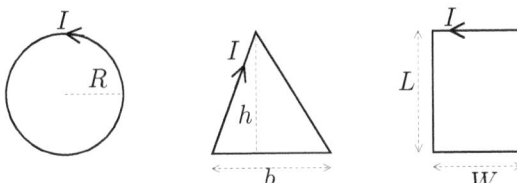

(a) Compute the magnetic moment associated with the current in each of the loops.

A uniform magnetic field,   (b) $\vec{B}_{(b)} = B_0$ [→]   and   (c) $\vec{B}_{(c)} = B_0$ [⊗],   is suddenly switched on. Determine the magnetic   (i) force acting on,   (ii) torque acting on, and (iii) potential energy of   each loop.

L.30 A current, $I = 3$ A, flows in the loop of wire exhibited in the nearby figure.

(a) Determine the magnetic dipole moment vector associated with this loop.

A uniform magnetic field,   (b) $\vec{B}_{(b)} = \frac{1}{6}$ T [↑] and (c) $\vec{B}_{(c)} = \frac{1}{3}$ T [⊙],   is suddenly applied. Compute the magnetic   (i) force   and   (ii) torque acting on the loop.   (iii) Calculate the potential energy of the loop in this field.

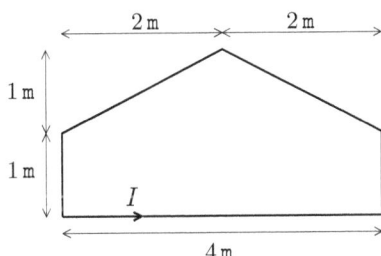

L.31 Consider the same planar region as was considered in L.26. Here, we shall compute the net force acting on a loop of current directed anticlockwise along the perimeter of the region in the presence of a magnetic field directed perpendicular to the plane with magnitude: $B(x,y) = B_0\, y(L - y)/x^2$.

Determine the magnetic force acting on the straight wire segment
(a) from $(a, 0)$ to $(b, 0)$,   (b) from $(b, 0)$ to (b, L),   (c) from (b, L) to (a, L), and (d) from (a, L) to $(a, 0)$.

(e) Determine the net magnetic force acting on the current loop. Comment.

L.32
A non-planar current loop consists of a semi-circular
section with radius r lying in the xy-plane, along
with a semi-elliptical section with semi-major and
semi-minor axes r, and $r/2$ respectively. A current
flows along the loop as indicated by the arrows in the
nearby figure.

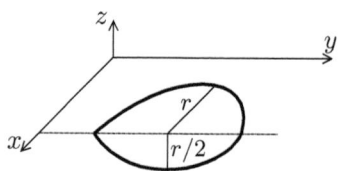

Determine the magnitude and direction of the unique magnetic moment associated with
this current loop.

L.33
A non-planar current loop consists of four
straight wire segments joining the points:
$\{(0, 0, 0), (2, 0, 0), (1, 1, -1), (0, 2, 0), (0, 0, 0)\}$,
as shown in the nearby figure. A current $I = \frac{1}{3}$ A flows
in the loop. Determine the loop's magnetic moment.

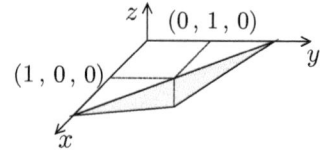

L.34 A wire loop in the shape of a triangular pennant, with edge lengths X_0, Z_0, and
$\sqrt{X_0^2 + Z_0^2}$, is bent to fit on the surface of a half-cylinder of radius $r_0 = X_0/\pi$.

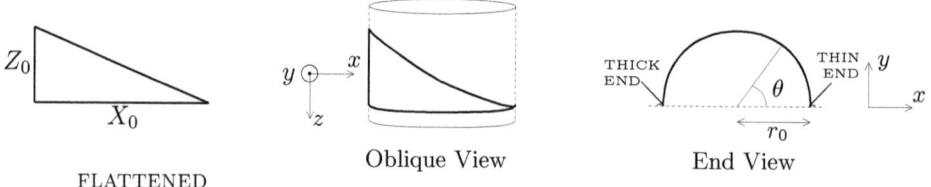

FLATTENED   Oblique View   End View

WRAPPED ONTO A HALF-CYLINDER

The wire carries current I. Determine, up to overall orientation, the magnetic moment
associated with the current in this bent loop of wire.   [HINTS: Partition the area on the
cylindrical surface (bounded by the loop) into thin strips. The height of the strip at angle θ is
$z(\theta) = Z_0\,\theta/\pi$. The differential area of the thin strip is $dA = z(\theta)\,r\,d\theta$. The radial unit vector
at angle θ has (x, y) components: $(\cos(\theta), \sin(\theta))$. Also, $\int u \cos(u)\,du = \cos(u) + u\sin(u)$ and
$\int u \sin(u)\,du = \sin(u) - u\cos(u)$, as may be verified using integration by parts.]

L.35 A long straight copper wire with rectangular cross-section has edge dimensions $0.5\,\text{mm}$
and $2.0\,\text{mm}$. A $1.5\,\text{A}$ current flows through the wire (with current density distributed uniformly
across its face area). An external magnetic field with magnitude $B_0 = 0.25\,\text{T}$ is applied to
the wire perpendicularly to both its length and its wider edge. Compute the Hall potential
across the thickness of the wire.

L.36 Repeat L.35 when the magnetic field is perpendicular to the length and narrower
edge and compute the Hall potential across the width of the wire.

L.37 Repeat L.35 for an aluminium wire.

B

Magnetic Field Production Problems

Useful Data

μ_0 The magnetic permeability of free space is $\mu_0 = 4\pi \times 10^{-7}$ T·m/A.

B.1 A thin straight wire of length 5 m carries current $I = 3$A.

(a) Determine the strength of the magnetic field produced by the current in the wire at a field point lying 10 cm directly away from the centre of the wire.

(b) Compare your result with the approximate value for the strength of the field when the wire is assumed to be infinitely long.

B.2 Repeat B.1 for a field point lying 1 m directly away from the centre of the wire.

B.3 Repeat B.1 for a field point lying 10 m directly away from the centre of the wire.

B.4 For the current-carrying wire in B.1 determine the magnitude of the magnetic field produced at a point lying 10 cm away from an endpoint,   (a) in a direction perpendicular to and   (b) parallel with   the wire.

B.5 For the current-carrying wire in B.2 determine the magnitude of the magnetic field produced at a point lying 1 m away from an endpoint,   (a) in a direction perpendicular to and   (b) parallel with   the wire.

B.6 For the current-carrying wire in B.3 determine the magnitude of the magnetic field produced at a point lying 10 m away from an endpoint,   (a) in a direction perpendicular to and   (b) parallel with   the wire.

B.7 Determine the strength of the magnetic field at the centre of a circular loop of current, $I = 3$A, when the radius of the loop is: (a) 0.1 m,   (b) 1 m, and   (c) 10 m.

B.8 A circular loop of conductor with radius 1 m bears a current of 3 A.

(a) Compute the magnetic dipole moment of the current loop.

(b) Determine the strength of the magnetic field produced along the symmetry axis of the loop at distance   (i) 0.1 m,   (ii) 1 m, and   (iii) 10 m   from its centre.

B.9
Consider the arrangement of a long straight wire and concentric circular loops illustrated in the adjacent figure. The radii of the inner and outer loops are 2 m and 4 m, respectively. The perpendicular distance from the centre of the loops to the long straight wire is 3 m. The inner loop carries a current of 2 A oriented anti-clockwise; the outer 3 A clockwise; and the straight wire 4 A from left to right.

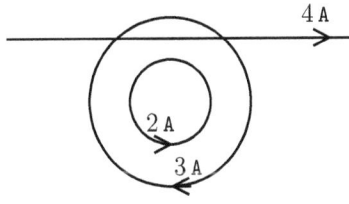

(a) Determine the magnetic field contribution at the centre of the concentric loops provided by the   (i) inner loop,   (ii) outer loop, and   (iii) straight wire.

(b) Sum these to obtain the net magnetic field at the centre of of the loops.

B.10 Repeat B.9 with the current in the long straight wire reversed.

B.11

Our goal is to determine the magnetic field produced at the point labelled $\mathcal{P}$ in the adjacent figure. The current in the very long horizontal wire is $I_h = 1.5\,\text{A}\ [\rightarrow]$, while that in the long vertical wire is $I_v = 1.2\,\text{A}\ [\uparrow]$. The inner loop, with radius $R_i = 0.3\,\text{m}$, bears a current of $I_i = 0.9\,\text{A}\ [\text{clockwise}]$, while the outer loop, with radius $R_o = 0.6\,\text{m}$, carries $I_o = 1.8\,\text{A}\ [\text{anti-clockwise}]$.

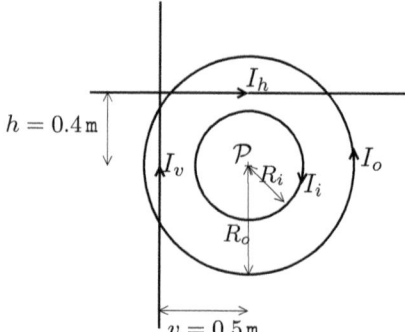

Compute the contribution to the magnetic field at $\mathcal{P}$ provided by the current   (a) I_h,   (b) I_v,   (c) I_i, and   (d) I_o.

(e) Sum these to obtain the net magnetic field at $\mathcal{P}$.

B.12 Repeat B.11 with the current in the inner loop reversed.

B.13 Repeat B.11 with the current in the vertical wire reversed.

B.14 A circular loop of radius $0.80\,\text{m}$, made of conducting wire, bears an electric current, $I = 0.4\,\text{A}$. A particle with charge $q = 1.2\,\mu\text{C}$ is launched with speed $v_0 = 500\,\text{m/s}$ directly toward the loop along its axis, from an initial point a distance of $z_0 = 0.60\,\text{m}$ away. Determine the magnetic force exerted on the particle at the instant it is launched.

B.15

An electric current, I, flows down the y-axis toward the origin, along a quarter-circular segment of radius R, and then outward along the x-axis. This path is illustrated in the adjacent figure. Determine the magnetic field produced by this current at the origin.

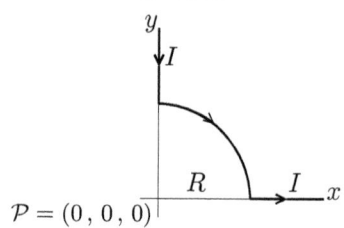

B.16 [See also B.17.]

A semi-infinite thin wire extending along the positive x-axis from the origin, $x = 0$, bears current I_0 toward increasing values of x. Determine the magnitude and direction of the magnetic field associated with this current at the point $(x_0, y_0, 0)$, located in the first quadrant of the xy-plane $[i.e., x_0 > 0$ and $y_0 > 0]$.

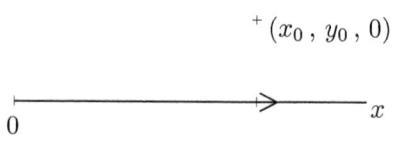

B.17 Let's consider two variations on B.16.

(a) Suppose that the field point lies in the fourth quadrant, $i.e., (x_0, -y_0, 0)$.

(b) Suppose that the current flows toward $x = 0$.

B.18 Two long thin wires consist of two straight segments joined by a semi-circular piece of radius R as shown in the nearby figures. A current, I_0, flows in the direction indicated by the arrows. The field point of interest lies at the midpoint of the line joining the endpoints of the semi-circular segment. Determine the net magnetic field produced by the current in the wires at the field point in (a) and (b).

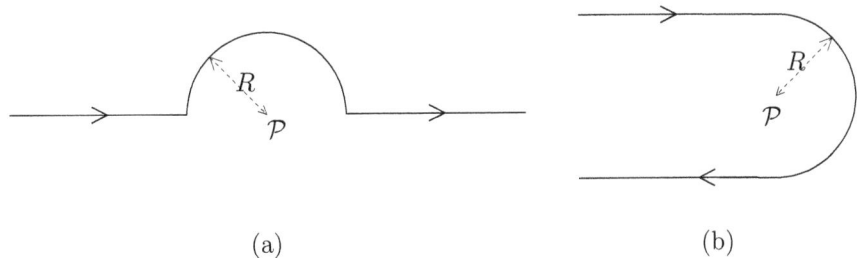

(a)   (b)

B.19 [See also B.22, B.23.] A very long wire is bent at a 90° angle and carries current I_0. In the plane containing the wire, the current flows down the y-axis [toward $y = 0$] and out along the x-axis [away from $x = 0$]. Express the net magnetic field, as a function of x, produced by the current in this wire at field points lying along the line $y = x$, for $x > 0$.

B.20 Repeat B.19 for points along the line $y = x$, with $x < 0$.

B.21 Repeat B.19 for points along the line $y = -x$, with $x > 0$.

B.22 Consider the scenario described in B.19. Suppose that a short thin straight wire carries current I_w from $\vec{r}_i = (2, 2, 0)$ m to $\vec{r}_f = (8, 8, 0)$ m. Compute the magnetic force acting on the short wire. [HINTS: Employ $d\vec{F} = I_w\, d\vec{L} \times \vec{B}$, using the expression for the field along the wire as determined in B.19, and integrate over the extent of the short wire. Although it is conventional to parameterise the integral in terms of arclength, here it is somewhat simpler to use x, with $ds = \sqrt{2}\, dx$.]

B.23 [See also B.19, B.24.] A very long wire is bent at a 90° angle and carries current I_0. In the plane containing the wire, the current flows down the y-axis and out along the x-axis. (a) Determine the net magnetic field [expressed as a function of x] produced by the current in this wire at field points lying along the line $y = \alpha x$, for $\alpha > 0$ AND $x > 0$. (b) Take the limit $\alpha \to 1$ of your result and comment. [HINTS: The cosine of the angle between the line of field points and the x-axis is $1/\sqrt{1 + \alpha^2}$; while the sine of this angle is $\alpha/\sqrt{1 + \alpha^2}$.]

B.24 Consider the scenario described in B.23. Suppose that a short thin straight wire carries current I_w from $\vec{r}_i = (2, 2\alpha, 0)$ m to $\vec{r}_f = (8, 8\alpha, 0)$ m. Compute the magnetic force acting on the short wire. [HINTS: As given in B.22, except that here $ds = \sqrt{1 + \alpha^2}\, dx$.]

B.25

A very long thin wire is bent at an angle α at a point we will consider to be the origin of coordinates. A current I_0 flows in the wire (as shown in the nearby figure).

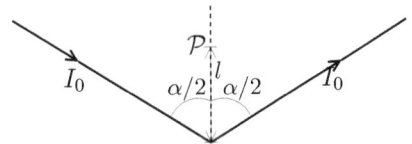

Our goal is to ascertain the magnetic field at the field point a distance l from the origin along the ray which bisects the bend angle.

(a) Determine the perpendicular distance from the field point to each straight section of wire [in terms of l and $\alpha/2$].

Determine the contribution to the magnetic field at $\mathcal{P}$ produced by the   (b) inbound and   (c) outbound   current in the wire.

(d) Sum the results from (b) and (c) to obtain the net magnetic field at $\mathcal{P}$.

(e) Take the limits: (i) $\alpha \to \pi/2$ and   (ii) $\alpha \to \pi$ of the magnitude of the field. Comment.

B.26 [See also E.104.]

An infinitely long thin wire bent in the shape of a parabola bears constant electric current I. The mathematical expression of the parabola which places its focus at the origin of coordinates is $y = \alpha x^2 - \frac{1}{4\alpha}$, where α is the parameter governing its width.

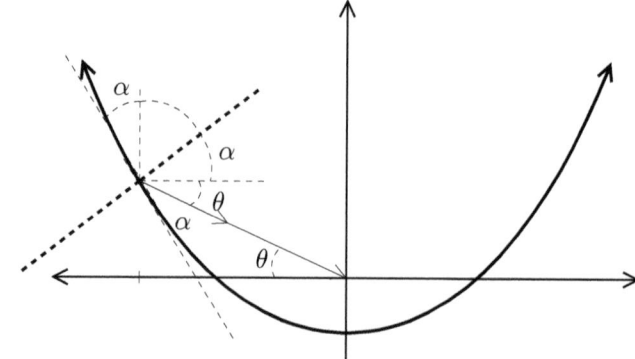

(a) Predict whether the magnitude of the magnetic field at the origin will be greater than, equal to, or less than that which would be produced by an infinitely long straight wire lying parallel to the x-axis and passing through the parabola's vertex, $(0, -\frac{1}{4\alpha})$.

(b) Compute the magnetic field at the origin [*viz.*, the parabola's focal point] produced by the current in the wire. [HINTS: See E.104.]

B.27

Two very long parallel wires are set $5\,\text{m}$ apart. A point of interest, $\mathcal{P}$, lies $2\,\text{m}$ from the first wire and $3\,\text{m}$ from the second.

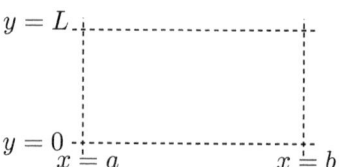

(a) Suppose that each wire carries current $I = 2\,\text{A}\ [\rightarrow]$. Determine the contribution to the magnetic field at $\mathcal{P}$ produced by the current in   (i) wire 1 and   (ii) wire 2. (iii) Determine the net magnetic field at $\mathcal{P}$.

(b) Argue whether the force of interaction between wires 1 and 2 is attractive or repulsive when their currents are flowing in the same direction.

(c) Suppose that wires 1 and 2 carry currents $I_1 = 2\,\text{A}\ [\rightarrow]$ and $I_2 = 2\,\text{A}\ [\leftarrow]$. Determine the contribution to the magnetic field at $\mathcal{P}$ arising from the current in   (i) wire 1 and (ii) wire 2.   (iii) Determine the net magnetic field at $\mathcal{P}$.

B.28   Two thick wires each carry current I_0 in the same horizontal direction. Each wire is of length L and weight $M g$. One wire is held in place by insulating support pieces, while the other floats, unsupported, directly beneath the first. The centres of the wires are a constant distance, a, apart. Assuming that a is greater than the diameter of the wires, and that $a \ll L$, determine the magnitude of I_0 in terms of M, g, L, a, and constants of nature.

B.29   Explicitly compute the value of the current in B.28 when $L = \frac{3}{4}\,\text{m}$, $a = 10\,\text{mm}$, $M = 75\,\text{g}$, and $g = 10\,\text{m/s}^2$.

B.30

A planar rectangular region is bounded by the lines $y = 0$, $y = L$, $x = a$, and $x = b$, with $0 < a < b$ and $0 < L$. Everywhere within the bounds of the rectangle, there is a magnetic field with magnitude given by $B(x,y) = \frac{B_0 L}{x}$, directed perpendicular to the plane, *i.e.*, along $\hat{k}$.

Compute $\int \boldsymbol{B} \cdot d\vec{s}$ along the segments:
(a) from $(a, 0)$ to $(b, 0)$,   (b) from $(b, 0)$ to (b, L),   (c) from (b, L) to (a, L), and
(d) from (a, L) to $(a, 0)$.

(e) Ascertain the circulation of the magnetic field around the perimeter of the rectangle.

B.31 Repeat B.30 when the magnetic field is: $\vec{B} = \frac{B_0\,L}{x}\,[\rightarrow]$.

B.32 Repeat B.30 when the magnetic field is: $\vec{B} = \frac{B_0\,L}{x}\,[\uparrow]$.

B.33 Repeat B.30 for (a) $\vec{B} = \frac{B_0\,L\,y}{x^2}\,[\rightarrow]$ and (b) $\vec{B} = \frac{B_0\,L\,y}{x^2}\,[\uparrow]$.

B.34 Repeat B.30 for (a) $\vec{B} = \frac{B_0\,y\,(L-y)}{x^2}\,[\rightarrow]$ and (b) $\vec{B} = \frac{B_0\,y\,(L-y)}{x^2}\,[\uparrow]$.

B.35 Repeat B.30 for $\vec{B} = B_0\,\frac{x^2\,(y+L)}{(b-a)^2\,L}\,\hat{\imath} + B_0\,\frac{x\,(y+L)^2}{(b-a)\,L^2}\,\hat{\jmath}$.

B.36 Consider the Amperian circuits [closed paths for computing the circulation of the magnetic field] shown in the nearby figure. The currents in the thin wires are: $I_1 = 1$, $I_2 = 2$, $I_3 = 3$, and $I_4 = 4$ (all in A). Determine the enclosed current for each of the Amperian paths (a–d).

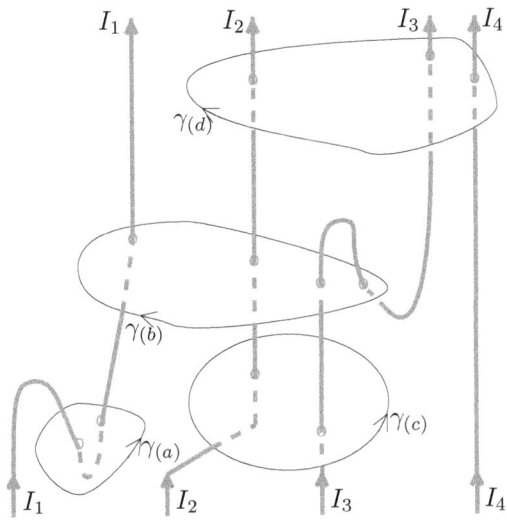

B.37
Three thin-walled very long and straight conducting cylinders are concentrically nested as illustrated in the figure. The radii of the cylindrical shells are R_i, R_m, and R_o for the inner, middle, and outer shells respectively. The middle shell carries an electric current, $I_m = I_0$, which is uniformly distributed about its circumference and directed into the page. The currents on the inner and the outer shells are $I_i = \frac{1}{3}I_0$ and $I_o = \frac{2}{3}I_0$, with both emerging from the page.

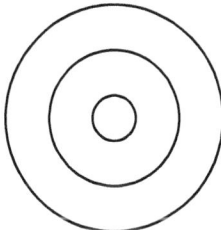

(a) Compute the circulation of the magnetic field about a closed circular path of radius r, concentric with the conductors and in a plane perpendicular to their axis of symmetry.

(b) Ascertain the current enclosed within the Amperian circuit with radius r, for (i) $R_o < r$, (ii) $R_m < r < R_o$, (iii) $R_i < r < R_m$, and (iv) $0 < r < R_i$.

(c) Invoke Ampère's Law to determine the magnetic field produced by the currents in these concentric cylindrical shells in the four separate regions of space: (i) $R_o < r$, (ii) $R_m < r < R_o$, (iii) $R_i < r < R_m$, and (iv) $0 < r < R_i$.

B.38 A long conducting rod is made by coating an insulating rod with a negligibly thin layer of conductor. The original insulating rod was 1 cm in diameter. Suppose that a current of 5 A flows [uniformly] along the rod. Determine the magnetic field produced (a) in the space exterior to and (b) within the rod.

B.39 An infinitely long thick wire with radius a bears electric current which is more densely distributed as one moves outward from the centre of the wire, according to

$$J(r) = J_0 \frac{r^4}{a^4} , \ r \le a , \qquad \text{and} \qquad J(r) = 0 , \ r > a .$$

The total current flowing through the wire is measured to be I_0.

(a) Determine the factor J_0 in terms of the total current and the radius of the wire.

(b) Determine the magnetic field produced by this current in the region of space (i) outside of and   (ii) within   the wire.

B.40 Repeat B.39 for the current density within the wire: $J(r) = J_0 \left(1 - \frac{r^2}{a^2}\right)$, $r \le a$.

B.41 Repeat B.39 for the current density within the wire: $J(r) = J_0 \frac{r\,(1-r)}{a^2}$, $r \le a$.

B.42 Repeat B.39 for the current density within the wire: $J(r) = J_0 \cos\left(\frac{\pi r}{2a}\right)$, $r \le a$.

B.43 Repeat B.39 for the current density within the wire: $J(r) = J_0 \cos^2\left(\frac{\pi r}{2a}\right)$, $r \le a$.

B.44 Repeat B.39 for the current density within the wire: $J(r) = J_0 \sin\left(\frac{\pi r}{2a}\right)$, $r \le a$.

B.45 Repeat B.39 for the current density within the wire: $J(r) = J_0 \sin^2\left(\frac{\pi r}{2a}\right)$, $r \le a$.

B.46 Two long straight thin parallel wires, separated by a negligible distance, carry currents of $12\,\text{A}$ and $6\,\text{A}$. Use Ampère's Law to estimate the magnitude of the magnetic field at the field point located $0.75\,\text{m}$ from the wires when the two currents are   (a) parallel and (b) anti-parallel.

B.47
A planar rectangular region is bounded by the lines $y = 0$, $y = L$, $x = a$, and $x = b$, with $0 < a < b$ and $0 < L$. Everywhere within the bounds of the rectangle, there is a time-dependent electric field given by $\boldsymbol{E} = E(t; x, y)\,\hat{k} = \frac{E_0\,x\,y\,t\,(8-t)}{(b-a)\,L}\,\hat{k}$, for times $0 \le t \le 8$ (in s).

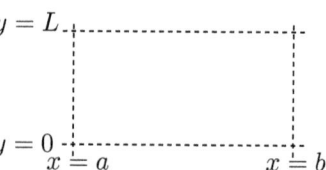

(a) Compute the electric flux passing through the planar rectangular surface at time t.

(b) Determine the displacement current at time t.

(c) What happens to   (i) the flux and   (ii) the displacement current   at $t = 4\,\text{s}$?

B.48 Repeat B.47 for $E(t; x, y) = E_0 \frac{x^2}{(b-a)^2} \sqrt{t\,(8-t)}$.

B.49 Repeat B.47 for $E(t; x, y) = E_0 \frac{y^2}{L^2} \sin(\pi t)$.

B.50 Repeat B.47 for $E(t; x, y) = E_0 \frac{x^2\,y^2}{(b-a)^2\,L^2} \sin\left(\frac{\pi t}{4}\right)$.

B.51 Repeat B.47 for $\boldsymbol{E} = E_{0x} \cosh\left(\pi\,(t-4)\right)\,\hat{i} + E_{0y} \sinh\left(\pi\,(t-4)\right)\,\hat{j} + E_{0z} \tanh\left(\pi\,(t-4)\right)\,\hat{k}$, where E_{0x}, E_{0y}, and E_{0z} are constants with appropriate units.

I

Faraday–Induction Problems

I.1
A long straight wire lying in a small groove on a smooth surface bears current I. As shown in the nearby figure, a circular loop of wire is moved from the point marked i to that marked f. Determine the orientation of the current induced in the loop [*i.e.*, clockwise or anti-clockwise as viewed in the figure] while the loop is moving past the point marked (a) and (b).

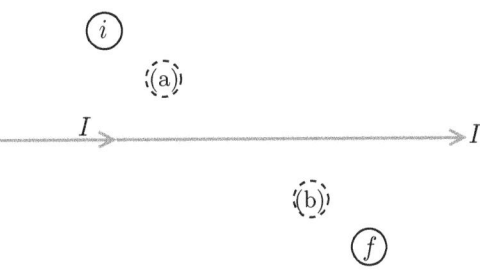

I.2 COAXIAL cable consists of a cylindrical wire concentric with a thin cylindrical conducting shell. The radius of the inner wire is R_i; the shell R_o. The current used to drive some device or to carry a signal from one piece of equipment to another passes along the inner wire and returns along the outer shell, completing the circuit. We suppose that the current, I, flows along the wire's outer surface, and that both surface currents are uniformly distributed.

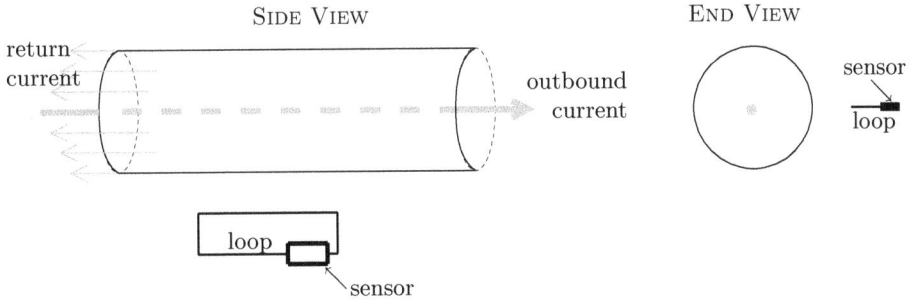

Consider a particular piece of coaxial cable alongside a loop of conductor connected to a (black-box) sensor which detects the presence or absence of electric current in the loop. We shall use Ampère's Law to determine the magnetic field produced by the current flowing in the coaxial cable at various points in space. We choose a circle of radius r centered on and perpendicular to the coaxial cable as the Amperian circuit.

(a) Apply and adapt the expression for the magnetic circulation to the present context, and simplify.

(b) Ascertain the net electric current enclosed by the Amperian circuit for (i) $R_o < r$, (ii) $R_i < r < R_o$, and (iii) $r < R_i$.

(c) Apply Ampère's Law to determine the strength of the magnetic field within the distinct regions of space: (i) $R_o < r$, (ii) $R_i < r < R_o$, and (iii) $r < R_i$.

(d) In particular, what contribution does the current flowing in the coaxial cable make to the net magnetic flux through the loop?

(e) A sudden failure in the device being powered causes the outbound current flowing along the inner cable to not return via the coaxial conducting shell. Explain the effect that this failure has on the magnetic   (i) field in the space surrounding the cable, and (ii) flux impinging on the loop.

[The black-box sensor associated with the loop triggers a circuit-breaker which stops the flow of current through the cable. In this way the risk of accidental electrocution is lessened.]

I.3 [See also I.4.]

A planar rectangular region is bounded by the lines $y = 0$, $y = L$, $x = a$, and $x = b$, with $0 < a < b$ and $0 < L$. Everywhere within the bounds of the rectangle, there is a magnetic field with magnitude given by $B(x,y) = \frac{B_0 L}{x}$, directed perpendicular to the plane, *i.e.*, along $\hat{k}$.

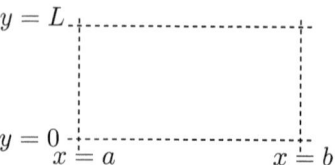

Compute the magnetic flux through the rectangle [using the planar [MINIMAL] surface].

I.4 Is it possible to obtain a different result for I.3 by computing the flux through a NON-MINIMAL surface bounding the rectangle?

I.5 Repeat I.3 for $\vec{B} = \frac{B_0 L}{x}$ $[\rightarrow]$.

I.6 Repeat I.3 for $\vec{B} = \frac{B_0 L y}{x^2}$ $[\odot]$.

I.7 Repeat I.3 for   (a) $\vec{B} = \frac{B_0 y (L-y)}{x^2}$ $[-\hat{k}]$ and   (b) $\vec{B} = \frac{B_0 y (L-y)}{x^2}$ $[\hat{j}]$.

I.8 Repeat I.3 for $\vec{B} = B_0 \frac{x^2 (y+L)}{(b-a)^2 L} \hat{i} + B_0 \frac{x (y+L)^2}{(b-a) L^2} \hat{j} + B_0 \hat{k}$.

I.9 Repeat I.3 for $\vec{B} = B_0 \frac{x (y+L)}{(b-a) L} \left[\frac{x}{b-a} - \frac{y+L}{L} \right] \hat{k}$.

I.10 Repeat I.3 for $\vec{B} = B_0 \frac{x (y+L)}{(b-a) L} \left[\frac{x}{b-a} + \frac{y+L}{L} \right] [-\hat{k}]$.

I.11 A circular loop of area $A = 0.24\,\mathrm{m}^2$ lies in the xy-plane. A uniform magnetic field with $B_0 = 0.25\,\mathrm{T}$, directed along $-\hat{k}$, bathes the loop. In each of the cases considered separately below, the change occurs in 3×10^{-3} s. Indicate the sense in which electric current is induced in the loop and estimate the magnitude of the EMF when   (a) the field strength increases to $0.30\,\mathrm{T}$,   (b) the field reverses direction [from $-\hat{k}$ to $+\hat{k}$], and   (c) the loop rotates $30°$ about the x-axis.

I.12 Repeat I.11 when the magnetic field originally points in the $-\hat{j}$ direction. [Under reversal in (b), it changes to point along $+\hat{j}$.]

I.13 The planar rectangular region shown in the figure adjacent to I.3 is bounded by the lines $y = 0$, $y = L$, $x = a$, and $x = b$, with $0 < a < b$ and $0 < L$. Everywhere within the bounds of the rectangle there is a time-dependent magnetic field given by $\vec{B} = B(t; x, y) \hat{k} = \frac{B_0 x y t (8-t)}{(b-a) L} \hat{k}$, for times $0 \le t \le 8$ (in s).

(a) Compute the magnetic flux passing through the planar rectangular surface at t.

(b) Determine the induced EMF at t.

(c) What happens to   (i) the flux and   (ii) the EMF   at $t = 4\,\mathrm{s}$?

I.14 Repeat I.13 for $B(t; x, y) = B_0 \frac{y^2}{L^2} \sqrt{t(8-t)}$.

I.15 Repeat I.13 for $B(t; x, y) = B_0 \frac{x^2}{(b-a)^2} t^2 (t^2 - 8t + 16)$.

I.16 Repeat I.13 for $B(t; x, y) = B_0 \frac{x^2 y^2}{(b-a)^2 L^2} \sin\left(\frac{\pi t}{4}\right)$.

I.17 Repeat I.13 for $B(t; x, y) = B_0 \frac{x (y+L)}{(b-a) L} \left[\frac{x}{b-a} t - \frac{y+L}{L} t^2 \right] \hat{k}$.

I.18 Repeat I.13 for $B(t; x, y) = B_0 \frac{x(y+L)}{(b-a)L} \left[\frac{x}{b-a} \sin(\pi t) + \frac{y+L}{L} \cos(\pi t) \right] [-\hat{k}]$.

I.19 Repeat I.13 for $\vec{B} = B_{0x} \cosh\left(\pi (t-4)\right) \hat{i} + B_{0y} \sinh\left(\pi (t-4)\right) \hat{j} + B_{0z} \tanh\left(\pi (t - 4)\right) \hat{k}$, where B_{0x}, B_{0y}, and B_{0z} are constants with appropriate units.

I.20 [See also I.21.]

A conducting bar is able to move without friction or drag along two parallel rails separated by 0.4 m. The rails are joined by a short length of wire and a 12 Ω resistor. The entire system of rails, bar, and resistor is immersed in a uniform magnetic field directed perpendicular to the plane formed by the rails, $\vec{B}_0 = 1.5\,\mathrm{T}\, [\odot]$. Suppose that the bar is in motion to the right at an initial speed of 0.8 m/s.

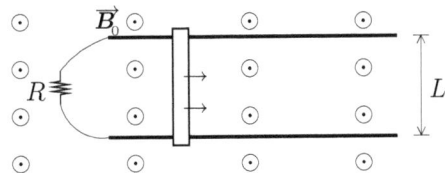

(a) In what direction does the induced current traverse the bar?

(b) What is the magnitude of the induced EMF?

(c) What is the magnitude of the induced current?

(d) What force must be applied to the bar in order that it maintain its forward velocity?

I.21 Suppose that the bar in I.20 has mass 2 kg. Determine the maximum distance through which it moves when no external force is applied to it.

I.22 Repeat I.20 when the magnetic field is $\vec{B}_0 = 1.5\,\mathrm{T}\, [\otimes]$.

I.23 Repeat I.20 when the magnetic field is $\vec{B}_0 = 1.5\,\mathrm{T}\, [\uparrow]$.

I.24 Repeat I.20 when the magnetic field is $\vec{B}_0 = 1.5\,\mathrm{T}$ [angled at 30° from the horizontal to the right and coming out of the page].

I.25 Referring to the figure in I.20, suppose that $R = 6\,\Omega$, that $L = 1.5\,\mathrm{m}$, that $B_0 = 2.5\,\mathrm{T}$, and that 24 W are being dissipated in the resistor. Determine   (a) the speed of the bar and (b) the force required to keep it in uniform motion.

I.26 Two parallel conducting rails on the surface of a plane inclined at angle θ [WRT horizontal] are separated by distance L and joined by a resistor, R. A conducting bar of mass M is able to move without friction along the rails. A uniform magnetic field directed straight down fills the space occupied by the rails.

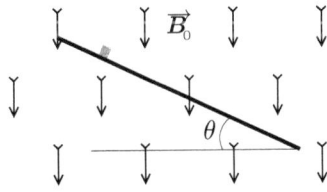 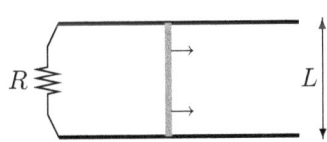

(a) Determine the component of the weight of the bar acting down along the incline.

(b) When the bar is moving at uniform speed [its terminal velocity] down the incline, the up-incline component of the magnetic force exerted on the bar precisely cancels the force determined in (a). Infer the current in the bar when this occurs.

(c) Confirm that the input power from the action of the weight force is matched by the power dissipated in the resistor.

(d) Discuss qualitatively what would happen should the initial speed of the bar be (i) less than   and   (ii) greater than   the terminal velocity.

I.27

Two parallel conducting rails are fixed in space, separated by distance L. Each pair of ends is joined by a resistor, R. A conducting bar is able to move without friction along the rails. A uniform magnetic field is directed perpendicular to the plane including the rails. An external force is applied so as to keep the bar in uniform motion at speed v_0 along the rails in the direction shown.

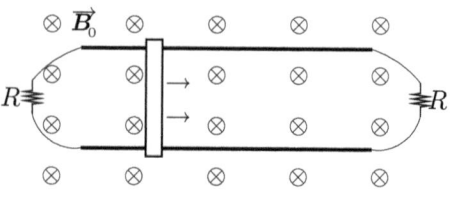

(a) Determine the current flowing through the resistor on the   (i) left   and   (ii) right   in the figure.

(b) Ascertain the current flowing through the moving bar.

(c) Determine the magnetic force acting on the bar.

(d) What applied force is required to keep the bar in uniform motion?

(e) Determine the electrical power being dissipated in the resistor on the   (i) left   and   (ii) right   in the figure.

(f) Ascertain the net electrical power being dissipated by the two resistors.

(g) Determine the mechanical power input by the external applied force which acts to keep the bar moving uniformly.

I.28   Repeat I.27 with $L = 0.30\,\text{m}$, $R = 20\,\Omega$, $B_0 = 0.75\,\text{T}$, and $v_0 = 3\,\text{m/s}$.

I.29

A long thin straight wire and a rectangular loop made of thin wire lie in the same plane as shown in the nearby figure. The dimensions of the loop are $L = 8\,\text{cm}$ and $W = 6\,\text{cm}$. The near edge of the loop is at distance $d = 4\,\text{cm}$ from the wire. A voltmeter, V, is inserted into the loop, with negligible effect on its geometry. The current in the wire increases linearly from zero according to $I = 5\,t$.

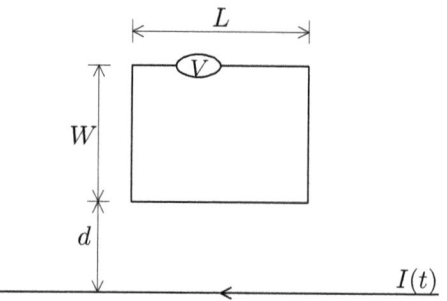

(a) Ascertain the voltage in the loop registered by the voltmeter [while the current is increasing].

(b) How does your result for (a) change if, rather than a single loop, the wire in the rectangle forms a coil with $\mathcal{N} = 12$ turns?

I.30   Repeat I.29 when the current in the wire is $I(t) = 2.0 \sin(120\,\pi\,t)$.

I.31

A coaxial cable consists of two long straight thin cylindrical tubes fashioned from conducting material. The radius of the outer tube is b; the inner a. Both tubes have length l, and $l \gg b$. Current I flows in one direction on the inner tube and returns in the opposite direction on the outer. The current in each tube is assumed to be uniformly distributed. A short section of the cable is shown in the adjacent figure.

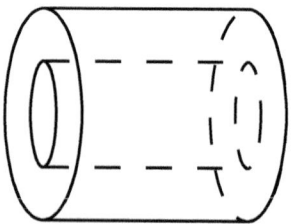

(a) Apply Ampère's Law to determine the magnetic field   (i) outside of the outer tube, (ii) between the tubes, and   (iii) within the inner tube.

(b) A planar surface from the axis of symmetry of the coaxial cable passes through both shells and extends toward infinity in a particular direction. Calculate the magnetic flux passing through this surface. [The cable, and hence also the surface, has length l.]

(c) The inductance of this particular coaxial cable is the ratio of the magnetic flux, computed in (b), to the current, I. Determine the flux per unit length for this cable.

I.32   Another coaxial cable is fashioned with the same dimensions as that considered in I.31, except that the inner cylinder is filled with conducting material. Suppose that the current carried on the inner wire is no longer confined to the surface of the tube but instead distributed throughout the bulk of the solid conductor. Without doing any computation, predict whether the inductance per unit length of this cable is less than, equal to, or greater than that of the hollow cable in I.31.

I.33   Repeat the analysis of I.31 for an inner current distributed uniformly across the face of the inner cylinder [as discussed in I.32.].

I.34

The RL circuit shown in the adjacent figure has $\varepsilon = 5\,\text{V}$, $R = 25\,\Omega$, and $L = 100\,\mu\text{H}$. The switch is closed at the instant $t = 0$.

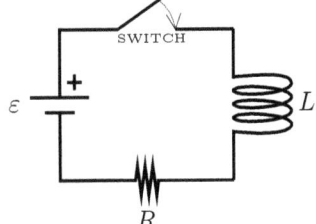

Compute   (a) the inductive time constant for   and (b) the asymptotic current in   this circuit

(c) Determine the instant at which the current is $\frac{1}{e}$ times its asymptotic value.

I.35   Repeat I.34 with the values: $\varepsilon = 12\,\text{V}$, $R = 10\,\text{m}\Omega$, and $L = 50\,\text{mH}$.

I.36

The RL circuit shown (with its switch open) in the adjacent figure was connected long in the past. At $t = 0$ the switch is thrown, opening a bypass of the inductor. Let $R_1 = 50\,\Omega = R_2$, $\varepsilon = 9\,\text{V}$, and $L = 5\,\text{mH}$.

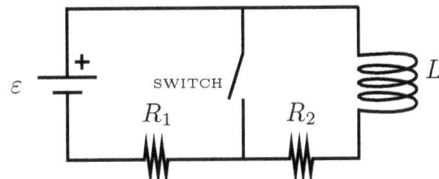

(a) Determine the current passing through the inductor just prior to $t = 0$.

(b) Ascertain an expression for the current through the inductor valid after $t = 0$.

I.37   Repeat I.36 with $R_1 = 100\,\Omega$, $R_2 = 50\,\Omega$, $\varepsilon = 5\,\text{V}$, and $L = 2\,\text{mH}$.

I.38   Repeat I.36 for unspecified R_1, R_2, ε, and L.

I.39   A battery powering a circuit failed suddenly at $t = 0$. At $t = 50\,\mu\text{s}$, the current in the circuit had dropped to half its initial value.   (a) Estimate the inductive time constant of the circuit.   (b) If the resistance in the circuit was $125\,\Omega$, compute its inductance.

I.40   A battery powering a circuit failed suddenly at $t = 0$. At $t = 3\,\text{ms}$, the current in the circuit had dropped to one quarter of its initial value.   (a) Estimate the inductive time constant of the circuit.   (b) If the resistance in the circuit was $125\,\Omega$, compute its inductance.

I.41   A battery powering a circuit failed suddenly at $t = 0$. At $t = 0.5\,\text{s}$, the current in the circuit had dropped to 80% of its initial value.   (a) Estimate the inductive time constant of the circuit.   (b) If the resistance in the circuit was $36\,\text{m}\Omega$, compute its inductance.

I.42   [See also R.19, S.41.] The resistor network analysed in R.19 was later found to possess self-inductance $L = 0.9\,\text{mH}$. Compute the value of the inductive time constant for this circuit.

I.43 [See also R.22, R.48, I.43, S.44.] The network of resistors considered in R.22 was later determined to have self-inductance $L = 11\,\mu\text{H}$. Compute the value of the inductive time constant for this circuit.

I.44 [See also R.27, S.42.] The network of resistors studied in R.27 was later found to possess self-inductance $L = 0.5\,\text{H}$. Compute the value of the inductive time constant for this circuit.

I.45 [See also S.57.]

A transformer consists of a ferrous core around which are wrapped two coils, designated the primary and the secondary. The core efficiently channels magnetic flux from one coil to the other. In an IDEAL TRANSFORMER, all of the magnetic flux generated by the primary passes through the secondary and dissipative effects are negligible.

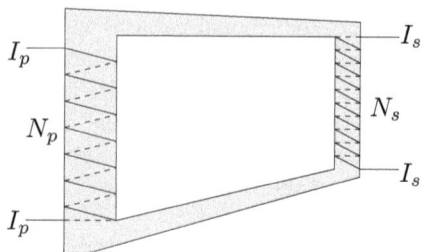

Let the number of windings in the primary be N_p; the secondary N_s. A current, I_p, passes through the ideal wire making up the primary coil. This current generates a magnetic field in the surrounding space—particularly within the transformer core. The field in the core is amplified [it's ferromagnetic] and assumed to be completely confined. Hence, a constant flux, $\Phi_{pp} = \Phi_0 = \Phi_{ss}$, passes through every loop of the primary and secondary coils.

(a) Express the self-inductance of the   (i) primary and   (ii) secondary   coil in terms of Φ_0 and the winding numbers.

(b) Express the mutual inductances   (i) M_{ps} and   (ii) M_{sp} in terms of Φ_0 and the winding numbers.

I.46 [See also C.25, S.39.] The network of capacitors studied in C.25 was later found to possess self-inductance $L = 0.25\,\mu\text{H}$. Compute the natural [angular] frequency of this circuit.

I.47 [See also C.33, S.40.] The network of capacitors studied in C.33 was later found to possess self-inductance $L = 0.05\,\text{H}$. Compute the natural [angular] frequency of this circuit.

I.48 Six capacitor–inductor pairings in LC circuits are presented in the nearby table.

System	C [nF]	L [μH]	System	C [nF]	L [μH]
A	400	20	B	20	200
C	8000	20	D	200	80
E	8000	400	F	400	400

Which pairs of systems, if any, are dynamically equivalent [*i.e.*, possess the same natural oscillation frequency, ω_0]?

I.49 An LC circuit has $L = 0.1\,\text{H}$ and $C = 100\,\text{mF}$ connected via a switch. Prior to $t = 0\,\text{s}$, the switch is open and a charge of $3.0\,\mu\text{C}$ resides on the capacitor. At $t = 0\,\text{s}$, the switch is closed. Determine the   (a) charge on the capacitor and   (b) current in the circuit   for times after $t = 0\,\text{s}$.

I.50 An LC circuit has $L = 0.1\,\text{H}$ and $C = 100\,\text{mF}$ connected via a switch. The switch was closed at some time in the past (prior to $t = 0\,\text{s}$). At the instant $t = 0\,\text{s}$, the capacitor has charge $q(0) = 3.0\,\mu\text{C}$, and current $I(0) = 1.5\,\mu\text{A}$ flows in the circuit [leaving the capacitor]. Determine the   (a) charge on the capacitor and   (b) current in the circuit   for times after $t = 0\,\text{s}$.

I.51 Repeat I.50 with the current flowing into the capacitor.

I.52

The tracing from an oscilloscope measuring current in an LC circuit is sketched in a the figure on the right. The horizontal axis shows time (in **s**), while the vertical axis shows current, measured in **mA**.
(a) From the oscilloscope tracing, determine the (i) amplitude, (ii) period, (iii) angular frequency, and (iv) cycle frequency of the oscillating current.
(b) Express the current in terms of a (i) **sine** and (ii) **cosine** function.

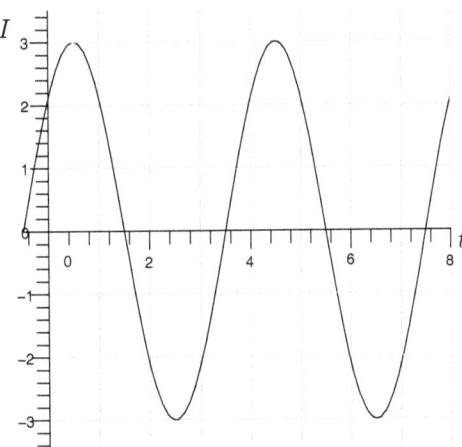

I.53 [See also I.73, S.45.] A $500\,\Omega$ resistor, a $500\,\text{nF}$ capacitor, and a $20\,\text{mH}$ inductor are connected in series by means of ideal wires.

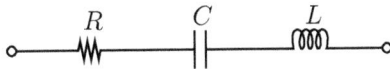

Compute the (a) natural frequency and (b) damping parameter of this circuit.
(c) Determine whether the circuit is under-/over-/critically damped.

I.54 [See also I.74, S.46.] Repeat I.53 with $R = 10\,\text{k}\Omega$, $C = 5\,\mu\text{F}$, and $L = 2\,\text{mH}$.
I.55 [See also I.75, S.47.] Repeat I.53 with $R = 10\,\text{k}\Omega$, $C = 5\,\mu\text{F}$, and $L = 50\,\text{mH}$.
I.56 [See also I.76, S.48.] Repeat I.53 with $R = 10\,\text{k}\Omega$, $C = 50\,\text{pF}$, and $L = 20\,\text{mH}$.
I.57 [See also I.77, S.49.] Repeat I.53 with $R = 200\,\Omega$, $C = 500\,\text{nF}$, and $L = 20\,\text{mH}$.
I.58 [See also I.78, S.50.] Repeat I.53 with $R = 500\,\Omega$, $C = 100\,\text{pF}$, and $L = 90\,\text{mH}$.
I.59 [See also I.79, S.51.] Repeat I.53 with $R = 400\,\Omega$, $C = 100\,\text{pF}$, and $L = 4\,\mu\text{H}$.
I.60 [See also I.80, S.52.] Repeat I.53 with $R = 400\,\Omega$, $C = 500\,\text{nF}$, and $L = 20\,\text{mH}$.
I.61 [See also I.81, S.53.] Repeat I.53 with $R = 10\,\text{k}\Omega$, $C = 20\,\text{pF}$, and $L = 0.5\,\text{mH}$.
I.62 [See also C.25, R.47.] The combined resistor and capacitor network analysed in R.47 was later found to possess self-inductance $L = 36\,\text{mH}$. Determine the (a) natural frequency and (b) damping parameter for this circuit. (c) Ascertain whether the circuit is under-/over-/critically damped.
I.63 Repeat I.62 with self-inductance $L = 180\,\text{mH}$.
I.64 Repeat I.62 with self-inductance $L = 90\,\text{mH}$.
I.65 [See also C.28, R.22, R.48.] The combined resistor and capacitor network analysed in R.48 was later found to possess self-inductance $L = 30\,\text{mH}$. Repeat the analysis of I.62 for this circuit.
I.66 Repeat I.65 with self-inductance $L = 120\,\text{mH}$.
I.67 Repeat I.65 with self-inductance $L = 4.8\,\text{mH}$.
I.68 A particular RCL circuit happens to be critically damped.

(a) Show that the circuit parameters must satisfy the relation $4\,L = R^2\,C$.

(b) Determine how the character of the circuit changes when (i) the resistance and inductance are both doubled and the capacitance halved, (ii) the resistance and capacitance are both doubled and the inductance halved, and (iii) the capacitance and inductance are both doubled and the resistance halved.

(c) Show that $(A + Bt)\,e^{-\gamma t}$ [for unspecified constants $\{A, B\}$] satisfies the dynamical equation when the condition expressed in (a) is in force.

I.69
A battery, resistor, capacitor, and inductor are joined in series, along with an open switch. The capacitor bears no initial charge. At $t = 0$, the switch is closed. Take $R = 60\,\Omega$, $C = 138.9\,\mu F$, $L = 80\,mH$, and $\varepsilon = 9\,V$.

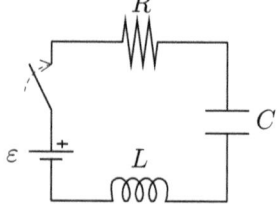

(a) Predict   (i) the amount of charge on the capacitor and   (ii) the current in the circuit   after a long time has elapsed.

(b) Determine   (i) the charge on the capacitor and   (ii) the current in the circuit at all times after $t = 0$.

I.70 Repeat I.69 with the parameters: $R = 1000\,\Omega$, $C = 40\,nF$, $L = 10\,mH$, and $\varepsilon = 9\,V$.
I.71 Repeat I.69 with the parameters: $R = 40\,\Omega$, $C = 16\,pF$, $L = 1\,\mu H$, and $\varepsilon = 9\,V$.
I.72 Repeat I.69 with the parameters: $R = 12\,k\Omega$, $C = 200\,mF$, $L = 30\,\mu H$, and $\varepsilon = 12\,V$.
I.73 An ideal battery with constant EMF $\varepsilon = 9\,V$ is connected to the circuit in I.53 at the instant $t = 0$. Determine the current in the circuit as a function of time for $t \geq 0$.
I.74 Repeat the analysis of I.73 for the circuit described in I.54.
I.75 Repeat the analysis of I.73 for the circuit described in I.55.
I.76 Repeat the analysis of I.73 for the circuit described in I.56.
I.77 Repeat the analysis of I.73 for the circuit described in I.57.
I.78 Repeat the analysis of I.73 for the circuit described in I.58.
I.79 Repeat the analysis of I.73 for the circuit described in I.59.
I.80 Repeat the analysis of I.73 for the circuit described in I.60.
I.81 Repeat the analysis of I.73 for the circuit described in I.61.
I.82 It is our intention to build an AC generator with RMS voltage output $V_{\mathrm{RMS}} = 120\,V$ at $60\,Hz$.   [This means that the instantaneous voltage varies sinusoidally with time between $\pm 120\sqrt{2}\,V$.] The magnet that we have at hand produces a uniform field of $0.20\,T$ throughout the region of space occupied by the coil. The coil has $\mathcal{N} = 200$ turns of wire. Assume that we are able to choose optimal geometry to maximise the flux changes through the coil as it rotates in the field.

(a) Ascertain the face area of the coil needed to meet the design specifications.

(b) Determine the length of wire needed to make the coil under the assumption that its cross-sectional shape is   (i) circular and   (ii) square.

S

AC Circuits Problems

S.1 Compute the magnitude and phase of (a) $2 + i$ and (b) $3 - 3i$.

S.2 Compute the magnitude and phase of $5892839 \times (1648868 + 190995\,i)$.

S.3 Express the complex numbers (a) $40 + 9i$, (b) $60 + 91\,i$, and (c) $12 - 35\,i$ as phasors (*i.e.*, in Euler form).

S.4 Simplify $\frac{5+5\,i}{3-3\,i}$.

S.5 Simplify $\frac{3+4\,i}{i(1-i)}$ and express the result in terms of magnitude and phase.

S.6 Simplify $\frac{7+24\,i}{4-3\,i}$ and express the result as a phasor.

S.7 Simplify $\frac{10+24\,i}{52+39\,i}$ and express as a phasor.

S.8 Represent $10\,\cos(\omega\,t)$ as a phasor.

S.9 Represent $10\,\cos(\omega\,t + \frac{\pi}{4})$ as a phasor.

S.10 Represent $10\,\cos(\omega\,t) + 10\,\sin(\omega\,t)$ as a phasor.

S.11 Represent $6\,\cos^2(2\,\pi\,t) - 3$ as a phasor.

S.12 Represent $8\,\sin(4\,\pi\,t)\,\cos(4\,\pi\,t)$ as a phasor.

S.13 Sketch a phasor diagram showing current leading voltage by $\pi/6$.

S.14 Sketch a phasor diagram showing current lagging voltage by $\pi/4$.

S.15 Sketch a phasor diagram showing current lagging voltage by $\pi/3$.

S.16

Careful observations of an AC source voltage were made and the data plotted on the nearby graph. The vertical scale is in mV; the horizontal in s.

Estimate the voltage source's
(a) amplitude, (b) period, and
(c) (i) cycle frequency and (ii) angular frequency.

(d) Write $v(t)$ in terms of a (i) sine and (ii) cosine function of time [employing suitable phase angles].

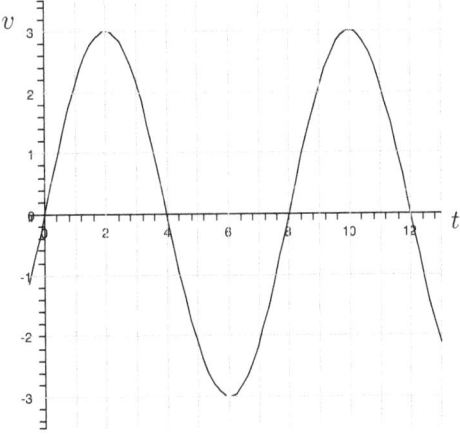

S.17 Determine the RMS voltage for the AC source in S.16.

S.18 A rectifier is a device that converts AC to DC. There are two common primitive forms: full-wave and half-wave. [Sophisticated ones also reduce the variation in the output by various smoothing techniques.] These are illustrated in the sketches below. In both cases, the input voltage was sinusoidal with amplitude V_0.

(a) Full-wave rectification effectively takes the absolute value of the input voltage, as illustrated just below. Compute V_{RMS} for the full-wave rectified signal and compare it with V_{RMS} for the input sinusoidal signal.

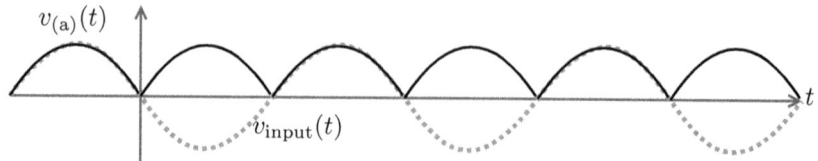

(b) Half-wave rectification takes only the positive part of the input voltage, as illustrated just below. Compute V_{RMS} for the half-wave rectified signal and compare it with V_{RMS} for the input sinusoidal signal.

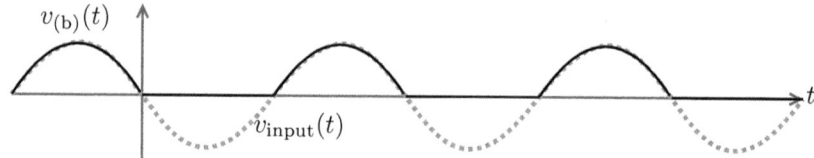

S.19   A triangular-wave voltage source is described in the formula and the figure below.

$$v(t) = \begin{cases} -4 - t\,, & -4 \le t < -2 \\ t\,, & -2 \le t \le 2 \\ 4 - t\,, & 2 \le t \le 4 \end{cases}$$

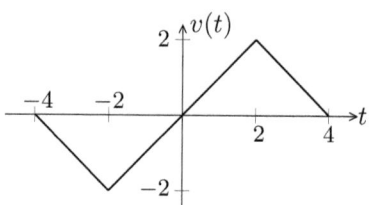

(a) Square the function, *i.e.*, determine $v^2(t)$.

(b) Plot $v^2(t)$ over one complete cycle of $v(t)$.

(c) Determine the average value of $v^2(t)$ through one cycle of $v(t)$.

(d) Determine the RMS value of $v(t)$.

S.20   Apply the full- and half-wave rectifiers of S.18 to the triangular voltage appearing in S.19. In each case, compute V_{RMS} and compare with V_{RMS} for the unrectified input signal.

S.21   In North America the standard parameters for household electric power are $V_{\mathrm{RMS}} = 120\,\mathrm{V}$ at $60\,\mathrm{Hz}$. In most of the rest of the world, the nominal supply voltage is $230\,\mathrm{V}$ at $50\,\mathrm{Hz}$. [Those parts of the Commonwealth that formerly used $240\,\mathrm{V}$ have switched to $230\,\mathrm{V}$.] Determine the peak-to-peak voltage differences for household electric services in   (a) North America and   (b) most of the rest of the world.

S.22   A capacitor has $C = 160\,\mathrm{nF}$ and an inductor has $L = 4\,\mathrm{H}$.

Compute their respective reactances when driven at angular frequency: (a) $10\,\mathrm{rad/s}$, (b) $1000\,\mathrm{rad/s}$, and   (c) $1.0 \times 10^5\,\mathrm{rad/s}$.

(d) Determine the driving frequency at which the reactances are equal.

S.23   Repeat S.22 when $C = 16\,\mu\mathrm{F}$ and $L = 40\,\mathrm{mH}$.

S.24   Repeat S.22 when $C = 1.6\,\mathrm{nF}$ and $L = 400\,\mu\mathrm{H}$.

S.25 [See also R.69.] Here we seek to improve the model for the current flowing through PK's toaster and coffee pot developed in R.69. Recall that the wall outlet is a source of AC voltage with $V_{\text{RMS}} = 120\,\text{V}$. A footnote in each of the product manuals clarifies that the power ratings, $1300\,\text{W}$ and $1100\,\text{W}$ respectively, are average values. Supposing that the toaster and coffee pot are plugged into the same outlet, ascertain the RMS current flowing from it. [Assume that the appliances are impedance matched for maximum efficiency. This is equivalent to setting the power factor equal to 1.]

S.26 [See also R.70.] Here we seek to improve the model for the current flowing through PK's waffle iron, microwave oven, and laptop developed in R.70. Recall that the average power consumptions for these three devices are: $P_{\text{wi}} = 1200\,\text{W}$, $P_{\text{mo}} = 1440\,\text{W}$, and $P_{\text{l}} = 240\,\text{W}$. Supposing that the appliances are plugged into the same outlet, ascertain the RMS current flowing from it. [Assume that the outlet is an AC voltage source with $V_{\text{RMS}} = 120\,\text{V}$, and that each device is impedance matched (at $60\,\text{Hz}$) to cancel inductive and capacitive effects. This has the effect of setting each power factor equal to 1.]

S.27
The nearby RL circuit is driven by an AC voltage source, $v(t) = 50\,\sin(120\,\pi\,t)$ V. Here, $R = 50\,\Omega$ and $L = 150\,\text{mH}$.

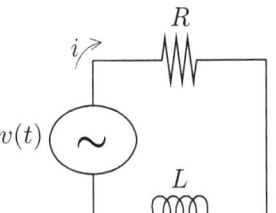

(a) Determine the (i) magnitude and (ii) phase of the current response in the circuit.

(b) Compute the average rate at which power is being dissipated in the resistor.

S.28 Repeat S.27 for $R = 5000\,\Omega$, $L = 250\,\text{mH}$, and $v(t) = 12\,\sin(40000\,t)$ V.

S.29 Repeat S.27 for $R = 3\,\Omega$, $L = 400\,\text{mH}$, and $v(t) = 0.5\,\sin(0.5\,t)$ V.

S.30
The nearby RC circuit is driven by an AC voltage source, $v(t) = 50\,\sin(120\,\pi\,t)$ V. Here, $R = 5000\,\Omega$ and $C = 150\,\text{nF}$.

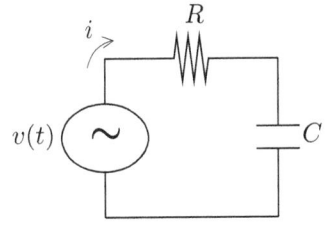

(a) Determine the (i) magnitude and (ii) phase of the current response in the circuit.

(b) Compute the average rate at which power is being dissipated in the resistor.

S.31 Repeat S.30 for $R = 20\,\Omega$, $C = 2.5\,\mu\text{F}$, and $v(t) = 12\,\sin(40000\,t)$ V.

S.32 Repeat S.30 for $R = 300\,\Omega$, $C = 5\,\text{mF}$, and $v(t) = 0.5\,\sin(0.5\,t)$ V.

S.33
The nearby LC circuit is driven by an AC voltage source, $v(t) = 50\,\sin(120\,\pi\,t)$ V. Here, $L = 40\,\text{mH}$ and $C = 160\,\mu\text{F}$.

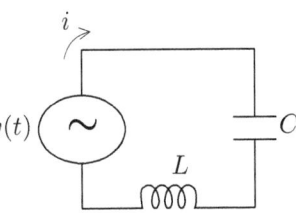

(a) Determine the (i) magnitude and (ii) phase of the current response in the circuit.

(b) Compute the average rate at which power is being dissipated in the circuit.

S.34 Repeat S.33 for $L = 500\,\mu\text{H}$, $C = 2.5\,\mu\text{F}$, and $v(t) = 12\,\sin(40000\,t)$ V.

S.35 Repeat S.33 for $L = 400\,\text{mH}$, $C = 200\,\mu\text{F}$, and $v(t) = 0.5\,\sin(0.5\,t)$ V.

S.36
The nearby RCL circuit is driven by an AC voltage source, $v(t) = 50\ \sin(120\,\pi\,t)$ V. Here, $R = 50\,\Omega$, $L = 40\,\text{mH}$, and $C = 160\,\mu\text{F}$.

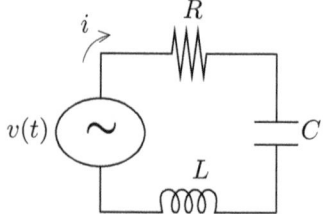

(a) Determine the   (i) magnitude and   (ii) phase   of the current response in the circuit.

(b) Compute the average rate at which power is being dissipated in the resistor.

S.37   Repeat S.36 for $R = 50\,\Omega$, $L = 2.5\,\text{mH}$, $C = 0.5\,\mu\text{F}$, and $v(t) = 12\ \sin(40000\,t)$ V.

S.38   Repeat S.36 for $R = 50\,\Omega$, $L = 400\,\text{mH}$, $C = 20\,\text{mF}$, and $v(t) = 0.5\ \sin(0.5\,t)$ V.

S.39   [See also C.25, I.46.]   The capacitor network introduced in C.25 was later found to have self-inductance $L = 0.25\,\text{mH}$. An AC voltage, $v(t) = 5\ \sin(2000\,t)$ V, is applied across this network. [It is assumed that $R \simeq 0\,\Omega$.] Compute the   (a) inductive and   (b) capacitive reactance and   (c) impedance   of the circuit.   (d) Determine the relative phase of the current response to the driving voltage.

S.40   [See also C.33, I.47.]   The capacitor network introduced in C.33 was later found to have self-inductance $L = 0.5\,\text{H}$. An AC voltage, $v(t) = 5\ \sin(200\,t)$ V, is applied across this network. [It is assumed that $R \simeq 0\,\Omega$.] Compute the   (a) inductive and   (b) capacitive reactance and   (c) impedance   of the circuit.   (d) Determine the relative phase of the current response to the driving voltage.

S.41   [See also R.19, I.42.]   The resistor network introduced in R.19 was later found to have self-inductance $L = 0.9\,\text{mH}$. An AC voltage, $v(t) = 5\ \sin(40000\,t)$ V, is applied across this network. Compute the   (a) inductive reactance and   (b) impedance   of the circuit. (c) Determine the relative phase of the current response to the driving voltage.

S.42   [See also R.27, I.44.]   The resistor network introduced in R.27 was later found to have self-inductance $L = 0.05\,\text{H}$. An AC voltage, $v(t) = 5\ \sin(300\,t)$ V, is applied across this network. Compute the   (a) inductive reactance and   (b) impedance   of the circuit. (c) Determine the relative phase of the current response to the driving voltage.

S.43   [See also C.25, R.19, R.47, I.62.]   An AC source, $v(t) = 5\ \sin(600\,t)$, drives the RCL circuit considered in I.62. Determine the properties of the long-term current response in the circuit.

S.44   [See also C.28, R.22, R.48, I.65.]   An AC source, $v(t) = 11\ \sin(12000\,t)$, drives the RCL circuit considered in I.65. Determine the properties of the long-term current response in the circuit.

S.45   [See also I.53, I.73.]   A $500\,\Omega$ resistor, a $500\,\text{nF}$ capacitor, and a $20\,\text{mH}$ inductor are combined in series by means of ideal wires, as shown in the figure accompanying I.53. An AC voltage source with amplitude $V_0 = 10\,\text{V}$ and angular frequency $\omega = 8000\ \text{rad/s}$ is connected across this network.

(a) Compute the cycle frequency of the AC source.

(b) Determine the RMS voltage of the source.

(c) Compute the   (i) inductive and   (ii) capacitive   reactance of the circuit.

(d) Determine the impedance of the circuit.

(e) Determine the   (i) amplitude and   (ii) relative phase   of the current response.

(f) State whether the current leads or lags the voltage.

(g) Compute the average power deposited into the resistor.

(h) Would the average power in (g) increase or decrease, if the frequency of the AC source were to be slightly increased?

S.46   Repeat S.45 for the components described in I.54.
S.47   Repeat S.45 for the components described in I.55.
S.48   Repeat S.45, with $\omega = 10^6$ rad/s, for the components described in I.56.
S.49   Repeat S.45, with $\omega = 12000$ rad/s, for the components described in I.57.
S.50   Repeat S.45, with $\omega = 320000$ rad/s, for the components described in I.58.
S.51   Repeat S.45, with $\omega = 5 \times 10^7$ rad/s, for the components described in I.59.
S.52   Repeat S.45, with $\omega = 9600$ rad/s, for the components described in I.60.
S.53   Repeat S.45, with $\omega = 1.1 \times 10^7$ rad/s, for the components described in I.61.
S.54   We have inductors, capacitors, and resistors with the parameters:

$$L \in \{0.5,\, 20,\, 100\}\,[\text{mH}], \qquad C \in \{10,\, 500,\, 10000\}\,[\mu\text{F}], \quad \text{and} \qquad R \in \{0.1,\, 100\}\,[\Omega].$$

(a) Using components from this set, construct a circuit that will react vigorously to driving frequencies in a narrow band about 1000 rad/s, and respond poorly to all other frequencies.   [This is a narrow-band filter.]

(b) Using components from this list, design a circuit that responds to a very broad range of frequencies centred on 50 rad/s.   [This is a broad-band filter.]

S.55   We have a box of 20 μF capacitors, a 10 mH inductor, and a pair of connectors with resistances $r = 5\,\Omega$ and $R = 500\,\Omega$.

(a) Design an RCL circuit which is strongly responsive to angular frequencies near $\omega = 2000$ rad/s and quite unresponsive otherwise.

(b) Design an RCL circuit which is most responsive to a broad range of angular frequencies centred on 1000 rad/s.

S.56   Determine approximate quality factors for the sinusoidally driven circuits whose average power frequency response curves appear below. [The angular frequencies on the left axis are quoted in kiloradians per second; the right, gigaradians per second.]

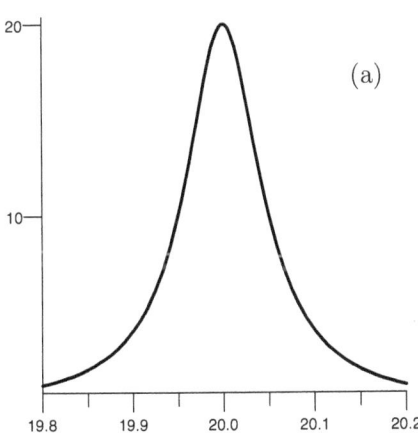

(a)

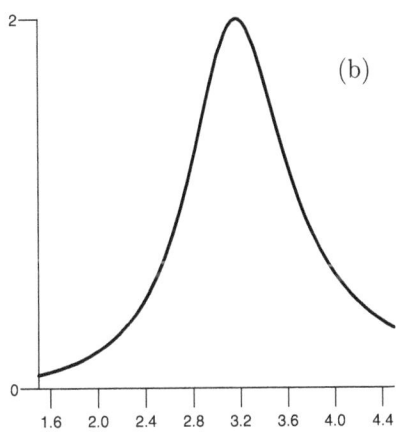

(b)

S.57   [See also I.45.] Suppose that the primary coil in the transformer studied in I.45 is connected to an AC power supply, $v(t) = V_p \sin(\omega t)$. Show that an AC voltage, with amplitude $V_s = \frac{N_s}{N_p} V_p$, is induced in the secondary coil.

S.58   [See also S.57.] Determine the secondary voltage in a transformer, V_s, when $N_p = 400$, $N_s = 30$, and $V_p = 120$ V. Here the transformer is said to "step-down" the voltage.

S.59 [See also S.57.] Determine the secondary voltage in a transformer, V_s, when $N_p = 690$, $N_s = 360$, and $V_p = 230\,\text{v}$.

S.60 [See also S.57.] Determine the secondary voltage in a transformer, V_s, when $N_p = 40$, $N_s = 200$, and $V_p = 12\,\text{v}$. Here the transformer is said to "step-up" the voltage.

M

Maxwell's Equations Problems

M.1 For each of Maxwell's Equations:

Name it

Express it mathematically as a formula

Explain (in words) what the formula means

List one or more of its direct physical implications

M.2 In Chapter 47, we used a very narrow rectangular plaquette lying in the xy-plane in analysis of [the vacuum form of] Faraday's Law. There, the electric and magnetic fields were taken to have non-zero projections onto the $\hat{\jmath}$ and $\hat{k}$ basis directions only. Argue that the results obtained are unchanged when　(a) the electric field has a non-zero z-component, $E_z(t, x)$, and　(b) the magnetic field has a non-zero y-component, $B_y(t, x)$.

M.3 Adapt the vacuum Faraday's Law analysis of Chapter 47 to the situation in which the electric and magnetic fields are $\vec{E} = E_z(t, x)\, \hat{k}$ and $\vec{B} = -B_y(t, x)\, \hat{\jmath}$, employing the plaquette with corners at:　$\{\, (x,\, 0,\, z_0),\, (x,\, 0,\, z_0 + l),\, (x + dx,\, 0,\, z_0 + l),\, (x + dx,\, 0,\, z_0) \,\}$.

M.4 In Chapter 47, we used a very narrow rectangular plaquette lying in the zx-plane in analysis of [the vacuum form of] Ampère's Law. There, the electric and magnetic fields were taken to have non-zero projections onto the $\hat{\jmath}$ and $\hat{k}$ basis directions only. Argue that the results obtained are unchanged when　(a) the electric field has a non-zero z-component, $E_z(t, x)$, and　(b) the magnetic field has a non-zero y-component, $B_y(t, x)$.

M.5 Adapt the vacuum Ampère's Law analysis of Chapter 47 to the situation in which the electric and magnetic fields are $\vec{E} = E_z(t, x)\, \hat{k}$ and $\vec{B} = -B_y(t, x)\, \hat{\jmath}$, employing the plaquette with corners at:　$\{\, (x,\, y_0,\, 0),\, (x,\, y_0 + l,\, 0),\, (x + dx,\, y_0 + l,\, 0),\, (x + dx,\, y_0,\, 0) \,\}$.

M.6 *Local Form of Vacuum Gauss's Law for Electric Fields*
Consider the rectangular parallelepiped with sides dx, dy, and dz. For simplicity, assume that the electric field is $\vec{E} = E_x(t,\, x)\, \hat{\imath} + E_y(t,\, x)\, \hat{\jmath}$.

Compute the electric flux impinging on the pairs of faces each with area:　(a) $dz \times dx$,　(b) $dy \times dz$,　(c) $dx \times dy$.

(d) Ascertain the net flux passing through the parallelepiped. [HINTS: Make similar approximations to those employed in Chapter 47.]

(e) In vacuum, the charge density is zero, and hence so too is the RHS of Gauss's Law. Identify the appropriate constraint on the local—derivative—behaviour of $E_x(t,\, x)$.

M.7 *Local Form of Vacuum Gauss's Law for Magnetic Fields*
Apply the analysis performed in M.6 to the magnetic field: $\vec{B} = B_x(t,\, x)\, \hat{\imath} + B_z(t,\, x)\, \hat{k}$.

M.8 [See also C.11.] A capacitor consists of two hollow thin-walled concentric conducting tubes of length L. The radius of the inner tube is a; the outer b; and $a < b \ll L$. The goal of this question is compare the macroscopic and microscopic [local] computations of the electric energy stored in the capacitor.

(a) Let's compute the capacitance of the system of conducting tubes.
(i) Assume that charges $\pm Q$ are uniformly distributed on the the cylindrical shells, and infer the local electric field everywhere in the interstitial region (between the shells).
(ii) Determine the potential difference between the shells [each is an equipotential region].
(iii) Determine the capacitance of the system.

(b) Compute the total amount of energy stored in the capacitor, via (i) $U = \frac{1}{2} C V^2$ and (ii) $U = \frac{Q^2}{2C}$, in terms of Q and geometrical factors, using results from (a).

(c) Construct the energy density in the interstitial region: $u = \frac{1}{2} \epsilon_0 E^2(r)$.

(d) Integrate the energy density throughout the volume of the interstitial region.

(e) Comment on your results for (b) and (d).

M.9 [See also I.31.] A straight section of coaxial cable consists of two hollow thin-walled concentric conducting tubes of length L, bearing uniformly distributed, oppositely directed currents, I_0. The radius of the inner tube is a; the outer b; and $a < b \ll L$. The goal of this question is compare the macroscopic and microscopic [local] computations of the magnetic energy stored in the capacitor.

(a) Let's compute the inductance of the system of conducting tubes.
(i) Assume that currents $\pm I_0$ are uniformly distributed on the the cylindrical shells, and infer the local magnetic field everywhere in the interstitial region (between the shells).
(ii) Determine the magnetic flux passing through a plane orthogonal to both shells and extending radially outward from the symmetry axis in a particular direction.
(iii) Determine the [self-]inductance of the system.

(b) Compute the total amount of magnetic energy stored in the system, via $U = \frac{1}{2} L I_0^2$, in terms of I_0 and geometrical factors, using results from (a).

(c) Construct the energy density in the interstitial region: $u = \frac{B^2(r)}{2\mu_0}$.

(d) Integrate the energy density throughout the volume of the interstitial region.

(e) Comment on your results for (b) and (d).

M.10 The electric component of an electromagnetic plane wave is $\vec{E} = E_0 \cos(k_z z - \omega t) \, \hat{\imath}$. The wave has amplitude $E_0 = 300\,\text{V/m}$ and oscillates with frequency $2.4\,\text{GHz}$.

Determine the (a) angular frequency and (b) wavelength of this wave.

(c) In which direction is the wave travelling?

(d) Determine (i) the magnitude of, (ii) the direction of, and (iii) an expression for the magnetic field portion of the plane wave.

(e) Write down the expression for the Poynting vector. In what direction is electromagnetic energy flowing?

M.11 Determine the (a) electric, (b) magnetic, and (c) combined energy intensity of the plane wave considered in M.10.

M.12 Repeat M.10 for $\vec{E} = E_0 \cos(k_z z + \omega t) \, \hat{\imath}$, where $E_0 = 500\,\text{V/m}$ and the wave is oscillating at $1.8\,\text{GHz}$.

M.13 Repeat M.11 for the wave in M.12.

List of Symbols

(ij...k) Series composition of the circuit elements labelled i, j, ..., k

[ij...k] Parallel composition of the circuit elements labelled i, j, ..., k

α Lineal coefficient of temperature-dependence of resistivity

$\vec{B_0}$ Uniform magnetic field

$\vec{B_i}(\mathcal{P})$ Magnetic field produced by a source labelled i at the field point $\mathcal{P}$; magnetic flux density

χ Magnetic susceptibility

χ_C Capacitive reactance

χ_L Inductive reactance

ΔH The quantity of heat deposited in a resistor and its environment via Joule heating

Δq A charge element; esp. as appearing in a partition of a body or other assemblage of charge

ΔV Electric potential difference between two points in space; voltage difference

Δx Displacement in 1-d; Cartesian x-component of displacement

$\vec{E}_\parallel, \vec{E}_\perp$ Parallel and perpendicular components of the electric field

$\vec{E_0}$ Uniform electric field

$\vec{E_q}(\mathcal{P})$ Electric field produced by a charge labelled q at the field point $\mathcal{P}$

ϵ_0 Permittivity of free space

η Number density of charge carriers within a substance

γ Damping parameter for an oscillator

$\hat{i}, \hat{j}, \hat{k}$ Unit vectors in the x, y, and z Cartesian basis directions

$\hat{r}, \hat{\theta}, \hat{\phi}$ Unit vectors in the polar basis directions, r, θ, and ϕ

$\hat{r}_{if}$ Unit vector in the direction of $\vec{r}_{if}$

$I_{\mathrm{disp.}}$ Displacement current

$I_{\mathrm{encl.}}$ Enclosed current: that piercing a surface bounded by a closed [non-self-intersecting] curve

λ, σ, ρ Charge densities for lineal, areal, and volume distributions of matter

$\mathcal{N}$ An integer, often taken to be large

μ_0 Permeability of free space

μ_b Bohr magneton

ν Cycle frequency

ω Angular speed; angular frequency

ω Driving (angular) frequency of an AC source

ω_0 Natural or resonant frequency for an oscillator

$\mathcal{T}$ Period: the time interval corresponding to a single oscillation

Φ_e; $\Phi_{e,S}$ Electric flux; electric flux passing through surface S

Φ_m; $\Phi_{m,S}$ Magnetic flux; magnetic flux passing through surface S

Ψ A wavefunction

ψ_R, ψ_L A rightmoving/leftmoving wave

$q_{\text{encl.}}$ Enclosed electric charge: that within a region bounded by a closed [non-self-intersecting] surface

ρ Electric resistivity of a substance

ρ_0 Base or reference value of resistivity

$\vec{S}$ The Poynting vector; electromagnetic energy intensity

σ Electric conductivity of a substance

τ A time constant characterising the exponential behaviour of RC and RL circuits

a   Constant (instantaneous) acceleration in 1-d

θ, ϕ, α, β Angles

θ_0, ϕ_0 Constant or initial angles

ε An EMF

φ Phase difference between AC driving voltage and current response

$\vec{F}_{\text{NET}}$ Net force acting on a particle or system

$\vec{F}$ A generic force

$\vec{F}_{\text{B}}$, $\vec{F}_{\text{M}}$ Lorentz force acting on a moving charged particle or electric current; magnetic force

$\vec{F}_{\text{C}}$ The Coulombic force acting between two point-like charged particles

$\vec{J}$; J Electric current density; its magnitude

$\vec{\tau}$; τ Vector torque; its magnitude

$\vec{\nabla} \cdot \vec{F}$ The divergence of the field $\vec{F}$

$\vec{\nabla} \times \vec{F}$ The curl of the field $\vec{F}$

$\vec{\nabla}$ The gradient operator

$\vec{a} \cdot \vec{b}$ The DOT PRODUCT, or scalar product, of vectors $\vec{a}$ and $\vec{b}$ [in n dimensions]

$\vec{a} \times \vec{b}$ The CROSS PRODUCT, or vector product, of three-dimensional vectors $\vec{a}$ and $\vec{b}$

$\vec{a}$; a Instantaneous vector acceleration; its magnitude

$\vec{a}_c$; a_c Centripetal acceleration; its magnitude

$\vec{k}$ Wave vector

$\vec{m}$, m Magnetic dipole moment vector; its magnitude

$\vec{p}$; p Electric dipole moment vector; its magnitude

$\vec{r}$ Instantaneous vector position

$\vec{r}_{if}$ Relative position vector, extending from i to f

$\vec{v}$; v Instantaneous vector velocity; its magnitude

$\vec{v}_{\parallel}$, $\vec{v}_{\perp}$ Parallel and perpendicular components of velocity

$|\vec{r}_{if}|$, r_{if} Magnitude of $\vec{r}_{if}$

V_{RMS}, I_{RMS} Root-mean-square values of voltage and current in an AC circuit

B_0 Magnitude of a uniform magnetic field

C Capacitance; a particular capacitive element; capacitor

c Speed of a wave; speed of light

$d\vec{A}$ Directed infinitesimal area element

$d\vec{s}$ An infinitesimal element of arc length along a [prescribed] path

e The elementary quantum of electric charge

E_0 Magnitude of a uniform electric field

g_l, g_s Orbital and spin gyromagnetic ratios

I Instantanous electric current

$i(t)$ The current response in an AC circuit

I_0 A constant or initial value of electric current; AC current amplitude

I_{∞} Asymptotic value of electric current

I_{av} Average electric current throughout a specified or implicit time interval

k Coulomb's constant

k Dielectric constant

k Wave number

k_B Boltzmann constant

L A fixed length: often that of a wire

L Inductance; a particular inductive element; inductor

m, M Mass of a particle

M_{ij} Mutual inductance

n, l, m, s Quantum numbers for electrons in atomic systems

P Power; rate of transfer, or flow, of energy

Q Quality factor

q_0, Q_0 Test charge; a constant or initial value of electric charge; charge amplitude

q_{∞}, Q_{∞} Asymptotic value of electric charge

q_a; Q_b Electric charge borne by particle a; by b

Q_{Total} Total charge

R Resistance; a particular resistive element; resistor

r Internal resistance of a battery

R, r Radial coordinate or parameter; a fixed value thereof

T Temperature; specified time

T_0 Reference temperature

T_C Curie temperature

U_{12} Electrostatic potential energy associated with the q_1–q_2 charge system

$U_{[C]}(\vec{r})$ A potential energy function associated with the conservative force $\vec{F_c}$

$u_E(\mathcal{P})$ Electrostatic energy density at a specified point in space

$u_M(\mathcal{P})$ Magnetic energy density at a specified point in space

$V(\vec{r})$ Electric potential; electrostatic potential energy per unit charge; voltage

$v(t)$ An AC voltage source

V_0 A constant or initial value of electric potential [voltage]; AC voltage amplitude

v_0 A specific [initial] value of the 1-d velocity

v_d Drift speed attributed to charge carriers within a conducting medium

v_t Tangential speed

$W_{if}\left[\vec{F}\right]_{\text{(path)}}$ Mechanical work performed by a force, $\vec{F}$, acting through a displacement along a prescribed path

x, y, z 1-d position; Cartesian component of position in 2- or 3-d

$\mathcal{A}$ Cross-sectional area

$\mathcal{O}, \mathcal{P}, \mathcal{Q}$ Points in space

$\text{Circ}(\vec{F})_{\gamma_0}$ Circulation of a field, $\vec{F}$, about the curve specified by γ

S   A [smooth, 2-d] surface; any such surface bounded by a particular curve

AC   Alternating current circuits are driven by oscillatory (often sinusoidal) EMFs

DC   Direct current circuits are driven by EMFs which do not oscillate

Z Impedance

EMF   Electromotive force; electric potential provided by a circuit element

FBD   Free body diagram

IRF   Inertial reference frame: a set of coordinates and clock(s) borne by a class of observers for whom Newton's First Law is evident.

K1, K2   Kirchoff's Rules

LHS, RHS   Left and right hand side of a mathematical expression

RHR   One of the various right hand rules used for assigning orientation to a vector quantity

RMS   Root-Mean-Square

WRT   With respect to

Battery   A constant source of EMF

Capacitor A device with a particular immutable capacitance and no resistance or inductance

Inductor   A device with a particular immutable inductance and no resistance or capacitance

LC   Circuits comprised exclusively of inductive and capacitive elements

QED   Quantum Electrodynamics

RCL   Circuits with resistive, capacitive, and inductive elements

RC   Circuits comprised exclusively of resistive and capacitive elements

Resistor A device with a particular immutable resistance and no capacitance or inductance

RL   Circuits comprised exclusively of resistive and inductive elements

Switch   A means of starting and stopping the flow of electric current or completing a circuit

Wire   A zero-resistance means of conveying current between two circuit elements

WLOG   Without loss of generality

Index

Milton Keynes UK
Ingram Content Group UK Ltd.
UKHW051943071024
449327UK00026B/2149